UNIVERSITY OF WINNIPEG
LIBRARY
515 Portage Avenue
Winnipeg, Manitoba R3B 2E9

Developments in Geotechnical Engineering 4A

COASTAL ENGINEERING, I

Further titles in this series:

1. *G. SANGLERAT*
THE PENETROMETER AND SOIL EXPLORATION

2. *Q. ZARUBA AND V. MENCL*
LANDSLIDES AND THEIR CONTROL

3. *E.E. WAHLSTROM*
TUNNELING IN ROCK

Developments in Geotechnical Engineering 4A

COASTAL ENGINEERING, I
Generation, propagation and influence of waves

by

RICHARD SILVESTER

*Department of Civil Engineering, University of Western Australia,
Nedlands, W.A., Australia*

ELSEVIER SCIENTIFIC PUBLISHING COMPANY
Amsterdam   London   New York   1974

ELSEVIER SCIENTIFIC PUBLISHING COMPANY
335 Jan van Galenstraat
P.O. Box 211, Amsterdam, The Netherlands

AMERICAN ELSEVIER PUBLISHING COMPANY, INC.
52 Vanderbilt Avenue
New York, New York 10017

Library of Congress Card Number: 72-97435

ISBN 0-444-41101-1

With 232 illustrations and 73 tables.

Copyright © 1974 by Elsevier Scientific Publishing Company, Amsterdam

All rights reserved. No part of this publication may be reproduced, stored in a retrieval system, or transmitted in any form or by any means, electronic, mechanical, photocopying, recording, or otherwise, without the prior written permission of the publisher,
Elsevier Scientific Publishing Company, Jan van Galenstraat 335, Amsterdam

Printed in The Netherlands

PREFACE

This book and its companion volume have evolved from courses developed at the Asian Institute of Technology (A.I.T.) in Bangkok, Thailand. These consisted of Coastal Engineering, Wave Hydrodynamics, Coastal Sedimentation, Tidal and Estuarine Hydraulics, and Coastal Modelling. It is envisaged, therefore, that similar courses at graduate or higher undergraduate level could be served by these tomes. Examples and problems have been included to aid in any such teaching programme.

However, another main purpose of the works has been to collect the latest information available in this fast developing field and to present it in a form for ready application. They should therefore serve as hand-books for the practising engineer who is concerned with coastal problems, either periodically or perennially. Even if a maritime project is considered too involved for him or his organisation, he can become aware of the complexities before calling in a consultant.

A third group of workers who could benefit from these books, particularly Volume II, are geologists and geographers who are interested in coastal geomorphology. Wave action can produce many submarine and shoreline features, which have here-to-fore been "explained" by tidal and other oceanic currents. To the author's knowledge, there is no aspect of submarine geology on which unanimous agreement has been reached respecting its origin. This may be due to the omission of this forceful factor of wave generation in the storm areas of the oceans and the concomitant spread of energy to their distant margins. A better understanding, therefore, of the principles of wave characteristics and beach processes should aid the descriptive scientist in his observations of nature and subsequent reporting of same.

Unlike many other treatises on hydraulics and wave mechanics, these omit the differential equations and manipulations leading to the final formulae. Where a researcher wishes to further the topic, and must go to the original works, ample references are supplied as a lead into the literature. It was felt that the practising engineer, always a busy person, wants to have a handy solution, preferably in the form of graphs or tables, which is suggested to him from the two or more generally available. The author has tried to digest the available information in order to recommend a course of action. This is a great responsibility, of which he is fully conscious, but is happy in the thought that his suggestions should be better (if only a little) than the apparently undisciplined guesses made by some engineers in the past and even the present.

The minimization of mathematics is made without remorse. The engineer's role, in any case, is to proceed where angles, scientists, and mathematicians alike, fear to tread. There are far too many practical problems for him to be overly concerned with deriving, or redeveloping or checking theoretical analyses. He must, of course, be aware of the premises upon which they are based, and know the simplifications used in obtaining a solution. There are occasions when he must substitute constants into a relationship which are not proven beyond reasonable doubt, but therein lies the "art" of engineering.

Acknowledgements

The author's interest in coastal engineering was initiated in 1954 by Professor K.L. Cooper, then Head of the Department of Civil Engineering at the University of Western Australia, who subsequently encouraged the author in his researches. During sabbatical leave from the University in 1956-7 the author spent six months at the National Institute of Oceanography in the United Kingdom, where the complexities of wave generation and other coastal processes were imbibed from the comprehensive N.I.O. library and discussions with oceanographers. Thanks are expressed to Dr. G.E.L. Deacon F.R.S. for his warm support. It was during this period that the importance of the crenulate shape of shoreline first struck the author, which led later to model studies at his own campus. During a second study leave (1963-4) the author spent twelve months at the University of California, Berkeley, whilst in receipt of a National Science Foundation fellowship. There he studied many geological phenomena which could be related to the action of ocean waves. His contact with such workers as Professors J.W. Johnson, R.L. Wiegel and H.A. Einstein helped his understanding of many facets of wave forces and sediment transportation.

In 1967 the University of Western Australia granted the author leave of absence to fulfil a SEATO appointment as Professor of Coastal Engineering at the SEATO Graduate School of Engineering (now an autonomous body called the Asian Institute of Technology). During his four years in Thailand the author was supported by the Australian Department of Foreign Affairs. It was here that a treatise was conceived, which ended up as two volumes. The thesis projects supervised by the author helped clear up some outstanding problems. Special mention should be made of Dr. Suphat Vongvisessomjai, who devoted much spare time with the author in modifying the latest wave forecasting formulae to the spectral form found in this volume. Dr. M.E. Bender, President of A.I.T., was extremely encouraging in the introduction of this new specialty, and provided funds for the conduct of a three-week "Institute" for practicing coastal engineers from a dozen Southeast-Asian countries. This helped the author learn of the problems facing men in the

Preface

professional front line and has, it is hoped, improved the presentation. After returning to Western Australia in 1971 the author was granted sabbatical leave on campus in order, among other things, to complete this work. To these various institutions and personnel in authority the author expresses his deep sense of gratitude.

Besides the many oceanographers and engineers with whom the author has been in direct contact, there are others to whom he has written for personal explanations of points not understood, or for data not contained in publications. Thanks are expressed to Professors N.H. Brooks of Caltech, A.T. Ippen of M.I.T., T. Hayashi of Chuo University, J.B. Herbich of Texas A & M, K. Horikawa of Tokyo University, W.J. Pierson Jr. of New York University and W.H. Munk of University of California, San Diego, and to Doctors D.E. Cartwright of National Institute of Oceanography (U.K.), B.D. Dore of University of Reading and S. Senshu of Central Research Institute of Electrical Power Industry, Tokyo, and Messrs. P. Donnelly of Department of Public Works, Canada, H. Sakai of Ministry of Construction, Japan, H. Yamashito of Ministry of Transport, Japan, and T. Yano of Ministry of Agriculture and Forests. These and many others have borne the brunt of the author's copious correspondence. Special thanks are due to Mr. L. Draper of the National Institute of Oceanography (U.K.) who supplied copies of wave records for analysis.

Richard Silvester

CONTENTS

Preface	V
Chapter 1. Introduction	1
What is coastal engineering?	2
Interest in the oceans	4
Costs and rewards in ocean engineering	6
Need for education of ocean engineers	8
Definitions	12
Sources of information	18
References	19
Chapter 2. Wave-generation processes	21
Mechanism of generation	21
Resonance	22
Shear flow	25
Resonance/shear-flow combination	27
Sheltering effect	29
Wave breaking	31
Wave characteristics	33
Storm waves	33
Swell	38
Meteorological conditions	43
Extra-tropical cyclones	44
Anticyclones	49
Tropical cyclones	50
Seasonal pattern	54
Problems	61
References	64
Chapter 3. Wave forecasting	65
Wave statistics	65
Sverdrup and Munk	66
Longuet-Higgins	67
Cartwright and Longuet-Higgins	68

History of wave prediction	70
SMB method	71
PNJ method	75
PM method	78
Darbyshire method	82
Fully arisen sea	84
The PM frequency spectrum	84
The PM period spectrum	85
Dimensionless presentation	88
Fetch and duration for FAS	93
Comparison of FAS formulae	95
Developing seas	98
Average spectra	105
Design data	109
Fully arisen sea	109
Developing sea	111
Optimum wave conditions	113
Wave dispersion	115
Waves from tropical cyclones	121
Examples	127
Problems	135
References	139
Chapter 4. Theory of waves	**143**
Progressive wave: linear theory	146
Water-particle motion	149
Wave energy	154
Progressive wave: finite height	156
Surface profile	157
Water-particle motion	160
Cnoidal theory	163
Hyperbolic theory	166
Stream function	169
Solitary wave theory	171
Comparison of progressive wave theories	173
Theoretical comparison	173
Experimental verification	176
Standing waves	179
Standing waves: linear theory	179
Standing waves: second-order theory	182
Partial clapotis	184
Clapotis gaufré	186

Contents　　　　　　　　　　　　　　　　　　　　　　　　　　　　　XI

　　Effects of viscosity 191
　　　　Viscous damping 193
　　　　Damping due to bottom permeability 195
　　　　Damping due to non-rigid bottom 196
　　　　Mass transport 196
　　Examples 206
　　Problems 212
　　References 214

Chapter 5. Effects of shoaling water 217
　　Wave refraction 217
　　　　Parallel contours 220
　　　　Non-parallel straight contours 222
　　　　Curved contours 224
　　　　Method when α_o approaches 90° 225
　　　　Tracing orthogonals seawards 225
　　　　Choice of depth increments 226
　　　　Alternative methods 227
　　　　Wave energy 229
　　　　Refraction of short-crested waves 233
　　　　Refraction of wave spectrum 236
　　　　Refraction by currents 237
　　Breakers and surf 240
　　　　Run-up procedure 240
　　　　Shoaling of wave spectrum 247
　　　　Breaking processes 248
　　Examples 249
　　Problems 256
　　References 260

Chapter 6. Wave recording 263
　　Analysis of a single record 265
　　Summation of records 269
　　Extreme values 271
　　Design waves 272
　　Types of recorder 274
　　　　Above surface 274
　　　　At surface 274
　　　　Below surface 275
　　Major difficulties 276
　　Wave direction 276

Spectral analysis 277
Errors arising 278
   Pressure attenuation 278
   Wave steepness 284
   Refraction and diffraction 286
   Currents 286
   Estuarine conditions 286
   Non-rigid bottom 287
Length of record 288
Record analysis 289
Recommendations 289
Examples 290
Problems 295
References 298

Chapter 7. Effect of structures 301
Wave diffraction 302
   Semi-infinite breakwater 303
   Breakwater gap 311
   Diffraction of wave spectrum 316
   Combined diffraction and refraction 318
Land-backed structures 319
   Reflection and dissipation 320
   Run-up 324
   Overtopping 327
   Seiching 330
   Complex harbour shapes 334
   Exclusion of resonant waves 335
Water-backed structures 337
   Permeable breakwaters 337
   Submerged breakwaters 338
   Pile arrays 344
Mobile breakwaters 346
   Pontoons 347
   Flexible rafts 351
   Miscellaneous 351
Pneumatic and hydraulic breakwaters 352
Examples 354
Problems 361
References 366

Contents

Chapter 8. Effect of waves on structures	371
Seawalls	372
Pressures due to standing waves	372
Pressures due to breaking waves	379
Pressures on sloping walls	388
Pressures on composite structures	388
Cylindrical structures	393
Slender piles	394
Horizontal and sloping members	403
Submerged pipelines	406
Submerged objects of large dimensions	408
Rubble-mound breakwaters	416
Examples	422
Problems	428
References	431
Appendix. Table of functions of d/L_o	435
Index	449

Chapter 1

INTRODUCTION

It has been considered ironical that man should have called our planet "Earth" when, in fact, 71% of it is covered with water. The presence of this unique substance water has initiated life on this planet, which has evolved to the stage of providing man an existence. Its presence also provides the moderate climate in which fauna and flora can flourish. The resultant rain has helped to flush sediment formed from the breakdown of basaltic rock to lower levels and into the sea. The waves and tides of the sea have then distributed this detritus along the coastal margins. Until such sediment was available vegetable matter could not exist, so inhibiting the development of creatures which depend upon green matter for their existence.

Not only is food supplied, but also oxygen is made available by photosynthesis. The percentage of the world's production of oxygen by the sea is a disputable point, but it ranges from 30 to 70%. This, together with the ameliorating effect on climate by the oceans being a container and distributor of heat from the sun, make the aquatic area of the globe an essential element of existence. It therefore behoves man to study it thoroughly, in order to conserve it in a natural state, work in harmony with it whilst utilizing its many bounties, and assist it in maximizing the products and influences man would desire of it. Much more food could be gained from the sea by better breeding methods and modes of harvesting. Only in recent years the techniques have been developed to exploit minerals on and under the sedimentary shelves adjacent to most continents.

On the other hand, the oceans insert an expensive obstacle to travel which man is continually trying to make more economical. For this purpose a greater understanding of the physical nature of waves and tides, plus sediment transport resulting therefrom, can aid in the overall reduction of the maritime burden. Oceanic storms can be catastrophic to coastal boundaries besides floating objects. Annual losses of life and livelihood due to the sea could well exceed the monetary value of benefits from it. This adverse balance of payments will be rectified the more we learn about the physics and chemistry of the oceans and the nature of their biota. Unlike the space race, the sea scramble can pay more immediate dividends, not only to its participants, but also to the less-developed nations, whose subsistence depends more upon marine activities than do the more affluent societies.

WHAT IS COASTAL ENGINEERING?

The study of any topic generally involves a scientific base (the *why* and *where* aspect) and an engineering component or practical application of scientific knowledge (the *how* aspect). The scientific erudition of the oceans is termed *oceanography*, whilst its applied brother is called *ocean engineering*. This latter terminology became evident during the 1960's when oil exploration extended further out on the continental shelves. Prior to this the term *coastal engineering* had been used for problems of a structural nature connected with the sea, with *marine engineering* and *naval architecture* being specialties related to mechanical and transportational operations. Man's concern with stability of installations on the sea bed will be restricted to continental shelves for the foreseeable future. These comprise about 5% of the ocean area, so that the term *coastal engineering* might well apply to this activity. However, the waves and tides involved in design are generated in the wide expanses of the oceans so that *ocean engineering* could equally apply.

Institutes of Oceanography have existed in academic circles since the early 1930's, and have concentrated on the physical, chemical, biological and geological aspects of the sea and its boundaries. Physics of the sea is concerned with water motions, such as tides, waves or currents, acoustics, heat balance, plus transfer of energy from the atmosphere to the sea surface by wind. Chemistry of the sea includes distribution of chemical compounds, stratification due to density differences and the mechanism of corrosion. The biological aspect deals with the ecology of marine plants and animals and has been in the forefront of controlling the killing of whales, which provide the most economical method for harvesting the substance of the sea. Marine geology is a sister to the terrigenous science and, in fact, should supply the answers to many of the latter's unknowns, since the bulk of sediments have been deposited under marine conditions.

The field of ocean engineering was first recognized in academic and professional circles during the 1960's. A definition was sought by the National Academy of Engineering [1][1], established in the United States in 1965, through its Committee on Ocean Engineering. The tentative attempt resulted in: "Ocean engineering is that activity which combines knowledge of the ocean and engineering to utilize the oceans, their contents or boundaries for the achievement of human objects." As noted by Kavanagh [1]: "In its proper technical context, the above definition is extended to include the estuarine, near-shore and coastal environment as well as large bodies of inland waters."

Silvester [2] believes that the above stresses too much the exploitation of oceanic resources, whereas the catastrophic forces generated in the ocean need to be

---

[1] Numbers between straight brackets refer to the numbered references at the end of each chapter.

understood and controlled by engineers equally as much as the utilization of its living or fossilized contents. As noted already, it would be interesting to compute worldwide whether the sea absorbs more wealth than it effuses each year. The loss of ships, oil rigs, buildings and beaches from wind-generated waves, tsunamis, and storm surges, could far outstrip the value of food and minerals harvested from the oceans and their boundaries. Perhaps a more concise definition might be acceptable as follows: "Ocean engineering is the application of scientific knowledge of the oceans to the economic welfare of man." The term *economic* is included intentionally as this is the essence of engineering.

A comparison with engineering *of* or *in* the land indicates immediately the breadth of cover in the term *ocean engineering*. It is more comprehensive than the title *civil engineering* in the land context, since it involves geological, petroleum, mining and marine engineering, plus naval architecture. The equivalent oceanic term to civil engineering could be *coastal engineering,* but activities in and under the sea demand a nautical emphasis to existing phases of engineering, including civil, mechanical, electrical and chemical, to name a few. Thus, whilst much engineering activity in the oceans depends upon existing knowledge in other branches of engineering, there is, as pointed out by Calhoun et al. [3]: "... an identifiable area of engineering activity which will not fall within any of the traditional engineering areas unless they are modified considerably to meet the demands of the new environment." The same authors present a diagram as in Fig. 1—1, which depicts the

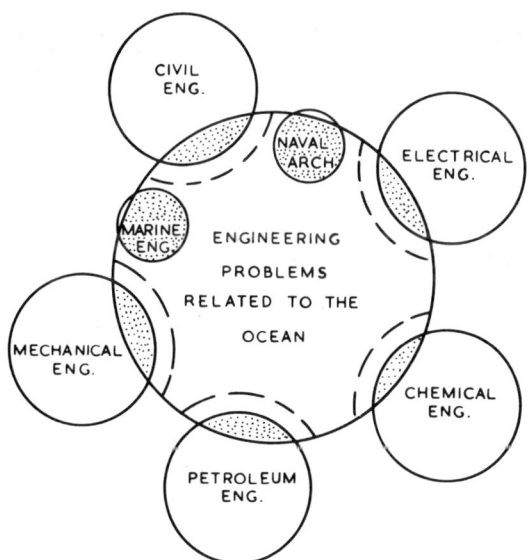

Fig. 1-1. Present overlap and probable expansion of existing disciplines into ocean engineering problems.

overlap of interest and the expansion of each branch into the centre as any group of workers widens its knowledge to encompass special oceanic conditions.

Similar concentration of effort has occurred in such fields as nuclear, space and bio-medical engineering, where research and subsequent teaching programmes have been modified to cope with the new environment. At this stage the title *ocean engineering* is probably too comprehensive on which to build a post-graduate or undergraduate programme, but its recognition as a field may help to integrate the activities of scientists and engineers working with coasts, the continental shelves, the deep ocean, the sea bed or the air mass above the surface.

INTEREST IN THE OCEANS

The demand for knowledge during World War II (1939–1945) on wave generation and beach processes provoked the first research efforts into topics now known as coastal engineering. In 1948, the first of the biennial conferences on coastal engineering was held, sponsored by the "Council on Wave Research of the American Society of Civil Engineers". Table 1-I illustrates the growing interest in this field, with a quadrupling of papers published per conference over the twelve held to 1970. This omits the special conferences conducted between times [4,5,6] and those now being held regularly on the topics of ocean engineering [7] and dredging [8]. Neither does it account for the new journals launched to cover marine topics [9–13].

The substantial increase in papers since the 1960's is indicative of the support given to research in this field. Countries in the forefront of this activity are the

TABLE 1-I

Papers published in Proceedings of Conferences on Coastal Engineering

Conference	Year	No. papers
1st	1948	35
2nd	1950	28
3rd	1952	26
4th	1954	28
5th	1956	43
6th	1958	54
7th	1960	59
8th	1962	45
9th	1964	53
10th	1966	86
11th	1968	101
12th	1970	138

United States, Japan, the United Soviet Socialist Republics, The Netherlands and France. Substantial economic and strategic interest in the oceans was exhibited by the U.S.A. President L.B. Johnson who provided the impetus for the United Nations Organisation to launch its "International Decade of Ocean Exploration". Prior to this he formed a "Panel on Oceanography" of his Science Advisory Committee, who submitted to him in 1966 a report entitled "Effective Use of the Sea" [14]. This was forthright in its assessment of the state of the art at this time, as can be gauged from the following quotation: "The nation needs to improve the technology for constructing coastal zone structures, which will make the national expenditure on break-waters, harbors, beach erosion, docks etc., more effective. The panel was distressed to find a high failure rate of construction projects in the surf zone and on beaches; the destruction of beaches by break-waters designed to extend the beaches, the silting of harbors and marinas as a result of construction designed to provide shelter; and the enhancement of wave action by the building of jetties supposed to lessen wave erosion are but a few examples of the inadequacy of our knowledge and practice in coastal engineering." It is to be hoped we have come a long way since 1965, but as a field for learning and research the abyssal depths are the limit, as distinct from the proverbial sky.

The report as noted above [14] was probably instrumental in the formation of the "Commission on Marine Science, Engineering and Resources" being established by the U.S. Congress in June 1966. This commission submitted its report [15] in 1969 entitled *Our Nation and the Sea* (300 pp.) with three complementary volumes of panel reports entitled: I. "Science and Environment" (340 pp.); II. "Industry and Technology" (309 pp.); III. "Marine Resources and Legal-Political Arrangements for their Development" (406 pp.).

These tomes are a comprehensive assessment of the present situation and future hopes for a full utilization of the seas surrounding the United States. Every other nation could well apply the findings to their own oceanic margins because, as stated in Chapter 1 [15]: "The Nation's stake in the uses of the sea is synonymous with the promise and threat of tomorrow. The promise lies in the economic opportunities the sea offers, in the great stimulus to business, industry, and employment that new and expanded sea-related industries can produce.... The threat lies in the potential destruction of large parts of the coastal environment, in the future deterioration of economically important ports, recreational facilities, coastal shell fisheries, and fisheries of the high seas."

At the same time as it set up the above commission the U.S. Congress established the "National Council on Marine Resources and Engineering Development", which was charged with the planning and coordination of current marine programmes and with advising and assisting the President. This involved the overseeing of the "National Sea Grant Programme" [16] in which the National Science Foundation dispersed funds for research, education and advisory services directed to the development of the nation's marine resources and uses of the sea.

In the United Kingdom, the Institution of Civil Engineers is equally concerned with marine matters. A research council reported in 1968 on research needs for the next two decades [17] many of which were nautical in character. In the same year a Research and Development Committee of I.C.E. reported on problems of undersea engineering [18]. The primary aspects considered were "...exploitation (and conservation) of the biological resources of the sea; and the exploitation of the mineral resources of the sea and, more particularly, the sea bed".

Many other countries and international organisations could be quoted in respect to their accelerated interest in the ocean and its resources. The potential for pollution from the mammoth tankers plying the high seas has also inserted a common concern for research and legal problems. The population explosion is another incentive to more fully utilize the food content of oceans in order to supplement the land supply of protein.

COSTS AND REWARDS IN OCEAN ENGINEERING

Engineering tasks associated with the ocean are vastly more expensive than similar activities on land. The hazards are greater and the organization more demanding. The further from shore, or the greater the distance from sources of supply, the more prodigious the cost of any maritime operation. The forces of weather must be contended with continually, and hence probable delays must be part of the price of the project.

Costs of maritime activities are no small proportion of total governmental expenditure on construction as instanced by the following figures.

Some values and costs in the *United States* back in 1950 have been published by Mason [19]. New Jersey spent 1 million dollars annually over 20 years protecting its beaches which were valued at U.S. $20,000 per foot length. The yearly cost of dredging navigation channels throughout the country was $48 million, which amount did not consider losses by interception of littoral drift. Mason added: "It is difficult to assess the total economic significance of shore control to the people of the United States. Perhaps a conservative guess might be of the order of 100 million dollars annually, considering only actual expenditures and damage costs, and disregarding the intangible and difficult to evaluate costs associated with loss of recreational benefits and loss of shore area without immediate resulting damage."

Schwartz [20] has quoted similar figures for 1968 which had reached an annual level of around 1,250 million dollars for civil works alone. The previous figure of approximately 150 million dollars could be doubled to cover inflationary trends up to 1968, so that overall a four-fold increase had occurred over two decades. With ever larger ships demanding access to ports, dredging costs are bound to increase, as will the costs of remedial measures if knowledge of shoreline processes at specific locations is lacking.

TABLE 1-II

Annual expenditure (million U.S. dollars) on various maritime activities

Activity	Japan, Ministry of:			Japan total	Canada Public Works
	Transport	Construction	Agriculture		
Sea defense	32	24	9	65	3.6
Harbour works	84	–	28	112	17.4
Dredging	57	–	9	66	18.5
Reclamation	216	–	2	218	3.3
Reparations	5	9	8	22	5.1
Total	394	33	56	483	47.9

Silvester [2] has quoted costs of marine structures, coast protection and land reclamation in *Japan* and *Canada*. These are contained in Table 1-II and are representative of mid 1960's. The three national totals approximate $2.5 per capita, which may be fortuitous, but the point to be made is that the figures are conservative (as they exclude private or defense expenditure on facilities) and are therefore illustrative of an expensive phase of engineering which requires specialist knowledge.

On the other side of the slate, the sea and its bottom boundaries have provided prizes valued in millions of dollars. Naturally this is not pure profit, the extraction

TABLE 1-III

Mineral production from oceans bordering the United States of America (million U.S. dollars)

Year	Magnesium metal and compounds, salt and bromine	From wells: petroleum, natural gas, and sulfur	Sand and gravel, zircon, feldspar, cement rock and limestone	Total
1960	69	424	47	540
1961	73	497	46	616
1962	89	621	44	954
1963	85	731	42	858
1964	94	820	44	958
1965	103	933	51	1087
1966	117	1178	52	1347
1967	114	1451	56	1621
Total	744	6655	382	7781
Average	93	832	48	873

of minerals has been, and still is, worth the effort. Dean [21] has published data on the effort to utilize the resources contained within the continental shelves of the United States of America. He quotes figures supplied by the U.S. Bureau of Mines for 1966, which are reproduced in Table 1-III, rounded off to the nearest million dollars.

Dean [21] states that the continental shelves of the Atlantic and Gulf coasts have produced 7,000 million dollars worth of oil and gas already. The landward zone of this wedge of sediment (the Mississippi Valley) has yielded altogether 85,000 million dollars worth of minerals and fuels. If the shelf can provide the same value as an equal area of coastal margin it has the potential to provide 160,000 million dollars worth of minerals and 80,000 million dollars worth of oil and gas. Dean concludes: "Although it is more costly to exploit marine minerals than comparable resources on land, nevertheless, the opportunity for greatly increased offshore production is bright." This can be considered the understatement of the year, as gauged by the volume of investment pouring into offshore exploration, not only on the U.S. coastal margins, but all over the world.

NEED FOR EDUCATION OF OCEAN ENGINEERS

With the mammoth tasks facing the engineering profession in the nautical field, the need for expertise at graduate, baccalaureate and technician levels becomes evident. Goodwin [16], in noting that an engineer might be able to transfer from a land specialty into a sea job, remarks also: "The marine environment is a lot tougher than land or air. It's a marvellous combination of a biological broth, a corrosive chemical soup, a weak electrolyte, an energy transfer medium of rather awesome proportions, and a realm of pressure that varies with depth at about 1 p.s.i. for each 2 ft. of depth until it reaches several tons per square inch in the abyssal depths." Outlining the main aims of the Sea Grant Programme already mentioned he states: "... ocean development depends on closer cooperation between scientists and engineers, with strong inputs from the social scientists. It depends on getting a feedback loop established between industry, the academic, scientific and engineering communities, and federal and state governments."

Any well balanced teaching programme should include, besides the basic technical courses, peripheral topics related to the medium and a research programme at various levels. Major engineering courses could be:

coastal hydraulics	sediment transport
wave hydrodynamcs	port and coastal structures
harbour hydraulics	dredging
tidal and estuarine hydraulics	reclamation
inland waterways	offshore structures

To these should be added basic courses in oceanography and fluid mechanics, plus core subject matter in ocean engineering or oceanology [3] such as:

environmental aspects
the human in the environment
materials in the environment
instrumentation and operations
communication and transfer

A more detailed subdivision of ocean engineering has been supplied by Calhoun et al. [3] as in Table 1-IV, with additions as gleaned from a contents page of the *Oceanic Citation Journal* [9].

TABLE 1-IV

Subdivisions of oceanology

Basic information	*Applications*
Ocean environment	Shore and estuaries
fluid flow	beach processes
waves	wave propagation
oceanography	wave forces
–, physical	coastal structures
–, chemical	foundations
–, geological	sedimentation
	corrosion
Material behaviour	desalination
properties	pollution
corrosion	waste disposal
structures	tidal power
vehicles	thermal power
Human factors	Offshore
diving	floating structures
habitation	fixed structures
performance	pipelines
maritime activities	dredging
Communication	undersea probes
acoustics	undersea structures
electro-acoustics	salvage techniques
optics	foundations
television	corrosion
surveying	Deep ocean
navigation	undersea installations
Operations and instrumentation	bottom stations
submersibles	tool development
installations	dredging
salvage	foundations
recording	corrosion
data processing	deep submersibles
economics	Measurements
legal aspects	recording devices
	Aquaculture
	fish farming

There is little doubt that the sea is a sphere of engineering activity for the coming decades. The rewards to be had will receive the investment, which in turn will demand countless engineers over a wide spectrum of specialties. These personnel cannot be provided at short notice, a lead time of 5–10 years is necessary. With the present expansion in marine activity, in spite of 1970 cutbacks in space and aircraft industries, it would appear that employment opportunities are more stable in the turbulent sea situation. The same warning in respect to training can be given for programmes of research, as noted by Dean [21] in 1969: "An aggressive and viable marine minerals research programme now is the only effective deterrent to a much more expensive and less efficient effort in the future, when the need will become critically apparent. Crash programs of that nature are highly undesirable in terms of social and economic costs and can only be avoided if sufficiently long-range plans are formulated and executed with adequate allowance for the required technological lead time."

In a report of an informal discussion by the "Research and Development Committee of the Institution of Civil Engineers" (London) in 1969 [22] the same sentiments were aired, as can be instanced by the following quotations: "While ocean engineers existed in Britain by evolution rather than training, there are at present in the U.S.A. 4400 enrolments of students and graduates in 65 institutional departments of marine sciences and ocean engineering, engaged in research and development projects valued at nearly 14 million pounds ..... The present annual government expenditure on oceanography and hydrography totals about £10 million, including about £4 million on fisheries. The corresponding figures in U.S.A. are about £196 million (civil) and £94 million (defence) with expenditure in the U.S.S.R. running at about £100 million ... Although the long-term commercial possibilities are attractive there appears no likelihood of adequate commercial investment on account of the complexity and variety of the technical problems to be solved. There is, therefore, a need for the Government to take a lead in awarding research and development contracts to industry until the ultimate goals could be more clearly focussed."

Sir Frederick Brundrett delivered the Graham Clark lecture to the Institution of Civil Engineers in November 1969 on the topic "The sea of opportunity – what the oceans have to offer the engineer" [23]. He reports: "Stimulated by complaints in the House of Commons and elsewhere that this country was falling behind other nations in the race to extract advantage from the development of ocean-based industries, and by the interest taken in the proceedings of the conference arranged by the Ministry of Technology at Harwell in April 1967, the Government set up in July 1967 an Interdepartmental Working Party on Marine Science and Technology. The report of this body was presented to Parliament as a White Paper, 'Report on Marine Sciences and Technology' Command 3992, in April 1969."

The following opportunities for engineers were discussed [23]:

ship design	ship's machinery	navigation
mapping oceans	moored stations	harbour engineering
tidal power	weather forecasting	pollution
minerals	naval defense	fishing

The above items emphasize the mechanical facets of an ocean engineering career. Openings for the civil-oriented graduate should be available as employees of or consultants to:

public works departments	irrigation departments
local councils	fisheries organisations
petroleum companies	defense departments
electric power authorities	public health departments
teaching institutions	research organisations

As already noted, marine activities are always expensive. Engineers associated with design and planning must be as conversant with the principles and processes involved as the technical literature will allow. When a maritime mistake is made it may remain hidden for years but by the time it is exhibited, the costs to remedy it may be exorbitantly high.

Technology of the ocean is a post-World War II development, as are many facets of physical oceanography on which it is based. Whilst the scientist can pursue his pet theories as his desires take him, the engineer must work in a dictated field and come up with answers, which he hopes are as soundly based in theory and experimental verification as is possible at the time. The scientist can also devote full time, for a long time, on a narrow field and therefore push out the frontiers of knowledge more readily than the engineer who must be concerned as much with economics as with principles of science. It is this large discrepancy in academic opportunities that probably prompted Munk to comment at the close of the Conference on Ocean Wave Spectra [24] in 1963 as follows: "I think it is a fair conclusion to say that those who are interested in the movement of sand, the disappearance of beach houses, the reflection of radar, the reason why a Texas Tower failed, and how sewage spreads into the sea have not really gotten tangible methods out of the great improvements that have taken place in our basic understanding of the physics of ocean waves. Though engineers have improved their own understanding a great deal and have adopted more sophisticated methods, the theorists have improved their approach even more, so that the two are further apart than ever."

There is a wide scope for any engineer who wishes to enter oceanology. He can serve as a liaison between the scientist and the designer, he can be a researcher into many aspects of theory still unknown, he can design or enter the contracting field. All the time he will be dealing with a fascinating medium which covers nearly three quarters of the globe. With this great attraction ahead of us, let us enter it with a few simple definitions of terms.

DEFINITIONS

A comprehensive glossary of terms related to coastal engineering is available [25] and another covering oceanography and oceanology [26]. The following brief outline of topics is meant to serve only as a basis for the text to follow, which will contain more definitive explanations.

Wave is a surface undulation of a liquid surface, generally at the interface with a gas. Unless otherwise specified, waves will be considered only at the interface of fresh or salt water and the atmosphere. Where layers of water with different densities are in contact waves can occur which undulate this interface and are known as *internal waves*. This latter phenomenon may, in the future, be found to be of some engineering significance.

Wave crest is the section of a wave profile where the water reaches its greatest height. It is a convex upward shape, the curvature of which is determined by many factors (see Fig. 1-2).

Wave trough is the section of a wave profile where the water is at its lowest level. It is concave upwards or may even appear horizontal over a reasonable length of the wave.

Wave height is the vertical distance from the crest level to the trough level. In a simple series of waves the height of each wave remains the same, but in a more complicated undulation the height will vary for successive waves, so that height must be consistently measured from a crest to a preceding or following trough (see Fig. 1-2).

Wave amplitude is always taken as half the wave height. For waves with small height in very deep water this amplitude is the distance vertically from the still-water level (SWL) to either the crest or the trough. For a simple series of waves of greater height, or shallower water, the half-way mark between crest and trough is termed the *mean water level* (MWL), which is not necessarily coincident with the SWL (see Fig. 1-2).

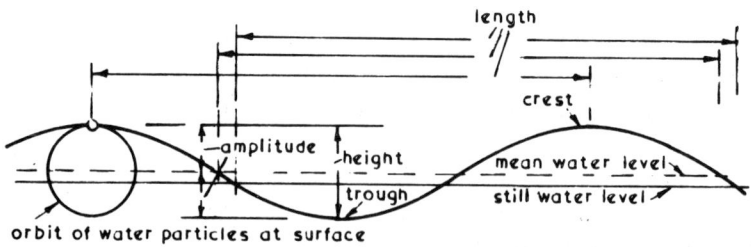

Fig. 1-2. Definition sketch of a wave.

DEFINITIONS

Wave length is the horizontal distance between successive crests or troughs. For a simple series of waves the wave length is also given by the distance to points where the water level is at a given datum (for example, MWL) and the surface is either rising or falling (see Fig. 1-2).

Wave period is the time for two successive crests (or troughs) to pass one point on the water surface. It is the one characteristic of a wave that remains constant at all times, no matter what changes occur in height or length. The period is normally expressed in seconds.

Wave frequency is the reciprocal of the wave period and represents the number of waves passing a location in some unit of time (generally seconds).

Wave train is a series of waves whose period is constant. When in water of constant depth, the height and length of successive waves in a train is the same. A train may be considered to contain an infinite number of waves or a finite number, in which case the total may be termed a *wave group*. This simple model for waves in a prototype situation is used extensively in engineering for computation of wave forces and other influences on fixed boundaries.

Wave celerity or wave speed is the velocity at which a wave travels across a liquid surface, so that by definition:
celerity $\equiv$ length/period
or length $\equiv$ celerity $\times$ period
or period $\equiv$ length/celerity

As already noted, the period remains constant, thus making celerity and length directly proportional. These two characteristics depend upon the ratio of (water-depth/wave-length) for many engineering problems.

Deep water is the depth beyond which the wave celerity is not affected by depth and is therefore dependent only upon length or wave period. This limit is given by a specific ratio of (water-depth/wave-length) = 0.84 but between this value and 0.5 the changes are so slight that engineers accept the latter proportion for many computational purposes.

Shallow water is the depth within which the wave celerity depends solely upon depth and is therefore independent of wave period. This limit also is given by a specific ratio of (water-depth/wave-length) which is chosen by engineers to simplify situations.

Transitional depths are those between deep water and shallow water, in which the wave celerity depends upon wave period and water depth. In most engineering problems the changes occurring in a wave train over the transitional zone are of extreme importance.

Wave steepness is the ratio (wave-height/wave-length) which for any wave train increases as it travels from deep water to transitional depths and then shallow water. There is a limiting value as the wave increases in height or decreases in length, either due to addition of energy in the generating process or decreasing speed in the shoaling process, when the wave becomes unstable and breaks. This limiting steepness varies from about 1/7 in deep water to about 1/10 in shallow water.

Sinusoidal wave is that wave in which the profile is assumed to be the shape of a sine or cosine function. This condition exists for waves in deep water when the wave height is very small, infinitesimal in mathematical language. Many computations are carried out for wave phenomena based on this assumption which provide relationships that are applicable to prototype situations where such restrictions do not apply. The sinusoidal wave form lends itself to a powerful mathematical tool known as the Fourier analysis.

Finite height refers to waves whose steepness is such as to disallow the sinusoidal assumption. However, by considering a wave to consist of an accumulation of sinusoidal components of equal period, it is possible to simulate finite height waves.

Trochoidal wave is one whose profile is that of a trochoid, which is a curve traced by a point within a circle which rolls along a horizontal datum (see Fig. 1-3). Because of the steepness limitation of $1/7 = 2r/2\pi R$, the maximum practical ratio of r/R is around 0.5, when the angle contained in the apex of the wave crest is around 120°, a criterion for breaking found from other mathematical procedures. If the point traced out were on the circumference of the circle, the curve is called a *cycloid,* but cycloidal profiles are impossible for water waves due to their instability. Trochoidal wave profiles occur only in deep water and when their steepness is small they tend to be sinusoidal. A number of workers [27] have derived wave relationships on the basis of the trochoidal shape, which are available through more realistic mathematical approaches.

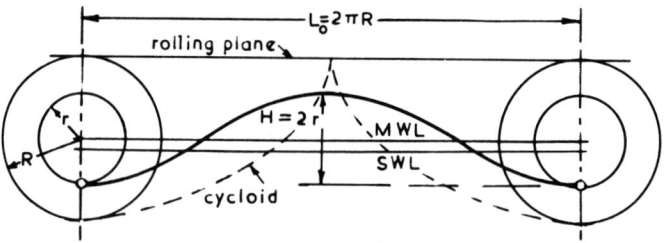

Fig. 1-3. Definition sketch of a trochoidal wave.

DEFINITIONS

Cnoidal wave is the profile assumed by waves in shallow water and is given by a mathematical series containing the Jacobian elliptic cosine designated "cnu". Besides the profile other characteristics are derivable, but the relationships are not readily presentable for design purposes.

Hyperbolic wave is the name given to cnoidal wave relationships in which certain functions are assumed constant. By this means [28] the various relationships can be graphed or tabulated. They apply only to waves in shallow water.

Solitary wave is a hypothetical shape of wave just prior to its breaking. It purports to rise directly from SWL so that MWL is (wave height/2) above SWL. Although unrealistic in this sense, the various relationships can indicate limits of action in controlled experiments.

Wind velocity refers to some mean wind speed occurring at some specified height above the sea surface (10 m, for example). Relationships between wave characteristics and wind velocity ignored the height factor for many years, but those presented later incorporate it. Conversion of velocities from one level to another is based upon a logarithmic distribution near the surface, which is yet to be proven conclusively.

Fetch is the area of ocean over which a wind has blown and generated waves. More specifically the term refers to the length of this zone as measured along the mean direction of wind near the sea surface. The upwind and downwind boundaries may be land masses or zones where sharp changes in wind direction are known to exist, or the wind reduces speed significantly. Besides a length, a fetch has a width, which can influence the growth of waves, but is not normally incorporated into forecasting formulae. As long as the fetch width is at least a quarter of the length the usual relationships should hold. In the ocean these dimensions are in the order of 100 – 1000 miles.

Duration is the time (generally in hours) for which the wind of constant velocity has been blowing across the specified fetch. Waves will continue to grow from the commencement of the wind up to a specific duration, after which no further growth at the given location is possible.

Fully arisen sea (FAS) is the condition when the fetch length and duration are long enough for a given wind velocity to produce the highest waves possible. This steady wave state requires a minimum fetch and duration which can be related to the wind velocity at a specific height above the sea surface.

Decay area is the term normally used for the zone downwind of a fetch into which the bulk of the waves propagate after leaving the fetch. A more appropriate term might be *dispersal area,* as it is here that the waves disperse longitudinally and transversally. Since their reduction in height is due to spreading of energy over an increasing area of ocean and not due to viscous or turbulent effects, the adjective decay appears misplaced.

Nautical mile (N.M.) is the circumferential distance along the equator subtended by 1 min of longitude. It equals 6080 ft., 1.85 km or 1.15 statute miles. The same angle will subtend a smaller distance at higher latitudes, but the above values should serve up to 50°N or S. This term introduces the term *knot* which is a speed of 1 N.M. per hour, and may be used for wind velocities or wave celerities.

Fathom is a depth measurement used in hydrographic surveys equal to 6 ft. or 1.78 m. Where fathoms are used on marine charts intermediate depths may be expressed in ft., such as 0^5, 2^4 etc.

Propagate is the term used to describe the passage of a wave across a liquid surface. It is only the wave form that moves since the water-particle motions creating it take place ostensibly in the same location (see Fig. 1-2). It is because of this modest motion transmitting the wave form that little or no energy is lost when one wave train passes through another, either in alignment or angled to it.

Dissipate refers to the loss of energy in any kinematic process, in this case the loss of height in a wave train. Energy can be computed per unit area of ocean, or per wave length for a unit length of crest. If energy is not lost or spread, as a wave travels towards shore, the reduction of length is balanced by an increase in wave height. This increases the steepness, as previously mentioned, until breaking occurs.

Attenuate refers to a reduction in some specific factor such as wave height, water-particle orbit, pressure fluctuation, etc. Attenuation can be effected without dissipation or loss of energy, as it can result from the physical nature of the phenomenon, for example, attenuation of orbital motion with depth.

Progressive wave is one in a train of waves which is propagating into areas of calm, or areas where other trains exist. It is the type of wave in which the coastal engineer is normally interested. The French term for it is "la houle".

Standing wave is produced when two progressive waves of equal period and height are propagating in exactly opposite directions. The resulting surface oscillation contains crests which form alternately at points called *antinodes,* which are one half the progressive wave length apart (see Fig. 1-4). At points mid-way between the antinodes the water level remains constant, forming nodal points. Such complete standing waves are termed "le clapotis". These are difficult to observe in nature or in the laboratory since opposing waves are likely to differ slightly in height or period, in which case particles at the nodes have some vertical motion. These waves are called "partial clapotis" and are also illustrated in Fig. 1-4, where envelopes of the maxima and minima are drawn.

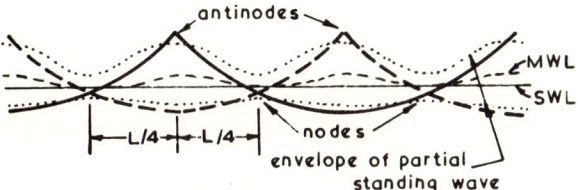

Fig. 1-4. Definition sketch of a standing wave and a partial standing wave.

Short-crested waves are formed when two progressive waves are angled to each other. At the intersection of the crests a hump is created and at the intersection of the troughs a cumulative depression occurs (see Fig. 1-5). The combined crests will pass along the diagonal formed by the successive crests of the longer wave and the crest alignments of the shorter waves, the interval between which is the same as the

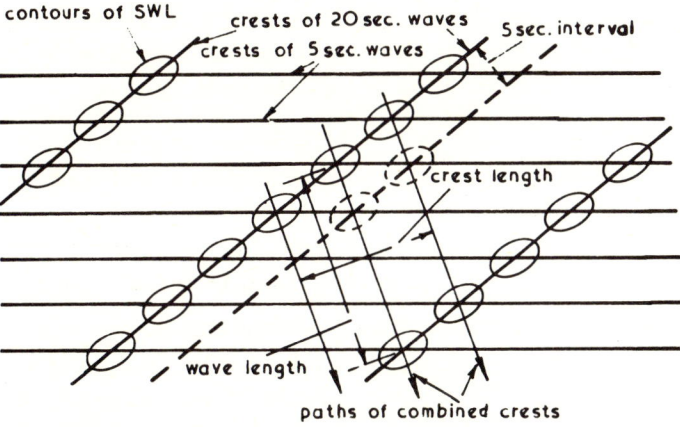

Fig. 1-5. Definition sketch of short-crested waves.

period of the longer wave. (For example, in Fig. 1-5 the successive 5-sec locations of the 20-sec wave will create combined crests as illustrated, which travel along the paths shown. The wave length is the distance between successive combined crests, whilst the crest length is the transverse distance between adjacent crests. Such short-crested patterns, on a much more complicated basis than depicted in Fig. 1-5, occur within fetches, where wave trains of similar and different periods are propagating over a variety of directions around the mean wind vector. Many other cases arise in which angled wave trains play a predominant role in coastal processes. The French term for this wave system is "clapotis gaufré".

SOURCES OF INFORMATION

Besides the texts already mentioned [25, 27] there are books on oceanographical topics [29—36], wave generation [37, 38], coastal processes [39—44], coastal structures [45—49], reclamation and dredging [50—52] and sources of general information pertaining to the sea and its margins [53—62].

Most of the above publications are based upon information originating in the technical literature. In the field of coastal engineering this has been spread over many journals, because of its variety of basic science, mathematics and applications. The reader would do well to peruse periodically the journals listed in Table 1-V.

TABLE 1-V

Journals containing information pertaining to ocean engineering

Journal	Frequency	Coverage
Proc. Am. Soc. Civil Engrs. (Waterways) (Hydraulics)	quarterly	all eng. topics
Proc. Inst. Civil Engrs. (London)	monthly	all eng. topics
J. Natl. Eng. Institutions (of various countries)	monthly	all eng. topics
Coastal Eng. Japan	annually	all eng. topics
Proc. Coastal Eng. Conferences	biennial	all eng. topics
Proc. Int. Assoc. Hydraul. Res.	biennial	all eng. topics
Coastal Eng. Res. Centre (U.S. Army)	sporadic	all eng. topics
La Houille Blanche	monthly	all eng. topics
Dock Harbour Auth.	monthly	all eng. topics
Shore and Beach	monthly	sediments
J. Geophys. Res.	monthly	scientific
J. Fluid Mech.	monthly	mathematical
Ocean Eng.	monthly	scientific
Ocean Industry	monthly	application
Oceanology	monthly	application
Deutsche Hydrogr. Z.	monthly	scientific
Cahiers Océanogr.	monthly	scientific
Oceanic Citation J.	monthly	abstracts
J. Geol.	monthly	sediments
Marine Geol.	monthly	sediments
Sedimentology	quarterly	sediments

Other useful sources of information will be gleaned from the list of references at the end of each chapter. Although these are not claimed to be comprehensive, they should aid as an introduction to the literature for the particular topics. Because of the great advances being made in the science and application of knowledge in ocean engineering, it behoves the coastal engineer to keep abreast of the literature.

REFERENCES

[1] T.C. Kavanagh, 1968. Ocean engineering and the American Society of Civil Engineers. *Civil Eng.*, 38(2): 55–56.
[2] R. Silvester, 1971. Australia's interest in ocean engineering. *Trans. Inst. Engrs. Aust., Civil Eng.*, CE 13: 71–75.
[3] J.C. Calhoun, C.H. Samson and J.B. Herbich, 1968, What is the core in ocean engineering? (Paper presented to Annual Meeting Am. Soc. Eng. Educ., June 1968; available from Texas A and M Univ.).
[4] *Proceedings Santa Barbara Specialty Conference on Coastal Engineering, 1965.* Am. Soc. Civil Engrs., New-York, N.Y., 1966.
[5] *Proceedings 1st Conference on Ships and Waves, 1954.* Am. Soc. Civil Engrs., New York, N.Y., 1955, 509 pp.
[6] *Proceedings on Ocean Wave Spectra.* Prentice Hall, Englewood Cliffs, N.J., 1963.
[7] *Preprints of Offshore Technology Conferences.* (Sponsered annually by many U.S. Engineering Societies since 1969.)
[8] *Proceedings World Dredging Conferences* (Sponsored annually by World Dredging Association since 1968.)
[9] *Ocean Citation Journal with Abstracts.* (Journal commenced in 1964.)
[10] *World Dredging and Marine Construction.* (Journal commenced in 1965.)
[11] *Oceanology.* (Journal commenced in 1965.)
[12] *Ocean Industry.* (Journal commenced 1966.)
[13] *Ocean Engineering.* (Journal commenced 1968.)
[14] Panel on Oceanography of the Presidents Science Advisory Committee, 1966. *Effective Use of the Sea.* U.S.G.P.O., Washington D.C., 144 pp.
[15] Commission on Marine Science, Engineering and Resources, 1969. *Our Nation and the Sea* (plus 3 volumes of Panel Reports). U.S.G.P.O. Washington, D.C., 305 pp.
[16] H.L. Goodwin, 1968. The role of technology in ocean development – the sea grant perspective. *Proc. Workshop Ocean Eng. Educ., Univ. Deleware*, 1968: 8–16.
[17] Anonymous, 1968. Future research: research and development report. *Proc. Inst. Civil Engrs.*, 39: 477–482.
[18] Anonymous, 1968. Undersea engineering: research and development report. *Proc. Inst. Civil Engrs.*, 39: 483–487.
[19] M.A. Mason, 1950. Geology in shore controls problems. In: P.D. Trask (Editor), *Applied Sedimentation.* Wiley, New York, N.Y., ch. 15.
[20] H.A. Schwartz, 1968. Ocean and marine related activities of the U.S. Army Corps of Engineers. *Proc. Workshop Ocean Eng. Educ., Univ. Delaware*, 1968: 116–122.
[21] G.W. Dean, 1969. A pragmatic look at the oceans' mineral resources. *Trans. New York Acad. Sci.* 31: 731–736.
[22] Anonymous, 1969. The advancement of undersea engineering; research and development committee – informal discussion. *Proc. Inst. Civil Engrs.*, 42: 601–602.
[23] Sir Frederick Brundrett, 1970. The sea of opportunity: what the oceans have to offer the engineer. *Brit. Eng.*, 1970: 2–4.
[24] W.H. Munk, 1963. Discussion. *Ocean Wave Spectra*, 1963: 345.
[25] Anonymous, 1966. *Shore Protection Planning and Design* (3rd ed.) – *U.S. Army, Coastal Eng. Res. Centre, Tech. Rep.*, 4.
[26] M.L. Hunt and D.G. Groves (Editors), 1965. *A Glossary of Ocean Science and Undersea Technology Terms.* Natl. Acad. Sci. – Compass Publ., New York, N.Y.
[27] R.L. Wiegel, 1964. *Oceanographical Engineering.* Prentice Hall, Englewood Cliffs, N.J., 532 pp.
[28] Y. Iwagaki, 1968. Hyperbolic waves and their shoaling. *Proc. 11th Conf. Coastal Eng.*, 1: 124–144.

[29] P.K. Weyl, 1970. *Oceanography: an Introduction to the Marine Environment.* Wiley, New York, N.Y., 539 pp.
[30] C. Eckart, 1960. *Hydrodynamics of Oceans and Atmospheres.*
[31] A. Defant, 1961. *Physical Oceanography, 1, 2.* Pergamon, London, Vol. 1: 745 pp., Vol 2: 606 pp.
[32] Von Arx, 1962. *An Introduction to Physical Oceanography.* Addison-Wesley, 422 pp.
[33] M.N. Hill, (Editor), 1962. *The Sea, 1. Physical Oceanography.* Wiley-Interscience, New York, N.Y., 864 pp.
[34] G. Neumann and W.J. Pierson Jr., 1966. *Principles of Physical Oceanography.* Prentice Hall, Englewood Cliffs, N.J., 555 pp.
[35] Natl. Bur. Standards, 1951. Gravity waves. *Natl. Bur. Std., Circ.*, 521: 287 pp.
[36] *Advances in Geophysics.* Academic Press, New York (since 1945 each year one volume).
[37] B. Kinsman, 1965. *Wind Waves.* Prentice Hall, Englewood Cliffs, N.J., 676 pp.
[38] W.J. Pierson Jr., G. Neuman and R.W. James, 1955. *Practical Methods for Observing and Forecasting Ocean Waves.* U.S. Hydrogr. Office, Publ., 603: 284 pp.
[39] E.C.F. Bird, 1968. *Coasts.* Austr. Natl. Univ. Press, 246 pp.
[40] J. Larras, 1957. *Plages et Cotes de Sable.* Eyrolles, Paris.
[41] C.A.M. King, 1959. *Beaches and Coasts.* Edward Arnold, London, 403 pp.
[42] F.P. Shepard, 1963. *Submarine Geology.* Harper and Row, London, 557 pp.
[43] N.N. Djounkovski and P.K. Bojitch, 1959. *La Houle.* Eyrolles and Gauthier-Villars, Paris, 404 pp.
[44] P. Bruun and F. Gerritsen, 1960. *Stability of Coastal Inlets.* North-Holland, Amsterdam, 123 pp.
[45] R.R. Minikin, 1950. *Wind Waves and Maritime Structures.* Charles Griffin, London, 216 pp.
[46] J. Chapon, 1966. *Travaux Maritimes, 1, 2.* Eyrolles, Paris, 1: 285 pp., 2: 276 pp.
[47] A.M. Muir Woods, 1968. *Coastal Hydraulics.* McMillan, London, 187 pp.
[48] H.F. Cornick, 1958–1962. *Dock and Harbour Engineering, 1–4.* Charles Griffin, London.
[49] A.J. Savory (Editor), 1969. *Tanker and Bulk Carrier Terminals – Proc. Conf. Inst. Civil Engrs.,* 1969: 109 pp.
[50] J. van Veen, 1952. *Dredge-Drain Reclaim.* Martinus Nijhoff, The Hague, 3rd ed.
[51] J. Huston, 1970. *Hydraulic Dredging – Theoretical and Applied.* Cornell Maritime Press, 318 pp.
[52] Anonymous, 1968. Dredging. *Proc. Symp. Inst. Civil Engrs.*, 1967: 119 pp.
[53] A.T. Ippen (Editor), 1966. *Estuary and Coastline Hydrodynamics.* McGraw Hill, New York, N.Y., 744 pp.
[54] J.F. Brahtz (Editor), 1968. *Ocean Engineering: Goals, Environment, Technology.* Wiley, New York, N.Y., 720 pp.
[55] W. Bascom, 1964. *Waves and Beaches – the Dynamics of the Ocean Surface.* Anchor Books and Doubleday, London, 267 pp.
[56] R.W. Fairbridge (Editor), 1966. *Encyclopedia of Oceanography.* Reinhold, New York, N.Y., 1021 pp.
[57] R.W. Fairbridge (Editor), 1968. *Encyclopedia of Geormorphology.* Reinhold, New York, N.Y., 1295 pp.
[58] B. Le Méhauté, 1969. *An Introduction to Hydrodynamics and Water Waves, 1, 2.* U.S. Dept. Commerce, ESSA, Tech. Rep., 1: 503 pp., 2: 725 pp.
[59] C.L. Bretschneider (Editor), 1970. *Topics in Ocean Engineering, 1, 2.* Gulf Publ. Co., New York, N.Y.
[60] G.M. Grant, 1971. *Oceanography.* C.E. Merrill, London, 2nd ed.
[61] R. Barton, (1970). *Oceanology Today: Man Exploits the Sea.* Aldus Books, London.
[62] M.S. Yalin, 1971. *Theory of Hydraulic Models.* Macmillan, London, 266 pp.

Chapter 2

WAVE-GENERATION PROCESSES

Coastal engineers are mainly interested in the larger waves that are generated by the wind in the storm areas of the oceans. Any particular storm may cover hundreds of square miles. But, to obtain an insight into the final product of this transfer of energy, it is instructive to account for the waves from the instant of their formation on the smooth water surface. Their development can then be followed through the stages of small waves, as encountered on rivers and lakes, to the complete series of waves, of long and short period, which combine at times to form waves of some tens of feet in height.

Waves on the ocean have distinct characteristics, depending upon whether they are still under the influence of the wind or are spreading out from the zone of generation. These two categories will be called, respectively, *storm waves* and *swell*. They may be present singly or in combination, since swell waves can travel thousands of miles before being dissipated on some shoreline. The characteristics of these two wave systems will be described, because of their importance in beach processes and in forces exerted on man-made structures.

It will be seen that even a simple fetch situation can produce complex wave conditions. The introduction of additional variables such as weather patterns complicates the problem to a seemingly insurmountable one, in respect to computing heights and periods of waves for design purposes. The influencial low-pressure systems, which produce the high-velocity winds, assume various forms dependent on their latitude. Their frequency of occurrence and general path of travel are of paramount importance to the coastal engineer in assessing a wave climate for a particular section of shoreline.

MECHANISM OF GENERATION

The development of waves will be traced in the order it is believed to occur, which is not the order in which scientists and mathematicians have applied themselves to the problem. However, it should be realized at the outset that these various phases of generation are simultaneous and continuous. Whilst the longer waves are slowly being built up, the shorter ones are growing to an unstable state, breaking, and new ones being generated. The breaking process disposes of much more energy, but some would appear to be transferred to the longer waves. For a given wind velocity there is a certain optimum wave energy expressable in statistical terms of wave height beyond which growth does not occur. This stage is termed a *fully arisen sea* (FAS), when the wind energy being transferred to the sea is being dissipated by wave breaking and other modes of loss at an equal rate. The wave conditions in a FAS are specific for each wind velocity.

There would appear to be four stages in the generation of waves that may be recognized as distinct mechanisms; these are: *(1) resonance* — produced by air turbulence on the water surface; *(2) shear flow* — arising from the profile of the mean wind velocity; *(3) sheltering effect* — as the waves influence the air flow near the surface; *(4) breaking* — after the waves have reached maximum steepness.

Resonance

The concept and associated theory of wave initiation by resonance of the water surface was developed in 1957 by O.M. Phillips [1]. This was a significant advance in the understanding of wave genesis. The author has developed this theory further over the years [2].

Movement of air in the form of wind is always associated with turbulence, or random fluctuations of pressure and velocity about some mean value. These are the result of swirls or eddies in the moving mass of air which, in essence, are revolving spheres of air varying in size from centimetres to hundreds of metres. The mode of this size distribution is determined by the wind velocity, as well as other atmospheric conditions and surface roughness. These eddies produce random fluctuations of pressure and of shear on the water surface, but Phillips' theory incorporates only the pressure fluctuations, which are considered normal to the surface.

The pressure pulse imparted on the water traverses the surface at the speed of the eddy in which it is contained, known as the convective speed. This speed will vary with the size of the eddy, because the smaller ones will travel at the wind velocity near the surface, whilst the largest ones will travel at the speed of the air some metres higher, which necessarily is much greater. Naturally these spheres of rotating air are relatively short lived, since they are attenuated by contact with other eddies and the air mass in general. Thus their influence on the water surface is transient, not only because of this variation in intensity, but also because of their change in level from the surface.

The depression caused by the pressure from the eddy is minute, but it creates a wave with a specific wave length. However, if the pressure can be maintained on this trough at the same speed of application as the natural speed of the wave it will grow deeper and so the wave will grow in dimension from crest to trough. One restriction applicable to this mechanism, therefore, is that the wave length is very much smaller than the travel distance of the eddy near the water surface. This condition is strictly only met for capillary and small gravity waves.

MECHANISM OF GENERATION

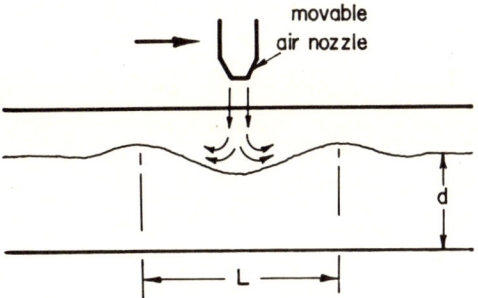

Fig. 2–1. Suggested flume test to illustrate wave generation by the resonance mechanism.

The matching of the speed of the wave first initiated to that of the eddy producing it introduces the concept of resonance. This can be demonstrated in a flume, as depicted in Fig. 2-1, by applying a jet of air vertically to the water surface. This produces a depression and associated crests either side, which forms a wave length. For the specific depth of water a wave of this length will have a given speed. If the air jet is moved along the flume at this particular speed the wave will grow in height. However, if it proceeds too quickly or too slowly the initial depression will be annulled and the only effect will be the temporary fluctuation occurring directly under the jet. The speed of the jet is not that critical, since velocities either side of the resonant value will also cause some wave generation.

Turbulence in any fluid is a three-dimensional phenomenon so that depressions of the water surface are random, spatially and temporally. The propagation of the waves from their origin is in all directions, like the circular ripples from impact of a solid body on a water surface. Thus, whilst in the two-dimensional experiment described above the jet velocity had to match the wave velocity for resonance to occur, this is not necessary in the three-dimensional case.

This might best be explained by considering Fig. 2-2, in which discrete locations of pressure cells X, Y and Z are considered. At some instant depressions X and Y occur and these proceed to propagate in all directions. The portion of Y that reaches Z at the same time as the depression from X has moved to Z, will receive extra impetus or be in resonance. As seen in the figure, the distance between Y and Z is $L(C)$ (associated with the velocity of the wave) and the length XZ is $L(U_c)$ (the distance traversed by the turbulent eddy in the same time at the convective speed U_c).

It is seen, therefore, that the trough travelling at an angle α to the wind direction is enlarged, whilst that travelling with the wind will be out of phase and so not receive full benefit from the eddy. The angle at which optimum generation occurs is known as the critical angle. Troughs moving at angles between α_{cr} and less will receive some resonant energy, but those outside this limit will be attenuated. Thus,

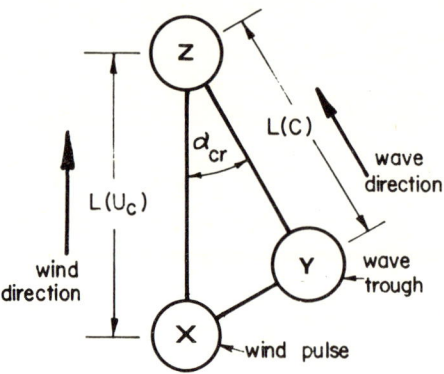

Fig. 2-2. Definition diagram for three-dimensional aspect of resonance model.

waves of a specific period will be moving in all directions between α_{cr} either side of the mean wind vector, but those propagating along the resonant or critical alignment will receive the most energy.

This so-called principal stage of wave development takes place when the application time is much greater than dissipation time of the eddies. This length of life of a pressure pulse cannot be gauged from the periods of pressure fluctuations measured at a point. If the probe were to travel at the convective speed appropriate to the eddy size a much longer sequence time would be recorded.

Another requirement for the application of the theory in this mechanism is that the waves so formed are small in height, otherwise the assumptions regarding smooth surface conditions no longer apply. The critical angle at this stage of development, when energy is being transferred linearly with time, is given by the equation:

$$\cos \alpha_{cr} = C/U = gT/2\pi U \tag{2-1}$$

where C is wave velocity, U is wind velocity, T is wave period and g is acceleration due to gravity. This relationship which is applicable to deep-water conditions, has been graphed in Fig. 2-3, where it is immediately obvious that the longer waves are more closely aligned to the wind than the shorter-period waves. The above holds so long as $U \geqslant C$, the limit of energy transfer by resonance being reached when the wave velocity equals the wind velocity. In the case of $U \gg C$, as in short-duration or limited fetches, the resonant maxima fail to materialize, other effects of shear flow, sheltering and breaking assuming importance.

The principal stage of wave generation by resonance discussed above is accompanied by the production of capillary waves. These are of little consequence to the coastal engineer, so will not be discussed here. Suffice it to say that they are swiftly attenuated by surface tension and have a wave length of around 1.7 cm.

MECHANISM OF GENERATION

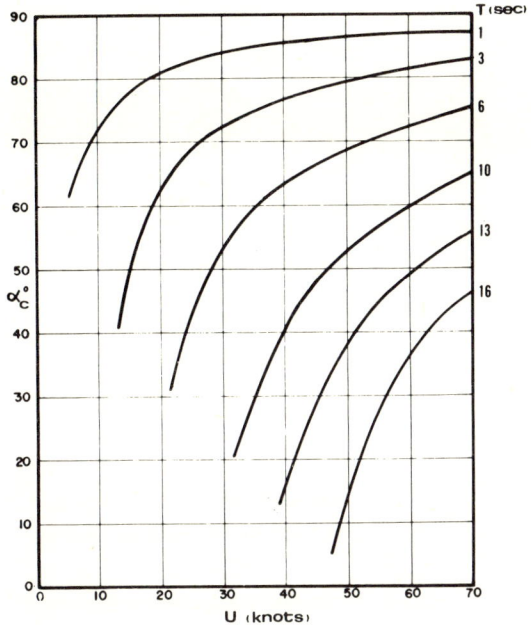

Fig. 2-3. Critical angle for resonance build-up of waves of various period for different wind velocities.

The resonance theory of Phillips excludes the action of air–sea coupling. It can only explain, therefore, the initiation of waves, after which other processes of generation take over. However, it provides an explanation for the initiation of even the longest waves in the spectrum. It is based upon certain assumptions which should be perused in the original papers of Phillips or in the convenient book form by Kinsman [3].

Shear flow

This theory of wave generation was devised by J.W. Miles in 1957 [4] almost simultaneously with that of Phillips. It is not meant to substitute for the resonance model, but rather to complement it. Many of the initial assumptions in the shear-flow model are similar to those of Phillips, the major additions being the exclusion of turbulent fluctuations and the inclusion of a logarithmic wind-velocity profile above the water surface. Transfer of energy is again assumed to be effected through pressures normal to the surface, since drag forces and drift currents are ignored.

The model assumes an initially disturbed surface consisting of sinusoidal waves of infinite crest length. This in turn induces waviness in the streamlines of air flow near the surface. This compression and spreading of the streamlines travels at the

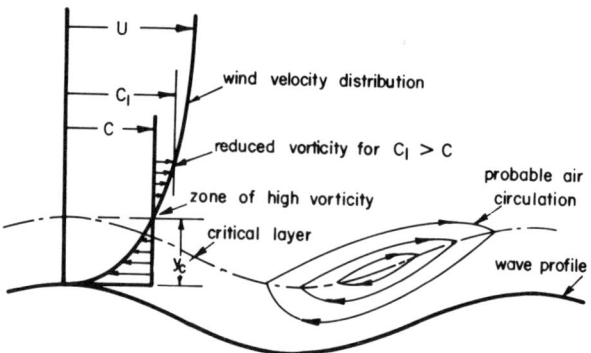

Fig. 2-4. Definition sketch for shear flow model of wave generation.

speed of the wave celerity C. If some distribution of wind velocity $U(y)$ is assumed as in Fig. 2-4, the subtraction of C from this distribution results in a zero point at a height y_c. This defines a critical zone below which the air velocity in essence is reversed in respect to the celerity of the streamline undulations. At the layer y_c above the water surface a strong vortex is established, which drains energy from the wind at the same rate as it is transferred to the wave.

The critical height y_c is a fixed distance above the undulating surface of the water. Vertical fluctuations at this critical height, however, are determined by the balance of a fluctuating vortex force and a fluctuating pressure force. If the waves were static the resultant pressure distribution due to a given wind velocity would be in phase with the wave (see Fig. 2-5), a positive pressure over the trough and a negative pressure over the crest in the case of sinusoidal waves of slight steepness. However, when the waves are propagating across the surface this pressure force lags in phase with the wave (see Fig. 2-5), a positive pressure over the trough and a surface particles of water in their downward motion. The wave thus receives energy and builds up. The rate of energy transfer varies as the square of the wave height so that wave energy grows exponentially as the wave steepens. This mechanism of

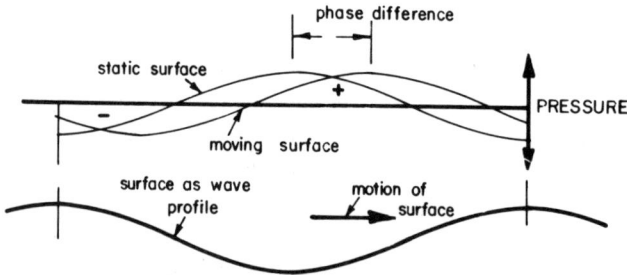

Fig. 2-5. Pressure distribution over an undulatory surface with and without waves in motion.

wave generation is limited to the stage before the streamlines break away from the wave crests. Some other mechanism then takes over, such as the sheltering effect, or breaking at a later stage.

The critical height y_c is in the order of $L/10$ and therefore is different for each component of the full-wave spectrum. As seen in Fig. 2-4, these intersect the wind profile at its various curvatures, so that the vorticity generated is lessened the greater the wave celerity length y_c. Thus, the shorter waves are built up more quickly by this mechanism, but provision is made for generation of waves with celerities approaching that of the wind. This is based upon the assumption that the wind distribution at the appropiate height of around $L/10$ is not affected greatly by the shorter waves already built up to their optimum steepness.

This coupling mechanism of wind and wave was derived initially from analytical methods, but has since been shown to apply in the laboratory and in prototype conditions. The logarithmic distribution of air velocity above the surface does not appear to be so critical as was at first believed. As emphasized by Stewart [5], the consumption of energy from the wind at the edge of the critical zone will greatly influence this distribution. Even if it is initially logarithmic it is not likely that it will remain so within the critical layer of the largest wave components.

Resonance/shear-flow combination

The two mechanisms just discussed complement each other, resonance from the air turbulence providing the initial undulation to the sea surface with a wide spectrum of frequencies, whilst the shear flow of the mean velocity wind profile selectively promotes the shorter waves. The two processes act simultaneously, but in the early stages of generation resonance predominates with linear build-up of energy, whereas later the coupling of wind and wave produces an exponential increase of energy.

A particular wave component may reach the limit where these mathematical models no longer apply, due to non-linearities and non-appliance of some of the restrictions, such as no breakaway of streamlines. Beyond this level of energy intake, the wave is generated further by mechanisms to be discussed later, but which have not as yet been formulated mathematically. A particular component may reach this stage by passing through both stages of generation, resonance and shear flow, but it may also reach it without suffering the coupling mechanism of the latter, especially if the pressure fluctuations are strong. The longer components, for which C/U tends to 0.8, will tend to have this linear energy growth, as illustrated in Fig. 2-6, even though time to reach the "mathematical" limit is much longer.

The shorter waves ($C/U \ll 1$) more quickly pass through both stages of generation, with the linear and exponential build-up separated by a transition zone, as depicted in Fig. 2-6. The time to reach this transition zone was computed by Phillips

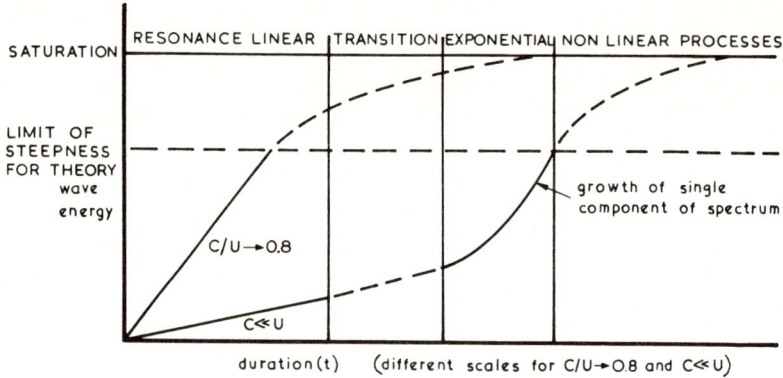

Fig. 2-6. Temporal energy increase in waves to point of saturation.

and Katz [6] and the order of time for waves of certain periods to reach this stage in a 35-knot wind in deep water are as follows:

wave period (sec)	12	10	8	6	4
time to transition (h)	5	2	0.7	0.25	0.1

As noted previously, the longer-period waves may have reached the non-linearity stage and be generated by other processes before passing through this transition into the shear-flow process of build-up. This particular wind speed requires 47 h to create a fully arisen sea, so that the transition stage is reached very early in the overall generation process.

Whilst interest has been directed to the time history development of specific components in the spectrum the normal mode of measurement is to record waves at a fixed location in the generation area. When steady conditions have been reached the spectrum of waves likely to exist is illustrated in Fig. 2-7. The energy of each component is drawn for the range of frequencies (in cycles or waves per second). The high-frequency end rises steeply in a 5-power law, which is predicted by the combined Phillips—Miles theory. The steep slope at the lower frequencies (longer waves) is associated with other generating mechanisms. Further down the fetch the steep rise in the spectrum occurs at lower frequencies since the longer waves have been under the influence of the wind long enough for them to reach their transition stage, after which a swift build-up in height ensues.

The significance of this exponential growth of waves by the coupling of wind and wave was brought out in the results of tests by the U.K. National Institute of Oceanography [7]. A mushroom-shaped buoy was placed in a fetch so that it could record a multitude of variables, including wave height, wave slope and surface air pressures, all being related to the wind direction. From the analysis of wave and wind fluctuations it was concluded that 90% of the spectrum of the air pulsations

MECHANISM OF GENERATION

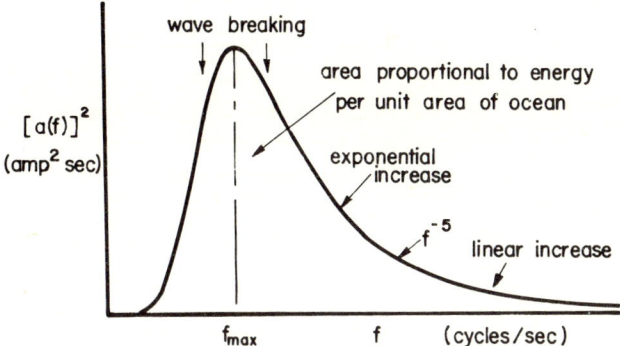

Fig. 2-7. Typical energy-frequency spectrum for ocean waves.

was coupled to the waves and only 10% to the general wind turbulence and, as stated: "...over much of the spectrum Mile's instability mechanism is probably responsible for much of the wave growth".

The theories of Phillips and Miles developed in the mid 1950's have certainly gone a long way to bringing the seemingly impossible complexity of wave generation within the hands of man to predict from basic theory. However, as Kinsman [3] states: "We are a long step forward, but the show is hardly over."

Sheltering effect

With the continual growth of any wave, it reaches a steepness or surface slope at which the air streamlines will break away from the water surface. At this stage, if not before, the Mile's mechanism will no longer apply. Even so, it is common observation that further generation occurs. This has been considered by some to result from a sheltering effect, much after the style of drag on a roughened surface.

Measurements of pressures, on profiles of static surface undulations similar to ocean waves being subjected to an air flow above them, have shown differentials over the upwind and leeward side of the crests. These could provide energy to the wave. It has been shown by experiment that this distribution differs when the same profiles are moving forward. As noted previously, the effect of the Mile's unstability at the critical air layer is to cause a phase lag of the pressure distribution. This is so located as to apply a positive pressure zone where it amplifies the surface undulation already existing.

Jeffreys in 1925 [8] promoted the thesis that the whole generation mechanism resulted from drag on the wave form. Stanton in 1937 [9] conducted flume experiments to evaluate the drag coefficient. His values were about one twentieth of those assumed by Jeffreys. Only pressures normal to the water surface were considered relevant, tangential shear stresses being ignored. These latter would appear to be effective when it is considered that at the crest, where maximum air velocities occur, so also do the maximum orbital velocities of the water particles in the same direction. Any impetus given to these particles must be influential in building wave height. It is known that an oil slick on water will inhibit wave growth to a substantial degree, not forgetting the attenuating effect this would have on the smaller wave crests through additional surface tension.

Another disability of the sheltering concept as previously devised was that it was applied to the "significant" or major wave only. In practice a rising and fully arisen sea consists of waves of many heights and lengths travelling together. Each one, as it steepens, has its own sheltering effect within the shadow of a larger wave.

Hino [10] has put forward a hypothesis on the energy budget near the sea surface based upon the premise that maximum energy is transferred to the waves with minimum loss. Of the wind energy available he concluded that an optimum of approximately 1/6 is transferrable to the waves, the remainder being consumed in air turbulence at the surface, or convected to higher levels, or per eddies in the wake of waves. He constructed a wave spectrum by combining the equations of Phillips and Miles, without utilizing any empirical factors. From the assumption of logarithmic distribution of wind velocity he determined the proportion of pressure drag to total drag force being exerted on the waves along the length of the fetch. His resulting graphs have been modified to the form in Fig. 2-8, where values for two wind speeds (16 and 32 knots) are included. The drag ratio reaches 100% for the 16-knot wind, since it is a FAS at the end of a fetch of 34 nautical miles. The

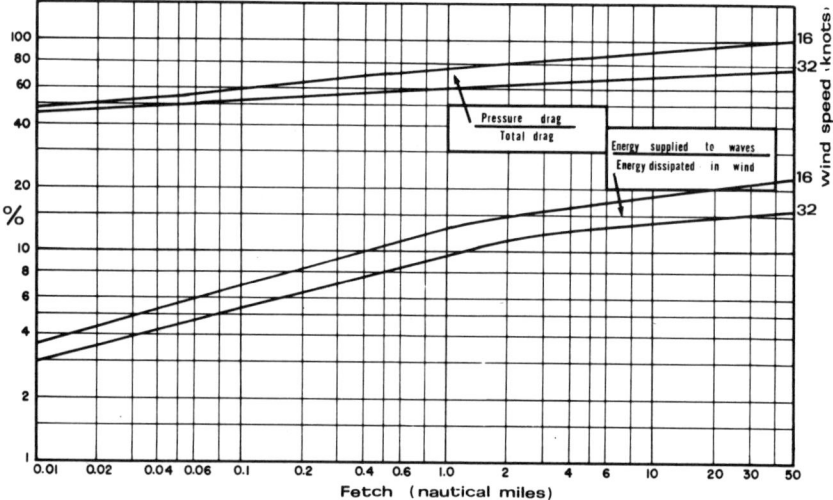

Fig. 2-8. Percentage of pressure drag to total drag exerted on waves by wind and efficiency of energy transfer to waves.

MECHANISM OF GENERATION

sea roughness increases along the fetch, thus providing projections for the wind to grip or press against. The steady state is reached when the input energy is being consumed at an equal rate by wave breaking, turbulence, viscous effects and drift currents at the surface. As noted previously, the optimum percentage of energy supplied to that dissipated is approximately 20% at the fully arisen state. At the commencement of generation it is around 3%, because at this stage the sea surface is dynamically smooth.

Wave breaking

Breaking is an integral part of wave generation. It cannot occur without it. Besides, growth continues long after the mathematical models are invalid. This instability and concomitant breaking can only be discussed at present in a qualitative manner. Notwithstanding, any rationalization must fit into the final pattern of observed wave characteristics in order to be acceptable.

Breaking of waves is considered by most mathematicians and oceanographers to be solely a dissipating mechanism. Little credence is given to it as a process for transferring energy from shorter to longer wave lengths. Whilst much energy is no doubt lost in the turbulence associated with the "white horses" of the sea, it is submitted that the growth of the longer waves could greatly depend upon the shorter waves being "fed" to them.

The smaller components of the spectrum, which are built up swiftly, reach a point of instability and then break. Accepting that the longer-period waves have commenced to form (by resonance action over the still relatively smooth sea) these shorter waves will tend to break at their crest locations. The reason for this is that they are stretched between a trough and the preceding crest and compressed (in length) between the trough and the following crest (see Fig. 2-9).

This is significant, because just at the time the small wave steepens to breaking point the wind velocity is maximum at the crest of the longer wave. The small wave is being carried forward at the orbital velocity of the water particles near this crest of the longer wave. Also, the wind pressure will cause the wave to break in the direction of wave advance. The body of water thrown forwards and downwards

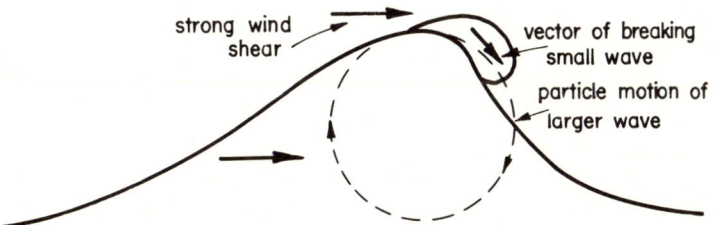

Fig. 2-9. Possible momentum transfer from breaking short wave to longer-period wave.

over part of the crest of the longer wave should add momentum to that already contained in the circular motion of the surface water particles in this wave (see Fig. 2-9). Further details of this mechanism will be given later.

Such a process could transfer energy so long as the wind speed were greater than the celerity of the longer wave. Once these became equal the smaller waves would be carried forward near the crest at the same speed as the wind. There would then be less of a tendency for the small wave to be forced forward on breaking, so that the momentum previously available would be substantially reduced.

By this mechanism wave energy could be transferred through the whole spectrum of waves, the smaller waves feeding longer ones and they, in turn, feeding longer ones still. Once the limit were reached, at which breaking occurred at the crest of the longest waves, with no forward motion, dissipation rather than transfer would be predominant. At this fully arisen state the feeding process would still be taking place amongst the higher-frequency components of the spectrum.

As noted previously, the resonance build-up of waves causes them to travel in directions oblique to the mean wind velocity. The longer its period, so a wave is more aligned to the wind, the shorter waves propagating at greater angles. Even when the shear-flow feed-back takes place on the shorter waves they still maintain their obliquity.

This complex three-dimensional surface undulation might be simplified to the case of two long-crested wave trains angled to each other. Such an occurrence is depicted in Fig. 2-10, where the intersection of the two sets of lines, or wave crests, will be the location of wave peaks. It is at these points of intersection that breaking is likely to take place. The act of breaking will cause this portion of the crest system to travel a little faster and so round off the diamond pattern in such a way as to straighten the crests normal to the wind (see Fig. 2-10). Thus the energy purportedly given to the longer waves from the smaller oblique components is conveyed in the downwind direction, along which energy is known to be concentrated, in the measurements mentioned previously [7].

Longuet-Higgins [11] has provided a theoretical basis for the transfer of energy

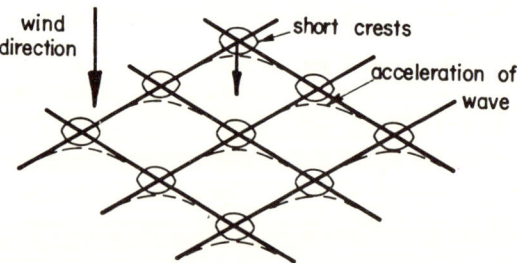

Fig. 2-10. Illustration of short-crested wave system becoming long-crested as breaking occurs at peaks.

from shorter waves to longer waves. If the decay of the short waves is by viscosity their momentum transfer is slight. "But when the short waves are forced to decay strongly by breaking on the forward slopes of the long waves the gain of energy by the lattter is greatly increased. Calculations suggest that the mechanism is capable of imparting energy to sea waves at the rate observed." This process Longuet-Higgins termed the "maser mechanism" since the crests of the longer waves sweep the energy from the smaller waves, which is then replenished during the passage of the following trough. The action would appear to "account for generation of waves travelling faster than the wind, and the observed damping of waves by an adverse wind".

WAVE CHARACTERISTICS

For the purposes of coastal engineering, waves can be divided into two categories, namely, *storm waves* (sea) and *swell*. The former are still being generated or maintained by the wind, they are therefore within the confines of the fetch or storm zone. The term "swell" is applied to waves that have left the fetch and are dispersing across the oceans to some distant shore. Each of these wave systems has distinctive characteristics, which have quite different influences on coasts, marine structures and vessels.

Storm waves

As noted in the discussion on wave generation, the sea surface undulation in a fetch consists of many waves of differing height and length. Each component, particularly the small-period ones, are being built-up to a maximum steepness until breaking occurs. Some energy is thus dissipated and some is fed into longer waves, near whose crests these events take place. The residue of the small wave is rebuilt to breaking point once more. Other waves of similar period are being formed and enlarged by wind pulsations, shear-flow or sheltering effects, besides energy transfer from the breaking of still smaller waves. The combination of all these wave components make up the random fluctuations of the stormy sea surface. The resultant major waves are therefore steep in character, since the ratio of height to apparent wave length is very great.

Some authors, in discussing the effect of wave steepness on beach profiles, distinguish a certain value above which erosion occurs and below which accretion takes place. These have been derived from model tests in which continuous single-wave trains have been used — in essence they are swell-type waves. Whilst such waves may vary somewhat in their steepness they should not be confused with the storm waves at present being considered. There is no comparison, where sediment motion is concerned, between the complex sea of a fetch situation with the simple form of a wave train arriving as swell, even where the latter may have a relatively large height to length ratio.

Another feature of storm waves is their asymmetrical shape. Due to the pressure of the wind, and the non-linearities in the water-particle motions as maximum steepness is reached, the wave becomes steep-fronted or tilted forward. The extra steepness, plus the momentum transferred through breaking, causes these waves to travel at a slightly higher speed than the theoretical value for the sinusoidal wave of very small height.

These two characteristics of instability (steepness and asymmetry) make storm waves prone to breaking. As they come into shore, shoaling or a contra-current will trigger them into breaking before they would otherwise. This has particular relevance to coastal processes where the ready dissipation of incoming storm waves prevents further beach erosion at a particular stage of shoreline degradation.

It will be recalled from previous sections that the initiation of wave motion, through resonance with random pressure pulsations of the air, results in waves with differing directions. The closer the celerity of a wave to the wind velocity the more aligned is the wave to the wind. The slower or shorter waves on the other hand will be more oblique. These shorter waves, however, are the ones to be built up more quickly. As they become steep enough for the wind to promote breaking they add momentum to the longer waves more-or-less in the direction of the wind. Thus whilst the shorter waves are angled greatly to the wind, the longer waves deviate only slightly from the wind or fetch alignment. Fig. 2-11 provides an impression of the planimetric shape of waves in a fetch.

This multi-directional nature of storm waves, particularly the short-period end of the spectrum, indicates the importance of fetch width in the generation process. Unless there is time for the shorter waves to develop, the longer will not grow so readily. It could be conceived that a fetch is so narrow that the shorter waves move out of the wind field as soon as they are generated. The long waves present will be those generated only by the wind pulsations and shear-flow mechanism. They

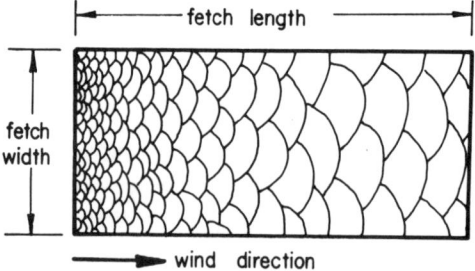

Fig. 2-11. Typical crest plan of waves along the length of a fetch.

WAVE CHARACTERISTICS

would be smaller in height than when the more complete spectrum is present, feeding energy along the line to the longest component. In this respect it might be envisaged that most flume tests of wind-generated waves are narrow in the above sense. However, waves in a flume are reflected from the vertical sides and so are equivalent to those in an infinitely wide fetch.

The propagation of waves at angles to each other provides a surface oscillation which is short-crested. This feature will be more pronounced at the upwind end of the fetch, since downwind the longer waves will predominate and are more aligned to each other and to the wind. Even so, the presence of the shorter-period end of the spectrum will make it difficult to discern where the next composite crest is likely to form — the bane of all ships' masters.

The waves of any given component in the spectrum, that is of a particular period, will be travelling in all directions of the fan contained by the critical angle, either side of the wind vector. If the energy content contained in any directional train were computed or measured its distribution would be as shown in Fig. 2-12. The smallest waves would have a low energy content, which is spread widely across the possible 180°-fan. As the wave period increases so the energy is concentrated in a narrower band of directions. The longest waves would contain almost all their energy in the wind alignment or 0° of the figure.

When the surface undulation at a point is recorded continuously it is the same no matter from how many directions the various component waves arrive. When such a record, or more particularly a short length of it, is analyzed to determine the height of the waves in the various trains, a distribution as illustrated in Fig. 2-13 is likely to result. The histogram would be obtained by averaging the heights (squared)

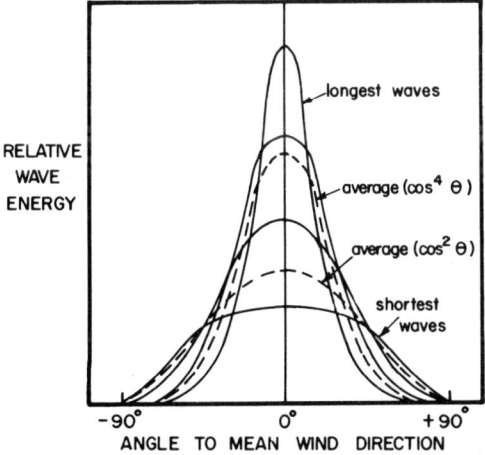

Fig. 2-12. Directional energy distribution in waves of various periods for a specific wind velocity.

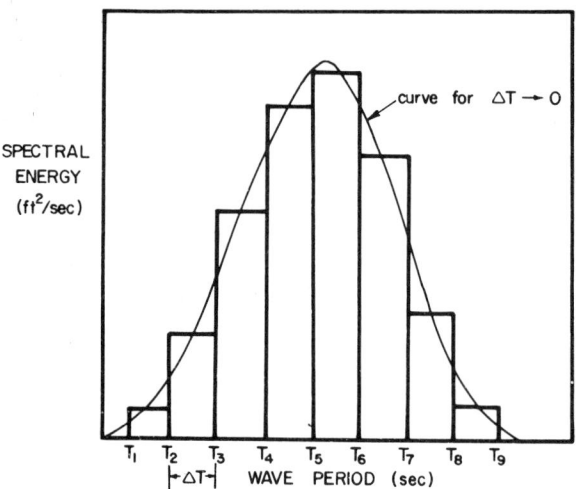

Fig. 2-13. Histogram of wave energy for increments of wave period.

of waves within each incremental period, for example, 0–1, 1–2, 2–3 sec, etc. If the increments of T tended to zero the curved line would best represent the wave energy distribution against period. As noted earlier such a spectrum of wave energy (amplitude squared) is generally graphed against wave frequency (see Fig. 2-7), with the ordinate in terms of squared length × sec., so that the area under the curve becomes total energy of the composite sea per unit area of ocean. However, for the present discussion the distribution of Fig. 2-13 is more readily pictured in terms of wave period and squared wave height. The shapes of the two curves differ besides being reversed along the abscissa.

The energy distribution curve (EDC) of Fig. 2-13 will represent conditions at a specific point in the fetch. If it should be that recorded at the downwind limit of the fetch after a steady wave state has been reached (for the particular wind velocity), it then represents the FAS. The waves measured at that location will provide a similar distribution no matter how long the wind blows.

At locations further upwind, the resulting spectrum may be limited by the distance over which the wind has had contact with the sea surface. Even when constant wave conditions have been reached they will be smaller in magnitude than further downwind. The generation of waves is then said to be fetch-limited, but the spectra bear a special relationship to that of the FAS as indicated in Fig. 2-14. The curves F_1, F_2, F_3 represent increasing distances along the fetch until at F_4 the maximum wave condition is reached for the existing wind velocity. Any further distance along the same fetch (F_5), the spectrum will be the same as F_4. The features to be noted about Fig. 2-14 are, firstly, the proximity of the rising curves to each other and, secondly, the similar steep slopes of the falling curves at the

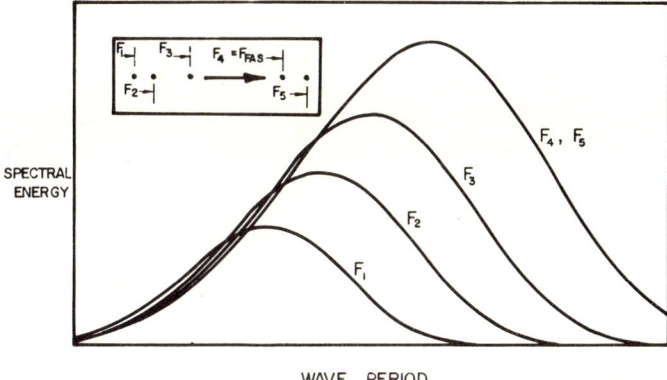

Fig. 2-14. Energy-period spectrum for points along the length of a fetch.

higher-period end of the EDC. As the fetch increases, so the low-period end of the EDC becomes insignificant and disappears. The sloping sides thus shift to the right (greater periods), whilst maintaining their relative shape.

The above discussion relates to the steady state reached at each point. Prior to this minimum duration the waves are still increasing, implying an enlarging spectrum at that location. It is readily conceived that near the beginning of the fetch the waves will reach their peak or steady state before those that are further downwind. During this early stage the spectra existing all over the fetch will be similar, because control of the wave size is exerted by duration of the wind.

Considering now the point in the fetch where the fully arisen sea can exist for the specific wind speed, its final spectrum is reached after some minimum duration t_4 (see Fig. 2-15). Stages of development are depicted by the curves t_1, t_2 and t_3

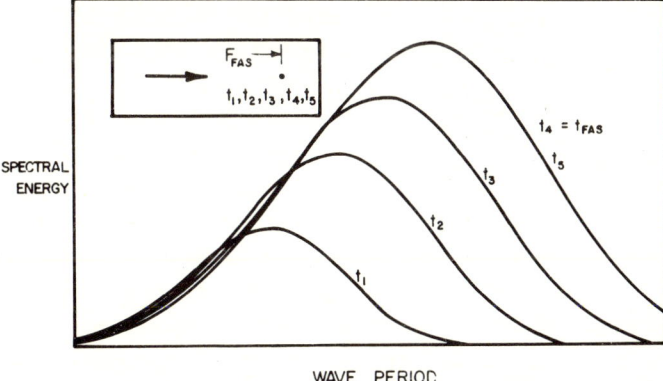

Fig. 2-15. Energy-period spectrum for durations less and greater than that for the FAS.

and can be considered similar to the fetch-limited condition of Fig. 2-14. After time t_4 has been passed (t_5) the spectrum remains the same.

Thus, even for the simple situation of a well-defined rectangular fetch, with a constant wind velocity blowing for an unlimited duration, the waves recorded at various locations at the same or other times will be substantially different. At each point it will take time to reach steady conditions, this minimum duration being different along the fetch, and being unique to the wind velocity present, as is the spectrum of the FAS. As will be seen later, the waves which emerge from such a fetch will also vary temporally and spatially as they propagate across the ocean.

Swell

As has already been noted, waves of any particular period in the fetch will be travelling in many directions. The shorter waves have a wider fan of operation than do the longer ones, which are more closely aligned with the mean wind velocity. For any given period there is a direction which contains the highest waves as depicted in Fig. 2-12.

Once generated, waves continue to move in the same direction and so traverse the oceans in *great circles*. These are lines of intersection of a plane through the centre of the earth. Thus, if no land masses intercepted, waves would proceed around the globe to their point of origin. These great-circle traverses can be considered as straight lines on map projections of the gnomonic and conical type, that is, those with curved latitudes and straight longitudes angled to each other. As noted

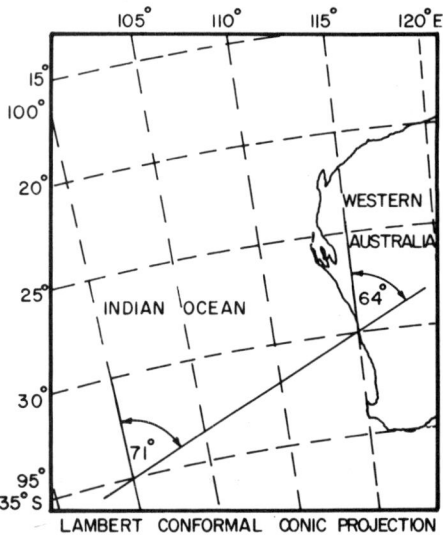

Fig. 2-16. Map of a conical projection showing straight wave ray.

WAVE CHARACTERISTICS

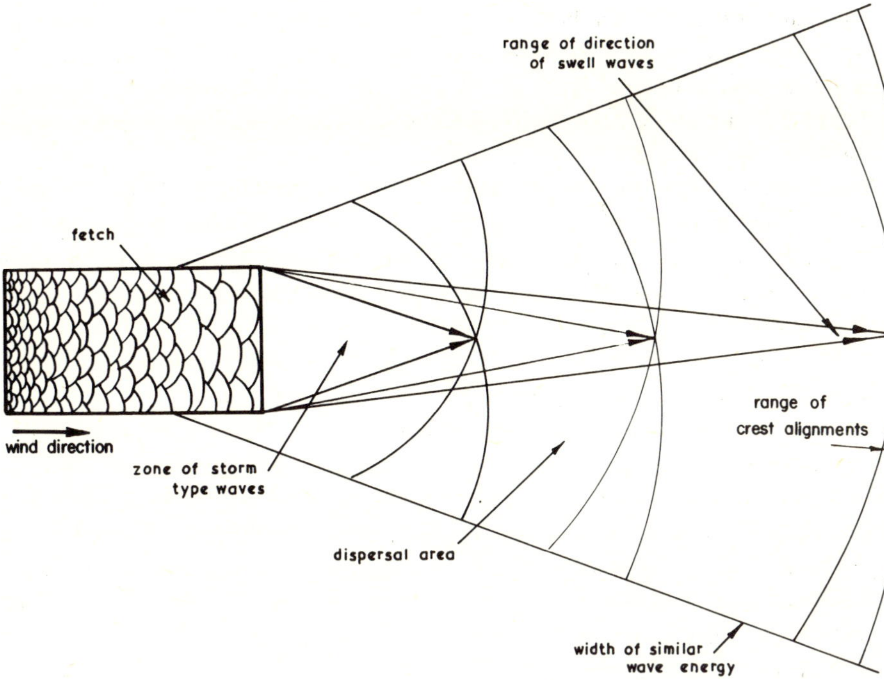

Fig. 2-17. Illustration of directional dispersion of waves outside the fetch.

in Fig. 2-16 a straight wave ray continually changes direction with respect to the azimuth. In tracing waves over long distances this change is of utmost importance when assessing obliquity of deep-water swell to a coastline.

In the downwind zone of the fetch where the FAS exists the energy concentration per unit area of ocean is high. When these waves, which are moving in a multitude of directions, disperse to zones outside the fetch, their energy is spread out thinly over the ocean. The waves follow paths which are essentially radial from a centre of high wave intensity inside the fetch. At any point away from the fetch boundary the only waves to arrive are contained within a small fan of directions as illustrated in Fig. 2-17, all other waves will by-pass this point of recording. The effect of this is to restrict the amount of wave energy that can be present at any time. Also those waves arriving are more closely aligned to each other. This latter tendency makes for a composite wave system which is longer crested than that inside the fetch.

The contained angle is smaller the further from the fetch. However, beyond a limit of two or three fetch widths, the reduction in possible wave directions is extremely small. Here again the width of the fetch exerts its influence on the wave climate in the dispersion area, as it did in the generation itself. This point is emphasized because it is a dimension that is not generally included in wave computations and may not be fully appreciated when using forecasting formulae or when analyzing wave records.

Not only are the waves spreading circumferentially, as discussed above, they are also dispersing radially. The longer-period waves travel the fastest, the medium-length waves follow, and the shortest-period waves trail them all. Interest here is concentrated on waves from 5 to 15 sec for open-ocean conditions. Waves less than 5 sec travel so slowly that climatic conditions could change before they can be of concern to the coastal engineer. Besides, they attenuate swiftly enough to make their influence negligible after a thousand miles of propagation.

The radial spreading of waves results in less wave trains being present concurrently at any one place in the dispersion area. The greater the distance the narrower is the band of wave lengths that could combine. This reduces the height of the maxima or any other height criterion of the composite undulation.

Unlike the FAS, where wave characteristics could remain relatively steady for some time, conditions outside the fetch are quite variable. During the period of growth to FAS the optimum waves of the developing sea (of medium period) may have reached the recording point. Later the longest waves of the FAS spectrum will pass through, they may even precede the medium group above if the point is a long distance away. These longest waves will be of very modest height, but as the period band with optimum energy arrives the swell grows to its peak in height. After this the waves slowly decrease both in height and period.

This transient situation is depicted in Fig. 2-18, where the original spectrum is seen to be reduced in the first place by circumferential spreading. Secondly, only a portion of this reduced wave energy can arrive simultaneously, as indicated by the vertical bands, the one on the left arriving last.

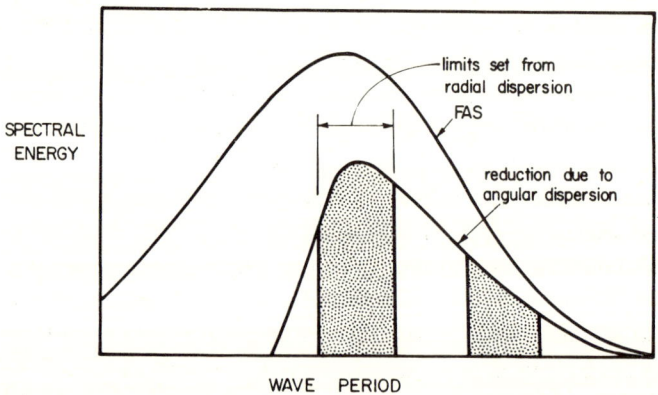

Fig. 2-18. Reduction of wave energy due to circumferential and radial dispersion.

The modified energy distribution resulting from circumferential spreading will vary according to the location in the dispersion area. A point directly downwind of the fetch will receive all waves, including the longest and highest, whereas a point oblique to this alignment will receive a greater proportion of the smaller-period waves and omit the longer components. The conditions in Fig. 2-18 more closely represent a location on the fetch axis where the long waves arrive with negligible reduction whilst the shorter waves may be missed altogether, because their energy is spread widely.

In a triangular section of the dispersion area adjacent to the downwind end waves can arrive from a wide angle, so that the resulting complex sea is not dissimilar to that inside the fetch (see Fig. 2-17). The major difference will be the absence of breaking or white caps. When such a zone is adjacent to a coast, the resulting effect could be the same as if the fetch were next to the coast itself. The variability in wave direction in waves arriving at a coast should indicate how close was the fetch that produced the particular set of waves.

Another characteristic of swell waves is their much longer duration of occurrence. Whilst storm waves in a fetch last little longer than the period of the storm, the full range of swell emanating therefrom may take three times as long to arrive at a distant shore. In 1964, Snodgrass et al. [12] recorded waves at five points across the Pacific Ocean. They were able to trace swell waves, generated in specific storm centres in the "roaring fourties" ($40°-50°S$) regions, around great circles to the coast of Alaska. This oceanographic experiment was of tremendous significance and the reader is encouraged to read the details of its execution, as well as the conclusions derived from it. It will be referred to on many occasions, herein, but the pertinent quotation at present is that regarding duration of wave incidence. Snodgrass et al. state: "Once or twice a week a wave train associated with a severe southern storm leads to an identifiable 'event' that can be traced across the entire ocean... A typical event begins at 30 mc/sec and ends at 80 mc/sec, lasting for 2 days at Tutuila (near Fiji) and for a week at Yakutat (Alaska), the progressive lengthening being attributable to dispersive stretching of the wave train."

Another feature of swell waves is their profile. Since the wind is no longer pressing them and there are no short-period elements present, they assume the stable profile of a sine wave, or a trochoid for the steeper samples. Swell can travel in this form with practically no loss of energy. The Pacific Ocean measurements indicated that almost the whole reduction of energy was caused by lateral spreading, particularly for the longest and highest waves of the spectrum.

It can be seen from the above description, that very little wave energy, particularly in the longer-period components, is lost. It is maintained and transported to distant shores, there to be applied to the sedimentary material resting on the continental shelves and beach profiles. As noted later, certain zones of the ocean contain the bulk of the storm energy, but the swell serves effectively to spread this around the world. Such distribution has taken place since the oceans were first formed, so that here is a source of energy of great significance when considered over geologic time.

One final feature of swell incidence is its constancy in approach direction at any particular section of coast. Even though related storm centres supplying swell to a distant shore may be a 1000 NM's apart, the waves from the two sources may differ in their deep-water direction by only a degree or two. This persistent nature of swell, both in occurrence and direction, has great significance in the transport of material along the shorelines of the world.

The Pacific Ocean swell report previously mentioned [12] contains a typical spectrum for a storm occurrence plus background waves for the month of July.

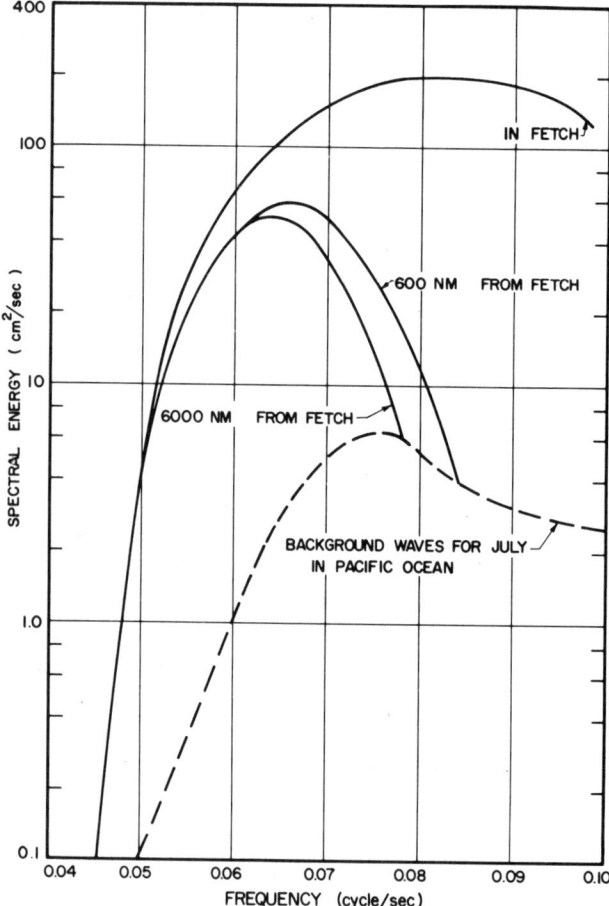

Fig. 2-19. Typical spectra measured at various distances from fetches in the southern Pacific. (Modified from Ref. 12.)

This is presented in slightly modified form in Fig. 2-19 and will be further discussed in the following chapter. However, two points should be observed at this stage: firstly, that the major reduction in wave energy occurred in the first tenth of this 6000-NM journey, and secondly, that most of this was effected in the higher frequency components. These recordings were made ostensibly on the great circle running from Yakutat (Alaska) through Honolulu to just east of New Zealand, and were directly downwind of the fetches which were located between 50° and 70° south.

METEOROLOGICAL CONDITIONS

From the discussion in the previous section of a simple fetch, in which a constant wind has blown indefinitely over a fixed rectangular section of ocean, it should now be appreciated that the wave incidence is anything but uniform. Two wave records taken simultaneously at points in close proximity would be hard to correlate. Waves within the fetch keep growing until FAS conditions exist. Outside the fetch the waves vary from place to place and also from hour to hour.

Now, to this hypothetical situation add the complexities of the natural environment. The fetch is likely to be changing in length, width, location and orientation. The wind velocity itself may be varying along the fetch and changing with time. It is little wonder that oceanographers have had great difficulty in obtaining steady enough conditions in order to correlate wave characteristics with conditions. In a comprehensive study of this nature Moskowitz [13] examined 460 weather charts in order to select 52 which were defined well enough for his use. The criteria for choice were: (*a*) duration and constancy of wind velocity and direction; (*b*) sufficient separation from other periods of high wind; (*c*) absence of swell in the area.

Whilst respect should be paid to the scientists for their tenacity of purpose in devising relationships for forecasting or hindcasting ocean waves, a tear of sympathy should also be shed for the engineers who are expected to use them. From such calculations, which can become extremely tedious, the engineer obtains wave data that he hopes are reliable enough for designing marine structures. From a series of figures, that he extracts from complex weather conditions, he must then judge if worse conditions could ensue for his particular port location and so amplify his values proportionately on some statistical basis.

Examination of typical wind structures gives an appreciation of the variables involved, and provides a qualitative if not a quantitative picture of the waves generated over wide areas. For a specific point on a coastline, details of the FAS can now be determined when a known wind has blown for a sufficient period of time over a great enough area of ocean.

Even so, the engineer must decide whether this particular storm situation could be worse than those recorded in the synoptic weather charts available to him. Some particularly severe storm located some distance from the coast during some past season may well recur closer to shore, for which due allowance must be made. The approach to the problem should be similar to that of the hydrologist who transposes his zone of precipitation over a catchment to give the worst flood condition at his dam site or other point of interest. The coastal engineer for some design purposes requires, in a similar manner, to maximize his significant or highest waves.

The winds that are important in ocean wave generation are those produced in low-pressure centres or *cyclones*. High-pressure systems or *anti-cyclones* contain very mild winds and are mainly important in separating the cyclonic centres with calm weather. These latter envelop large tracts of ocean (and land masses), whereas cyclones are concentrated over much smaller zones, although their centres move more rapidly than the ill-defined high-pressure regions.

Cyclones vary in size and intensity depending upon their geographic location, particularly in relation to their latitude of occurrence. Those generated between the equator and 30°N and 30°S are termed *tropical cyclones* and those initiated between 30° and the poles are called *extra-tropical cyclones*. They vary significantly in surface-wind structure and hence in their mode of generating ocean waves. In the following discussion of their features emphasis will be placed on sea-surface wind characteristics, as distinct from the three-dimensional motion of air that maintains and feeds such systems.

Extra-tropical cyclones

Since these cyclones in the 30° − 60° latitude band provide the fetch conditions implied in the previous discussion of wave generation it is fitting to treat them before the more unique tropical version. Besides, the extra-tropical cyclones are more frequent and produce the bulk of the swell across the oceans of the world.

When an area of low atmospheric pressure is created (in the order of 2 inches mercury, 26 inches water or 1 lb./sq. inch, approximately), the surrounding air immediately moves into it. The air arriving at the centre is drawn up vertically, so that an inward flow is maintained over a number of days. This initially radial motion is soon turned into a circular path due to rotation of the earth and its atmosphere, which provides a greater peripheral speed near the equator than at higher latitudes. The force applied to the moving air mass is known as the *coriolis force* and results in a clockwise rotation in the Southern Hemisphere and an anti-clockwise rotation in the Northern Hemisphere.

The velocity of the wind in any area depends upon the pressure there, or more strictly the pressure differential between it and the adjacent areas. The meteorologist records this pressure distribution by what are termed *isobars,* which are lines on a map joining points of equal atmospheric pressure. He produces synoptic weather charts with such isobaric patterns and other information at intervals of 3 or 6 h.

METEOROLOGICAL CONDITIONS

The closer the spacing of the isobars the greater the differential in pressure, since the same difference in pressure exists over a shorter distance of sea. Thus the closer the isobars, the stronger the wind represented by them. An increase in wind velocity is experienced as the low-pressure centre is approached, until the zone is reached where the air commences to turn upwards. The aligment of the isobars is that of the wind occurring some 2000 ft. from the sea surface, essentially at a level beyond the frictional effect of the earth's surface. This is known as the *geostrophic wind* and its velocity at any point can be computed from the isobar spacing, by the use of a scale normally reproduced on the synoptic weather map. If this is not present the relevant equation is:

$$U_{gs}/g = (0.52/\sin\theta)\Delta p/\Delta n \qquad (2\text{-}2)$$

where θ = latitude in degrees, Δp = pressure differential in millibars, Δn = spacing of isobars in degrees latitude, U_{gs} = geostrophic wind velocity in dimensions commensurate with g (acceleration due to gravity). This equation can be put in the form of a graph as in Fig. 2-20 for specific isobar differentials.

The wind velocity of concern in the generation of waves is, of course, the mean value of that occurring near the sea surface itself. This is called the *gradient wind*. It can be derived from the geostrophic wind when due allowance is made for latitude, curvature of the isobars and air-mass stability. The relationship for the first two of these factors is:

$$U_{gr} = U_{gs} \pm \frac{U_{gr}^2}{2\omega r \sin\theta} \qquad (2\text{-}3)$$

where U_{gr} = gradient wind velocity in knots, U_{gs} = geostrophic wind velocity in knots, ω = angular velocity of the earth (= 0.729 rad./sec), r = radius of curvature of isobars in nautical miles, θ = latitude in degrees. The minus sign is used for cyclones and the plus for anticyclones. This relationship has been graphed in Fig. 2-21 for wind speeds in knots and radius of curvature in nautical miles. Note that for straight isobars $U_{gr} = U_{gs}$.

The sea-surface velocity also varies for the same geostrophic wind as the sea and air temperatures differ. This influences the vertical distribution of wind velocity, it being greater at the sea surface the colder the wind in respect to that of the sea (see Fig. 2-22). The temperature difference here considered applies over the fetch as a whole. It would need to be ascertained from monthly or seasonal figures, available in atlases covering the ocean areas. The correction is applied to the geostrophic wind after allowance has been made for isobaric curvature. In spite of the difficulty in its assessment, this temperature differential has a significant influence on the gradient wind.

Another effect of this friction or drag effect on the geostrophic wind is a change

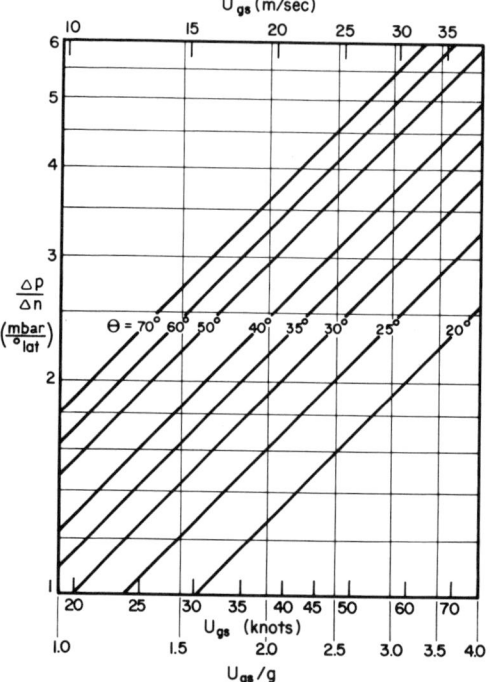

Fig. 2-20. Geostrophic wind velocity for given isobar differentials and latitude.

in direction from the isobar alignment. In the cyclonic system (which is of the greatest relevance) the gradient wind is deflected about 15° towards the low-pressure centre. This must be taken into consideration when forecasting waves towards a specific section of shoreline.

The air circulating around and towards a low-pressure centre will consist of batches of warm and of cold air. The latter, being denser than the former, will tend to wedge under it and so force it upwards. This constitutes a *cold front* and is normally associated with the formation of clouds and precipitation. The location of these cold fronts can be determined by the meteorologist from such information as a swift change in wind direction, or humidity and temperature changes. In a similar manner *warm fronts* can be formed and *occluded fronts*, which result from the combining of a cold and warm front.

These several features of cyclones are depicted in Fig. 2-23, which illustrates air circulation for the Southern Hemisphere. Also shown is the symbolic representation of sea-surface wind speeds, the number of tails to the arrow indicating the speed in knots or some other scale.

The various fronts of a cyclone rotate around the centre in the same rotational direction as the winds of which they are essentially composed. They move at the

METEOROLOGICAL CONDITIONS

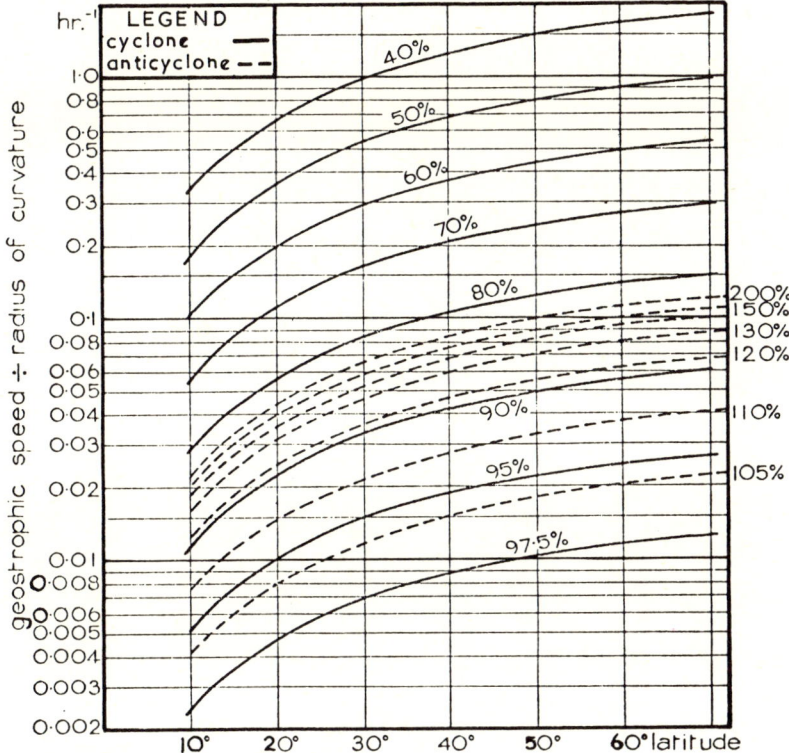

Fig. 2-21. Gradient wind velocity allowing for isobar curvature and latitude.

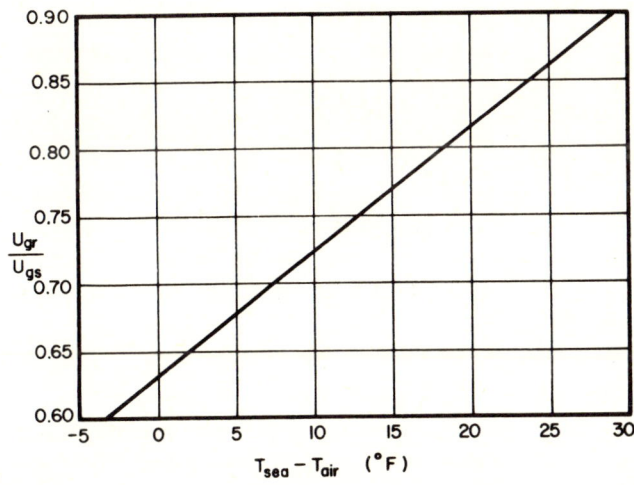

Fig. 2-22. Surface wind velocity as proportion of gradient wind, allowing for air–sea temperature difference.

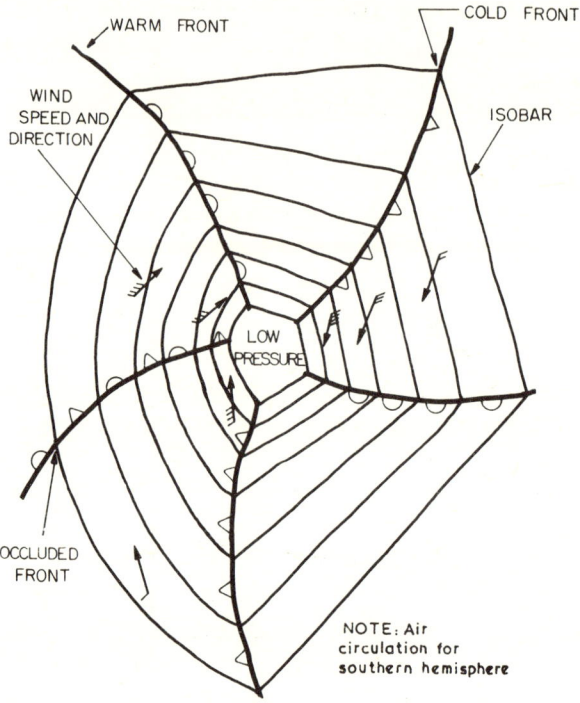

Fig. 2-23. Typical pattern for extra-tropical cyclone, Southern Hemisphere circulation.

speed of the gradient wind, the distribution of which produces a curvature in the front as indicated in Fig. 2-23. The centre itself is travelling across the sea surface at about three quarters of the gradient speed near the centre.

The accuracy in the delineation and positioning of cyclonic centres on synoptic charts depends greatly upon the experience of the meteorologist, plus the correctness, profuseness and distribution of the data available. It can be appreciated that some ocean areas have few sources of information, such as those off the main trade routes where few vessels visit. Needless to say, ships' masters try to avoid storm areas, where some of the most important wave-generation information is located.

The arrows of Fig. 2-23 would probably be wind measurements made aboard ships. Sometimes this is taken from an anemometer, or it may even be assessed by an officer viewing the state of the sea, in particular the proportion of "white horses" on the sea surface. As will be seen later, the height at which an instrument reading is taken can be all important in calculating waves. This information is not readily available and calls for some standardization in procedures of an international character. The meteorologist will, of course, fit his isobar spacing and orientation to suit this known surface wind speed at this point in space and time. If sufficient of these wind velocities are available along a fetch clearly defined by the isobars and cold fronts these anemometer readings should be used in preference to the gradient wind, whose value is subject to the imponderables of isobar curvature and air—sea temperature differentials, as noted already.

It should have become apparent from the above discussion that the fetches experienced in nature are far from simple, even when their accuracy is acceptable. In any one cyclone there could be four or five fetches sending waves out in an equal number of directions. Admittedly those waves being generated towards a specific point on the coastline might have a fetch in each 3-hourly synoptic chart very clearly defined. Methods have been devised for treating the case of moving fetches whose wind velocity and direction are essentially constant.

Waves generated in one zone may proceed into another wind field, which may or may not continue the process. Where the wind is in same direction as the waves coming under its influence the fetch and duration are, in essence, increased. If the sea is not already fully arisen this added energy will continue to build up the waves. There is thus one region of the cyclonic centre, the one in which the winds are in the same direction as the storm centre itself, where the largest waves will be created. In the Southern Hemisphere this is on the left and in the Northern Hemisphere it is on the right, when viewed along the cyclone path. In the other regions of the cyclone, winds are very transient. Zones experience winds for short periods, the waves from which move swiftly out of its influence. The most persistent paths for the passage of cyclones is thus seen to be of prime importance to the coastal engineer in assessing the most destructive waves possible for a port, or other point on the coast.

Anticyclones

Before outlining the features of the tropical cyclone, just a word or two about the anticyclone. As noted before, it consists of a high-pressure centre with an outward flow of air, which results in a circulation the direction of which is opposite to that for the cyclone in the same hemisphere. The resulting winds are generally mild, making the isobars widely spaced and more difficult to define. Few if any cold or warm fronts are recognizable so that fine weather is the order of the day. The anticyclones occurring in the same regions as the extra-tropical cyclones proceed, like them, from west to east. They change latitude throughout the year in concert with the north and south trend of the sun about the equator.

During these fine-weather periods, of little or no wind, other wind-producing mechanisms of modest character can exert an influence on the wave climate of a coast. One of these is the sea breeze, which is produced by the daily heating of a land mass adjacent to a section of relatively cold sea. The hot mass of air over the land rises and is replenished by a sea breeze which commences from midday to late afternoon and may last till early evening or midnight. This breeze may reach 20 knots and affect some 20 to 50 miles of ocean adjacent to the coast. This is sufficient for short choppy waves to be generated towards the shore, sometimes at quite an oblique angle to some sections of shoreline. When they break these waves can still be angled greatly and so can create a strong longshore current. This, in association with swell waves rolling in and helping to disturb sediment in the shallow zones, can have a significant effect on the daily profiles of beaches and even the transient balance of longshore drift. The coastal engineer should not discount the influence of these breezes, even though their presence may not be indicated in synoptic weather maps, except as a short-term wind arrows at some shoreline stations.

Tropical cyclones

The low-pressure centres generated in the 0–30° degree latitude zone north and south of the equator are characterized by their great intensity of wind and their unique structure. They produce the highest wind speeds known over land or sea, and the air mass follows an almost circular path. Even though the centres move continually, and hence limit fetches and durations for wave-generation purposes, they cause great harvoc with shipping and coastal facilities alike.

Different terms are used for these destructive cyclones according to their location. On the east coast of America they are termed *hurricanes*. In the near-east and Japan they are termed *typhoons*. In the vicinity of India they are known as *monsoons*. The Australians use the simple title of *tropical cyclone,* perhaps due to the modest size of centres in that area.

The tropical version of the cyclone differs greatly from the extra-tropical cyclone, it being initiated and fed by different forces. The major characteristics are listed on Table 2-I.

TABLE 2-I

Characteristics of cyclonic wind structures

Tropical	Extra-tropical
Occur in low latitudes 5°–35°	Occur between 30° and 60°
Isobars nearly circular	Isobars mainly straight
No cold fronts present	Many fronts circulate
Pressure gradients intense	Moderate gradients exist
Wind speeds 50–100 miles/h near centre	Winds rarely exceed 40 knots
Centre is circular calm zone (eye)	Centre not readily defined
Precipitation intense overall	Rain moderate at fronts
Heat budget source (summer)	Pressure source (winter)
Centre travels at 10–40 knots	Centre travels about 30 knots
Normally north–south movement	Direction usually west to east

METEOROLOGICAL CONDITIONS

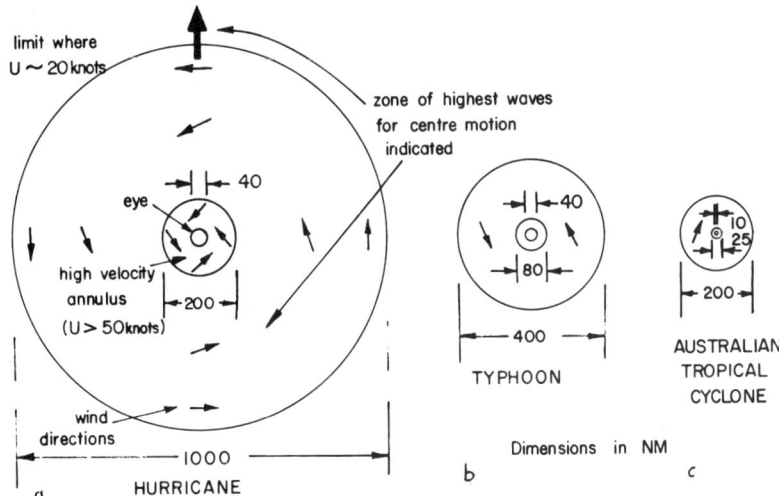

Fig. 2-24. Plan of sea surface wind conditions in a hurricane, a typhoon, and in a smaller tropical cyclone.

Like any meteorological phenomenon, tropical cyclones have a life cycle, which entails an immature, mature and declining stage. Even in maturity, or greatest intensity, these cyclones may differ greatly in stature. Because of this it is difficult to talk in terms of averages, but rather as maxima for the particular class of centre. In studying the zones of generation, and tracing the paths of tropical cyclones, the coastal engineer should seek knowledge of their variable intensity, to see if his specific port location is in a particularly vulnerable position.

The surface-wind structure in the hurricane-typhoon-species of tropical cyclone is illustrated in Fig. 2-24, a, b. This is a plan of the wind distribution and direction for the Northern Hemisphere. Fig. 2-24, c is a similar plan for the tropical cyclone occurring in the Australian region. The clockwise circulation of this hemisphere is indicated. Fig. 2-25 provides a graphical distribution of wind speed versus radius from the centre for the three samples depicted in Fig. 2-24. The variety of cross-sectional values has been drawn from figures supplied by Syono [13].

From these illustrations it can be seen that the eye of the cyclone appears well defined. Inside this eye there is practically no surface wind, the air being drawn upwards in an extremely strong vortex and the atmospheric pressure dropping swiftly to its minimum. At the edge of the eye the wind speed rises almost instantaneously from near calm to the maximum for the cyclone. The maximum for the hurricane-typhoon type of centre can average 120 miles/h with fluctuations to 150 miles/h. However a reasonable strong cyclone will produce 100 miles/h at the edge of the eye, which varies in diameter from 30 to 50 miles.

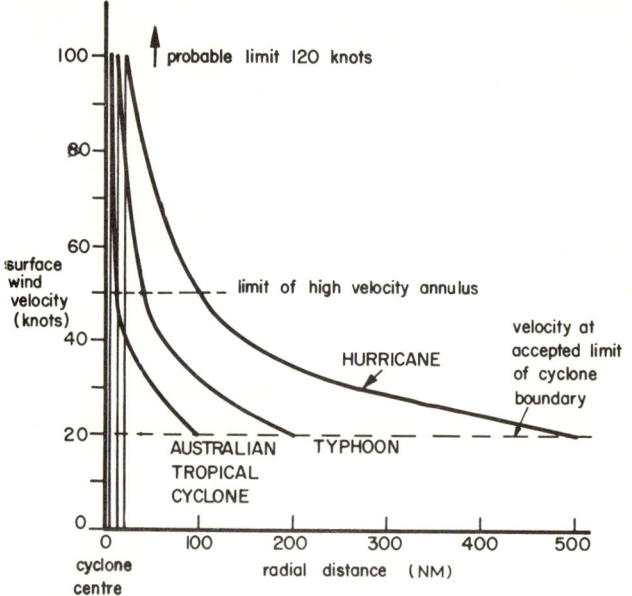

Fig. 2-25. Radial distribution of wind velocities in a hurricane, a typhoon and a smaller tropical cyclone.

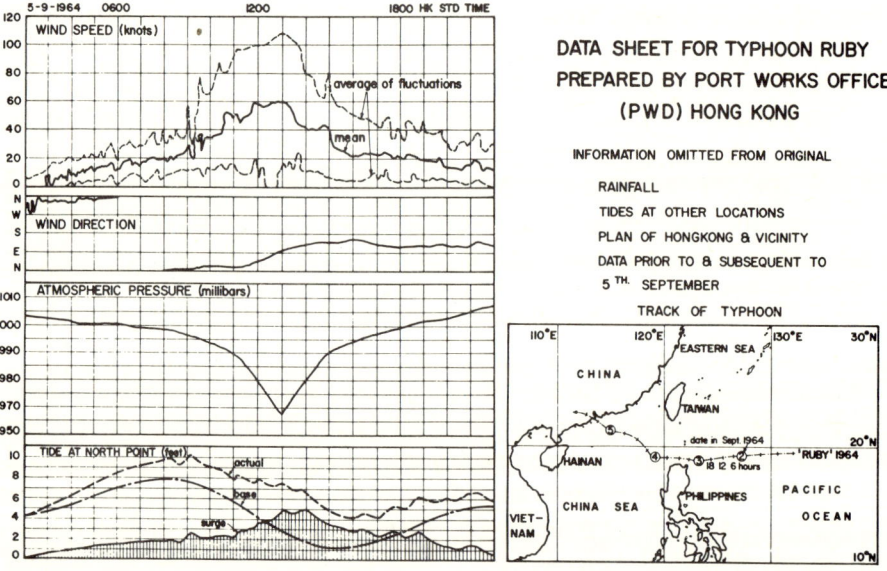

Fig. 2-26. Part of a comprehensive typhoon data sheet as constructed at the Public Works Department, Hong Kong.

METEOROLOGICAL CONDITIONS

In spite of its modest diameter the Australian version can still produce 100 miles/h winds at the margin of the 10-mile diameter eye. As will be seen later the strength of a cyclone can be gauged from the product of the eye radius and differential between the minimum pressure in the eye and normal atmospheric pressure.

The wind swiftly reduces speed to around 50 miles/h from the eye edge outwards. This annular zone of high wind velocity may vary in width from 10 miles to 70 miles in the smallest to largest species of tropical cyclone. During the passage of such a zone over an area of ocean or of land, extremely dangerous conditions always ensue. Large volumes of water and spray are hurled horizontally over the sea surface, whilst even large pebbles can become projectiles in the land case.

The outer zone, from the 50-miles/h circle to the boundary of the cyclone, contains a decreasing wind speed. The reduction to 20 miles/h (considered the limit of the cyclonic effects) occurs over a width of about 400 miles for the hurricane (see Fig. 2-24 and 2-25) and 90 miles for the tropical cyclone of Australia. Although more moderate in character than the winds in the annulus around the eye, this region probably provides the bulk of the ocean waves for the cyclone. This is because the waves are blowing over a reasonable fetch for a reasonable length of time, so that the FAS state is approached. Also the extremely high winds of the annulus tend to break waves and so assist in the production of spray. This and other aspects are to be discussed more fully in the section on wave prediction.

Another important feature of tropical cyclones, especially in respect to wave-generating capabilities, is the radial change in wind direction. As seen in Fig. 2-24 the wind blows at 45° to radii for most of the cyclonic area. Thus on opposite sides of the eye the wind directions are opposite.

Observation of Fig. 2-24, which indicates a specific direction of travel for the centre, would indicate that waves generated in the right rear quadrant will run with the cyclone. This will allow them to be under prolonged influence of the wind and hence permit them, perhaps, to reach their optimum size for the respective wind velocities. In a similar manner the left rear quadrant of a tropical cyclone in the Southern Hemisphere is the most dangerous for shipping and coastal works. This emphasizes once more the importance of knowing the general passage of cyclonic centres when assessing the wave climate for a particular point on the coast.

In this regard it should be realized that tropical cyclones are a lot less predictable in their movements than their extra-tropical sisters. It requires but a slight increase in atmospheric pressure in an adjacent area to deflect a hurricane or typhoon. In this way the immediate future passage of a centre may be forecast [14] but generalization regarding flow paths must be used with caution. Fig. 2-27 shows the record of tropical cyclones along the northwest coast of western Australia for the months of December to March 1962–1967. Later it will be shown how this information can be integrated by cyclonicity charts. However, this does not preclude a solitary

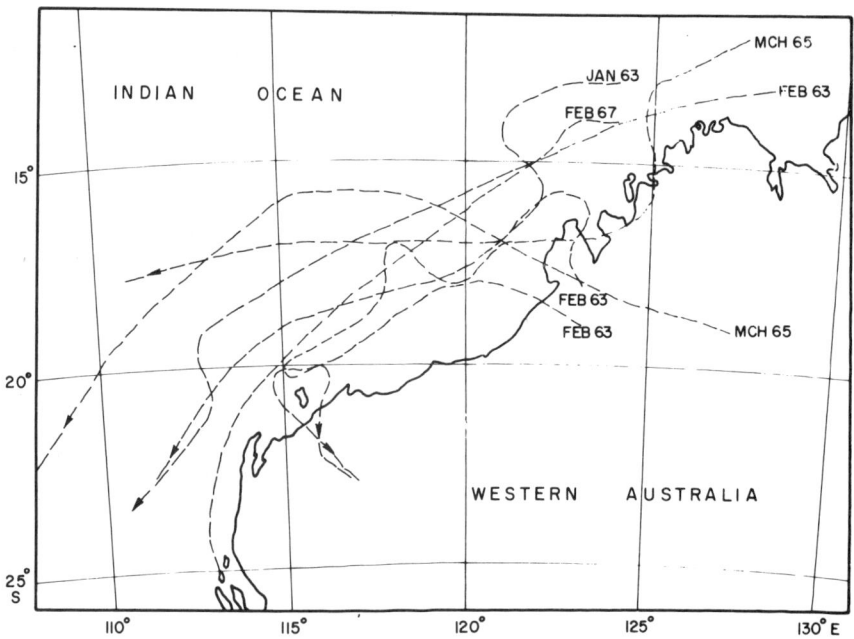

Fig. 2-27. Paths followed by tropical cyclones over the northwest coast of Western Australia from 1962 to 1967.

centre deviating from this average alignment and passing directly over, or in close proximity to, any port site in the whole region.

The speed of travel for a centre may vary along its life path, but generally the major variations are between sea and land travel. When tropical cyclones reach a continent, they tend to accelerate and dissipate. The normal cyclone speed for a region should be known, so that critical conditions for wave generation or storm surge can be predicted. Optimal wave generation occurs when the wind field maintains itself over the largest waves already produced. Optimal surges are determined by the ratio of the cyclone speed to the average speed of long shallow-water waves across the continental shelf.

Seasonal pattern

It is obvious from the previous discussion that the repetition of cyclonic centres (either tropical or extra-tropical), can be of the greatest importance to the engineer. He needs to know the annual cycle of events in the region when calculating the wave climate for certain sections of coast. For this purpose it is essential to study cyclone movements during all seasons, including summer and winter.

Meteorologists, or more specifically climatologists, have devised diagrams known

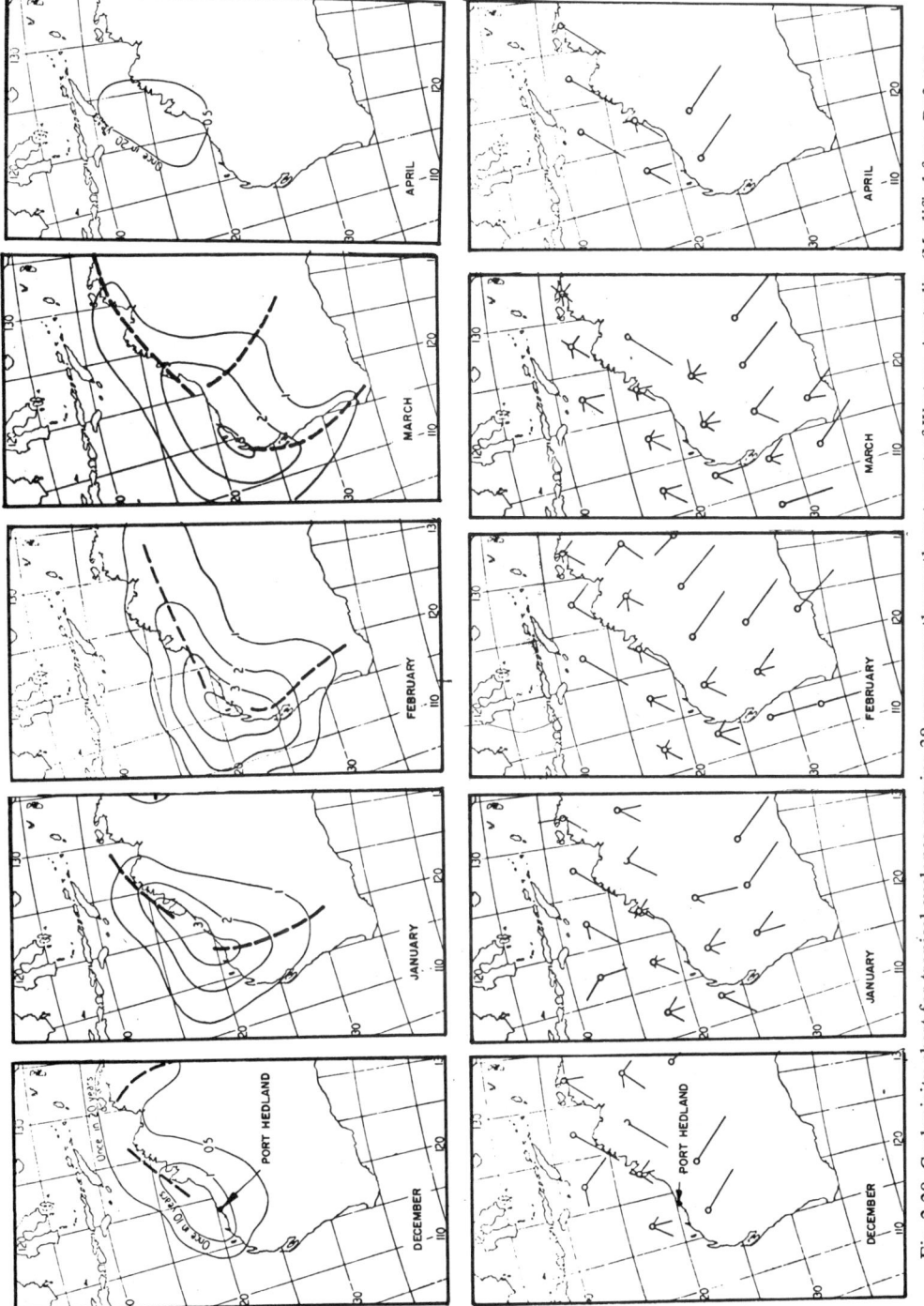

Fig. 2-28. Cyclonicity chart for tropical cyclones covering 38 years over the northwest coast of Western Australia. (Modified from Ref. 15.)

as *cyclonicity charts* and *anti-cyclonicity charts* to present such information. On a map of the region under study the presence of high- or low-pressure centres is noted as hours duration per month within 1°- or 5°-squares of latitude and longitude. These data are gleaned from the 3-hourly synoptic weather charts. By drawing isopleths through squares of equal numbers of hours the general path of the centres can be ascertained.

Such a chart is illustrated in Fig. 2-28 where the frequency and axis of travel for tropical cyclones is shown for the coast of western Australia [15]. Included in the figure are arrows which add the occasions on which centres in each square were moving in a given direction (concentrated at 45°-intervals). These may help in the forecasting of future cyclone movements. The data in Fig. 2-28 are based upon 38 years of record.

Other information can be summed in a similar manner, such as lowest pressure values, radii of eyes, speed of travel. The position of other relevant data on cyclones is now being recorded on punched cards in many centres so that information is readily available to the coastal engineer for specific seasons, years, or decades.

The wind pattern as recorded in atlases, covering vast stretches of ocean, is the resultant of all the cyclonic and anticyclonic events throughout the year, together with the air circulation of global character. Conditions for the extreme conditions of January and July are generally presented, although monthly values are available in special oceanic charts [16]. Besides general wind vectors the paths of cyclons are recorded, as in Fig. 2-29, so that the location of extreme winds can be noted.

In the charts just referenced additional information on wave climate is included. For each month of the year are recorded wave characteristics as observed by marine personnel over many years. These data are concentrated in 45°-directional bands for each 5°-square of latitude and longitude. Increments of wave height are recorded in the same manner as the original marine observation. They indicate also the general direction of movement of waves and this is probably of more use to the coastal engineer than the average wave heights recorded. A typical sea and swell rose from the British charts [16] is illustrated in Fig. 2-30. The charts of the United Kingdom and the United States differ a little in the scale of wave heights used (see Table 2-II) and in the fact that the former use data from certain types of vessel in which officers have been trained in wave observation, whereas the latter accept information from all sources willing to supply it.

Whilst such charts can give only qualitative data on wave height, and none on wave period, the directional nature of the ocean waves can be extremely useful. When such roses are inspected over the whole ocean a pattern is observed which is necessarily consistent with the wind pattern for that particular oceanic basin.

In employing sea and swell charts, the nearest 5°-square to the coast should be used, noting any consistency or otherwise with adjacent squares. When the more local wind pattern is added to this picture of wave height and direction an overall

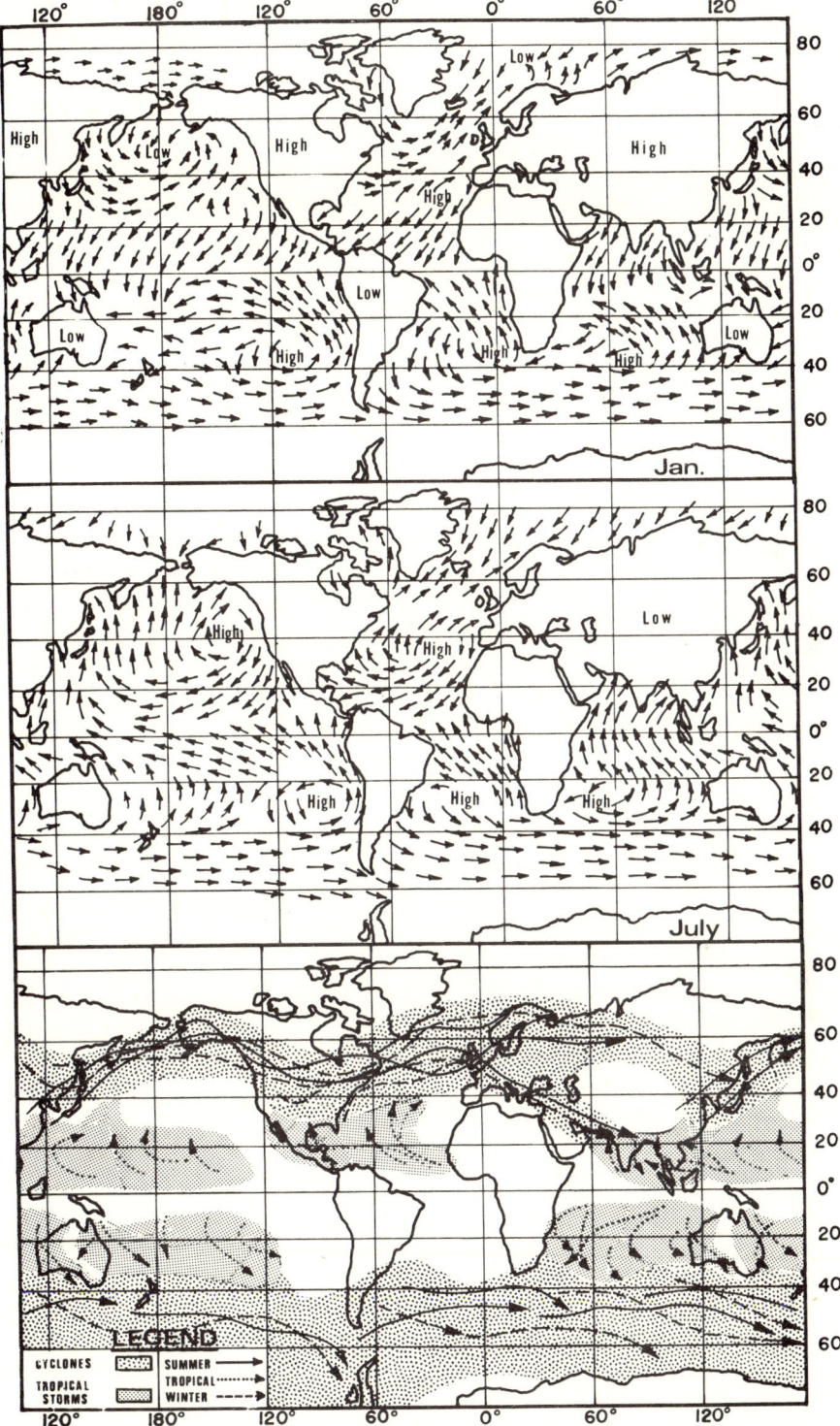

Fig. 2-29. Global distribution of winds and paths of cyclones for months of January and July.

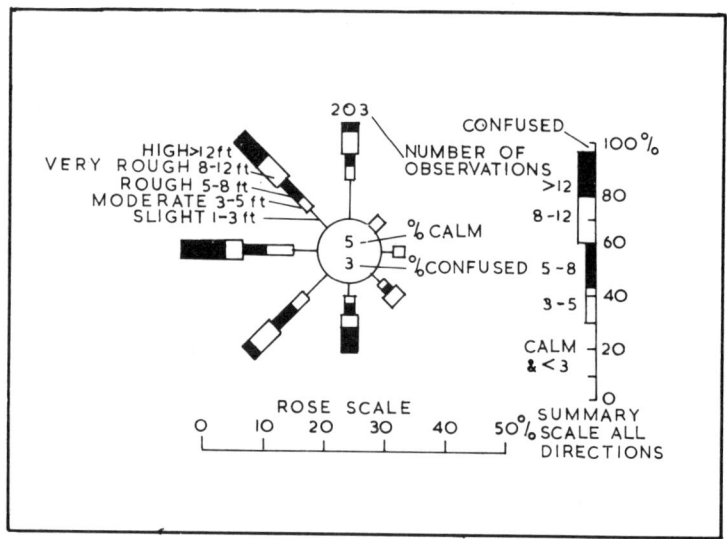

Fig. 2-30. Typical wave rose in British Admiralty sea and swell charts.

wave pattern for that section of coast can be derived. This information, it should be remembered, relates to the deep-water conditions beyond the continental shelf. As the waves travel towards the shore they will be modified somewhat by the shoaling depths, in both height and direction.

Besides the general wave distribution as noted above, it is useful to know where the zones of high wind are located over the oceans. From these will emerge the swell waves which traverse the largest of the oceanic basins and, in fact, can traverse more than one basin. Fig. 2-31 provides such a picture, with storm zones marked for each quarter of the year.

It will be observed in Fig. 2-31 that the regions of 40°– 60° latitude in both hemispheres are zones of high-intensity wind. As the year proceeds so do the

TABLE 2-II

Scale of wave heights used in reporting state of the sea

Sea scale	United States	United Kingdom
Slight	2– 4 ft.	<3 ft.
Moderate	4– 8 ft.	3– 5 ft.
Rough	8–13 ft.	5– 8 ft.
Very rough	13–20 ft.	8–12 ft.
High	20–30 ft.	>12 ft.
Very high	30–45 ft.	–
Phenomenal	>45 ft.	–

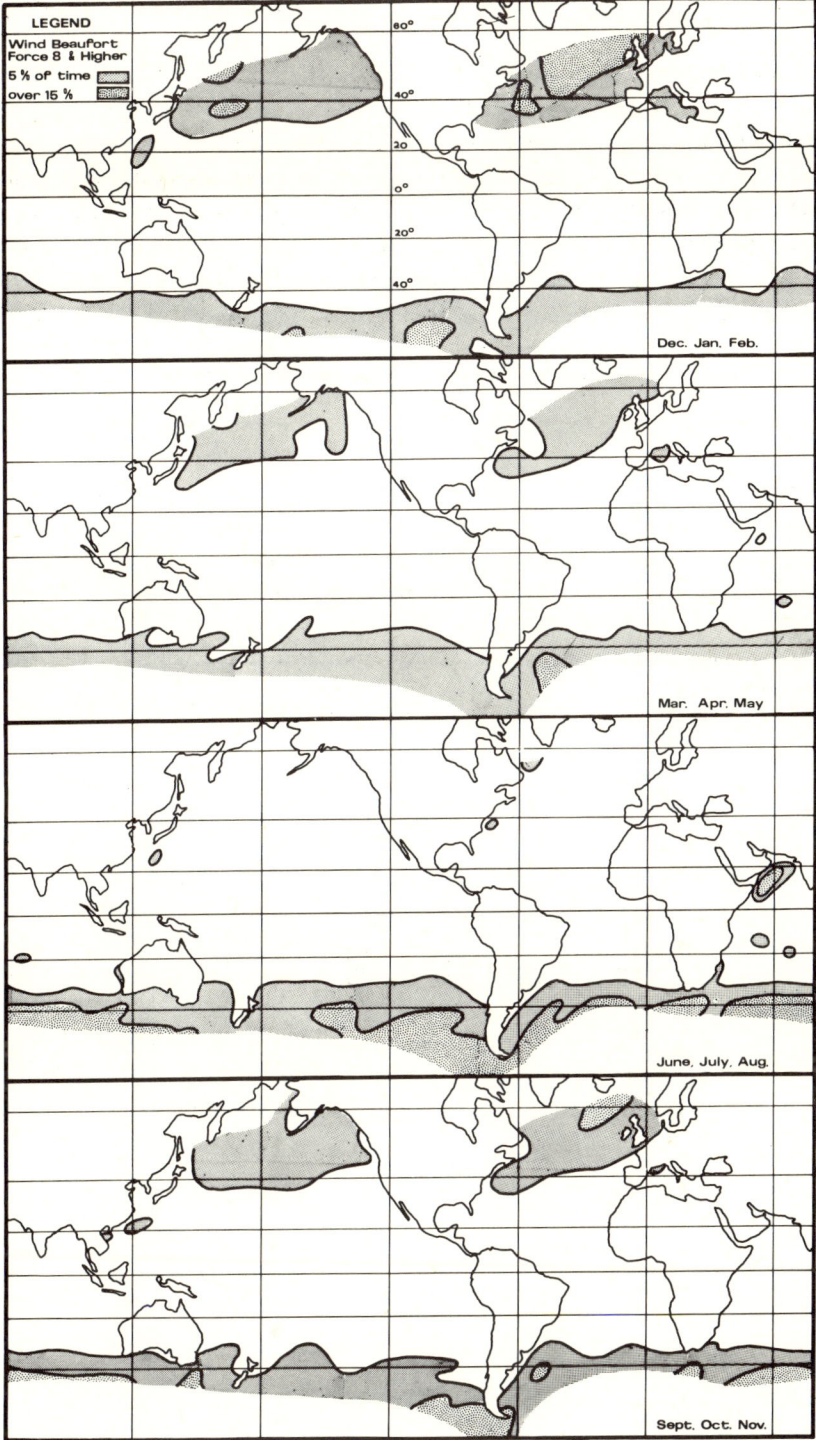

Fig. 2-31. Global distribution of high wind velocity zones for quarters of the year.

latitudinal limits of these storm zones vary. In the Northern Hemisphere, during June, July and August winds of Beaufort force 8 and higher almost disappear, although for the remaining nine months they occur at least 5% of the time.

Besides the major sources of swell noted above there are smaller storm areas which are worthy of note. One of these is the Arabian Sea where, in the June–August quarter, high winds are experienced over 15% of the time. Reference to Fig. 2-29 would indicate that these winds are from the southwest direction. Another small storm zone is located off the continent of China and south of Japan. This

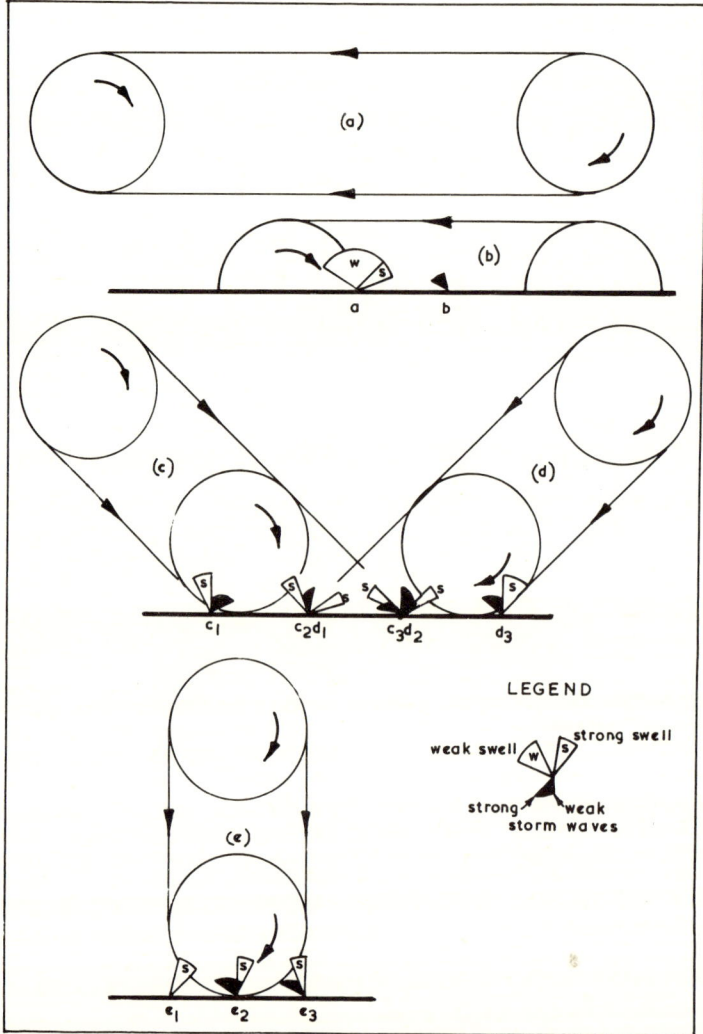

Fig. 2-32. Wave incidence from cyclones in proximity to a coast.

remains a storm-prone segment for most of the year, it being completely free in the March—May quarter only.

When deriving a wave climate for an area it is essential to consider the direction of rotation of winds in the cyclonic centres, besides the zones of most frequent occurrence. This, together with knowledge of the centre movement, shows from which direction the largest waves can be expected. The variation in waves at particular points on a coastline during the passage of cyclones along and across it is depicted in Fig. 2-32.

Referring to that figure it can be seen that for passage (a), along a coast offshore, point a will receive swell only, but this will be strong in a narrow directional band. For passage (b), with the centre line coincident with the coast, no waves will arrive at point b until the centre is just to the left of it. No swell will be experienced, only storm waves. For passage (c) points c_1, c_2 and c_3 will receive swell from slightly different directions and storm waves from totally different directions. In the same way passage (d), from the opposite quadrant, provides points d_1, d_2 and d_3 with specific wave conditions, which are different again from those at the c points. Passage (e) is normal to the coast and provides points e_1, e_2 and e_3 with swell from varying directions and storm waves over a wider band at two of the three locations.

In this simple model no allowance has been made for variable wind velocities within the cyclone as discussed previously. But it serves to illustrate the paramount influence of centre-alignment on waves experienced at a port site on the coast. In the Southern Hemisphere worse wave conditions ensue when the cyclone hits the coast to the right of the port, and to the left in the Northern Hemisphere (as gauged from the directional vector of the centre).

PROBLEMS

1
Discuss the various mechanisms by which waves are generated from their initiation to the FAS. Which of these provides the major energy?

2
Compare the angle to the mean wind direction at which 10-sec waves will receive maximum development from resonance in wind fields of 30 and 40 knots, respectively.

3
For a wind velocity of 35 knots list the critical angles, for optimum development from resonance with atmospheric pulses, of waves with periods 5, 7, 9, 11, and 13 seconds. Comment on the result.

4
List references on laboratory verification of wave generation and summarize their findings. What general conclusions can be drawn from the research work completed so far?

5
Discuss the several ways in which breaking of smaller waves may add energy to the longer waves in a developing sea.

6
List the main characteristics of storm waves and give reasons for their existence. List similarly the qualities of swell waves, indicating their source.

7
Considering an energy distribution curve (EDC) of quadrated amplitude versus wave period is an isosceles triangle for a FAS, draw the EDC at the end of the minimum fetch for such conditions when the wind has been blowing for only half the minimum duration. Draw also the EDC for a point half way along the fetch even when FAS conditions exist.

8
In what way will an EDC beyond, but directly downwind of, a fetch differ from one on the same radius from the fetch but located to one side of it? Consider the wave recordings to be made simultaneously.

9
How is the width of a fetch important in the waves generated by it, either inside or outside the fetch?

10
Why are waves in the dispersal area much smaller in height than inside the fetch?

11
Can the swell waves from a particular storm at sea, measured at one particular short period of time, provide much useful information of the generating conditions? Discuss the likely temporal variations of wave height and period.

12
In Fig. 2-33 a fetch is depicted in which a FAS has existed for some time. The EDC has been simplified to a triangle and that at the downwind end of the fetch (point C) is hatched. Draw the EDC's at the other locations (A to J inclusive) at the instant when the first waves arrive at point J. The diagram is not to scale but relevant dimensions are given.

13
In respect to wave generation how do tropical and extra-tropical cyclones differ?

14
As a consultant you are required to provide a wave climate for a certain port site. List the headings and sub-headings in the report you would prepare. Consider two locations one in latitudes $0°-30°$ and the second in latitudes $30°-50°$.

15
Curved isobars on a synoptic weather chart represent 5 millibar spacing and in a fetch area are $1.25°$ longitude apart. If the latitude of the fetch is $36°S$, find the velocity of the geostrophic wind. If the mean radius of curvature of the isobars is $6°$ and the sea-air temperature difference is $5°F$, compute the gradient wind speed.

16
When computing waves from wind data, why is it preferable to use anemometer readings from ships at sea rather than isobar spacing from synoptic weather maps?

17

Should formulae derived for forecasting waves from extra-tropical cyclones be applied without modification to tropical storms? Give reasons for your answer.

18

Why is the speed vector of a cyclonic centre important in its capacity for generating waves? How can a coastal engineer derive a picture of the general path of cyclones in a specific ocean area?

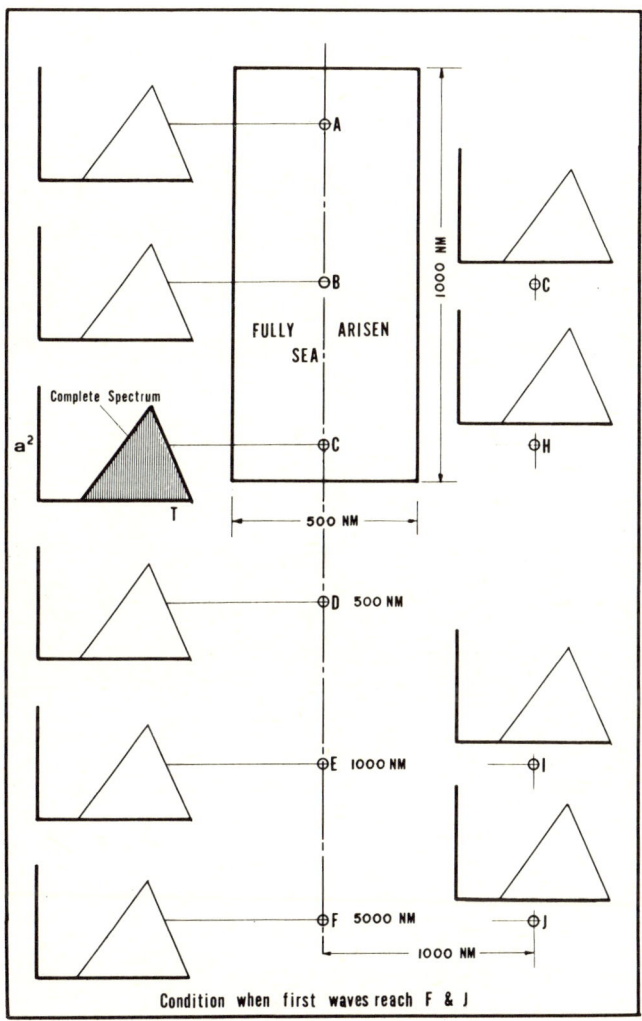

Fig. 2-33. Fetch information for Problem 12.

19

The "roaring fourties" of the Southern Hemisphere are notable for their influence on history. What are the longer-term effects of this phenomenon?

20

Do storm waves at a coastal area necessarily arrive from the same direction as swell from the same cyclonic centres? What is the relation of their approach to that of swell from more distant storm zones?

21

Where in the world's oceans do you expect the greatest storm energy? On what coasts of the world do you expect the greatest swell energy? On the minimum swell coasts, what is the general nature of the wave conditions?

REFERENCES

[1] O.M. Phillips, 1957. On the generation of waves by turbulent winds. *J. Fluid Mech.*, 2: 417–445.
[2] O.M. Phillips, 1960, 1961. On the dynamics of unsteady gravity waves of finite amplitude, 1. The elementary interactions. *J. Fluid Mech.*, 9: 193–217; 2. Local properties of a random wave field. *J. Fluid Mech.*, 11: 43–155.
[3] B. Kinsman, 1965. *Wind Waves*. Prentice Hall, Englewood Cliffs, N.J., 676 pp.
[4] J.W. Miles, 1957–1960. On the generation of surface waves by shear flows. *J. Fluid Mech.*, 3: 185–204; 6: 568–582; 7: 469–478.
[5] R.W. Stewart, 1961. The wave drag of wind over water. *J. Fluid Mech.*, 10: 189–194.
[6] O.M. Phillips and E.J. Katz, 1961. The low frequency components of the spectrum of wind-generated waves. *J. Mar. Res.*, 19: 57–69.
[7] M.S. Longuet-Higgins, D.E. Cartwright and N.D. Smith, 1963. Observations of directional spectrum of sea waves using the motions of a floating buoy. *Proc. Conf. Ocean Wave Spectra, 1963*: 111–132.
[8] H. Jeffreys, 1925–1926. On the formation of water waves by wind. *Proc. R. Soc., Ser. A*, 107: 189–206; 110: 241–247.
[9] T. Stanton, 1937. The growth of waves on water due to the action of wind. *Proc. R. Soc., Ser. A*, 137: 283–293.
[10] M. Hino, 1966. A theory on the fetch graph, the roughness of the sea and the energy transfer between wind and wave. *Coastal Eng. Japan*, 9: 11–26. (Also *Proc. 10th Conf. Coastal Eng., 1966*: 18–37.)
[11] M.S. Longuet-Higgins, 1969. A non-linear mechanism for the generation of sea waves. *Proc. R. Soc., Ser. A* 311: 371–389.
[12] F.E. Snodgrass et al., 1966. Propagation of ocean swell across the Pacific. *Phil. Trans. R. Soc., Ser. A*, 259: 431–497.
[13] S. Syono, 1963. Structure of typhoons. *Proc. Int. Seminar Trop. Cyclones, 1962, Tokyo – Meteorol. Agency, Tech. Rep.*, 21: 121–131.
[14] R.J. Renard, and W.H. Levings, 1969. The Navy's numerical hurricane and typhoon forecast scheme: application to 1967 Atlantic storm data. *J. Appl. Meteorol.*, 8: 717–725.
[15] A.T. Brunt and J. Hogan, 1956. The occurrence of tropical cyclones in Australian region. *Trop. Cyclone Symp., 1956, Brisbane, Pap.*, 1: 1–14.
[16] *Monthly Meteorological Charts*, H.M.S.O., London, or *Atlas of Sea and Swell Charts*. U.S.G.P.O., Washington D.C. (data listed for individual oceans).

Chapter 3

WAVE FORECASTING

In this chapter there are three distinct sections, although the subject matter is completely interrelated. There is firstly a discussion on wave statistics, secondly an outline of the history of wave forecasting, and thirdly a presentation of formulae and procedures for ready application to engineering problems. In the bulk of this discussion the FAS wave characteristics are paramount, but in the last section means are provided for forecasting waves in developing seas and those propagating outside the fetch.

Whilst the historical development of wave prediction would appear to take precedence over other topics, it is more readily appreciated after some information is presented on the statistics of waves. Until Longuet-Higgins in 1952 [1] applied some statistical theory, previously employed to acoustic and electronic phenomena, to ocean waves, the procedure known as wave forecasting or hindcasting did not really get off the water as it were. All methods considered acceptable by oceanographers today are soundly based on statistical theory, as must be the analysis of wave records used to verify them.

WAVE STATISTICS

The surface undulations of the sea, when waves are propagating across it, can be considered as a random process – the word "process" introducing the concept of time, and the term "random" inferring its representation by a probability law. It has already been noted that two wave records taken reasonably close together may in detail look extremely unlike. However, their statistical properties would indicate the similarities of their source. It is from the perspective of statistics that some order emerges from the seeming chaos of the sea surface.

The introduction of statistics into any phenomenon tends to add an air of accuracy to data, which may be unwarranted. The simplifications required in order to apply these mathematical models should always be kept in mind. No result is more accurate than the data on which it is based. An added complication is that geophysical observations are like records of history; they are hardly ever repeated exactly.

With generation of waves it is impossible to have a steady wind of infinite duration. Neither can an infinitely long record of waves be suitably analysed. In formulating a geophysical law of this nature, from a unique experience in time and space, the assumption is made that it represents other records should the measurement be repeatable. This assumption is that the process is *ergodic,* or that Nature is not freakish.

The mathematical model should really be applied only to waves generated over an infinite ocean for an infinite time. It is not strictly applicable to transient conditions where waves are still developing. But even in these circumstances approximations are possible for fetches and durations of finite magnitudes. Again, the statistical theory applies strictly to sinusoidal waves, which imply very low-amplitude waves. But real waves are fortunately catered for by having a Gaussian probability distribution for their crest heights above or below the mean sea level. This is certainly the case for the FAS state where waves are travelling in a multitude of directions.

There are three reasons for the Gaussian distribution not being strictly applicable, but for engineering purposes these are of slight consequence. For example:

(1) Ocean waves have a limited steepness and so the infinitely high waves implicit in this distribution cannot be present. Such peak waves in the marine situation constitute a negligible portion of the total record.

(2) The wave motion must satisfy the Bernouilli equation as well as the free-surface boundary conditions, even with all the short-period components. This short-period chop is of no engineering consequence.

(3) Waves of finite height are more trochoidal in shape than sinusoidal. A true Gaussian record could be read upside down with no error. Many wave records can be similarly inverted to provide equal statistical parameters. Such deviations from the sinusoidal shape, however, cause but a slight offset of the Gaussian probability peak from the mean position.

Three distinct advances can be recognized in the development of statistical relationships of ocean waves. The first, by Sverdrup and Munk [2], was the initial scientific attack on the problem of wave forecasting, whilst the later two stages emerged from the knowledge that ocean wave spectra were similar to those of acoustic wave systems.

Sverdrup and Munk

During World War II two groups of scientists were concerned with wave prediction from wind fields, in order to forecast conditions on enemy beaches at the time of future landings with amphibious vehicles. The group in the United Kingdom undertook the recording of waves by pressure transducers on the ocean floor in some 40 ft. of water and then subsequently the analysis of the output for the energy content in its many components. As will be explained later, this procedure emphasized the longer swell waves.

The group in the United States was headed by H.V. Sverdrup and W.H. Munk. The necessity of using visual data made them relate the wind speeds to the higher and more distinct oscillations of the complex sea surface. These waves were termed "significant waves" and their characteristics were called *significant height* and *significant period* [2]. It is fortunate that when the statistical distribution of wave heights should be determined later this "significant" value equated to the average of the highest one third of the waves (symbolized as $H_{1/3}$ or H_{33}). Also, the average period between the water surface rising through the still-water level at any point was found to approximate the value of the significant period as observed by Sverdrup and Munk.

At this stage of wave forecasting or hindcasting the significant wave was treated as a single-sine wave, which purportedly contained the same energy per unit area of ocean as the complex sea-surface undulations. For the pragmatism of the period it probably served its purpose. The unforeseen complexities at the landing beaches of Europe under storm conditions were possibly due to observing the actions of landing craft in the swell waves of the Californian beaches and not allowing for the three-dimensional nature of waves within storm zones as discussed earlier.

Longuet-Higgins

In the United Kingdom after the war, M.S. Longuet-Higgins became interested in the ocean wave records available and their spectra as provided by the harmonic analyser at the National Institute of Oceanography (NIO). The theory applied was that already used by Rayleigh in 1880 [3] for sound waves, and Rice [4] and Echart [5] later, when studying noise in electronic circuits. The mathematical result was then compared to the NIO spectra.

Since crest-to-trough height was not amenable to mathematical manipulation, resort was made to variance from some fixed datum (still-water level for example). The fluctuation of the water surface at some fixed location in the ocean varies about a mean, the amplitude of which can be measured at a multitude of points along a record of limited length. If N such points are chosen, a root mean square value can be calculated:

$$a_{rms} = \sqrt{\Sigma a^2 / N} \qquad (3\text{-}1)$$

where a is the successive amplitude of the motion from the mean position.

By accepting that the wave energy is being supplied to the recorder from a number of storm sources, and is in random phase, the Rayleigh probability distribution is applicable. Such a distribution has been found to apply to swell waves which have commenced to separate according to their component periods. This implies a narrow spectrum of waves, or a portion of the FAS passing the recording station at any one time.

With this restriction Longuet-Higgins in 1952 [1] was able to relate $H_{ave}, H_{1/3}$ and $H_{1/10}$ to a_{rms} or E, where:

$$E = 2\Sigma a^2 / N \qquad (3\text{-}2)$$

which from eq. 3-1 gives:

$$\sqrt{E} = \sqrt{2}\, a_{rms} \qquad (3\text{-}3)$$

Transposing from amplitude (to which a_{rms} refers) to the more useful crest-to-trough height, Longuet-Higgins doubled the value. This assumption is correct for a sinusoidal wave, which the narrow-spectrum condition represented reasonably well. The results obtained by application of the statistical theory were:

$$H_{ave} = 1.77\sqrt{E} = 1.77\sqrt{2}\, a_{rms} \qquad (3\text{-}4)$$

$$H_{1/3} = 2.83\sqrt{E} = 2.83\sqrt{2}\, a_{rms} \qquad (3\text{-}5)$$

$$H_{1/10} = 3.60\sqrt{E} = 3.60\sqrt{2}\, a_{rms} \qquad (3\text{-}6)$$

Longuet-Higgins also found a relationship for the probable maximum wave from any record, but this involves the number of waves, or period of coverage being considered, and will be discussed in the chapter on wave recording.

Thus if a_{rms} can be determined from a wave record, or E from specified wind conditions in a fetch, a means has been provided to find useful wave heights for engineering design purposes. Methods for computing both a_{rms} and E are to be described later.

Cartwright and Longuet-Higgins

The next advance in the statistical story was that made in 1956 by Cartwright and Longuet-Higgins [6]. They considered the statistical distribution of crest heights above SWL for a sea surface consisting of an infinite number of sine waves. This allowed analysis of the FAS condition, with its wide spectrum, except for the wave-shape restriction. Because crests can occur above and below the SWL, especially in a wide spectrum of waves, negative values must enter the analysis. As in the previous case, the mean or SWL must be determined in the wave record.

The distribution of crest heights (positive and negative) about the SWL was found analytically to depend upon two parameters, namely, a_{rms} (as previously) and ϵ, a measure of the relative width of the spectrum. For the single sinusoidal wave $\epsilon = 0$, whereas for the FAS ϵ tends to unity. As ϵ varies so does the distribution vary from the Rayleigh to the Gaussian as illustrated in Fig. 3-1.

A means for determining ϵ from a wave record will be discussed in the chapter on wave recording, but it should be noted here that this parameter applies to the sea-surface undulation. A pressure record taken at some depth will omit many of

WAVE STATISTICS

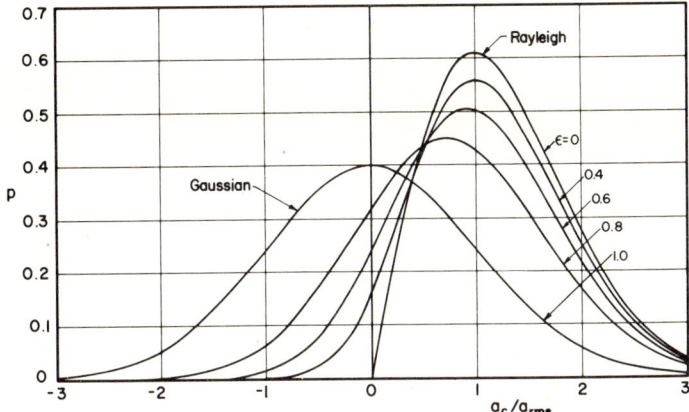

Fig. 3-1. Variation of probability distribution of wave crest heights above SWL with width of spectrum. (Modified from Ref. 6.)

the minor crests and so significantly influence the value of ϵ obtained. But using data from ship-borne wave recorders, which omit only waves of negligible proportions, it has been found for the FAS that ϵ varies from 0.6 to 0.8, independently of the wind velocity. Swell waves have provided an ϵ value of 0.3.

The relationship of a_{rms} to ϵ or more particularly $H_{1/n}/\sqrt{2}\, a_{rms}$ ($= H_{1/n}/\sqrt{E}$) to ϵ is shown in Fig. 3-2. It is seen that from the narrow-spectrum condition to the

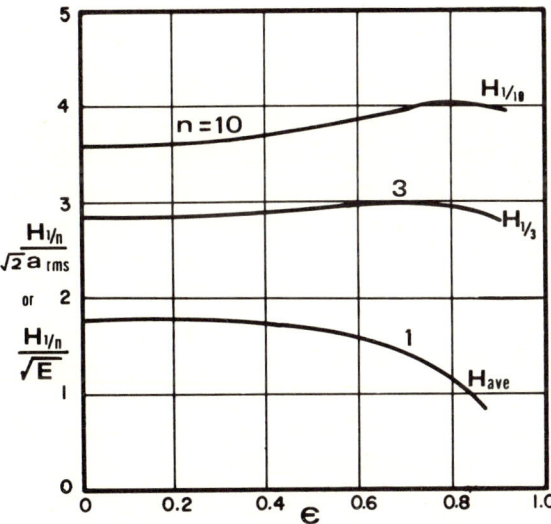

Fig. 3-2. Variation of $H_{1/n}/\sqrt{E}$ with width of spectrum.

FAS the ratio $H_{1/n}/\sqrt{E}$ varies as follows:

$H_{1/10}$ from $3.60\sqrt{E}$ to $4.0\sqrt{E}$ \hfill (3-7)

$H_{1/3}$ from $2.83\sqrt{E}$ to $2.95\sqrt{E}$ \hfill (3-8)

H_{ave} from $1.77\sqrt{E}$ to $1.1\sqrt{E}$ \hfill (3-9)

Eq. 3-8, from eq. 3-3, results in:

$$H_{1/3} = 2.95\sqrt{2}\, a_{rms} = 4.13\, a_{rms} \tag{3-10}$$

However, a value of 4 is generally accepted, so that:

$$\text{variance} = \Sigma a^2/N = (\tfrac{1}{4}H_{1/3})^2 \tag{3-11}$$

From eq. 3-7 to 3-9 and Fig. 3-2 it is seen that for the same a_{rms}, widening of the spectrum increases the height of the higher waves, but decreases the value of the average wave height. This supports the view that the average wave height is of little engineering value, because in the wide-spectrum situation, in which higher waves are encountered, the many more minor waves reduce the average. These smaller waves are not important to the design of engineering structures, particularly in the presence of the components with large magnitude.

Fig. 3-2 has been derived from the paper by Cartwright and Longuet-Higgins [5] on the assumption that all crest-to-trough measurements are twice the amplitudes from still-water level, for all values of ϵ. This is approximately correct, but the exact relationship is not known either mathematically or experimentally.

HISTORY OF WAVE PREDICTION

The forecasting, or hindcasting (computing after the event), of ocean waves dates essentially from World War II. Although some relationships were derived prior to this period, a consistent method of visually measuring wave heights had not been developed. Any equations so quoted should be used with extreme caution. Generally, they relate height with wind speed alone, omitting any reference to duration or fetch. As will be seen, this simplification is applicable to a FAS, but whether the conditions observed were just that, or intermingled with oceanic swell cannot be guaranteed.

As noted earlier, the war-time demand for information on wave conditions at enemy beaches promoted research on both sides of the Atlantic. From the United States emerged the Sverdrup-Munk relationships [1] and in the United Kingdom the wave recording data were being gathered, later to be put into equation form by Darbyshire [7]. Subsequent to Longuet-Higgins deriving the statistical relationships for the narrow wave spectrum, Neumann [8] at New York University derived theoretically an equation representing the spectrum of waves present, which could be related to the "significant" height and period previously used. Pierson et al. [9] then employed this spectral presentation to derive tables and graphs for ready application by engineers. Their dissertation on the subject should be read for its comprehensive treatment of the subject.

Many comparisons of these various formulae showed disconcerting disparities which were difficult to explain. However, Pierson [10] pursued the topic from the aspect of the wind speeds being used and found that the wind profile near the ocean was a critical element not previously appreciated. By noting the heights above the sea surface at which the stated wind speeds had been recorded during wave measurement, he was able to reduce these speeds to a common height. This greatly reduced the divergencies in the formulae. As the wind profile close to the sea surface can be predicted reasonably well, wind velocities recorded at one level can be readily converted to that at some other level.

At the same time, Moskowitz [11] undertook a comprehensive spectral analysis of data obtained by ship-borne wave recorders. This led to a modification of the spectral form previously employed by Pierson et al. [9] and for the FAS was in better agreement with other relationships available. Pierson and Moskowitz [12] placed the wind profile concept and the new spectral shape in the dimensionless form suggested by Kitaigorodskii [13]. From this a single spectrum was devised which could represent the FAS for any wind velocity, as long as the representative height above the sea surface could be specified for the wind.

SMB method

The work of Sverdrup and Munk [1] was later revised and added to by Johnson and Bretschneider [14], the latter having his name attached to the relationship [15], so that it became known as the SMB method of forecasting.

Since wave celerity in deep-water is given by:

$$C = gT/2\pi \tag{3-12}$$

where dimensions are consistently chosen, it is seen that gravity is an essential feature of wave propagation. This should therefore be considered with the other variables likely to influence wave speed (C) and wave height (H), namely, wind velocity (U), fetch length (F) and duration of wind (t). Thus:

$$C = f(U,F,t,g) \tag{3-13}$$

and:

$$H = f(U,F,t,g) \tag{3-14}$$

Dimensional analysis of the above results in:

$$\frac{C}{U} = f'\left(\frac{gF}{U^2}, \frac{gt}{U}\right) \qquad (3\text{-}15)$$

and:

$$\frac{gH}{U^2} = f''\left(\frac{gF}{U^2}, \frac{gt}{U}\right) \qquad (3\text{-}16)$$

In deep water, where all wave generation of this discussion is considered to take place, the ratio C/U (sometimes referred to as *wave age*) can be replaced by $gT/2\pi U$. Thus, four dimensionless parameters can relate the energy source to the energy sink, linked by the important variable of wind velocity:

source	fetch	gF/U^2
	duration	gt/U
sink	wave height	gH/U^2
	wave speed (period)	C/U or $gT/2\pi U$

Many data have been plotted over the years for various fetches and durations in order to determine the constants of proportionality. Separate graphs should be presented for these two variables each of which should contain two curves, one for wave height and the other for wave period. Since similar parameters apply to both fetch and duration the one graph generally serves (see Fig. 3-3), in which two sets of curves can provide height and the other two wave period.

It is readily seen that the growth of the significant wave can be controlled either by fetch length or by duration of wind. Thus, when two values of height ($gH_{1/3}/U^2$) result from a given fetch (gF/U^2) and a given duration (gt/U) the smaller of the two must be selected, as either fetch or duration will control the size. For any specified $H_{1/3}$ and U (i.e., $gH_{1/3}/U^2$ = const.) there will be a fixed value of gF/U^2 (fetch) and of gt/U (duration) to produce it. These may be considered the minimal requirements.

The growth of $H_{1/3}$ along a fetch is depicted in Fig. 3-4, where each intercept of F on the curve is associated with a duration t. At some particular distance along the fetch (F_1) the waves will grow with time. For example, at t_1 the height H_1 applies, not only to F_1 but all along the fetch from the point t_1, it being controlled by duration of the wind. At some later time t_2 the height H_2 exists from this point onwards. Upwind of this point the wave height is less as it is limited by the fetch available. When duration t_3 is reached this produces the maximum value for $H_{1/3}$ at F_1 and therefore this time may be considered the minimal value (t_{min}) to produce the optimum wave height at this particular fetch. For the given wind speed U and fetch F_1, a height of H_3 results, but if the duration is t_4 a greater height H_4

HISTORY OF WAVE PREDICTION

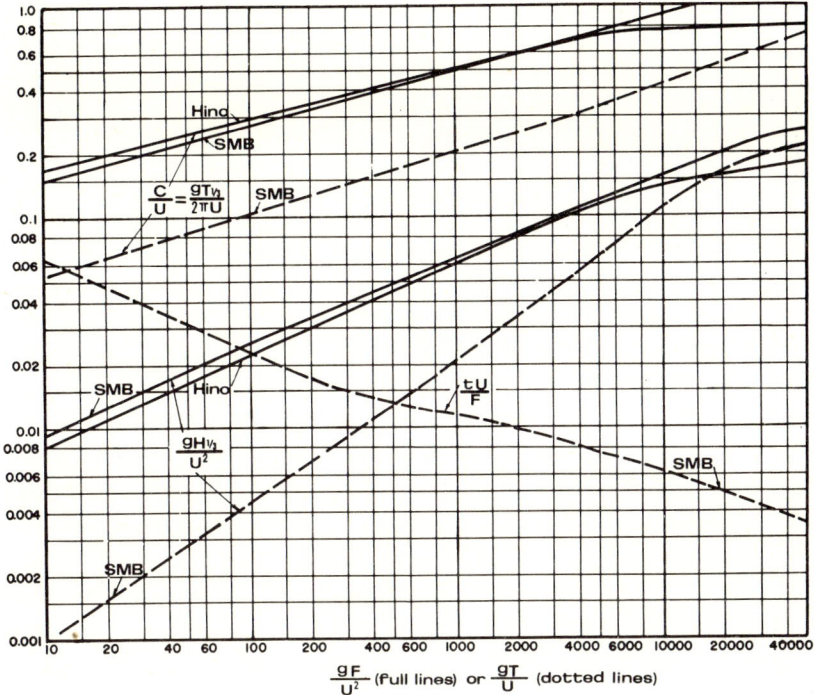

Fig. 3-3. Dimensionless relationships of the SMB formulae.

is given by Fig. 3-3. This value is not applicable since fetch F_1 controls the growth.

Thus, in the use of Fig. 3-3, compute $H_{1/3}$ for both F and t and choose the smaller of the two values. An alternative is to use the tU/F curve (shown dotted) to determine whether the known duration meets the minimum duration $t_{\min}$; if it

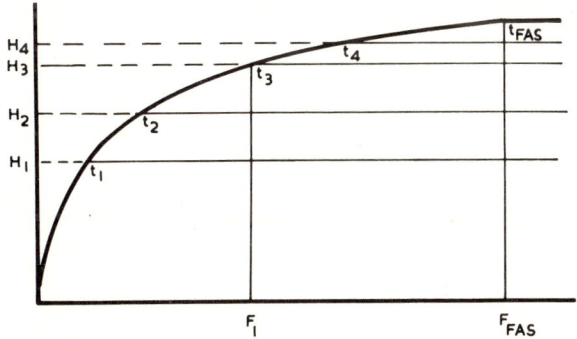

Fig. 3-4. Growth of significant wave height with distance along fetch for a specific wind velocity.

does not, use the duration curve instead of that for the fetch. The $t_{min}U/F$ relationship is derived by finding the gt/U value which provides the same height $(gH_{1/3}/U_2)$ as some specified fetch (gF/U_2) and finding the ratio:

$$\frac{gt/U}{gF/U^2} = \frac{t_{min}U}{F} \tag{3-17}$$

As illustrated in Fig. 3-4, there will be a fetch length and concomitant minimum duration when a wind of given velocity cannot generate higher waves. This is the FAS condition previously discussed. In Fig. 3-4 this is represented by t_{FAS} and F_{FAS} (the subscript FAS being reserved for the fully arisen sea), the heights beyond F_{FAS} being similar to H_{FAS}. At this stage neither the fetch nor the duration is required in the assessment of $H_{1/3}$, only the wind velocity U. This is exemplified in Fig. 3-3 where the curves of $gH_{1/3}/U^2$ are tending to the horizontal at high values of gF/U^2 and gt/U. In fact, the asymptote to both curves gives $gH_{1/3}/U_2 = 0.263$, or:

$$H_{1/3} = 0.263 \ U^2/g \tag{3-18}$$

$H_{1/3} = 0.00818 \ U^2 \qquad H_{1/3}$ (ft.), U (ft./sec) \hfill (3-19)

$H_{1/3} = 0.0233 \ U^2 \qquad H_{1/3}$ (ft.), U (knots) \hfill (3-20)

$H_{1/3} = 0.0268 \ U^2 \qquad H_{1/3}$ (m), U (m/sec) \hfill (3-21)

It will be noted in Fig. 3-3 that the curve C/U versus gF/U^2 rises above the value of 1.0 and, in fact, is purported to reach an asymptotic value of 1.37. This implies that the significant waves are travelling at a speed greater than that of the wind. Whilst some data appear to confirm this hypothesis [15], it is possible that winds in some parts of the fetch may have been greater than those recorded or some extraneous waves were present at the time.

Hino [16] has carried out a theoretical analysis which combines the theories of Phillips and Miles. He presented the result in the SMB format of dimensionless parameters. These are included in Fig. 3-3 and show the close similarity of these semi-empirical and theoretical approaches. The point to be noted here is the trend of the Hino curve to the limit of $C/U = 0.8$, rather than 1.37 of the SMB results.

The SMB method has been publicized extensively and consequently is in wide use today. Mathematicians, however, frown upon its semi-empirical approach now that the more sophisticated stochastic tools have been applied to the wave problem. In fact Munk, the central figure in the SMB method, has had occasion to publish a burial notice in 1957 [17]:

"I think the SMB method, at least in its earlier SM version, deserves retirement. It was first used in 1942 during the invasion of North Africa and published in 1947,

HISTORY OF WAVE PREDICTION

and it is amazing that it should have survived so long. Its accomplishments were to organize multitudes of scattered data into a few dimensionally consistent empirical relations. I feel now that in the various SM papers we should not have tried so hard to clothe these relations into a theoretical straightjacket."

In spite of this desertion from an old but reliable hulk, engineers might still utilize it, especially for conditions of limited fetch or duration. A later comparison will indicate the value of this work, carried out initially under pressure of time, but later verified by many wave measurements.

PNJ method

The historical record of the development of the PNJ method, plus the mathematical manipulations associated therewith, are amply treated by Kinsman [18], which the reader is recommended to read. Following is an extremely brief resumé, as the main interest is centered on the end result and its application.

In 1952 at New York University, Pierson recognized the problem of ocean waves as one of statistics [19]. From knowledge of energy spectra in electrical circuits he appreciated the type of data necessary in order to formulate a wind/wave-height relationship. At this time no analyses of wave records were available by which to test his hypothesis.

Neumann [8] simultaneously devised an energy spectrum of waves from visual observations taken aboard a German light vessel. This, together with an assumption of Gaussian distribution of wave heights in the FAS condition (the concept of which was born at this time), permitted an immediate attack on the problem of a wave forecasting formula based on stochastic processes.

By noting the height and period of many components in the complex sea of a fetch, Neumann plotted values of H/T^2 against $(T/U)^2$ which were equivalent to H/L_o against $(C/U)^2$. These he found to fill an area of the graph up to a boundary line which was defined by:

$$H/T^2 = \exp[-(gT/2\pi U)^2] = \exp[-(C/U)^2] \qquad (3\text{-}22)$$

For the shorter waves ($C/U \to 0$) the wave steepness tended to the limit for waves in deep water (i.e., 1/7). The longest waves had the least steepness since the wind had difficulty in transferring energy to them because their celerity approached that of the wind speed.

Since the wave energy per unit area of ocean is proportional to the square of the wave height of the composite waves, the energy for a single train of some specified period T (or narrow band of periods) can be expressed as:

$$H^2 = KT^4 \exp[-2(gT/2\pi U)^2] \qquad (3\text{-}23)$$

This can be converted to infinitesimal increases in period or frequency and inte-

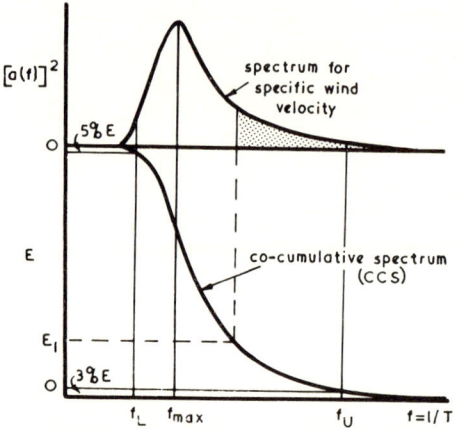

Fig. 3-5. The spectrum and co-cumulative spectrum as presented in Ref. 9.

grated giving the shape of spectrum shown in Fig. 3-5, the area under which represents the total energy per unit area of ocean. The equation for this spectrum, as derived by Neumann was:

$$E = C\,3(\pi/2)^{3/2}(U/2g)^5 \tag{3-24}$$

In order to evaluate the coefficient C, Neumann utilized the relationship of Longuet-Higgins [4] of $H_{1/3} = 2.83\sqrt{E}$. Thus, by using values of $H_{1/3}$ for known wind velocities U, he could determine C which varies with the measuring system employed. It should be recalled that the above relationship applied strictly to a narrow wave spectrum, whereas the conditions under study are the FAS. Cartwright [20] has discussed the error arising in C and hence in the evaluation of E in the PNJ method. However, in re-using E to find $H_{1/3}$ this error was annulled, so that the $H_{1/3}$ values predicted by the PNJ method can be accepted.

As seen in Fig. 3-5 the integral value of area throughout the infinite frequency range can be drawn as a curve of E versus f, or more correctly $H_{1/3}$ versus frequency. This has been termed by PNJ the co-cumulative spectrum (CCS).

It can be seen from the above equation for E that $H_{1/3}$ varies as $U^{5/2}$. This power of 5/2 has been criticized, as has been the shape of Neumann's spectrum [21]. As noted earlier, the SMB relationship was $H_{1/3} = KU^2$, derived on the basis of dimensional homogeneity. Pierson has since derived a new spectral shape [12] which is fitted by the exponent 2. Thus the tabular and graphical information on wave heights presented by the PNJ method [9] is no longer applicable. The characteristics of the PNJ spectrum will now be outlined so that the later developments of Pierson and Moskowitz can be appreciated.

The CCS in Fig. 3-5 presents the area under the curve from $f = \infty$ to the frequency being considered. Thus E, $H_{1/3}$ or a_{rms} for that portion of the spectrum can be

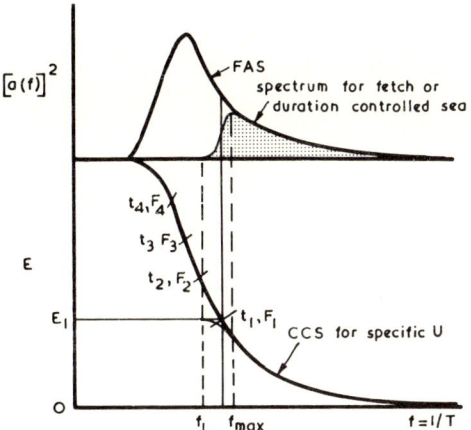

Fig. 3-6. Details of the CCS when fetch or duration limited.

found. The total E for the FAS is the maximum reached. As noted in Fig. 3-5, the build up of wave energy with fetch or duration is by the spectrum spreading from the high frequency end across to the lowest frequency value.

The energy E_1 represents the stippled area under the curve in the figure. In the PNJ forecasting method cut-off values at the upper and lower frequency limits, of 3 and 5% of the total energy, were used to define a boundary of engineering significance.

Each limited area of the spectral curve represents a control by fetch or duration, which can be identified as an F or t value on the CCS curve. These are represented by $t_1 F_1$, $t_2 F_2$ etc. in Fig. 3-6. At the top of the CCS curve the pertinent values represent F_{FAS} and t_{FAS} to produce the FAS condition for the given wind velocity. A different spectrum and CCS results for any other wind speed, becoming wider and higher for increased values.

The spectrum of a partially arisen sea is shown stippled in Fig. 3-6 in which the low-frequency slope follows the pattern of the FAS curve. The line intercept to give the equivalent area to that shown in Fig. 3-5 must bisect this slope approximately. When transferred down to the CCS this intercept gives the energy content E_1. The frequencies at peak energy and the lowest limit (f_{max} and f_L, respectively) are seen to differ from the frequency designated by the energy intercept. The point on the CCS curve at f_{max} is the steepest slope, since it is the point at which the asymptote to the E_1 value commences.

In the comprehensive presentation of the PNJ method, charts were prepared which covered a wide range of wind velocities and were separated into duration or fetch limitation. These included isopleths through the several CCS curves for equal values of F or t. This permitted the evaluation of $H_{1/3}$ for the FAS or developing sea conditions. These charts are not extremely easy to read as the duration or fetch isopleths recurve sharply on each other. Since results from the Pierson and Moskowitz equation [12] are to be recommended later, it would be confusing to reproduce figures or tables from the PNJ presentation.

A relationship arising from the PNJ spectra [9] is that of:

$$T_{max} = 2\pi U/\sqrt{3/2}\, g \tag{3-25}$$

which can be converted to:

$$U/C_{T_{max}} = \sqrt{3/2} = 1.22 \tag{3-26}$$

Thus the celerity of the wave group containing the maximum energy in this FAS spectrum ($C_{T_{max}}$) is slightly less than the mean wind speed $U(C/U = 0.8)$. This is the same ratio obtained by Hino [16] for his wave group with period $T_{1/3}$. It will be remembered that the SMB method indicated a FAS value of 1.37. When the wind velocity is specified for a given height above the sea-surface a more definite value of $C_{T_{max}}/U$ may be derived.

PM method

In 1964 Pierson and Moskowitz published a combination of three papers in the one issue of the *Journal of Geophysical Research* [10,11,12], which clarified some important issues respecting wave forecasting. This and previous work has been collected into a book by Neumann and Pierson [22]. The reader is advised to read the original sources to see the painstaking care which must be exercized in order to exclude inconsistencies. Also to be found are recommendations for taking measurements of waves and wind, which even for an engineer applying formulae can provide a caution in the acceptance of data supplied by meteorologists and others.

Moskowitz [11,23] selected 460 records made on ocean weather ships operating in the Atlantic Ocean. Synoptic weather charts of the areas were then inspected in order to choose suitable fetch conditions for correlating wind and wave characteristics. For certain wind velocities between 20 and 40 knots he was able to obtain subsets from which the averages were analyzed for their energy spectra. Statistical tests on these subsets confirmed that they were all for FAS conditions. To relate $H_{1/3}$ to the area of the spectrum or a_{rms}, he used the relationship $H_{1/3} = 4\, a_{rms} = 2.83\, \sqrt{E}$ as was employed in the PNJ method. After several statistical tests a final relationship $H_{1/3} = 0.0182\, U^2$ was adopted, where $H_{1/3}$ is in feet and U is in knots. Allowance was made for waves recorded to be too high at the low-period end of the spectrum as recorded by the ship-borne wave recorders.

Pierson and Moskowitz [12] used the subsets of the above project to test the dimensionless variables suggested by Kitaigorodskii [13] for the FAS. In this way the variable of wind velocity is eliminated and a spectral shape obtained which is

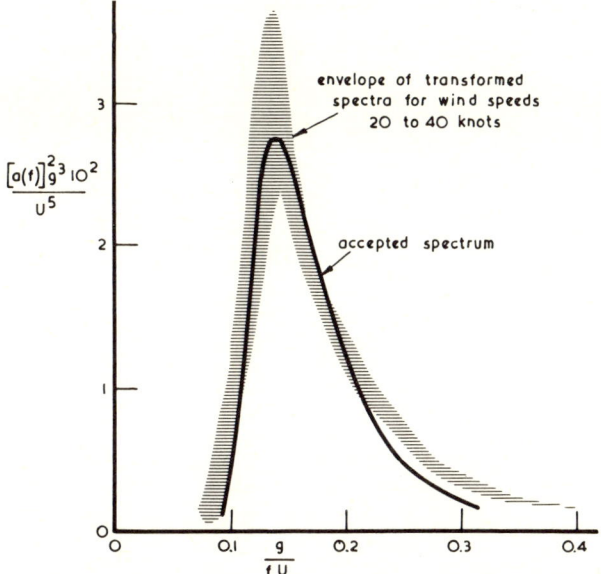

Fig. 3-7. Dimensionless PM spectrum showing range of spectra from which it was derived.

almost universal in its application. The equation derived after many statistical tests was:

$$\frac{[a(f)]^2 g^3}{U^5} = 8.10 \cdot 10^{-3} \, (g/fU)^5 \exp[-0.74(g/fU)^4] \qquad (3\text{-}27)$$

which was plotted with variables as displayed in Fig. 3-7. Also shown is the range of results from which the final shape was evolved. Since the height at which the wind velocity was measured had to be specified, the above equation applies to U values at 19.5 m above the water surface $(U_{19.5})$, as this was the anemometer level on the weather ships from which the records were supplied.

Pierson [10] has examined in detail the effect of the wind profile above the sea surface. To transpose a velocity taken at one height to that at another height (to 19.5 m, for example), the profile shape must be known accurately. Since Miles' theory of wave generation involves a critical layer, which approximates one tenth the wave length of the longest-period component, velocity variations to about 30 m height are of concern. This profile could depend upon the state of the sea (particularly when the significant waves reach 5–10 m height), and also upon the stability of the air near the sea surface, which is determined by temperature differentials of the air and water. A plea is made for more detailed observation of winds, values at 3 heights for example, with longer periods of measurement for computing averages (for the length of the wave record, in fact).

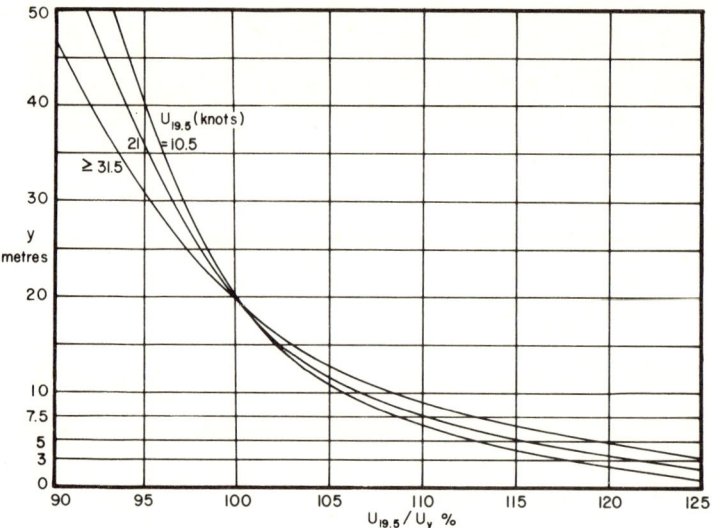

Fig. 3-8. Wind distribution near the sea surface.

Pierson [10] has shown the importance of taking into account the wind velocity distribution near the sea surface. The PNJ spectra [9] were based upon winds at 7.5 m, whilst the SMB relationships [2,15] employed 10–12 m high wind speeds. The wind used by Moskowitz [11] was measured 19.5 m from the sea surface. The selection of a standard height is of little importance so long as the wind distribution near the sea surface is known. In the relationships to be presented later the 19.5 m high values will be used, denoted by $U_{19.5}$.

Pierson compared the wind drag coefficients suggested by a number of workers and selected that of Shepard [24] for its ability to reduce discrepancies in wind profiles and its better application to open sea conditions. The equation recommended by Pierson was:

$$U_y/U_{10} = 1 + (C_{10}^{1/2} \ln y/10)/K \tag{3-28}$$

where U_y is the wind velocity at height y m; U_{10} is the wind velocity at height 10 m; C_{10} is the drag coefficient varying with U_{10}; and K is Karman's constant (= 0.4).

The drag coefficient C_{10} has been recorded by many workers and summarized by Wu [25]. By appropriate conversions the ratio $U_{19.5}/U_y$ has been graphed in Fig. 3-8 against height y in m and listed in Table 3-I. The wind velocity at any specified height can thus be transposed into that at $U_{19.5}$, or by suitable substitution into velocity at any other height. Since drag coefficients of differing values have been used it is necessary to utilize the curve appropriate to the final value of

TABLE 3-I

Respective wind velocities (knots) at various heights above the sea surface

U_3	U_5	$U_{7.5}$	U_{10}	U_{12}	$U_{19.5}$	U_{30}	U_{40}	U_{50}
4.25	4.41	4.60	4.74	4.80	5	5.15	5.26	5.38
8.50	8.82	9.20	9.48	9.60	10	10.3	10.5	10.8
12.5	13.2	13.8	14.1	14.4	15	15.5	15.9	16.2
16.4	17.3	18.2	18.8	19.1	20	20.7	21.3	21.8
20.2	21.2	22.4	23.2	23.8	25	26.0	26.9	27.6
23.9	25.1	27.6	28.3	28.9	30	31.4	32.6	33.7
27.9	29.3	32.2	33.0	33.7	35	36.7	38.0	39.3
31.9	33.5	36.9	37.8	38.5	40	41.9	43.5	45.0
35.9	37.6	41.5	42.5	43.4	45	47.1	49.0	50.6
39.8	41.9	46.0	47.2	48.1	50	52.4	54.4	56.2
44.8	46.0	50.6	51.9	53.0	55	57.5	59.8	61.9
47.8	50.2	55.3	56.6	57.8	60	62.9	65.2	67.5

$U_{19.5}$ (e.g., 10, 20, and 30 knots or greater, approximately). For example, a wind velocity of 35 knots recorded at 10 m height is equivalent to a velocity of 1.08 × 35 = 38 knots at 19.5 m height. At a height of 15 m the value is 38/1.03 = 37 knots. (See also Table 3-I.)

By modifying the various forecasting formulae to give $H_{1/3}$ for velocities at the same height, Pierson reduced discrepancies between these from 5 to 20%. This remaining error is possibly due to the difficulty of defining wind speeds or the heights to which they refer. Even an anemometer fixed at a certain level above the deck of a ship will suffer large fluctuations in height and horizontal position as the ship rides a stormy sea. The sensitivity of the wave data to errors in wind speed from any cause is illustrated in Table 3-II [10].

It should be noted once more that the Pierson-Moskowitz (PM) spectrum applies specifically to U values for 19.5 m above the surface, for which Fig. 3-8 should be used to transpose velocities recorded at any other level.

TABLE 3-II

Errors in percent produced by errors in wind velocity U

U	$H_{1/3}$	Area of spectrum	Location of T_{max}
5	10	22	28
10	21	46	61

Darbyshire method

Darbyshire in England had the assistance of many electronic and other devices for measuring and analyzing waves as they first became available. Whilst this would appear to be a distinct advantage, it can be appreciated that this can include the teething and calibration troubles of such novel instruments. Darbyshire's final relationships will not be presented here, as even the latest ones give much lower wave heights than other formulae. However, a discussion of his activities, and the reasons given for the errors arising, can provide some useful advice for others embarking on a wave recording programme.

During World War II Darbyshire recorded waves at Peranporth on the Cornwall coast. A piezo-electric pickup was used, which sat on the ocean bed in some 40 ft. of water and fed a signal by cable to a recorder on the shore. It was possible to measure the spectral energy distribution by means of a newly developed analyser [26].

Several sources of error became apparent as oceanographers solved the mysteries of wave generation, propagation and attenuation. Firstly, the wave heights recorded at the bed had to be amplified in order to provide the equivalent surface fluctuations. For this purpose linear wave theory was employed, using a common factor for all waves, based upon a coefficient for the significant wave period. As will be noted later, this gross simplification could have given wave heights which were too low. Besides this, the whole spectrum could have been influenced by reflection from shore and omission of the smaller-period elements completely.

Darbyshire referred his measured wave characteristics to generation factors in the eastern Atlantic Ocean. This required the computation of wave dispersion and attenuation from the fetch to the recording site, a procedure which was not highly developed at the time. It was also proved later that the waves arriving at Peranporth were greatly influenced by refraction, diffraction and tidal streams [27]. For all these reasons, it is not surprising that the 1952 formulae presented by Darbyshire [7] gave much lower wave heights than the SMB method, which was based upon visual records of surface undulations within their fetches.

When records became available from weather ships [28], Darbyshire was able to correlate waves and winds within the fetch. As previously, Darbyshire used the gradient wind value, as measured from synoptic weather charts [29]. From the previous discussion on wind profiles near the sea surface it can be appreciated that a comparison of his formulae with those of other workers using measured values is very difficult. Another factor influencing the spectra obtained by Darbyshire from the weather ships was a calibration error of the wave recorders. This appeared to be minor and hardly detectable in the scatter of the results.

Darbyshire also concerned himself with wave generation in the Irish Sea and so-called coastal waters [30,31]. Differences were then distinguished between limited fetches and FAS conditions [32]. The implication of this is not very clear since the depths concerned are not likely to retard even the longer components of the spectra significantly. Longuet-Higgins [33] has suggested that dynamics of wave generation, either by the resonance or the shear-flow mechanism, would be different for water of finite depth. However, this opinion was expressed without elaboration, and proof awaits further wave and wind data.

Darbyshire dealt with partially arisen seas as well as the FAS. In respect to fetch distances he found the minimum required for a FAS was in the order of 100 NM for all wind velocities [32]. This is distinctly different from the hundreds of miles proposed by the PNJ method and the thousands of miles suggested by the SMB formulae [34]. Phillips [35] has proposed an explanation of the anomaly, based upon his and Mile's mechanism of wave build-up. Considering one component of the wave spectrum emerging from a completely calm sea, the time or distance travelled for successive additions of wave energy will be as depicted in Fig. 3-9. After a lengthy period (or distance) of linear increase in energy a transition is reached after which an exponential build-up occurs (possibly associated with breaking) to the point of saturation. Now, if the sea already has some wave motion when the major wind structure of the fetch is applied, most of the linear energy time increase is by-passed, as indicated by x in Fig. 3-9. Waves could thus be taken to the FAS limit much sooner, as found by Darbyshire in the eastern Atlantic. This has evoked the suggestion that forecasting formulae should be employed only in the areas where they were evolved, with areas having their own specific empirical equations. Whilst this may be possible in the future, when data are available from all types of wave climates, it would appear a little unrealistic at present.

However, one precaution indicated to coastal engineers by these uncertainties in

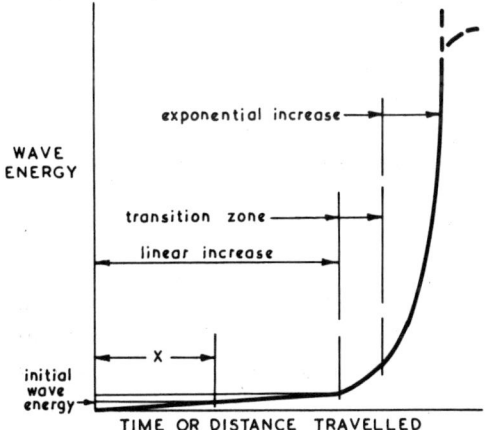

Fig. 3-9. Growth of a specific wave component in a fetch.

duration control, is to consider that the FAS exists on all occasions, except only when the fetch is limited by a land mass, or by the wind structure itself (as, for example, in a tropical cyclone). It is difficult to imagine, in an oceanic situation, that a 30–40 knot wind could be applied suddenly to a calm sea. As discussed in the section on meteorological conditions, a cyclonic centre will contain many fetches and will itself be travelling across the ocean. By the time the full force of a fetch is felt in an area, both short waves will be present from the prior wind build-up and from long swell emerging from adjacent areas. Under these conditions the FAS sea will be generated very quickly and could be limited only by distance over which the higher winds could be exerted.

FULLY ARISEN SEA

Since the evolution of the Neumann spectrum [8] and its application in the Pierson-Neumann-James (PNJ) forecasting technique [9], much discussion has taken place on its shape. The work of Moskowitz, resulting in the new spectral distribution by Pierson and Moskowitz [12], would appear to supercede this presentation. Although Bretschneider has put the Sverdrup-Munk-Bretschneider (SMB) relationships into spectral form [34,36], certain constants still rely on the original dimensionless parameters, for which the height of the wind velocities is not strictly known. For these reasons it is considered judicious to proceed from the Pierson-Moskowitz (PM) spectrum.

When a condition of fully arisen sea (FAS) exists, the variables of fetch and duration can be excluded from the calculation, so that wave height and period can be determined solely from the knowledge of wind velocity. It is the form and constants in this equation which Pierson and Moskowitz evolved from an analysis of many ocean wave records [23].

The PM frequency spectrum

The form of the FAS spectrum presented by Pierson and Moskowitz [12] is:

$$S_H 2(\omega) d\omega = \frac{\alpha g^2}{\omega^5} \exp[-\beta(g/U_{19.5}\omega)^4] d\omega \qquad (3\text{-}29)$$

where $S_H 2(\omega)$ is the energy per unit of ω, or dω; ω is the angular frequency = $2\pi f$ = $2\pi/T$; α is a dimensionless constant = $8.10 \cdot 10^{-3}$; β is a dimensionless constant = 0.74; $U_{19.5}$ is the wind velocity at 19.5 m height; g is acceleration due to gravity (commensurate dimension to $U_{19.5}$).

Integration of eq. 3-29 for ω from 0 to ∞, or the area under the curve of $S_H 2(\omega)$ versus ω, results in the total variance or the wave energy per unit area of

FULLY ARISEN SEA

ocean in terms of (quadrated) wave amplitude. This can be related to significant wave height by:

$$\text{variance} = (\tfrac{1}{4}H_{1/3})^2 \tag{3-30}$$

where $H_{1/3}$ is the average of the highest third of the waves.

Pierson and Moskowitz [12] did not graph the spectrum in this form, but converted it to a frequency spectrum by substituting $d\omega = 2\pi df$ so that:

$$S_{H^2}(f) = df \, \frac{\alpha g^2}{(2\pi)^4 f^5} \, \exp\,-\beta \left(\frac{g}{2\pi U_{19.5} f}\right)^4 \, df \tag{3-31}$$

from which:

$$S_{H^2}(f) = \frac{\alpha g^2}{(2\pi)^4 f^5} \, \exp\,-\beta \left(\frac{g}{2\pi U_{19.5} f}\right)^4 \tag{3-32}$$

Eq. 3-32 has been graphed in Fig. 3-10 for $U_{19.5}$ from 20 to 45 knots and the full range of effective frequencies. This is the usual presentation of wave spectra and is similar to the PNJ curves from which co-cumulative spectra were derived [9]. The changing location and height of the peak is to be noted, as well as the substantial increase in area (or $H_{1/3}$) as the wind velocity increases.

The PM period spectrum

Roll and Fischer [21] introduced the concept of a spectrum based upon a period abscissa as distinct from a frequency base. Bretschneider's application [36] of this raised the question of the meaning of a "spectrum" [37], so that to distinguish the two presentations perhaps the term "energy distribution curve" (EDC) should be used for the period presentation.

As noted by Roll and Fischer [21]:

$$d\omega = -\frac{2\pi}{T^2} \, dT \tag{3-33}$$

Substitution into eq. 3-29, neglecting the negative sign, gives:

$$S_{H^2}(T)\,dT = \frac{\alpha g^2 T^3}{(2\pi)^4} \, \exp\,-\beta \left(\frac{gT}{2\pi U_{19.5}}\right)^4 \, dT \tag{3-34}$$

from which:

$$S_{H^2}(T) = \frac{\alpha g^2 T^3}{(2\pi)^4} \, \exp\,-\beta \left(\frac{gT}{2\pi U_{19.5}}\right)^4 \tag{3-35}$$

Eq. 3-35 retains the same values of α and β as in eq. 3-32 and has been graphed in Fig. 3-11 for the same range of $U_{19.5}$ (in knots) as in Fig. 3-10. Because of the

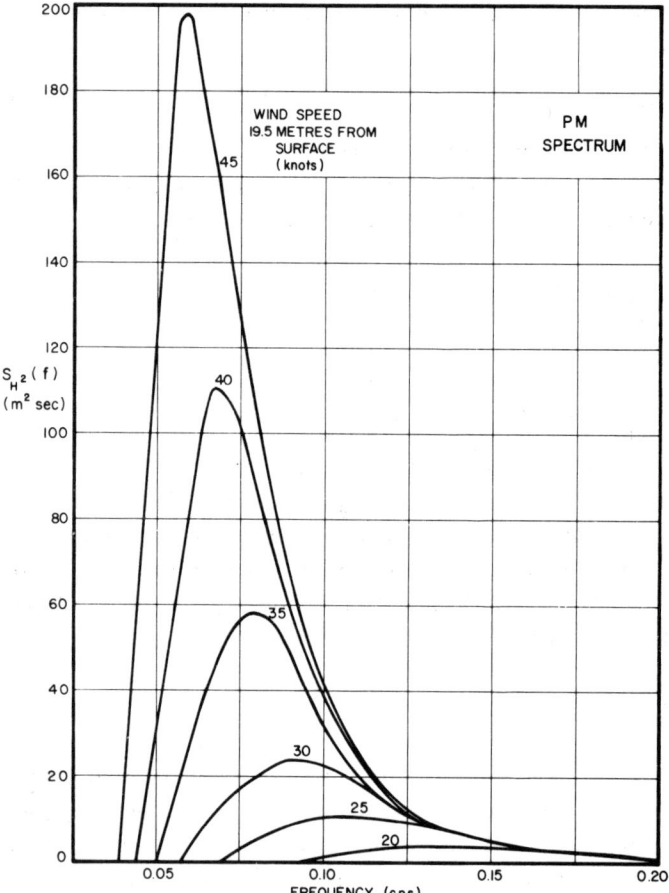

Fig. 3-10. Pierson-Moskowitz (PM) spectra for a range of wind velocities ($U_{19.5}$).

conversion noted above the area under the curve is the same as for eq. 3-32 and Fig. 3-10. However, there are notable differences between the EDC and the frequency spectrum. The shape of the EDC is almost triangular, with the straight rise and fall from the peak being almost linear. For purposes of discussing energy distribution and perhaps designating energy in period bands it may be advantageous to equate the shape to a triangle of equal area. Also, the EDC has an origin for both the abscissa and the ordinate, which helps to exhibit the shift of the EDC to higher periods as the wind velocity increases.

A form similar to eq. 3-35 was derived by Roll and Fischer [21], except for the constants and exponents, as follows:

$$S_{H^2}(T) = \Phi \frac{g^2 T^4}{(2\pi)^2} \exp[-2(gT/2\pi U)^2] \tag{3-36}$$

FULLY ARISEN SEA

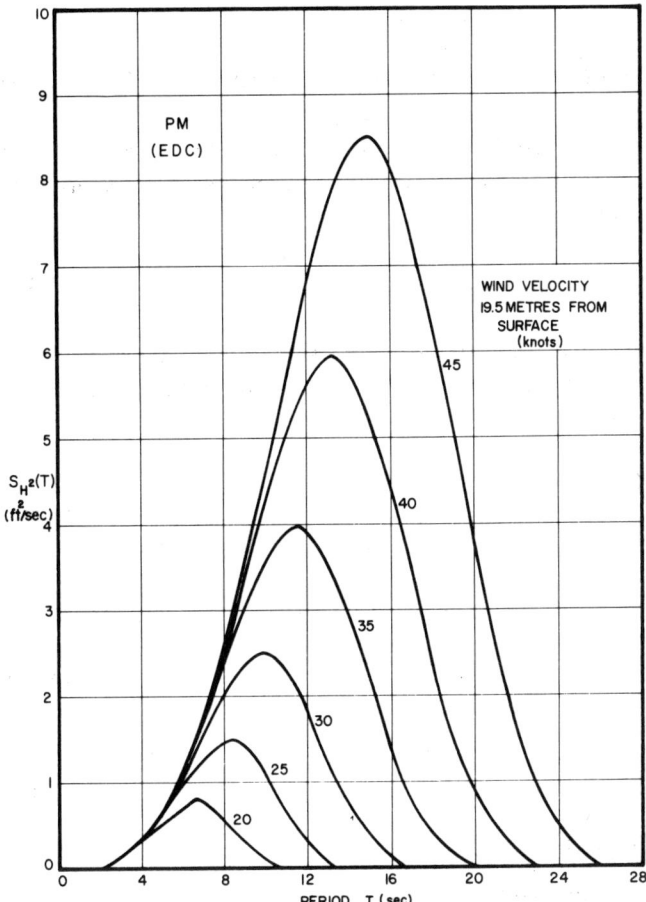

Fig. 3-11. Energy distribution curve (EDC) based upon PM spectral constants.

The index of 2 in the exponential term of the frequency spectrum (eq. 3-32) was tried by Pierson and Moskowitz, together with 3 and 4. Selection of 4, which spreads the base of the spectrum most, has been widely accepted by oceanographers [6].

The peak energy occurs at a defined period (T_{max}) in Fig. 3-11 which is different from the reciprocal of the frequency (f_{max}) at which this maximum occurs in the spectrum of Fig. 3-10. This can be shown by differentiating eq. 3-32 with respect to f, and eq. 3-35 with respect to T, and equating each to zero. This results in the following:

$$\frac{g}{2\pi U_{19.5} f_{max}} = (5/4\beta)^{1/4} = 1.14 \qquad \text{from eq. 3-32 and } \beta = 0.74 \qquad (3\text{-}37)$$

and:

$$\frac{gT_{max}}{2\pi U_{19.5}} = (3/4\beta)^{1/4} = 1.00 \qquad \text{from eq. 3-35 and } \beta = 0.74 \qquad (3\text{-}38)$$

The round figure of 1.00 in eq. 3-38 has no significance (the next decimal figure is 3) since it occurs only for the wind height of 19.5 m. In this latter case also, the following relationship is obtainable:

$$T_{max} = \frac{2\pi}{g} U_{19.5} \; (= 0.33 \, U_{19.5} = U_{19.5}/3) \qquad (3\text{-}39)$$

when T_{max} is in sec and $U_{19.5}$ is in knots for the bracketed expressions.

By integrating eq. 3-32 and 3-35 from 0 to ∞ for f and T respectively, the following relationship is obtained:

$$\text{area} = (\tfrac{1}{4} H_{1/3})^2 = \frac{\alpha U_{19.5}^4}{4\beta g^2} \qquad (3\text{-}40)$$

from which:

$$H_{1/3} = \frac{2U_{19.5}^2}{g} \sqrt{\alpha/\beta} \qquad (3\text{-}41)$$

or:

$$H_{1/3} = 0.0185 \, U_{19.5}^2 \qquad (3\text{-}42)$$

when $H_{1/3}$ is in ft. and $U_{19.5}$ is in knots.

Substitution of eq. 3-39 into 3-42 gives:

$$H_{1/3} = 0.17 \, T_{max}^2 \quad \text{or} \quad T_{max} = 2.43 \sqrt{H_{1/3}} \qquad (3\text{-}43)$$

for T_{max} in sec and $H_{1/3}$ in ft.

The relationships of eq. 3-43 are independent of wind velocity, but apply only to FAS conditions.

Dimensionless presentation

Pierson and Moskowitz presented the frequency spectrum in dimensionless form. Eq. 3-32 can be expressed as:

$$\frac{S_{H^2}(f) g^3}{U_{19.5}^5} = \frac{\alpha}{(2\pi)^4} \left(\frac{g}{f U_{19.5}}\right)^5 \exp - \frac{\beta}{(2\pi)^4} \left(\frac{g}{f U_{19.5}}\right)^4 \qquad (3\text{-}44)$$

from which $S_{H^2}(f) g^3 / U_{19.5}^5$ was graphed against $f U_{19.5}/g$ as in Fig. 3-12. This was

FULLY ARISEN SEA

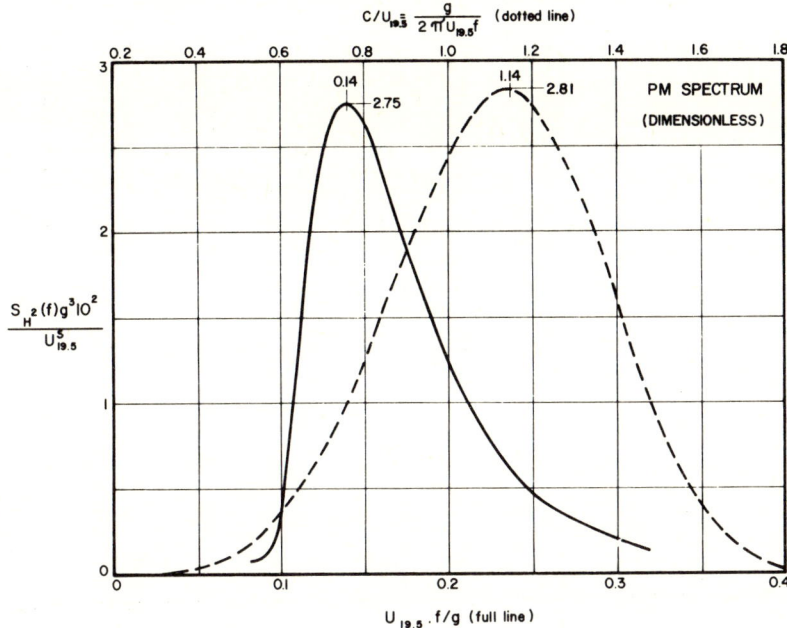

Fig. 3-12. The PM dimensionless spectrum based upon $U_{19.5}f/g$ or $C/U_{19.5}$. (Note the more exact peak value $2.81 \cdot 10^{-2}$ for values of α and β quoted in eq. 3-29.)

derived after much manipulation and matching of wave spectra, the details of which the reader is referred to [12].

Another method of obtaining a dimensionless frequency spectrum is to express eq. 3-44 as follows:

$$\frac{S_{H^2}(f)g^3}{U_{19.5}^5} = 2\pi\alpha \left(\frac{C}{U_{19.5}}\right)^5 \exp[-\beta(C/U_{19.5})^4] \qquad (3\text{-}45)$$

in which the "wave age", or ratio of wave celerity (C) to wind velocity ($U_{19.5}$), becomes a dimensionless parameter. Eq. 3-45 is graphed in Fig. 3-12. The frequencies for the peak energy content are the same when converted from the respective parameters in Fig. 3-12. The optimum value of $S_{H^2}(f)g^3/U_{19.5}^5$ is the same for the two diagrams and the areas under the curves still represent the total variance or $(\frac{1}{4}H_{1/3})^2$.

In a similar manner, eq. 3-35 can be expressed as:

$$\frac{S_{H^2}(T)g}{U_{19.5}^3} = \frac{\alpha}{2\pi} \left(\frac{C}{U_{19.5}}\right)^3 \exp -\beta\left(\frac{4}{U_{19.5}}\right)^4 \qquad (3\text{-}46)$$

which has been graphed in Fig. 3-13. As in the dimensional form of Fig. 3-11, this

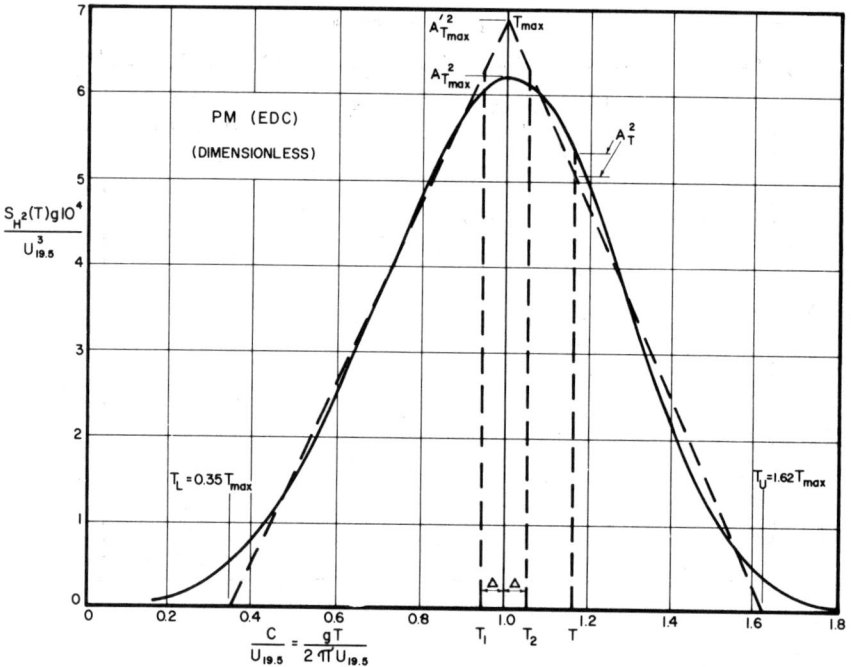

Fig. 3-13. The PM dimensionless EDC, with equivalent triangular distribution for FAS.

EDC exhibits a shape which can be readily equated to a triangle of almost isosceles form. The apex of both the curve and the triangle occurs at $C/U_{19.5} = 1.00$.

From Fig. 3-13 it is seen that the equivalent triangle gives:

$$\frac{T_U}{T_{max}} = 1.62 \; ; \quad \frac{T_L}{T_{max}} = 0.35 \; ; \quad \frac{T_U - T_L}{T_{max}} = 1.27 \qquad (3\text{-}47)$$

This gives the engineering bounds of the periods contained in the EDC. As $U_{19.5}$ increases so does T_{max} and so do T_U and T_L. There is thus a widening of the period band present and a shift in the EDC to higher periods as wind velocity increases.

The significant height ($H_{1/3}$) of any wave band with known period limits is a useful variable when computing wave height in the dispersal area outside the fetch. This can be derived from the following definite integral:

$$\int_{T_1}^{T_2} S_{H^2}(T)\,dT = -\frac{\alpha U_{19.5}^4}{4\beta g^2} \exp\left[-\beta\left(\frac{gT}{2\pi U_{19.5}}\right)^4\right] \Bigg|_{T_1}^{T_2} \qquad (3\text{-}48)$$

For convenience the ratio of such a height or wave energy can be obtained as a ratio

FULLY ARISEN SEA

of $H_{1/3}$ from eq. 3-41 so that:

$$\frac{\int_{T_1}^{T_2} S_{H^2}(T)\,dT}{H_{1/3}^2} = -\frac{1}{16}\exp\left[-\beta\left(\frac{gT}{2\pi U_{19.5}}\right)^4\right]\bigg|_{T_1}^{T_2} \quad (3\text{-}49)$$

To determine the amplitude ($A_{T_{max}}$) for the wave band centered on T_{max}, take $T_1 = T_{max} - \Delta$ and $T_2 = T_{max} + \Delta$ (see Fig. 3-13), where $T_{max} = 2\pi U_{19.5}/g$ from eq. 3-38. The ratio $A_{T_{max}}/H_{1/3}$ will vary with the band width (2Δ) chosen, but as $2\Delta \to 0$ an optimum value is reached. Alternatively it can be shown from the ratio of eq. 3-35 to eq. 3-41 for conditions at T_{max} that:

$$\frac{A_{T_{max}}}{H_{1/3}} = \frac{\frac{1}{2}(1.003)^2\{\beta\exp[-\beta(1.003)^4]\}^{1/2}}{T_{max}^{1/2}} = \frac{0.3}{T_{max}^{1/2}} \quad (3\text{-}50)$$

Substitution of $H_{1/3}$ from eq. 3-43 gives:

$$A_{T_{max}} = 0.051\, T_{max}^{3/2} \quad \text{or} \quad H_{T_{max}} = 0.102\, T_{max}^{3/2} \quad (3\text{-}51)$$

assuming the height $H_{T_{max}}$ is twice the amplitude $A_{T_{max}}$ (ft./sec$^{1/2}$). Eq. 3-39 then gives:

$$H_{T_{max}} = 19.26 \cdot 10^{-3}\, U_{19.5}^{3/2} \quad \text{for } H_{1/3} \text{ in ft. and } U_{19.5} \text{ in knots} \quad (3\text{-}52)$$

From eq. 3-50 and 3-51, graphed in Fig. 3-14, it is seen that the ratio of energy contained in the peak of the EDC ($H_{T_{max}}$) to the total energy per unit area of ocean ($H_{1/3}$) decreases as T_{max} or $U_{19.5}$ increases. This occurs due to widening of the spectrum, inspite of $H_{T_{max}}$ increasing with T_{max} or $U_{19.5}$.

The height of the apex of the equivalent triangle $A'_{T_{max}}$, as in Fig. 3-13, can be obtained by equating the area to that under the curve of the EDC and the knowledge that $(T_U - T_L)/T_{max} = 1.27$ from eq. 3-47. Thus:

$$\text{area of triangle} = \frac{(A'_{T_{max}})^2\, T_{max}\, 1.27}{2} = (\tfrac{1}{4}H_{1/3})^2$$

from which:

$$\frac{A'_{T_{max}}}{H_{1/3}} = \frac{0.314}{T_{max}^{1/2}} \quad (3\text{-}53)$$

or:

$$H'_{T_{max}} = 20.22 \cdot 10^{-3}\, U_{19.5}^{3/2} = 0.108\, T_{max}^{3/2} \quad (3\text{-}54)$$
(dimensions as in eq. 3-42)

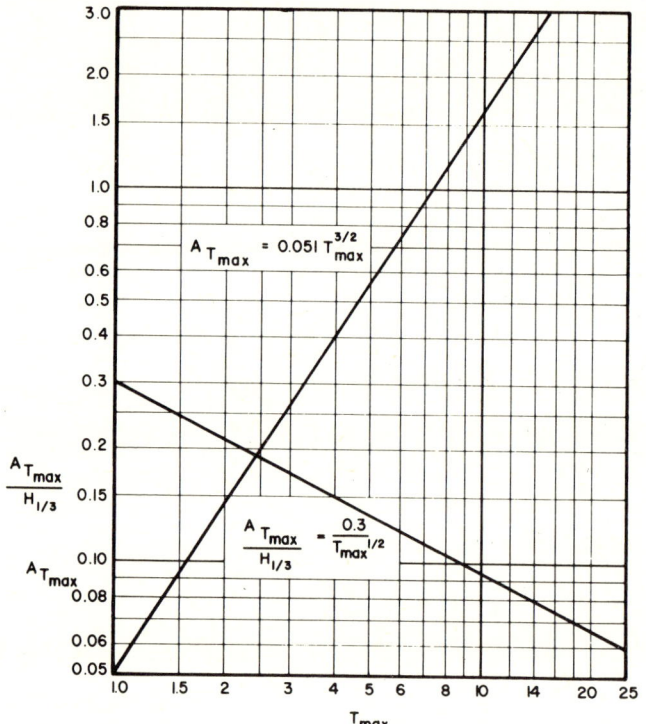

Fig. 3-14. Amplitude of wave component in EDC with peak energy ($A_{T_{max}}$).

The value of $A'_{T_{max}}$ is slightly higher than $A_{T_{max}}$ given by eq. 3-50, as is obvious from Fig. 3-13. However, its computation can serve to approximate amplitudes (A_T) of other wave trains with period T, as depicted in Fig. 3-13, where alternative values on the triangular or true EDC are shown. Depending upon whether $T \lessgtr T_{max}$, so A_T is given by:

$$\frac{A_T^2}{A_{T_{max}}'^2} = \frac{T_U - T}{T_U - T_{max}} \quad \text{or} \quad \frac{A_T^2}{A_{T_{max}}'^2} = \frac{T - T_L}{T_{max} - T_L} \tag{3-55}$$

To convert a PM spectrum for the FAS in terms of $S_{H^2}(f)$ versus f to an EDC in terms of $S_{H^2}(T)$ versus T, it is necessary to apply ratios of eq. 3-37 and 3-38 plus ratios of eq. 3-32 and 3-35. Having obtained T_{max}, values of T_U and T_L are known from eq. 3-47, and the EDC can be sketched from a knowledge of the triangular form (Fig. 3-13) and of $A_{T_{max}}'^2$, in eq. 3-53 for example.

FULLY ARISEN SEA

FETCH AND DURATION FOR FAS

Implied in the discussion on developing seas is a knowledge of minimum fetch or duration required to reach a FAS state. All other proportions will be related to the ratio of actual fetch length to the FAS value (i.e., F/F_{FAS}) and, therefore, an accurate assessment of F_{FAS} is necessary. Similarly, t_{FAS} needs to be known accurately, but this is closely related to F_{FAS}.

These minima have been derived by all workers who have produced forecasting formulae. As with the formulae for wave height and wave period, these values have varied greatly, perhaps more so than with the other variables. For example, as quoted by Bretschneider [34], the fetches and durations in Table 3-III have been specified for a 30-knot wind. Even when allowance is made for wind velocities at different heights in these two forecasting systems (7.5 m for PNJ and 10–12 m for SMB), the figures are nowhere near compatible.

TABLE 3-III

Fetch and duration quoted for 30-knot wind

Relationship		F_{FAS} (NM)	t_{FAS} (h)
PNJ [16]		280	23
SMB [21]	100%	7850	307
SMB	90%	795	54
SMB	80%	470	32.7

Inoue [38], from his predicted growth pattern, drew curves of $H_{1/3}$ versus F to the point where they reached the $(H_{FAS})_{1/3}$ values given by the PM [12] relationship. From these the FAS fetches could be measured, as were the durations. From the six wind velocity curves available [38], the points as in Fig. 3-15 were derived. These represent fetch and duration values for 70, 80, 90 and 100% of $(H_{FAS})_{1/3}$. The lower percentages helped determine the slope of the lines, since the FAS values had to be extrapolated beyond the figures. In spite of the drafting and measuring errors involved, the fetch data in Fig. 3-15 fit well the relationship:

$$F_{FAS} = 3.19 \, U_{19.5}^{1.5} \tag{3-56}$$

where F_{FAS} is in nautical miles when $U_{19.5}$ is in knots.

The duration data in the same Figure are related to the corresponding fetches, since the minimum duration for any stage of build-up is the time taken for some representative group of waves to traverse the fetch. For this purpose some workers have used the group velocity of the train with the significant period $(T_{1/3})$ existing

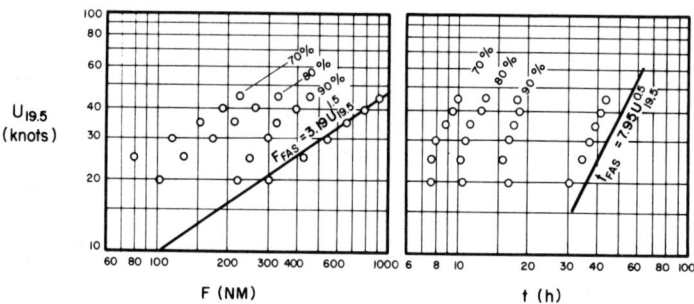

Fig. 3-15. Data points taken from Ref. 38., with suggested relationships for F_{FAS} and t_{FAS}.

at the end of the fetch. Others have used T_{max} as did Inoue. Substituting for this FAS condition of $(C_g)_{T_{max}} = \frac{1}{2}(C_o)_{T_{max}} = \frac{1}{2}(1.003\,U_{19.5})$ from eq. 3-38 into:

$$t_{FAS} = \frac{F_{FAS}}{(C_g)_{T_{max}}} \qquad (3\text{-}57)$$

we have from eq. 3-56:

$$t_{FAS} = 6.36\,U_{19.5}^{0.5} \qquad (3\text{-}58)$$

This would pass through the FAS points of the duration graph of Fig. 3-15, but has not been drawn.

However, it is submitted that the passage of this $(T_{FAS})_{max}$ train through the fetch is not the correct criterion for determining duration. It will be shown later that this train $(C_{T_{max}}/U_{19.5} = 1.0)$ has virtually no energy until 0.075 F_{FAS} from the upwind end of the fetch. Thus, if $(T_{FAS})_{max}$ is to be used it should only be applied for 92.5% of the F_{FAS} in order to derive t_{FAS}. But a more realistic approach is to consider the speed of the T_{max} group throughout the FAS fetch, from its zero group velocity at the outset to its optimum at the downwind end of the minimum fetch. This acceleration is depicted in Fig. 3-16 in terms of $C_{T_{max}}/U_{19.5}$ versus F/F_{FAS}. By using an average velocity:

$$\frac{\bar{C}_{T_{max}}}{U_{19.5}} = \frac{\int_0^{F/F_{FAS}} (C_{T_{max}}/U_{19.5})\,d(F/F_{FAS})}{F/F_{FAS}} \qquad (3\text{-}59)$$

over any given proportion of F_{FAS} the appropriate duration can be ascertained. This concept was used by Bretschneider in 1958 [15] in his revision of the Sverdrup-Munk relationships [2].

The above average is also shown in Fig. 3-16. At the end of the FAS fetch it

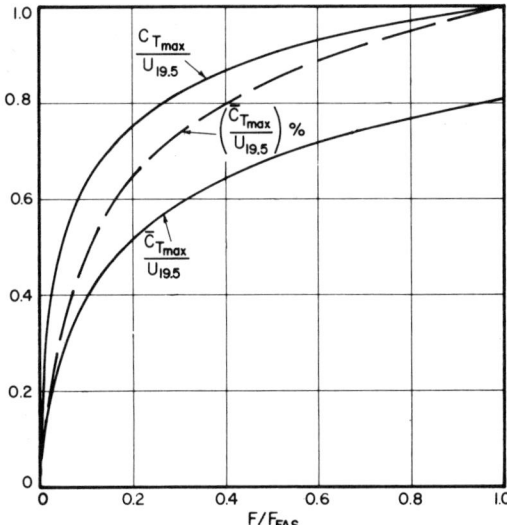

Fig. 3-16. Wave group velocities along the fetch.

reaches a value of 0.8, so that a more reasonable evaluation of t_{FAS} is obtained as follows:

$$t_{FAS} = \frac{F_{FAS}}{(\bar{C}_g)_{T_{max}}} = \frac{2(3.19\, U_{19.5}^{1.5})}{0.8(1.003\, U_{19.5})} = 7.95\, U_{19.5}^{0.5} \tag{3-60}$$

Eq. 3-60 has been included in the duration graph of Fig. 3-15 and represents t_{FAS} in hours for any wind velocity $U_{19.5}$ expressed in knots.

Expressing $\bar{C}_{T_{max}}/U_{19.5}$ as a percentage of the FAS value (see Fig. 3-16), the relationship between t/t_{FAS} and F/F_{FAS} can be derived from:

$$\frac{t}{t_{FAS}} = \frac{(F/F_{FAS})100}{[\bar{C}_{T_{max}}/U_{19.5}]\%} \tag{3-61}$$

Comparison of FAS formulae

At this stage it is worth comparing the formulae that have been devised for FAS conditions. This comparison encompasses significant wave height ($H_{1/3}$), minimum fetch (F_{FAS}) and minimum duration (t_{FAS}).

The $H_{1/3}$ (ft.) for a given wind velocity U(knots) at specific heights above the sea surface, as subscripts of U (m), has been derived by several workers as in Table 3-IV.

TABLE 3-IV

Formulae derived for $H_{1/3}$ (ft.) at FAS conditions with U (knots) specified at various heights (m)

Workers	References	Formula
SMB	2, 15	$H_{1/3} = 0.0233\ U^2_{10-12}$
PNJ	9	$H_{1/3} = 0.0443\ U^{5/2}_{7.5}$
Darbyshire	39	$H_{1/3} = 0.0133\ U^2_{grad}$
PM	12	$H_{1/3} = 0.0185\ U^2_{19.5}$

This group of equations is graphed in Fig. 3-17A, where it is seen that for any given wind velocity the SMB and Darbyshire values are above and below the PM curve. When the data of Table 3-IV are modified by Fig. 3-8 to provide relationships for winds at similar heights, as depicted in Fig. 3-17B, the various curves come much closer together. For this purpose the gradient wind used by Darbyshire has

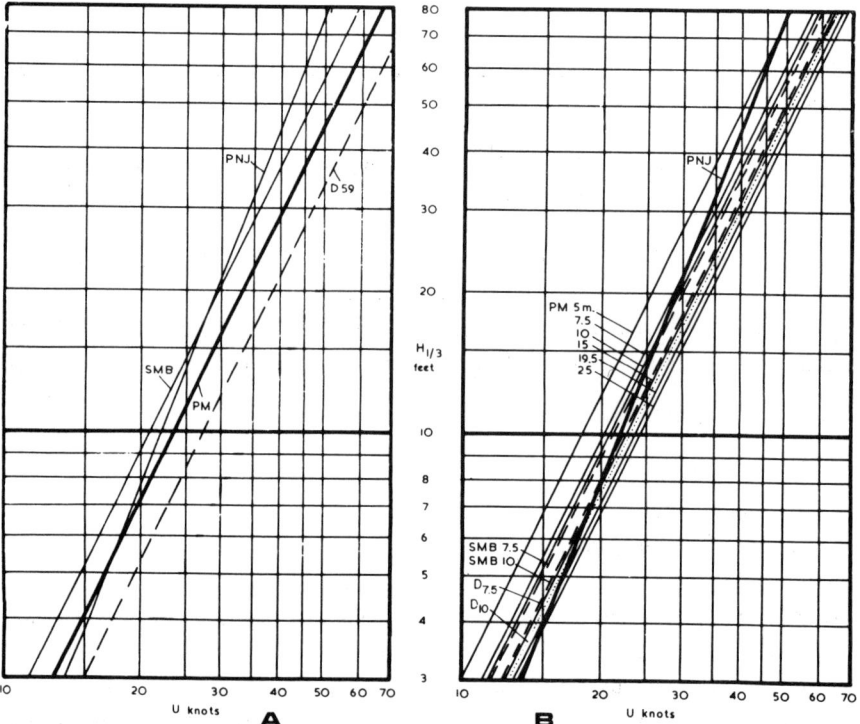

Fig. 3-17. Formulae from Table 3-IV for winds: A. at various heights; B. referred to common heights above sea surface.

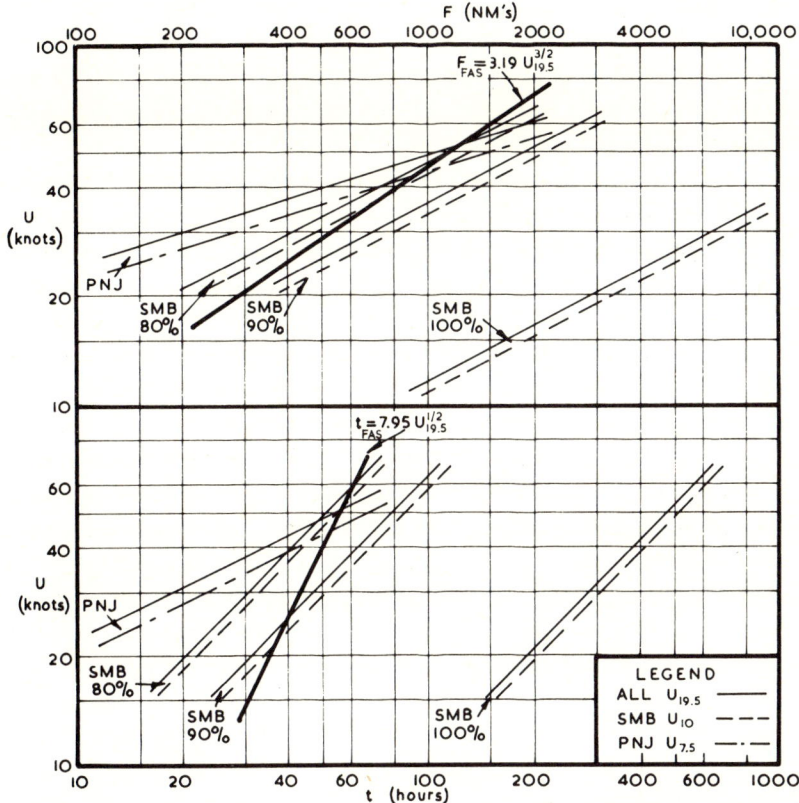

Fig. 3-18. Fetches and durations specified for FAS in various forecasting relationships.

been transposed from that at 50 m height above the sea surface; the $D_{7.5}$ values correspond to the SMB-10 line shown dotted. It is seen that the PNJ line at the steeper slope, associated with wind velocities at 7.5 m height, encompasses the PM curves for heights from 5 to 25 m. This is of only academic interest since the PM relationship should now be used in preference to that of the PNJ method.

The fetch and duration specified in the SMB and PNJ relationships, and listed by Bretschneider [36], have been graphed in Fig. 3-18. It is seen that there are three sets of curves for the SMB method, for FAS, as well as 90% and 80% of this value, as gauged by significant wave height. Although differences are apparent, there is an overlapping of the PNJ, 80% SMB and the Inoue (PM) curve (eq. 3-56) in the 40–50 knot range (all referred to the 19.5 m high wind). At lower wind velocities, which can more readily reach FAS conditions, the Inoue (PM) curve is close to the 90% SMB trace, whilst at higher wind velocities, which seldom have sufficient fetch or duration in nature, it approximates the 80% SMB line. The significance of these limitations will be discussed later.

DEVELOPING SEAS

Prior to the existence of the fully arisen sea, to which all previous relationships refer, the waves in the fetch are increasing both in period and in height. Control may be exercised by duration of the wind or by the length of fetch available.

Inoue [38] has developed a growth relationship to be used with the PM spectra in order to predict waves in a developing sea. He has presented curves for wave growth to the FAS condition, covering wind velocities from 20 to 45 knots (values at 19.5 m height). Two sets of these curves cover duration and fetch control respectively.

The controlling equations cannot be evaluated readily from the data presented. Values of f_{max} and $H_{1/3}$ were therefore taken from the graphs and tested in the original form of the spectrum as given in eq. 3-37 and 3-40. From eq. 3-37 it is seen that β is related directly to f_{max}, so that by reading f_{max} for the developing spectra for a given $U_{19.5}$, the value of β in the spectral form of eq. 3-32 could be computed. From eq. 3-40 it is seen that $H_{1/3}$ for a given wind velocity is related to α/β. Inoue provided growth curves of $H_{1/3}$ against fetch F, and also duration t, for a number of wind velocities $U_{19.5}$. The FAS values for the fetch and duration graphs do not match absolutely and the curvatures on some do not appear uniform. The asymptotic values in the fetch curves were measured in order that F_{FAS} and hence F/F_{FAS} ratios could be determined. From eq. 3-40 the changes in α/β could be assessed, which provided variations of α with F/F_{FAS}, from knowledge of β at these fetch ratios. Smoothing of the variables resulted in the curves of Fig. 3-19, for which the following equations apply:

$$\alpha = 0.0081 \, (F/F_{FAS})^{-0.194} \tag{3-62}$$

$$\beta = 0.1 \exp[(\ln 7.4)(F/F_{FAS})^{-0.284}] \tag{3-63}$$

The limiting FAS values are as noted in eq. 3-29.

The use of eq. 3-37 and 3-40 as above implies that the index of -5 for f and $+4$ for $(g/2\pi U_{19.5})$ in eq. 3-31 or 3-32 remain the same in developing seas as for the FAS condition. Measurements made in the ocean by Barnett and Wilkerson [40] and in the laboratory by Sutherland [41], plus the theory of Phillips [42] and Miles [43], support the f^{-5} growth of waves during the development stage. Hence eq. 3-37 would appear to be applicable. Changes in the index (4) for the C/U term would necessarily alter both the values of β and of α in the developing sea. However a comparison of f_{max} versus F/F_{FAS} derived from β above, with values obtained from Barnett and Wilkerson's ocean measurements [40] show good correlation, as illustrated in Fig. 3-20. From this reference it is difficult to assess the actual wind velocity at 19.5 m height but the several measurements from anemometers in the fetch indicate an average value very close to 35 knots. A value of $F_{FAS} = 660$ NM has been used for this wind velocity, which will be discussed later.

DEVELOPING SEAS

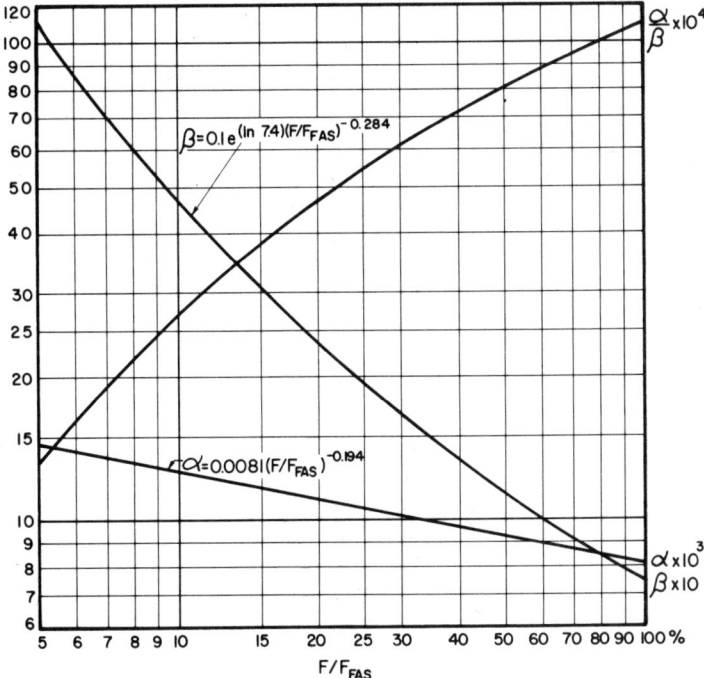

Fig. 3-19. Variations of α and β in a developing sea.

When the progressive values of α and β are substituted into eq. 3-32 to provide the developing frequency spectrum it is found that these disagree in shape with Inoue's curves. Whilst the optimum values occur at similar f_{max} and the areas under the curves are similar, the curves derived as above are flatter and the peak values smaller. This causes the energy of the lower-frequency components to exceed those of Inoue as illustrated in Fig. 3-21. Inoue graphed spectra for developing seas that had been recorded and hindcasted. He observed that the lower frequencies were undervalued by his curves.

Having established values of α and β for various stages of the fetch, it is possible to substitute into any of the relevant equations to obtain the progressive spectrum or EDC, in either dimensional or dimensionless form as in Fig. 3-22, 3-23 and 3-24. Since the equations retain the same form, the proportions of the triangular equivalent to the EDC will remain the same as that for the FAS condition. Thus the ratios T_U/T_{max}, T_L/T_{max} etc. remain constant as in eq. 3-47. This implies a similarity in the generation process during development to that when the sea is fully arisen. This again appears a reasonable supposition since there could be no drastic change in the procedure of smaller waves becoming unstable, breaking and being "fed into" the longer ones.

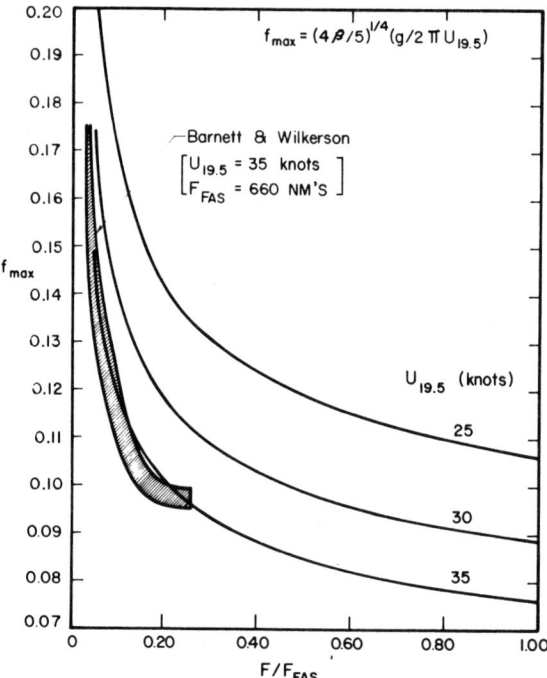

Fig. 3-20. Comparison of f_{max} in eq. 3-37 with recorded values in a limited fetch (Ref. 40).

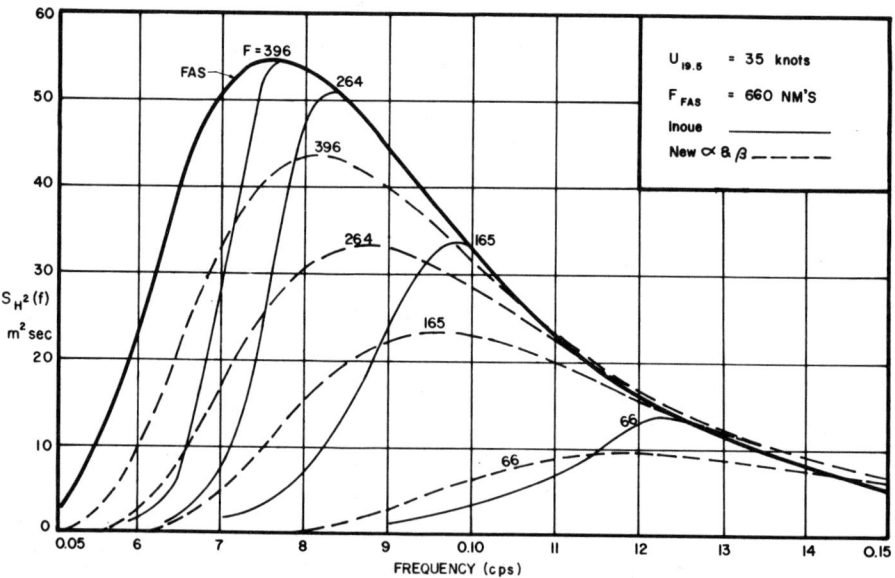

Fig. 3-21. Comparison of eq. 3-32 and curves from Ref. 38 for a developing sea with $U_{19.5}$ = 35 knots.

DEVELOPING SEAS

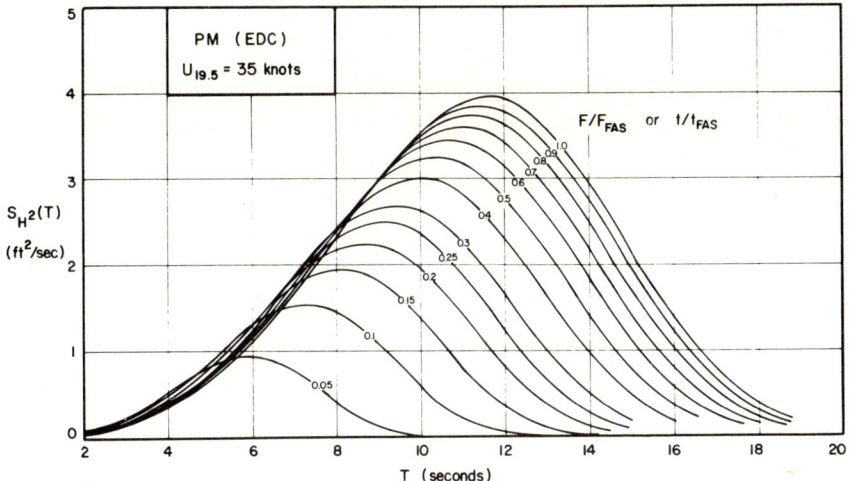

Fig. 3-22. Dimensional EDC for developing sea when $U_{19.5}$ = 35 knots.

The only difference between the FAS conditions and the earlier stages is the "overshoot effect", where waves with periods less than T_{max} grow initially to a height in excess of their "equilibrium" value in the FAS spectrum or EDC. Inspection of Fig. 3-22–3-24 would indicate that the form of the equations, and the values of α and β used for the developing sea, express the existence of this initial

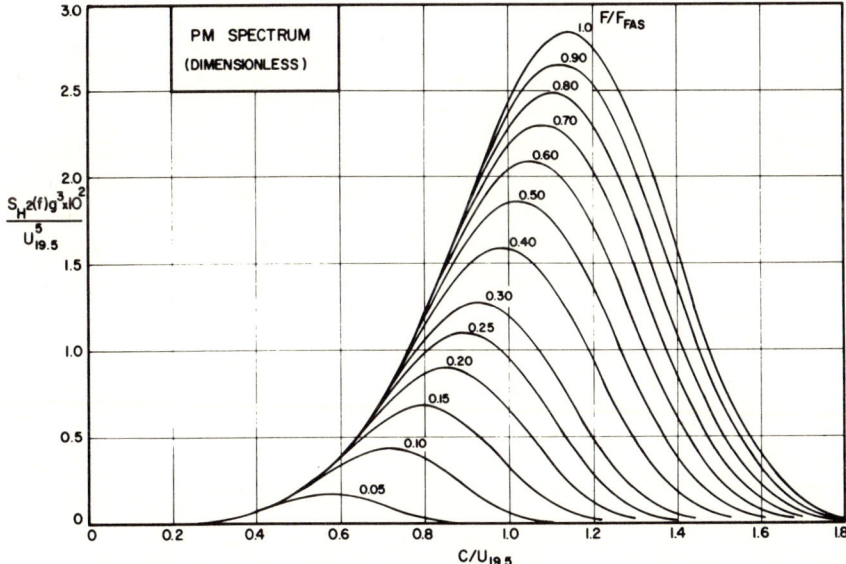

Fig. 3-23. Dimensionless PM spectrum for developing sea.

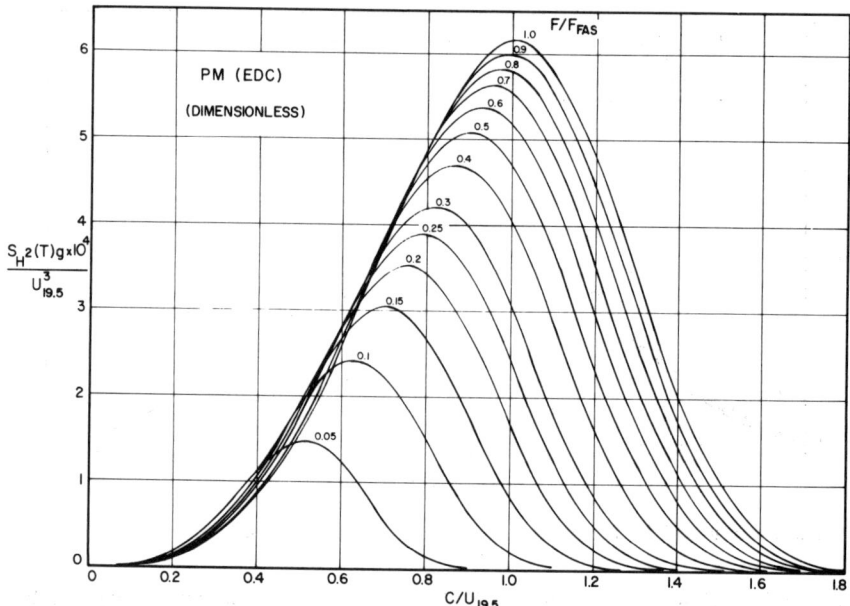

Fig. 3-24. Dimensionless EDC for developing sea.

exceedance, even close to the stage of the FAS. Inoue omitted these portions of the curves even though he discussed the "overshoot effect" [38]. Barnett and Sutherland [44] have reported on this phenomenon and have presented results from the laboratory and the ocean which exhibit it. They found that present theories were unable to explain it, leading to the conclusion that whitecapping or wave breaking "cannot be ruled out as a potential cause of the overshoot" [44].

Unna [45] and later Longuet-Higgins and Stewart [46] have shown analytically that small-length waves are steepened by longer waves. This is effected by contraction of the particles in the longer wave between the trough and the following crest. It is also caused by the work done by the longer waves against the radiation stress of the short waves, which is converted into short-wave energy. The resulting influence is an increase in the height of the short waves and a decrease in their length. The respective ratios of modified height H'_s and length L'_s to the mean height H_s and length L_s for deep water are given by:

$$\frac{H'_s}{H_s} = 1 + \pi \frac{H_L}{L_L} \tag{3-64}$$

and:

$$\frac{L'_s}{L_s} = 1 - \pi \frac{H_L}{L_L} \tag{3-65}$$

where subscript L refers to variables of the longer wave. The resulting increase in steepness is given by:

$$\frac{H'_s}{L'_s} \bigg| \frac{H_s}{L_s} = \frac{1 + \pi H_L/L_L}{1 - \pi H_L/L_L} \tag{3-66}$$

Eq. 3-66 is graphed in Fig. 3-25.

Using the EDC curve for $U_{19.5}$ = 35 knots in Fig. 3-22, at the stage of F/F_{FAS} = 0.2, then T_{max} = 8.5 sec and $H_{1/3}$ from computation is 15 ft. This combination of components gives a long-wave steepness of 0.04 which, from Fig. 3-25 is seen to increase the steepness of smaller waves by a factor of 1.3. Any short waves approaching a steepness of 1/10 would thus be forced to the critical value of 1/7 and be

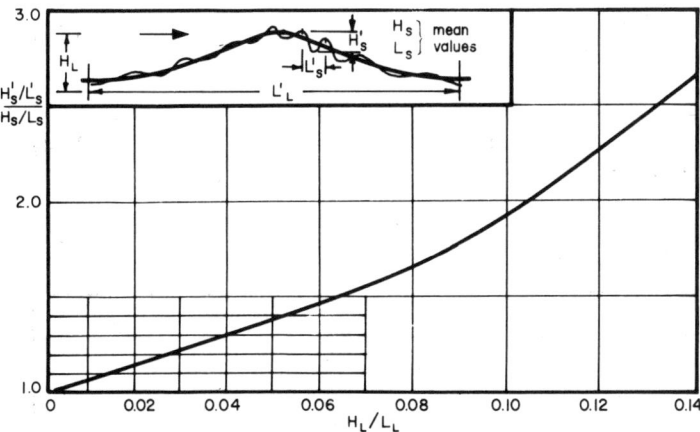

Fig. 3-25. Proportionate steepening of short waves on crests of long waves.

broken. As the FAS conditions are approached, so are the longer components in the short-period end of the EDC curve steepened and broken, whereas previously they may have remained stable at steepnesses just below the critical. Although $H_{1/3}$ has been taken for the height of the long wave, this occurs frequently enough for it to have a significant effect on the average energy of the small wave components.

With a knowledge of the variation of α and β over the full range of F/F_{FAS}, all previous relationships containing these two factors can be determined from their relationship to this ratio. Thus:

from eq. 3-38 $\quad \dfrac{C_{T_{max}}}{U_{19.5}} = m...$ (dimensionless) $\tag{3-67}$

from eq. 3-39 $\quad \dfrac{T_{max}}{U_{19.5}} = n...$ (sec/knots) $\tag{3-68}$

from eq. 3-41 $\dfrac{H_{1/3}g}{(U_{19.5})^2} = p$ ... (dimensionless) (3-69)

from eq. 3-42 $\dfrac{H_{1/3}}{(U_{19.5})^2} = q$ ... (ft./knot^2) (3-70)

from eq. 3-43 $\dfrac{T_{max}}{(H_{1/3})^{1/2}} = r$ ... (sec/ft.$^{1/2}$) (3-71)

from eq. 3-50 $\dfrac{A_{T_{max}}(T_{max})^{1/2}}{H_{1/3}} = s$ ... (sec$^{1/2}$) (3-72)

from eq. 3-51 $\dfrac{A_{T_{max}}}{(T_{max})^{3/2}} = v$ ... (ft./sec$^{3/2}$) (3-73)

from eq. 3-52 $\dfrac{A_{T_{max}}}{(U_{19.5})^{3/2}} = y$ ... (ft./knot$^{3/2}$) (3-74)

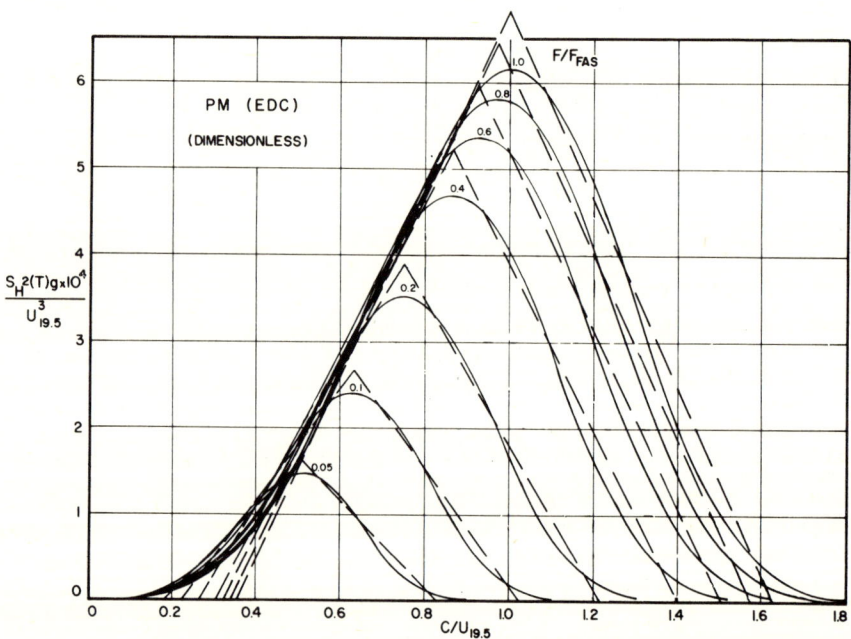

Fig. 3-26. Dimensionless EDC for developing sea illustrating the equivalent triangular distribution.

AVERAGE SPECTRA

Because the same form of eq. 3-34 and 3-35 is used for the developing sea as for the FAS condition, only α and β changing concurrently as per Fig. 3-19 or Table 3-VIII, so the proportions of the EDC curves remain the same. Thus the peak at T_{max} has the same relationship to T_U and T_L (eq. 3-47 and Fig. 3-13) as for the FAS. This is illustrated in Fig. 3-26 where the EDC curves and their triangular equivalent are presented for specific values of F/F_{FAS}. This implies similar generation and dissipation processes during development as for the final equilibrium stage. The only difference is the "overshoot effect" previously discussed. Values of T_U and T_L can thus be derived from T_{max} existing along the fetch, or occurring at some point over time.

AVERAGE SPECTRA

Values of $H_{1/3}$ can become very high if winds of around 50 knots are allowed to develop a FAS. Fortunately this is not generally the case. When high winds occur in some meteorological complex they are normally associated with short fetches or short duration or both. Moskowitz in his selection of wave records with their related fetches, refers to Walden's observations [47] and states that "the fetch required for lighter winds (up to about 30 knots) to produce fully developed seas occurs frequently in most areas. Wind speeds greater than 30 knots are rarely associated with fetches great enough to produce fully developed seas".

From FAS values of $H_{1/3}$ it is possible to compute optimum heights and periods for wind velocities which are fetch- or duration-limited. An arbitrary, but it is thought reasonable, scale of fetch limitation is listed in Table 3-V, together with the resulting $H_{1/3}$ and T_{max} for $U_{19.5}$ up to 55 knots. This indicates that optimum

TABLE 3-V

Heights and periods for probable maximum fetches

$U_{19.5}$	knots	30	35	40	45	50	55
F_{FAS}	NM	526	660	806	956	1130	1290
F_{act}	NM	700	600	500	400	300	200
F_{act}/F_{FAS}	%	100	91	62	42	26.5	15.5
$(H_{FAS})_{1/3}$	ft.	16.6	22.6	29.6	37.5	46.2	55.9
H/H_{FAS}	%	100	99	91	82	72	60
$H_{1/3}$	ft.	16.6	22.4	27	30.8	33.3	33.5
$(T_{FAS})_{max}$	sec	9.9	11.6	13.2	14.9	16.5	18.2
T/T_{FAS}	%	100	99	94	88	80	70
T_{max}	sec	9.9	11.5	12.4	13.1	13.2	12.8

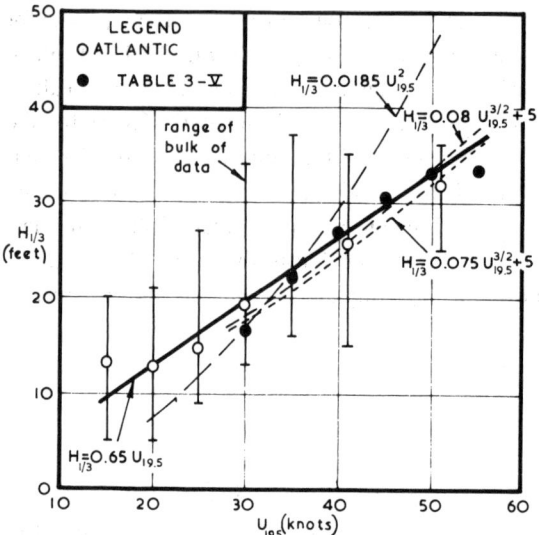

Fig. 3-27. Optimum wave heights from fetch control of higher wind velocities.

values of $H_{1/3}$ = 33 ft. and T_{max} = 13 sec could serve for design purposes when meteorological or other data are not available or are unreliable. Silvester has suggested such figures previously [48].

Scott [49] has analysed wave data from a number of sources which cover the Irish Sea [30,50] and Atlantic Ocean [23,51]. By grouping the significant wave heights for increments in wind speed, Scott obtained mean values which have been plotted in Fig. 3-27. This relates $H_{1/3}$ to $U_{19.5}$ and includes the values derived in Table 3-V. The closeness of the two sets of data supports the choice of fetch lengths used in this Table.

Scott derived a relationship of:

$$H_{1/3} = 0.08\, U_{19.5}^{3/2} + 5 \tag{3-75}$$

where $H_{1/3}$ is in ft. for $U_{19.5}$ in knots. Another equation:

$$H_{1/3} = 0.075\, U_{19.5}^{3/2} + 5 \tag{3-76}$$

had the high-frequency bias of individual spectra extracted from it.

Also indicated in Fig. 3-27 is the range of the bulk of the wave data from which the means were derived. The optimum of around 36 ft. should be noted. The curve of $H_{1/3} = 0.0185\, U_{19.5}^2$ (eq. 3-42) illustrates clearly that beyond 35 knots this FAS relationship over-estimates wave height. Likewise, it can be seen that below 35 knots it under-estimates; for example, at 20 knots it predicts $H_{1/3}$ = 7.4 ft. whereas the Atlantic data has a mean of 13 ft. The line given by:

$$H_{1/3} = 0.65\, U_{19.5} \tag{3-77}$$

AVERAGE SPECTRA

for $H_{1/3}$ in ft. and $U_{19.5}$ in knots, appears to express the optimum values of $H_{1/3}$ throughout the range of $20 < U_{19.5} < 55$, and is suggested for design purposes until further reliable data become available.

Scott [49] also grouped $H_{1/3}$ against zero crossing period (T_z), the latter as measured from the records of ship-borne wave recorders [52]. This has been converted to T_{max} from the relationship of T_z to T_f by Darbyshire [39]. The period T_f is the reciprocal of the frequency at which the peak energy occurs in the frequency spectrum. From eq. 3-37 and 3-38 it follows that:

$$T_{max} = T_f/1.14 \tag{3-78}$$

The data so analysed by Scott is illustrated in Fig. 3-28. The average Atlantic (T_z) data were obtained from summing spectral values included in the report of Moskowitz et al. [23], which contained a spurious hump in the high-frequency zone of the spectra. This was considered an error of the instrument plus an influ-

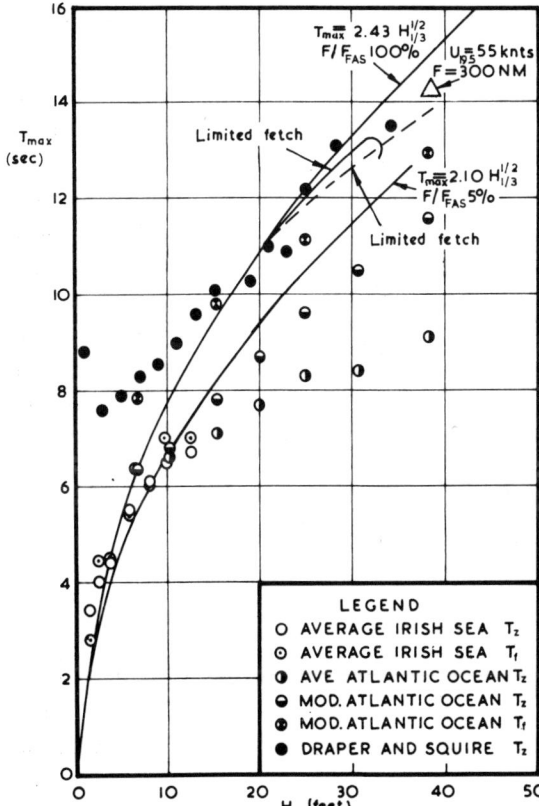

Fig. 3-28. Characteristics of optimum waves recorded on the sea and ocean.

ence of the ship. From Fig. 3-24 it can be seen that the "overshoot" effect, already alluded to, could partly explain the presence of such a hump. However, by smoothing this section of the spectrum the *modified Atlantic* (T_z) data points in Fig. 3-28 were obtained. The averages of 2,400 wave records analysed by Draper and Squire [51] are also included.

Scott also presented graphs of spectra from the seven Irish Sea and four Atlantic Ocean records. From these T_f and hence T_{max}, could be measured. The equivalent T_{max} for the *Irish Sea* (T_f) are plotted in Fig. 3-28. It can be seen that they differ very little from the T_{max} derived from the T_z quoted values. The Atlantic T_{max} values are plotted as *modified Atlantic* (T_f). It is seen that these T_{max} are significantly greater than those derived from the listed T_z periods, indicating a variable association between T_z and T_{max} over the full range of possible spectra or EDC's.

Eq. 3-43 provides a relationship between T_{max} and $H_{1/3}$ for the FAS condition. From the variations in α and β illustrated in Fig. 3-19 it can be shown that the ratio $T_{max}/H_{1/3}^{1/2}$ varies from 2.43 at FAS to 2.10 at 5% F_{FAS}. These two limiting relationships are depicted in Fig. 3-28 and are seen to follow the recorded data reasonably well. The deviation of the Draper and Squire points from the theoretical curve in the low period zone may be explained by Scott's statement: "The periods from the Draper and Squire data are not corrected for instrument response, and, due to high-frequency attenuation, are longer than the periods which would have been obtained had it been possible to apply such correction."

The results given by the limited fetch assumption made in Table 3-V are also drawn as a curve in Fig. 3-28. This falls a little below the observed results of Draper and Squire, although the optimum average height of 34.2 ft. and period 13.5 sec noted by them are very close to the 33.5 ft. and 12.8 sec computed in Table 3-V. If a fetch of 300 NM were used as optimum for $U_{19.5}$ = 55 knots, in place of the 200 in Table 3-V, a maximum of $H_{1/3}$ = 38.5 ft. would result. This represents 23% of F_{FAS}, and is plotted as a single point in Fig. 3-28. The *modified Atlantic* (T_f) data points cross over from the 100% F_{FAS} curve to the 5% F_{FAS} curve as the heights and periods increase. This indicates that the higher-velocity winds are operating under more limited fetch conditions.

By using eq. 3-77 it is possible, as illustrated in Table 3-VI, to compute $(H/H_{FAS})_{1/3}$ as a percentage, from which F/F_{FAS} and $T_{max}/H_{1/3}^{1/2}$ ratios can be derived from tables and graphs to be presented later. Respective values of $H_{1/3}$ and T_{max} are plotted in Fig. 3-28 as a dotted curve, which merges into the T_{max} = 2.43 $H_{1/3}^{1/2}$ line below $H_{1/3}$ = 20 ft., which is where FAS conditions exist for wind velocities below 30 knots. This curve would seem to serve the purpose of finding optimum values of T_{max} once $H_{1/3}$ was determined from eq. 3-77 for winds in excess of 35 knots.

An interesting comparison to the above average values is given by Mayeçon [53] who has analyzed the visual wave records from 500 ships in the North Atlantic over the years 1953 to 1961. The main ocean area was subdivided into eleven zones, but

DESIGN DATA

TABLE 3-VI

Average optimum wave characteristics for high wind velocities

$U_{19.5}$	knots	35	40	45	50	55
$H_{1/3}$ *	ft.	22.6	26	29.2	32.5	35.8
$(H/H_{FAS})_{1/3}$	%	100	88	78	70	64
F/F_{FAS} **	%	100	60	35	25	18
$T_{max}/H_{1/3}^{1/2}$ ***	sec/ft.$^{1/2}$	2.43	2.36	2.31	2.27	2.24
T_{max}	sec	11.6	12.0	12.5	13.0	13.5

* Eq. 3-77; ** Table 3-V or Fig. 3-30; *** Table 3-V or Fig. 3-29.

for the present purpose these can be combined into four latitudinal bands. From probability curves covering each zone the following annual values of wave height were obtained:

latitude (degrees)	20–30	30–40	40–50	50–60
$H_{1/3}$ (ft.)	26	35	40	45

For the 30°–35° latitudes, for which the data listed by Scott [49] applies, the range of height (35–40 ft.) fits the recorded points in Fig. 3-28 extremely well. Similar information is desirable for all oceans, for ready application by coastal engineers.

Thom [59] has applied the Frechet extreme value distribution to wave heights measured at 3-h and 1-h intervals on twelve ocean station vessels over 7 to 10 years. These stations were located between 30° and 50° latitude in the North Pacific and 30°–52° in the North Atlantic. The average of the significant wave for the twelve stations with a mean recurrence interval of two years was 33 ft. For 10, 25 and 50 year recurrence intervals the averages were 44, 51 and 57 ft., respectively. Extreme wave heights on the same basis are obtained by multiplying the above $H_{1/3}$ by 1.8 [59]. Thus the extreme height on average in these two areas of ocean that can be experienced once in 50 years is 1.8 × 57 = 103 ft. Two stations on longitude 20°W and latitudes 52°30' and 59°00N could both experience extreme heights in this period of 136 ft.

DESIGN DATA

Fully arisen sea

The above discussed relationships for FAS conditions can be summarized as in Table 3-VII for design purposes. The various graphs supplied can be useful when interpolating between the values listed. It should be noted that this Table applies to

TABLE 3-VII

Wave characteristics for FAS and optimum conditions

1	$U_{19.5}$	knots	20	25	30	35	40	45	50	55
2	$U_{19.5}$	mile/h	23.0	28.7	34.6	40.2	46.1	51.7	57.5	63.2
3	$U_{19.5}$	km/h	37.0	46.1	55.6	64.6	74.1	83.1	92.5	101.6
4	U_{10}	knots	18.5	23.2	27.8	32.4	37.0	41.6	46.3	51.0
5	F_{FAS}	NM	284	396	526	660	806	956	1130	1290
6	F_{95}	NM	213	297	395	495	605	717	850	970
7	t_{FAS}	h	35.6	39.8	43.5	47.0	50.2	53.4	56.4	59.0
8	t_{95}	h	28.5	36.9	34.8	37.6	40.2	42.7	45.1	47.2
9	$H_{1/3}$	ft.	7.4	11.5	16.6	22.6	29.6	37.5	46.2	55.9
10	$H_{1/10}$	ft.	9.4	14.6	21.1	28.8	37.7	47.8	58.9	71.1
11	H_{max}	ft.	13.8	21.4	30.8	42.0	55.0	70.0	86.0	104
12	$H_{T_{max}}$	ft.	1.7	2.4	3.1	4.0	4.8	5.6	6.8	7.8
13	$H'_{T_{max}}$	ft.	1.8	2.5	3.3	4.2	5.0	5.8	7.2	8.2
14	T_{max}	sec	6.6	8.25	9.9	11.6	13.2	14.9	16.5	18.2
15	T_U	sec	10.7	13.4	16.0	18.8	21.4	24.2	26.7	29.4
16	T_L	sec	2.3	2.9	3.5	4.1	4.6	5.2	5.8	6.4
17	$T_{1/3}$	sec	6.1	7.6	9.2	10.7	12.2	13.8	15.3	16.8
18	F_{opt}	NM	284	396	526	660	480	320	280	230
19	t_{opt}	h	35.6	39.8	43.5	47.0	33.0	24.0	20.0	17.0
20	$H_{1/3}$	ft.	13.0	16.0	19.5	22.6	26.1	29.5	33.0	36.4
21	$H_{1/10}$	ft.	16.6	20.4	24.8	28.8	33.2	37.6	42.0	46.4
22	H_{max}	ft.	24.2	29.8	36.3	42.0	48.5	54.9	61.4	67.6
23	$H_{T_{max}}$	ft.	2.6	3.0	3.5	4.0	4.4	4.8	5.2	5.6
24	$H'_{T_{max}}$	ft.	2.7	3.2	3.7	4.2	4.6	5.1	5.5	6.0
25	T_{max}	sec	8.7	9.7	10.7	11.6	12.0	12.5	13.0	13.5
26	T_U	sec	14.1	15.7	17.3	18.8	19.4	20.2	21.0	21.8
27	T_L	sec	3.0	3.4	3.7	4.1	4.2	4.4	4.5	4.7
28	$T_{1/3}$	sec	8.0	9.0	9.9	10.7	11.1	11.6	12.0	12.5

Rows 9–17: FAS. Rows 20–28: OPT.

extra tropical cyclones where fetches contain winds not exceeding 55 knots and which are essentially uniform in velocity and direction.

The following notes will be useful when employing Table 3-VII. The items are listed as in the table.

1. Winds should be referred to the height 19.5 m above the sea surface. See Fig. 3-8 for necessary conversions. Note that equalities including $U_{19.5}$ below refer to wind velocity in knots at 19.5 m above the sea surface.

2, 3. Wind velocities in mile/h and km/h equivalent to item *1*.

4. $U_{10} = U_{19.5}/1.08$.

DESIGN DATA

5. F_{FAS} is the minimum fetch length required to generate a FAS ($= 3.19\, U_{19.5}^{3/2}$).

6. F_{95} is the fetch length required for 95% of the FAS value of either $H_{1/3}$ or T_{max} to be developed ($= 0.75\, F_{FAS}$).

7. t_{FAS} is the minimum duration time required to generate a FAS ($= 7.95\, U_{19.5}^{1/2}$).

8. t_{95} is the duration time for 95% of the FAS value of either $H_{1/3}$ or T_{max} to be developed ($= 0.80\, t_{FAS}$).

9. $H_{1/3}$ is the significant wave height which is the average of the highest third of the waves ($= 0.0185\, U_{19.5}^2 = 0.17\, T_{max}^2$).

10. $H_{1/10}$ is the average of the highest one tenth of the waves [1] ($= 1.275\, H_{1/3}$) also equal to the highest $H_{1/3}$ in 10 years [53].

11. H_{max} is the highest wave in every 1000 waves ($= 1.86\, H_{1/3}$) [1].

12. $H_{T_{max}}$ is height of wave component with period T_{max} ($= 0.102\, T_{max}^{3/2} = 0.38\, H_{1/3}^{3/4}$).

13. $H_{T_{max}}'$ is height of wave component with period T_{max} based on equivalent triangular EDC ($= 0.108\, T_{max}^{3/2} = 0.40\, H_{1/3}^{3/4} = 1.06\, H_{T_{max}}$).

14. T_{max} is the period of the component with maximum energy in the EDC or its triangular equivalent ($= 2.43\, H_{1/3}^{1/2} = 0.33\, U_{19.5}$).

15. T_U is the upper limit of periods in the EDC of engineering significance ($= 1.62\, T_{max}$).

16. T_L is the lower limit of periods in the EDC of engineering significance ($= 0.35\, T_{max}$).

17. $T_{1/3}$ is the significant period [2, 15], or that period closely associated with waves represented by $H_{1/3}$ ($= 0.929\, T_{max}$) [39].

Developing sea

The ratios from eq. 3-67 to 3-74 are graphed in Fig. 3-29 and listed in Table 3-VIII. Also included in that Table are the extremely useful ratios of $(H/H_{FAS})_{1/3}$ and $(T/T_{FAS})_{max}$ as they vary along the fetch. For clarity they have been plotted separately in Fig. 3-30. It is seen that both these two ratios reach 95% of their FAS value at approximately 75% of F_{FAS}, which fact is noted in Table 3-VIII and used in Table 3-VII. Table 3-VIII also contains the variations in α and β, as displayed in Fig. 3.19. To aid application also, Fig. 3-31 has been provided, which shows the growth in $H_{1/3}$ for various wind velocities up to the FAS values. As noted previously, these differ a little from the curves presented by Inoue [38] due to the smoothing carried out.

The conversion of t/t_{FAS} to F/F_{FAS} from eq. 3-61 is illustrated in Fig. 3-32 as well as being listed in Table 3-VIII. Thus in all cases where F/F_{FAS} has been used as a parameter the corresponding values of t/t_{FAS} are applicable. For example, a spectrum or EDC representing conditions at $t/t_{FAS} = 50\%$ also represents conditions at $F/F_{FAS} = 40\%$. Thus, all the graphs which contain F/F_{FAS}, can be re-

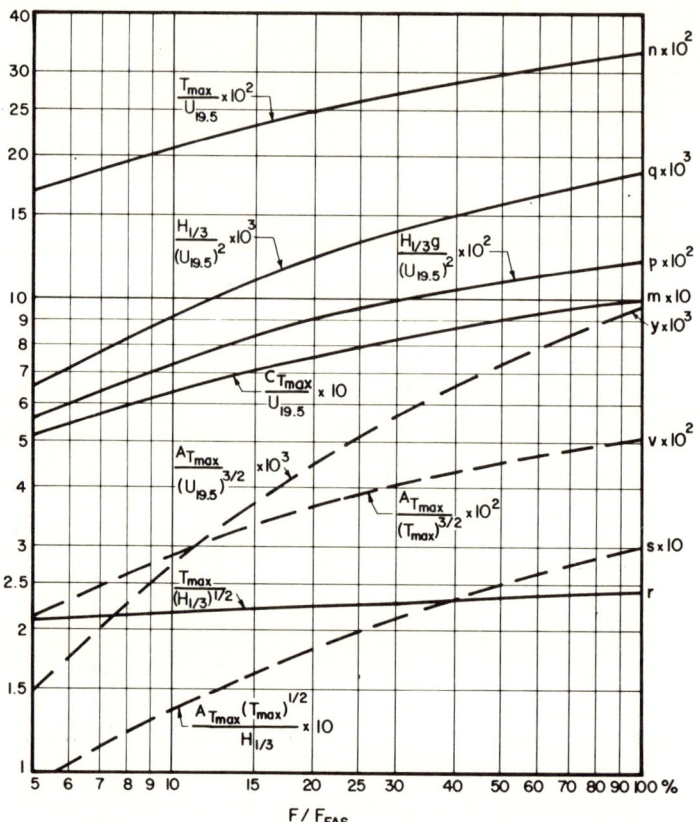

Fig. 3-29. Wave characteristics for limited fetch or duration conditions.

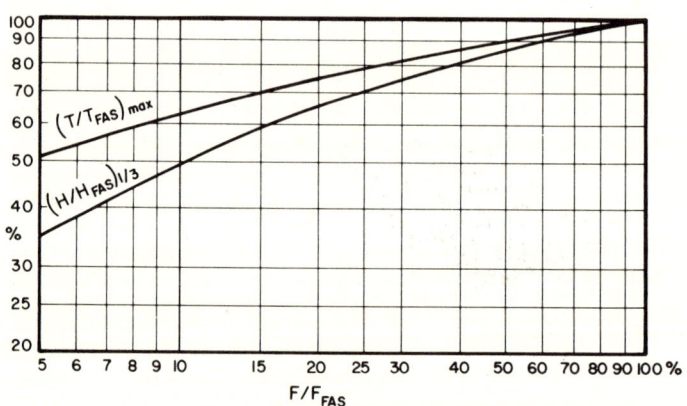

Fig. 3-30. Ratios of $H_{1/3}$ and T_{max} along fetch to values at F_{FAS}.

ferred to a limited duration condition by noting the t/t_{FAS} equivalence to the fetch ratio. Since 95% of the FAS significant wave height exists at 75% of the FAS

DESIGN DATA

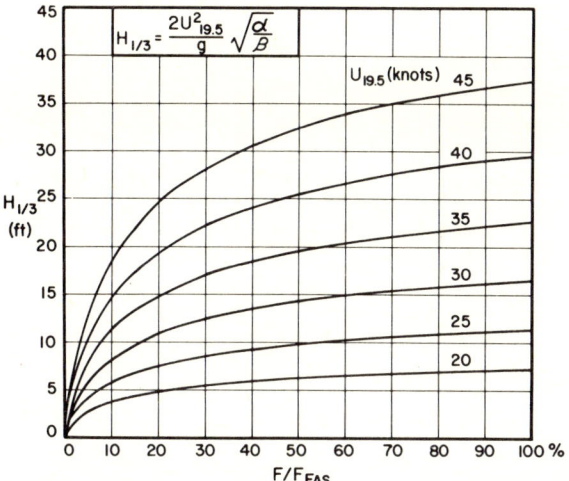

Fig. 3-31. Values of $H_{1/3}$ for a range of wind velocities with limited fetch.

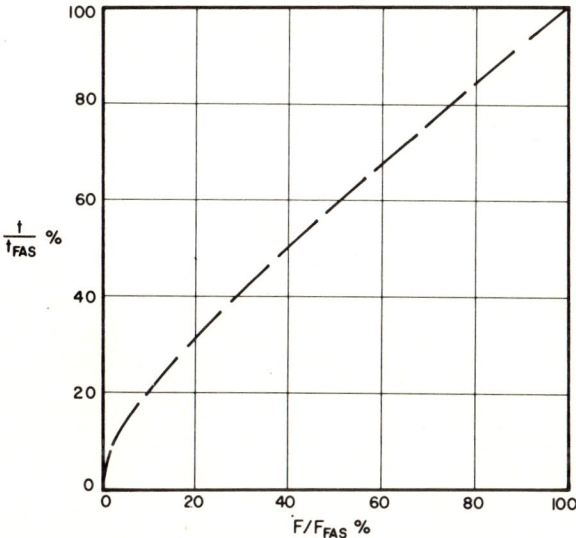

Fig. 3-32. Relationship between t/t_{FAS} and F/F_{FAS}.

fetch, it is seen that the same conditions exist at 80% of the FAS duration. Whilst the sea is developing the ratios of T_U and T_L to T_{max} remain the same as those for the FAS, as given in eq., 3-47.

Optimum wave conditions

From the discussion on average spectra it is reasonable to consider some optimum conditions for any particular section of coast, particularly for wind veloci-

TABLE 3-VIII

Relationships for limited fetch or duration conditions

$F/F_{FAS}\%$	$(\frac{H}{H_{FAS}})_{1/3}\%$	$(\frac{T}{T_{FAS}})_{max}\%$	$\frac{c_{T_{max}}}{U_{19.5}} = m$	$\frac{T_{max}}{U_{19.5}} = n$	$\frac{H_{1/3}g}{U_{19.5}^2} = p$	$\frac{H_{1/3}}{U_{19.5}^2} = q$
5	35	51	0.51	0.169	0.073	0.0065
10	49.5	63	0.63	0.208	0.104	0.0091
15	59	70	0.70	0.231	0.123	0.0109
20	65.5	75	0.75	0.247	0.137	0.0121
30	74.5	31.5	0.82	0.270	0.156	0.0139
40	81	86	0.86	0.285	0.170	0.0150
50	86	89.5	0.90	0.296	0.180	0.0160
60	90	92.5	0.93	0.306	0.188	0.0166
70	93	95	0.95	0.312	0.195	0.0172
80	95.5	97	0.97	0.320	0.200	0.0177
90	98	98.5	0.99	0.326	0.205	0.0181
100	100	100	1.00	0.330	0.209	0.0185

ties in excess of 35 knots. This has been carried out as illustrated in Tables 3-V and 3-VI and in Fig. 3-28. The resulting design data are listed in the lower part of Table 3-VII under section OPT. The following notes show the origin of these values, the item numbers referring to those in the table.

18. F_{opt} is the optimum fetch likely for any $U_{19.5}$, this is FAS up to 35 knots and proportions as in Table 3-VI for greater velocities.

19. t_{opt} is the optimum duration equivalent to F_{opt} as per Table 3-VIII.

20. $H_{1/3}$ is the significant wave height under limiting conditions imposed by items *18* or *19* for $U_{19.5} > 35$ knots or average values recorded in the Atlantic for $U_{19.5} < 35$ knots (= $0.65 U_{19.5}$).

21–24. As for *10–13* above, respectively.

25. T_{max} for item *20* as per ratio equation (3-71) and Fig. 3-29 or Table 3-VIII.

26–28. As for *15–17* above, respectively.

The choice of FAS or optimum conditions at any particular site will depend upon the general weather patterns experienced in the adjacent ocean. Where cyclonic centres travel towards the relevant section of coast there is greater possibility of even high winds having the minimum fetch or duration required for a FAS. Where such centres only pass across or away from the coast generation will be much more limited. This knowledge of cyclonicity patterns has been emphasized in Chapter 2.

The choice of average or FAS wave characteristics will also be dictated by the risks involved to life and property. In respect to coastal structures it must be remembered that waves discussed here are those occurring in deep water. When they propagate across the continental shelf they may be refracted or dissipated by

WAVE DISPERSION

$\dfrac{r_{max}}{H^{1/2}_{1/3}} = r$	$\dfrac{A_{T_{max}}}{H_{1/3}} = s$	$\dfrac{A_{T_{max}}}{T^{1/2}_{max}} = v$	$\dfrac{A_{T_{max}}}{T^{3/2}_{max}} = y$	$\alpha \cdot 10^3$	β	t/t_{FAS}
.10	0.093	0.021	0.0015	14.48	10.87	14.5
.17	0.136	0.029	0.0027	12.66	4.70	20.5
.22	0.163	0.033	0.0037	11.70	3.09	26
.25	0.184	0.036	0.0045	11.07	2.36	31.5
.29	0.213	0.040	0.0057	10.23	1.67	41
.32	0.233	0.043	0.0066	9.67	1.34	50
.35	0.250	0.045	0.0073	9.27	1.14	59
.37	0.263	0.047	0.0079	8.94	1.01	67.5
.39	0.274	0.048	0.0084	8.68	0.92	76
.40	0.283	0.049	0.0089	8.46	0.84	84
.42	0.291	0.050	0.0093	8.27	0.79	92
.43	0.299	0.051	0.0096	8.10	0.74	100

breaking before reaching the relevant depths of the structure. However, in computing limited heights due to wave instability, maximum tide and storm surge depths should be considered.

WAVE DISPERSION

Waves that have traversed the fetch will propagate out of it, in the same direction as they were initially generated and in which they received their maximum energy. Trains of different period will be moving at many angles to the mean wind vector, and even waves of similar period will be propagating over a range of directions. The theory of generation processes, plus measurements made at sea [54] have shown the orthogonals of longer waves to be more closely aligned with the wind than the smaller waves. This directional distribution of wave energy depends mainly upon the "wave age" (i.e., C/U), the smaller this ratio the more angled are the waves. As it tends to, or exceeds unity, the waves are propagating more-or-less directly down fetch. Thus, energy of waves in a certain period band will have a different directional distribution for each wind velocity that generated them.

In the area outside the fetch, generally termed the "decay area" (the writer prefers "dispersal area") waves spread out both circumferentially and radially. This reduces the energy content per unit area of ocean, which must be computed when determining wave conditions in this zone. It has been shown that during this dispersion very little energy is dissipated by viscosity or eddy effects, particularly for

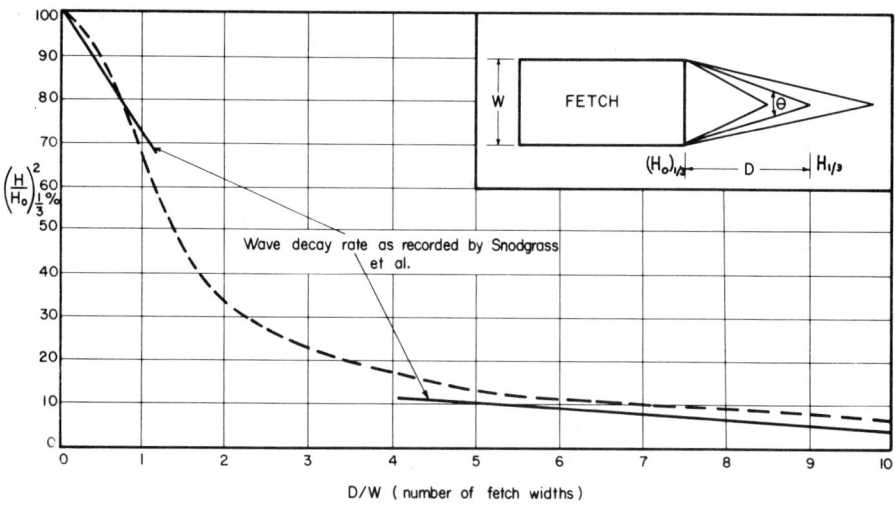

Fig. 3-33. Energy reduction with distance directly downwind of a fetch.

wave periods of 5 sec or more, which are the minimum of engineering significance.

From the diagram of the fetch in the inset of Fig. 3-33, it can be seen that the waves arriving at a point in the dispersal area, somewhere down the alignment of the fetch, are restricted to a band of directions contained by angle θ. All other waves will by-pass this point. In this respect the width of the fetch (W) assumes importance outside the fetch, more so than it does inside it. The further from the fetch (i.e., the greater D/W) the smaller the angular content of the wave energy. The ratio of energy along the centreline to that at the downwind end of the fetch is shown as the dotted line in Fig. 3-33. This should be compared to the energy reductions as recorded by Snodgrass et al. [55] of waves traversing the Pacific Ocean from the Antarctic to Alaska. It would seem, therefore, that the major reduction in height, when wave trains propagate thousands of miles across the ocean, is due to angular spreading of energy.

Fig. 3-33 is based upon a $\cos^4\theta$ energy distribution, which is an average value for a range of distributions around the mean wind direction for $C/U \geqslant 1.0$. The shorter-period components have a less peaked angular distribution, as discussed above, and tend more closely to a $\cos^2\theta$ curve [55]. Integration across such an angular distribution from $-90°$ to $+90°$ results in curves as in Fig. 3-34 which represent the two means of $\cos^2\theta$ and $\cos^4\theta$.

To use this graph, draw two arcs from the side boundaries of the fetch to the point at which wave heights are required (see inset Fig. 3-33). The angle of these to the fetch alignment is then marked on the relevant curve in Fig. 3-34. The difference in the two percentage values is the proportion of the energy at the end of the

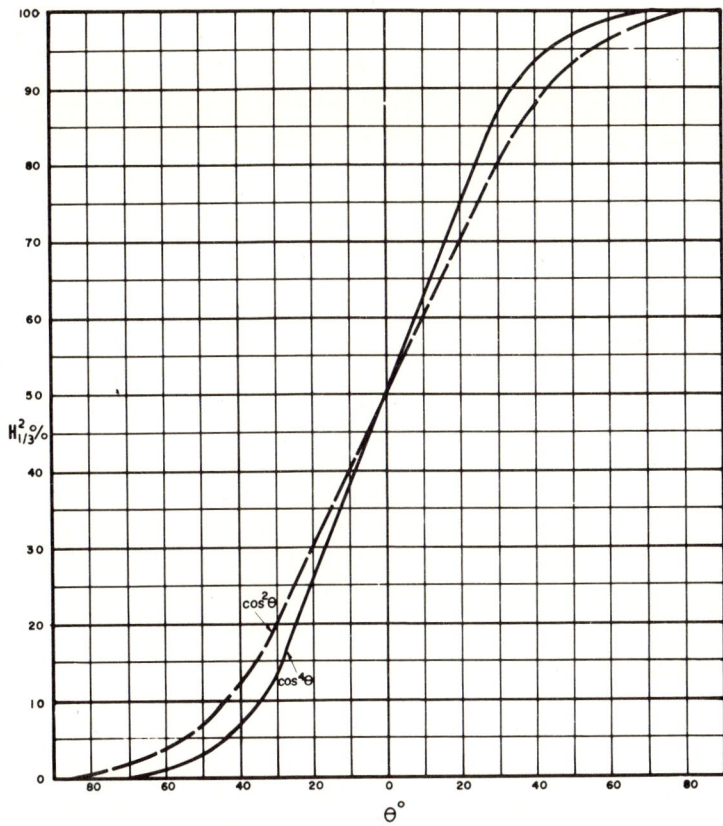

Fig. 3-34. Angular distribution of wave energy according to $\cos^2\theta$ or $\cos^4\theta$.

fetch. In the case of points on the fetch centre line the angle of each ray either side of 0° is the same; for points to one side of the centre line these angles are different, as illustrated in Fig. 3-35 where angles θ_1 and θ_2 can be read for ranges of deviation angle α and distances D/W. It can be seen from the shape of the two traces in Fig. 3-34 that for any enclosed angle θ the energy content will not be greatly reduced until the curves sections are encountered. This occurs when the point is located on a line angled some 30° from the fetch centre line. Thus the bulk of the wave energy is restricted to a 30° fan either side of this alignment. It is suggested that the $\cos^2\theta$ curve be used for waves of period from T_L to $T_{max}/2$ and the $\cos^4\theta$ curve for periods $T_{max}/2$ to T_U.

Besides spreading in this circumferential fashion the waves disperse radially, the higher period waves racing ahead of the medium and lower period components. To assess the band width of waves arriving concurrently at a point in the dispersal area, it is necessary to know the duration time t_e in excess of that required for the FAS.

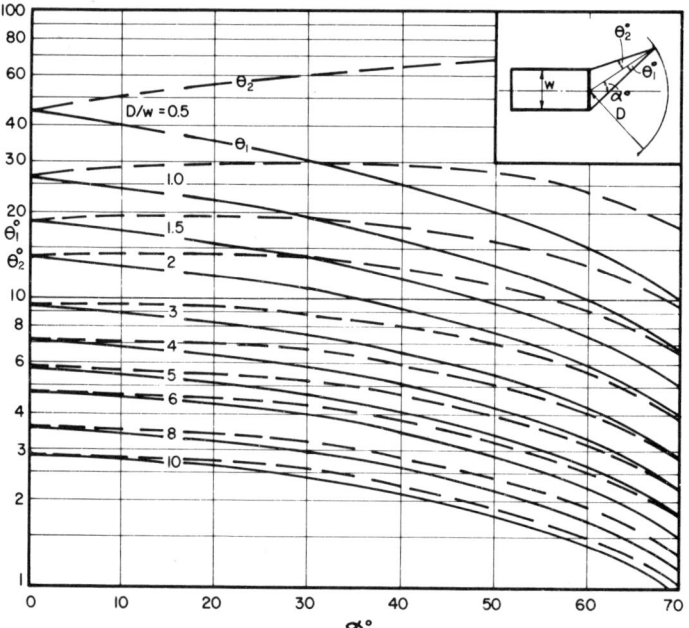

Fig. 3-35. Fan angles to points downwind of a fetch.

Since 95% of the FAS value of $H_{1/3}$ and T_{max} is produced in 80% of the FAS duration, it would seem more reasonable to compute t_e after $t_{95} = 0.80\, t_{FAS}$, even though $(H_{FAS})_{1/3}$ is used for the height calculation.

Each component wave train traverses the ocean at its group-wave-velocity, which for the deep-water conditions considered is $C_o/2$. At a given distance (S) from the downwind end of the fetch, at some time t *after* t_{95}, the period T_1 of the train which is just arriving is given by [9]:

$$t = \frac{S}{1.52\, T_1} \tag{3-79}$$

where t is time in hours; S is distance in nautical miles; T_1 is wave period in seconds.

The period T_1' of the train that has just finished arriving, which is a faster train travelling for a time $t - t_e$, is given by:

$$t - t_e = \frac{S}{1.52\, T_1'} \tag{3-80}$$

Thus the band width present $T_1' - T_1$, depicted in Fig. 3-36, encompasses a certain proportion of the energy existing at the end of the fetch. This may or may not be FAS conditions, as illustrated by equations for $H_{T_{max}}^2$ and $H_{T_{max}}'^2$ in Fig. 3-36.

WAVE DISPERSION

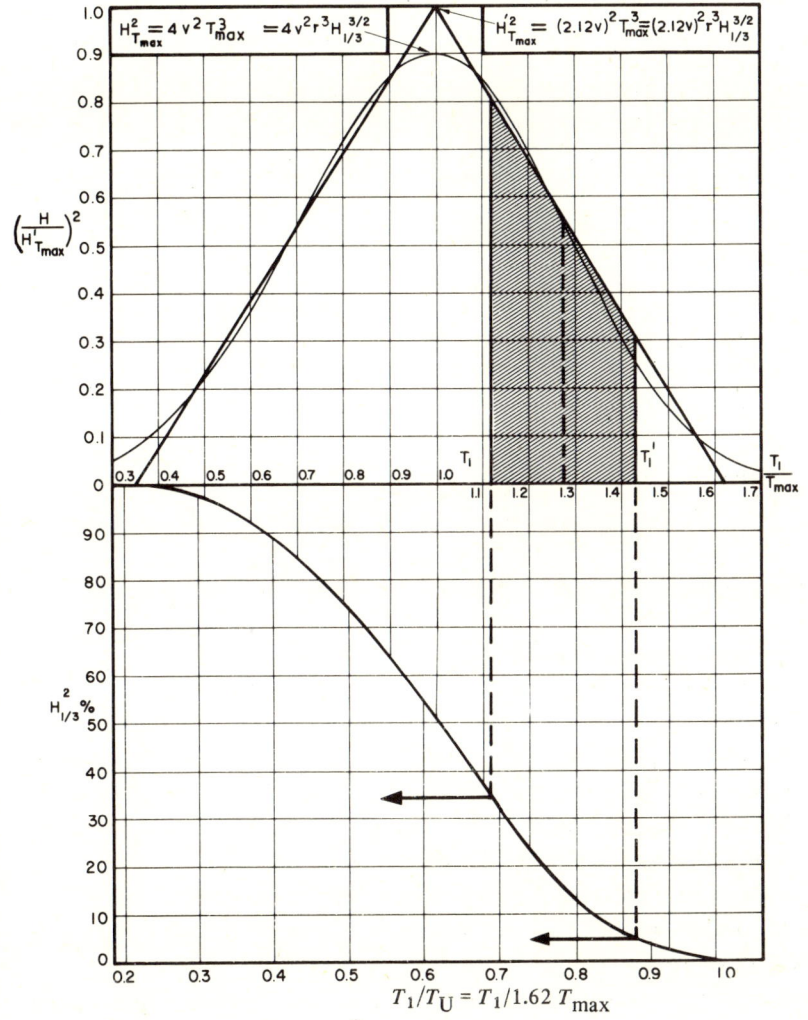

Fig. 3-36. Dimensionless EDC and $H_{1/3}^2$ as percentage of that at end of fetch.

To facilitate the calculation of this proportion, the cumulative area curve is provided in Fig. 3-36, the utilization of which requires T_1 and T_1' to be expressed either as a fraction of T_{max} or of T_U. If T_1' exceeds T_U (= $1.62\,T_{max}$), then only the triangular portion of the equivalent EDC, with a base $T_U - T_1$, is used.

The EDC curve of Fig. 3-36 is based upon a dimensionless ordinate $(H/H'_{T_{max}})^2$ so that when $T_1' - T_1 \to 0$, due to $t_e/t \to 0$, or S being large (making t large in comparison to t_e), the height of the single-period swell wave can be readily determined. Either the triangular equivalent value can be used or the EDC ordinate itself as illustrated in the figure.

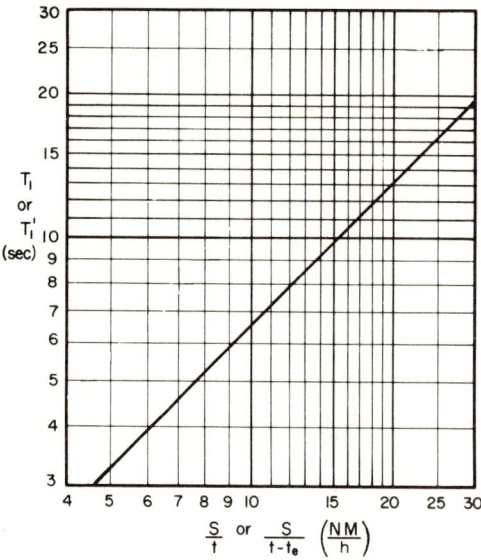

Fig. 3-37. Periods of waves arriving at points distance S from the fetch, t hours after $0.80\ t_{FAS} = t_{95}$.

Eq. 3-79 and 3-80 have been graphed in Fig. 3-37 for ready assessment of T_1 or T_1' for a given distance S from the downwind end of the fetch, at a time t hours after $0.80\ t_{FAS}$ (for T_1) or $t-t_e$ hours after $0.80\ t_{FAS}$ (for T_1'). Should the value of T_1 so derived exceed T_U then no waves have yet arrived at the site. If T_1' is less than T_L, then all energy of engineering significance has already arrived. The range of periods from 4 to 30 sec should be adequate for all purposes.

From the previous discussion of average wave conditions it was seen that $H_{1/3}$ = 33 ft. is a reasonable optimum in extra-tropical-cyclone regions. At long distances from the storm centres of the oceans the T_{max} components given by $H_{T_{max}}$ = $0.38\ H_{1/3}^{3/4}$ = 5.1 ft. These will be reduced in energy by angular spreading to around 10% of the FAS value of 1/3 of the wave height. Thus the biggest swell from a distant storm has a deep-water height = 5.1/3 = 1.7 ft. When this is amplified due to shoaling, its breaking height will be two to three times this, depending upon its angle of approach to the shore and the steepness of the beach. Thus swell waves arriving at a beach from distant storm centres are optimized at heights 3.4—5.1 ft. and periods of 13 sec. The more frequent maxima will emanate from a FAS with $U_{19.5}$ = 35 knots, in which case $H_{T_{max}}$ = 4 ft., so that breaker heights will approximate 2.7—4 ft.

Much of the above material has been published by the author in the technical literature [62,63]. It has been integrated herein in order to provide practical applications besides the bases for conversion from the spectrum (wave frequency) to the energy distribution curve (wave period).

WAVES FROM TROPICAL CYCLONES

In Chapter 2 the wind structure of tropical cyclones was discussed. It was seen that size and intensity of centres varied geographically and even fluctuated in any oceanic location throughout the summer season. The waves issuing from such low-pressure centres are therefore likely to vary greatly, making any correlation of wind velocities and wave characteristics particularly difficult. On the occasion of such meteorological events conditions are extremely dangerous, so that recording of waves within tropical cyclones is generally fortuitous.

Because of the exceptionally high winds in the annulus surrounding the eye it is doubtful whether the normal generation principles apply. However, Bretschneider [56] has presented an analysis, based upon the SMB formulae, for small fetch lengths, which will be outlined below. It is suggested that this be used until more information is available on wave generation inside these intense circular wind fields.

Inspection of Fig. 3-3, showing the SMB relationships, indicates that for values of $gF/U^2 < 10^4$ the traces of gH/U^2 and $gT/2\pi U$ are straight, with slopes of ½ and ¼, respectively. Since these velocities refer to measurements at 10–12 m from the sea surface the resulting equations are:

$$H_{1/3} = 0.0555 (U_{10}^2 F)^{1/2} \tag{3-81}$$

and:

$$T_{1/3} = 0.5 (U_{10}^2 F)^{1/4} \tag{3-82}$$

where H is in ft. for U_{10} in knots and F in nautical miles. This results in the ratio:

$$T_{1/3} = 2.13 (H_{1/3})^{1/2} \quad \text{or} \quad T_{max} = 2.29 (H_{1/3})^{1/2} \tag{3-83}$$

which, from Table 3-VIII, is seen to correspond to $F/F_{FAS} = 30\%$ or $t/t_{FAS} = 41\%$. These appear to be reasonable proportions for this purpose.

As noted previously, the winds in a tropical cyclone are circular in nature, not having cold fronts to straighten the isobars, as in the extra-tropical situation. Also, the wind is directed towards the centre, particularly in the high-wind annulus, so that it is angled at 45° to the radii in the region. A generating zone, considered as a finite width of sea aimed in some direction (see Fig. 3-38), will have longitudinal wind components which vary in magnitude along the length. As illustrated, the velocity distribution varies along chords at different distances from the centre. The main interest is centred on the fetch band passing through the high-wind annulus. In his analysis, Bretschneider integrated the wind components along the chord length, to obtain the $H_{1/3}$ and $T_{1/3}$ values.

Normally wind velocities are not available, so that a relationship must be known between pressure differentials and the highest sustained wind (as distinct from peak fluctuations). Holliday [57] has summarized relationships available of U_{max} (geo-

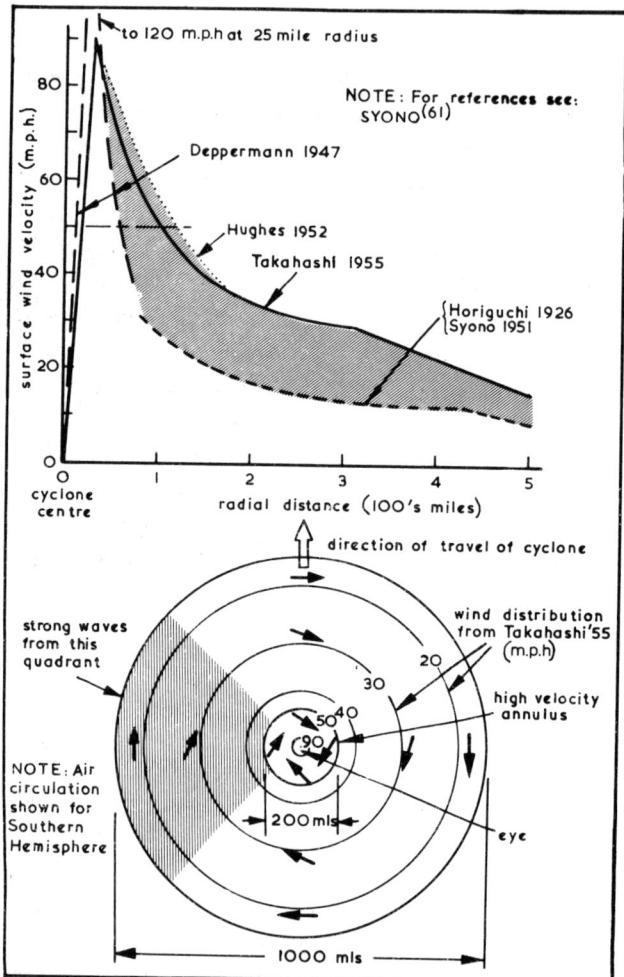

Fig. 3-38. Fetches within a tropical cyclone.

strophic value) and Δp the difference from normal atmospheric pressure (= 29.7 inches mercury, or 1013 mbar, i.e., $1''$ Hg = 34 mbar) and that at the centre of the cyclone. The maximum value of Δp likely to be experienced is $2.5''$ Hg or 85 mbar, but a coastal engineer should obtain such climatological data for the areas of ocean with which he is concerned. An example of such information is given in Fig. 3-39, which should be read in conjunction with Fig. 2-28. Where average values of centre pressures are not readily available in publications, they should be sought from the meteorological office covering the ocean area. Such pressure differentials vary throughout the life of a cyclone and some zones are prone to higher values than others.

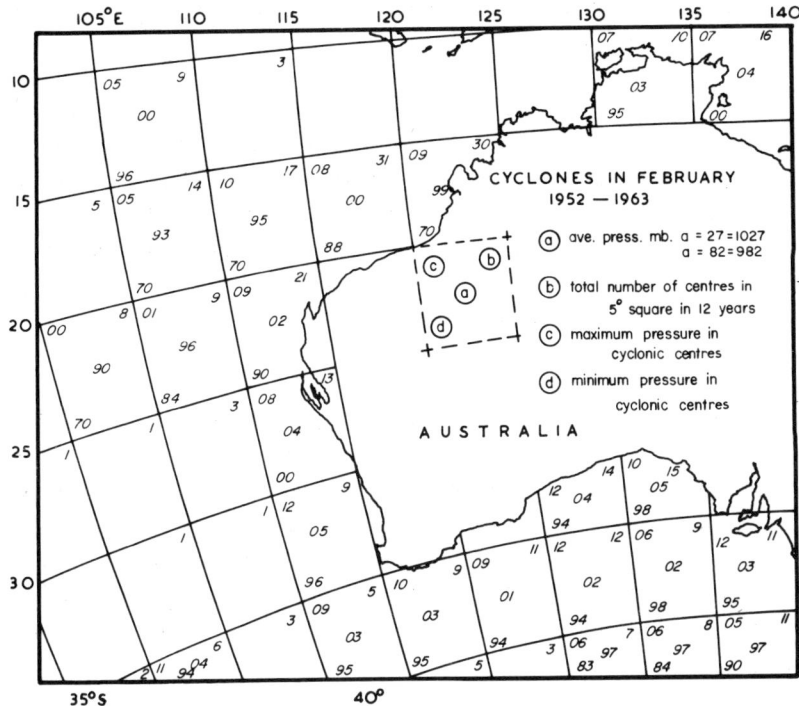

Fig. 3-39. Climatological information available for the northwest coast of Australia.

Besides Δp, another variable indicating the energy in a tropical cyclone is the radius (R) of the eye. This can vary from 5 to 50 NM, without relation to the maximum sustained wind (U_{max}). The larger the eye the longer is the diameter of the complete cyclone. It can be realized, therefore, that the wave-generating capacity can vary tremendously for the same U_{max} value at the edge of the eye. Since the analysis to be presented below is based upon wind structures and cyclone dimensions of Atlantic hurricanes, the resulting wave characteristics may exaggerate conditions for the more modest typhoon, monsoon or tropical-cyclone experienced in other parts of the world. But until wave data are available from these regions the computations will err on the side of safety.

To determine U_{max}, Bretschneider [56] used the relationship:

$$U_{max} = 0.868[73(\Delta p)^{1/2} - R(0.575f)] \qquad (3\text{-}84)$$

where Δp is in inches of mercury, R is in NM and f is a Coriolis parameter varying from 0.2 to 0.3 for latitudes from 5° to 30°. Since the second term in eq. 3-84 is about 5% of the first it could be ignored, thus giving:

$$U_{max} = 60(\Delta p)^{1/2} \qquad (3\text{-}85)$$

Of the data presented by Holliday [57], the equation given by Kraft [58] appears to be the most suitable, namely:

$$U_{max} = 80(\Delta p)^{1/2} \tag{3-86}$$

for Δp in inches of mercury. This gives a greater U_{max} than eq. 3-85, but it is the maximum sustained wind at the edge of the eye. If a reasonable width of fetch is considered the wind velocity at the outer edge will be around half U_{max} as in eq. 3-86. Thus the value of U_{max} as given in eq. 3-85 should be accepted. It is graphed in Fig. 3-40. Since this U_{max} represents the gradient wind (height about 35 m), the value at 10 m height from the sea surface (see Fig. 3-8) is given by:

$$U_{10} = 0.865 \, U_{max} \tag{3-87}$$

Tropical cyclones are noted for their mobility and it can be seen from Fig. 3-38, for example, that where a fetch follows waves already generated these will be enhanced still further. It is for this reason that waves emerging from the periphery of a tropical cyclone will vary greatly in height and length. The largest waves will occur in the right rear quadrant of cyclones in the Northern Hemisphere and in the left rear quadrant in the Southern Hemisphere. The effective maximum velocity in the direction of cyclone motion, when its speed is V_F in knots, is given by [56] as:

$$U_R = 0.865 \, U_{max} + 0.5 \, V_F \tag{3-88}$$

The characteristics of the largest waves within the cyclone are then given by [56]:

$$H_{1/3} = 16.5 \exp \frac{R \Delta p}{100} (1 + 0.208 \, V_F / U_R^{1/2}) \tag{3-89}$$

$$T_{1/3} = 8.6 \exp \frac{R \Delta p}{100} (1 + 0.104 \, V_F / U_R^{1/2}) \tag{3-90}$$

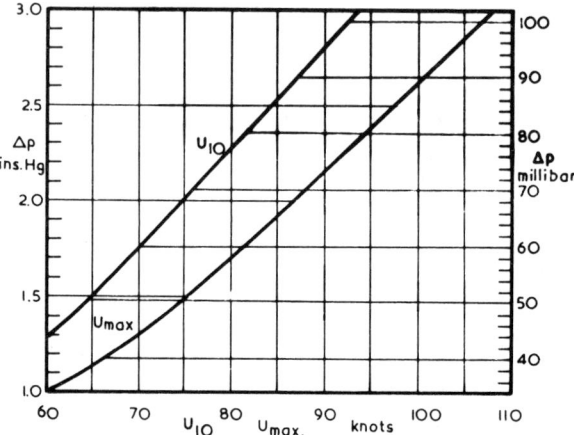

Fig. 3-40. Values of U_{max} for Δp in Hg inches or millibars.

WAVES FROM TROPICAL CYCLONES

where Δp is in inches of mercury, R is in NM and both V_F and U_R are in knots, to give $H_{1/3}$ in ft. and $T_{1/3}$ in sec.

When $V_F = 0$, the second term in the brackets can be omitted and when V_F equals the wave group velocity of the largest waves (V_{cr}), values of $H_{1/3}$ and $T_{1/3}$ should be greatest. Theoretically $V_{cr} = 1.52 T_{1/3}$ (knots) but measurements made in a number of hurricanes [56] indicate $V_{cr} = 1.9 T_{1/3}$. Under these critical conditions the wave height and period are given by [56]:

$$(H_{1/3})_{cr} = 25.8 \exp(R\Delta p/100) \tag{3-91}$$

and:

$$(T_{1/3})_{cr} = 10.7 \exp(R\Delta p/200) \tag{3-92}$$

Assuming that the difference between V_F and V_{cr} has similar effects on $H_{1/3}$ and $T_{1/3}$, whether V_F is greater or less than V_{cr}, curves as in Fig. 3-41 can be used to determine $H_{1/3}$ and $T_{1/3}$. As noted previously (Table 3-VII) $T_{max} = T_{1/3}/0.929$. To aid in the calculation of V_{cr}, a scale is inserted in Fig. 3-41 ($= 1.9 T_{1/3}$).

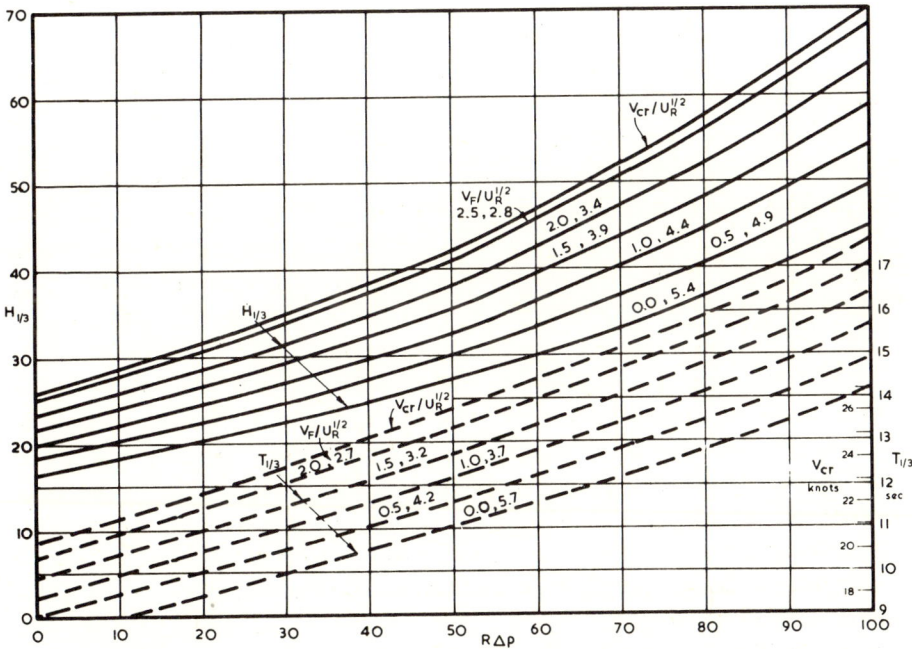

Fig. 3-41. Heights and periods of waves generated in tropical cyclones.

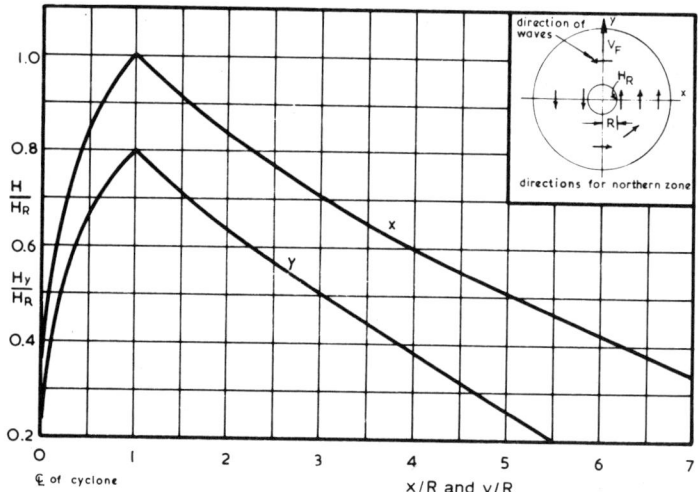

Fig. 3-42. Radial variation of wave height.

The height and period above refers to waves at the periphery of the eye travelling in the direction of the centre itself (see inset of Fig. 3-42). The reduction in height along the radius normal to this direction is given by the x curve in the figure, in terms of H_x/H_R. Also shown is the distribution along the radii aligned with the cyclone path (curve y); even at the eye periphery the height is less than H_R due to the absence of V_F in the magnitude of U_R. The periods for these waves are given by eq. 3-84.

The empirical relationships presented above require verification for the large variety of cyclonic conditions existing in tropical areas. Research is required into such topics as: influence of interaction of waves from many directions, the effects of relatively shallow water on such wave generation, influence of fetch widths on the height and period of waves, and the attenuation of waves through breaking under conditions of high wind velocity. The reader is referred to the work of Ijima et al. [60] and also to the recent work of Bretschneider [64].

The coastal engineer is advised to collect as much data as possible on tropical cyclones in seas adjoining his coast. This should include cyclonicity data which indicate the general path followed by cyclones, although it must be realized that the path of any particular storm cannot be predicted more than an hour or so ahead of its occurrence. Values of radius R, depression Δp and speed V_F should be noted throughout the life history of a tropical cyclone. A frequency curve for cyclone intensities $(R.\Delta p)$ should aid in the calculation of wave characteristics which have a known statistical repetition period. Where possible any wave records taken on a coast or aboard vessels caught in cyclones should be kept, and if possible related to the factors noted above.

EXAMPLES

1
A 12-min wave record has been analyzed and the root mean square amplitude was found to be 3.2 ft. Compute the values of $H_{1/3}$, $H_{1/10}$ for swell and storm wave conditions, respectively.

a_{rms} = 3.2 ft.
for swell conditions (eq. 3-5): $H_{1/3} = 2.83\sqrt{2}a_{rms}$ = 12.8 ft.
$H_{1/10} = 3.60\sqrt{2}a_{rms}$ = 16.3 ft.
for storm conditions (eq. 3-6): $H_{1/3} = 2.95\sqrt{2}a_{rms}$ = 13.3 ft.
$H_{1/10} = 4.0\sqrt{2}a_{rms}$ = 18.1 ft.

2
A wind, whose velocity at 25 m height above the sea is 35 knots, blows across 400 NM of ocean for 18 h. Determine $H_{1/3}$ and $T_{1/3}$ by the SMB relationships provided.

Given U_{25} = 35 knots.
From Fig. 3-8: $U_{19.5}/U_{25}$ = 0.975 and $U_{19.5}/U_{10}$ = 1.06
hence $U_{10} = U_{19.5}/1.06 = 35 \times 0.975/1.06$ = 32.2 knots = 62 ft./sec.

$gF/U^2 = 32.2 \times 400 \times 6080/62^2 = 20,400$
$gt/U = 32 \times 18 \times 3600/62 = 33,700$
from Fig. 3-3: $gH_{1/3}/U^2$ = 0.2 for fetch control = 0.195 for duration control
$gT_{1/3}/2\pi U$ = 1.1 for fetch control = 0.62 for duration control.
Using the smaller value for duration control:
$H_{1/3} = 0.195 \times 62^2/32.2$ = 23.2 ft.
$T_{1/3} = 0.62 \times 2\pi \times 62/32.2$ = 7.5 sec
To check from Fig. 3-3 $t_{min}U/F$ = 0.004
so that t_{min} = (0.004 × 400 × 6080)/62 = 157 h which confirms duration control.
To check from Fig. 3-17, for U_{10} = 32 knots, $H_{1/3}$ = 22 ft.

3
From Fig. 3-10 to 3-13 compare FAS values of $S_{H^2}(f)$ and $S_{H^2}(T)$ for $U_{19.5}$ = 40 knots at a frequency of 0.075 c.p.s. and a period of 12 sec, respectively. From the areas under the curves determine $H_{1/3}$ and compare with values in Table 3-VII.

From Fig. 3-10, f = 0.075 gives $S_{H^2}(f)$ = 100 m²/sec
from Fig. 3-12, $U_{19.5}f/g = 40 \frac{1855}{3600} 0.075/9.81 = 0.157$,
which gives $S_{H^2}(f)g^3 10^2/U_{19.5}^2$ = 2.5.
Likewise $C/U = g/2\pi U_{19.5}f$ = 1.01 giving 2.5 as above
so that $S_{H^2}(f) = 2.5(40\frac{1855}{3600})^5/9.81^3 10^2$ = 100.5 m²/sec.
From Fig. 3-11, T = 12 gives $S_{H^2}(T)$ = 5.6 ft²/sec.
from Fig. 3-12, $gT/2\pi U_{19.5} = 32.2 \times 12/2\pi \, 40 \frac{6080}{3600} = 0.91$
which gives $S_{H^2}(T)g \, 10^4/U_{19.5}^3$ = 5.9
so that $S_{H^2}(T) = 5.9(40\frac{6080}{3600})^3/32.2 \times 10^4$ = 5.65 ft.²/sec.
From Fig. 3-10 area under 40-knot curve equals approximately: $110(0.13-0.04)/2 = (¼H_{1/3})^2$
hence $H_{1/3}$ = 8.9 m, c.f. 29.6 ft. = 9.03 m in Table 3-VII
From Fig. 3-11 area under 40-knot curve equals approximately:
$6.3(22-4.5)/2 = (\frac{1}{4}H_{1/3})^2$
hence $H_{1/3}$ = 29.7 ft., c.f. 29.6 ft. in Table 3-VII.

4

Using the PM relationships, determine $H_{1/3}$ and T_{max} for $U_{19.5} = 15$ m/sec, if FAS conditions are known to exist.

$H_{1/3} = 2\, U_{19.5}^2 (\alpha/\beta)^{1/2}/g$ (eq. 3-41) where for FAS conditions $\alpha = 8.10 \cdot 10^{-3}$ and $\beta = 0.74$ so that $H_{1/3} = 215^2 (\frac{8.10}{0.74 \cdot 10^3})^{1/2}/9.81 = 4.81$ m

$T_{max} = 2\pi U_{19.5}/g$ (eq. 3-39) = $2\pi\, 15/9.81 = 9.62$ sec.
Cf. Table 3-VII, 15 m/sec = 30 knots, giving $H_{1/3} = 16.6$ ft. = 5.05 m, and $T_{max} = 9.9$ sec.

5

From wave records taken at intervals over a number of days the swell waves recorded were of greatest height when the period was 12.5 sec. What can be stated about the FAS conditions known to produce it and the other records taken?

$T_{max} = 12.5$ sec; since $T_{max} = U_{19.5}/3$ (eq. 3-39) the 19.5-m high wind velocity in the fetch = 37.5 knots.

The largest period perceptible in the batch of records will be $T_U = 1.62\, T_{max} = 20$ sec. The shortest waves to arrive later will be $T_L = 0.35\, T_{max} = 4.5$ sec. The fetch required for FAS is $F_{FAS} = 3.19\, U_{19.5}^{1.5}$ (eq. 3-56) so that $F_{FAS} = 3.19\, (37.5)^{1.5} = 730$ NM.
The duration required for FAS is $t_{FAS} = 7.95\, U_{19.5}^{0.5}$ (eq. 3-60) so that $t_{FAS} = 7.95(37.5)^{0.5} = 48.6$ h.

Cf. Table 3-VII by interpolation $F_{FAS} = 738$ NM and $t_{FAS} = 48.6$ h.

6

A wind $U_{19.5} = 30$ knots generates a FAS. Determine the height of wave trains at 1-sec intervals across the EDC and determine the resulting wave from the interaction of these.

Since $U_{19.5} = 30$ knots, $T_{max} = 0.33\, U_{19.5} = 10$ sec (eq. 3-39)
$A_{T_{max}} = 0.051\, T_{max}^{3/2}$ (eq. 3-51) = 1.58 ft./sec$^{1/2}$, $A_{T_{max}}^2 = 2.50$
$A'_{T_{max}} = 0.054\, T_{max}^{3/2}$ (eq. 3-54) = 1.68 ft./sec$^{1/2}$, $A'^2_{T_{max}} = 2.81$
$T_L = 0.35 \times 10 = 3.5$ sec, $T_U = 1.62 \times 10 = 16.2$ sec.

TABLE 3-IX

Computations for Example 6

T (sec)	4	5	6	7	8	9	10	11	12	13	14	15	16
$T-T_L$ (sec)	0.5	1.5	2.5	3.5	4.5	5.5	–	–	–	–	–	–	–
T_U-T (sec)	–	–	–	–	–	–	5.2	4.2	3.2	2.2	1.2	0.2	
A_T^2 (ft.²/sec) *	0.22	0.65	1.08	1.52	1.95	2.38	2.50	2.36	1.91	1.45	1.0	0.55	0.09
H_T (ft./sec$^{1/2}$)	0.93	1.61	2.08	2.46	2.79	3.08	3.16	3.06	2.76	2.40	2.0	1.49	0.06

* Employing eq. 3-55, see Fig. 3-13.

Area of histogram = $\Sigma A_T^2 \times 1.0 = 17.66 = (\frac{1}{4}H_{1/3})^2$
so that $H_{1/3} = 16.8$ ft., cf. Table 3-VII, $H_{1/3} = 16.6$ ft.

7

For a FAS produced by $U_{19.5} = 30$ knots, check the time for a wave group with period T_{max} at the end of the fetch to pass through the length of the fetch. Also compute the time for a batch of waves that represent T_{max} at all sections of the fetch. Use Fig. 3-16 and add the values to read for each 10% of F_{FAS}.

EXAMPLES

T_{max} = 9.9 sec, C_g = 1.52, T_{max} = 15.06 knots, t = 526/15.06 = 35 h.
In Fig. 3-16 we have $C_{T_{max}}/U_{19.5}$ versus F/F_{FAS} from which Table 3-X results.

TABLE 3-X

Computations for Example 7

F/F_{FAS}	0.1	0.2	0.3	0.4	0.5	0.6	0.7	0.8	0.9	1.0
$C_{T_{max}}/U_{19.5}$	0.65	0.75	0.83	0.87	0.91	0.935	0.955	0.97	0.985	1.0
t	5.5	5.11	4.53	4.21	4.02	3.89	3.81	3.72	3.65	3.61

$$t_T = \Sigma t = \Sigma \frac{F_{FAS}/10}{C_{T_{max}}/2} \text{ where } F_{FAS} = 3.19\, U_{19.5}^{1.5} = 536 \text{ NM}$$

and $C_{T_{max}}$ interpolated = 42.05 h which is a little less than t_{FAS} = 43.5 h from Table 3-VII, since the time taken over the first 5% of the F_{FAS} will be slower than the 5.5 computed.

8
A wind which has a velocity of 30 knots at 19.5 m from the sea surface generates a FAS. Compare the heights and periods of the waves as computed by the SMB, PNJ, PM and Darbyshire forecasting relationships.

From Fig. 3-7, $U_{19.5}/U_{7.5}$ = 1.13, and $U_{19.5}/U_{10}$ = 1.085.
Using curves in Fig. 3-17 for $U_{7.5}$ and the knowledge that $H \alpha U^2$ for all relationships except the PNJ, in which case $H \alpha U^{2.5}$, the $H_{1/3}$ values in Table 3-XI below result.
Values of T_{max} can be used as provided in the original references or from the general relationship $T_{max} = 2.43\, H_{1/3}^{1/2}$ (eq. 3-43). For the former we have:
SMB: from Fig. 3-3 for FAS $gT_{1/3}/2\pi U_{10} = 1.37$ (known extreme [15]), so that:

$$T_{max} = \frac{1.37 \cdot 2\pi \times U_{19.5} \times 6080}{0.929\, g \times 1.085 \times 3600} = 13.5 \text{ sec}$$

PNJ: from Ref. [9] $T_{max} = U_{7.5}/2.476$ $\quad T_{max} = \dfrac{U_{19.5}}{1.13 \times 2.476} = 10.7$ sec

PM: from eq. 3-39: $\quad T_{max} = 0.33\, U_{19.5} = 9.9$ sec

Darbyshire: from Ref. [32] $T_{1/3} = 1.5\, U_{50}^{1/2}$, so that: $T_{max} = \dfrac{1.5\, U_{19.5}^{1/2}}{0.929 \times 0.89^{1/2}} = 9.4$ sec

TABLE 3-XI

Computations for Example 8

	SMB	PNJ	PM	Darbyshire
$H_{1/3} - U_{7.5}$	20	21	21	18
$H_{1/3} - U_{19.5}$	15.7	15.6	16.6	14.1
T_{max}	13.5	10.7	9.9	9.4
$T_{max} = 2.43\, H_{1/3}^{1/2}$	9.6	9.6	9.9	9.1

Thus these relationships give very close results in spite of the varied modes of recording and analyzing wind and wave data.

9

For a FAS generated by $U_{19.5} = 35$ knots, compute the height of the wave components with periods 6, 9, 12 and 14 sec throughout the length of the fetch. Determine also their percentages of the FAS value. What is their optimum steepness?

From Fig. 3-22 the following table can be drawn up (wave heights in ft.):

TABLE 3-XII

Computation for Example 9

F/F_{FAS}	0.1	0.2	0.3	0.4	0.5	0.6	0.7	0.8	0.9	1.0
$2[S_{H^2}(6)]^{1/2}$	2.38	2.38	2.24	2.19	2.14	2.10	2.10	2.10	2.10	2.10
$2[S_{H^2}(9)]^{1/2}$	2.10	2.96	3.22	3.34	3.46	3.46	3.46	3.46	3.46	3.46
$2[S_{H^2}(12)]^{1/2}$	0.45	1.79	2.61	3.04	3.34	3.52	3.68	3.79	3.9	3.97
$2[S_{H^2}(14)]^{1/2}$	–	0.63	1.41	2.0	2.37	2.76	2.93	3.16	3.32	3.43
% FAS (6)	113	113	107	104	102	100	100	100	100	100
% FAS (9)	61	85	93	96	100	100	100	100	100	100
% FAS (12)	11	45	66	77	84	89	93	95	98	100
% FAS (14)	–	18	41	58	69	80	85	92	97	100

From this table it may be concluded that:

(a) the components with smaller period reach maturity more quickly than do the larger ones;

(b) the smaller-period components quickly reach a peak and then decrease in height slowly to their FAS values;

(c) the longer-period components do not become apparent until some distance along the fetch, e.g., 14-sec wave reaches 1/2 of its FAS value about $F_{FAS}/3$;

(d) there is some medium-period train which contains the most energy;

(e) optimum steepness = $2[S_{H^2}(T)]^{1/2}/5.12\,T^2$, see Table 3-XIII.

TABLE 3-XIII

Results for Example 9

T (sec)	6	9	12	14
$H_{opt.}$ (ft.)	2.38	3.46	3.97	3.43
H/L_o	0.013	0.0083	0.0054	0.0034

10

A wave record within a fetch indicates $H_{1/3} = 30$ ft. and $T_{max} = 12$ sec. Determine the wind conditions that produced such a result.

From eq. 3-72 $T_{max}/H_{1/3}^{1/2} = r = 12/30^{1/3} = 2.2$. Table 3-VIII or Fig. 3-7 indicates $F/F_{FAS} =$ 15%, and hence from eq. 3-70 $H_{1/3}/U_{19.5}^2 = q = 0.0109$ from Table 3-VIII or Fig. 3-7, so that $U_{19.5} = 52.5$ knots.

Similarly, from eq. 3-68 $T_{max}/U_{19.5} = n = 0.231$ giving $U_{19.5} = 52.5$ knots. Thus a 52.5-knot wind blowing for $0.15 \times 1210 = 180$ NM or for $0.26 \times 57.8 = 15$ h could produce this sea.

EXAMPLES

11

Compare the growth of 10-sec waves in FAS fetches where $U_{19.5}$ = 20, 30 and 40 knots, respectively.

To employ Fig. 3-24 values of $C/U_{19.5}$ are determined, so that:

$$\frac{C}{U_{19.5}} = \frac{g}{2\pi} \frac{10 \times 3600}{U_{19.5} \, 6080} = 1.52, \ 1.01 \text{ and } 0.76, \text{ respectively.}$$

At each of these ratios, for different percentages of F_{FAS}, values of $\dfrac{S_H{}^2(10) g \, 10^4}{(U_{19.5} \, 6080/3600)^3}$ can be read, from which $2[S_H{}^2(10)]^{1/2}$ gives the height of the component.

TABLE 3-XIV

Computations for Example 11

$U_{19.5}$	F/F_{FAS}	0.1	0.2	0.3	0.4	0.5	0.6	0.7	0.8	0.9	1.0
20	$2[S_H{}^2(10)]^{1/2}$	–	–	–	0.15	0.22	0.31	0.44	0.54	0.62	0.69
30	$2[S_H{}^2(10)]^{1/2}$	0.18	1.66	2.22	2.56	2.75	2.89	3.00	3.06	3.11	3.15
40	$2[S_H{}^2(10)]^{1/2}$	2.70	3.66	3.96	4.05	4.10	4.10	4.10	4.10	4.10	4.10

From Table 3-XIV it is seen that the generation of a 10-sec wave component is greatly dependent upon the wind velocity generating it. At low velocities it is slow to build up and at the FAS condition is still small in height. At greater velocities it attains its maximum height quickly along the length, e.g., in a 40-knot wind the 10-sec component is at its optimum half way along the F_{FAS}. In terms of distance the height 3.15 ft. is reached at F_{FAS} for 30-knot wind (= 526 NM, Table 3-VII) but is reached at 0.15 F_{FAS} for 40-knot wind (= 0.15 × 806 = 121 NM).

12

An exposed section of coast at latitude 35° S is known to suffer from extra-tropical cyclones, the magnitudes of which have not been measured accurately. Determine the heights and periods of deep-water waves for the design of structures on the coast to withstand the onslaught of the worst of these.

From Fig. 3-28 the average maximum $H_{1/3}$ for the Atlantic Ocean is 38 ft. If the coast under study is known to not have such intense storms as the Atlantic, a more modest value, as based on Table 3-V, might be used of 33 ft.

The optimum value of T_{max} as per Table 3-V is 13 sec, which is also applicable to the data presented in Fig. 3-28.

The $H_{1/3}$ so derived must then be modified for refraction, amplification, breaking, etc. as the waves propagate across the continental shelf to the site of the structure. When the $H_{1/3}$ value close in shore has been obtained, ratios should be applied to determine $H_{1/10}$ or H_{max} in 1000 waves. The same T_{max} is applicable, although many of the shorter components will have disappeared.

13
Determine $H_{1/3}$ and T_{max} for the following conditions:

$U_{19.5}$ (knots)	F (NM)	t (h)
25	400	50
30	400	50
40	800	40
35	132	9.4
20	30	36

From Table 3-VII determine FAS values of F and t and % FAS where necessary. The proportion of FAS values of $H_{1/3}$ and T_{max} can then be assessed from Table 3-VIII or Fig. 3-30. Results are listed in Table 3-XV.

TABLE 3-XV

Results for Example 13

$U_{19.5}$	F	t	F/F_{FAS}%	t/t_{FAS}%	$(H_{1/3})_{FAS}$	$(T_{max})_{FAS}$	$H_{1/3}$	T_{max}
25	400	50	> 100	> 100	11.5	8.25	11.5	8.25
30	400	50	76	> 100	16.6	9.9	15.8	9.5
40	800	40	≐ 100	80 *	29.6	13.2	28.1	12.7
35	132	9.4	20	20 **	22.6	11.6	11.3	7.3
20	30	36	10.5	= 100	7.4	6.6	3.7	4.2

* Equivalent to F/F_{FAS} = 75% (Table 3-VIII or Fig. 3-32); ** equivalent to F/F_{FAS} = 10% (Fig. 3-32).

14
A 30-knot wind at 19.5 m height generates waves in a fetch which is 400 NM long and 300 NM wide (see Fig. 3-43). Determine $H_{1/3}$ at a point directly downwind of the fetch at a distance 900 NM from the upwind limit of the fetch 60, 80, 100 and 120 h after the commencement of the wind, if its duration is 40 hours.

F_{FAS} = 526 NM (Table 3-VII), F = 400/526 = 76%, t_{FAS} = 43.5 h, t = 40/43.5 = 92%.
Equivalent F/F_{FAS} = 90% (Fig. 3-32) hence use 76% value.
$(H_{1/3})_{FAS}$ = 16.6 ft., $(H_o)_{1/3}$ = 15.8 (Fig. 3-30) $(T_{max})_{FAS}$ = 9.9 sec, T_{max} = 9.5
D/W = 500/300 = 1.67, $(H/H_o)^2_{1/3}$ = 40% (Fig. 3-33)   Equivalent t/t_{FAS} to 76% F_{FAS} = 80% (Fig. 3-32) so that time to reach optimum sea = 0.8 × 43.5 = 35 h.
  In eq. 3-79 and 3-80 at 60 h after storm commencement, t = 60−35 = 25 and $t-t_e$ = 60−35−5 = 20, since the extra time t_e = 40−35 = 5 h.
For this time S/t = 500/25 = 20 and $S/(t-t_e)$ = 25
From Fig. 3-36 T_1 = 13.2 and T'_1 = 16.5 sec, so that T_1/T_{max} = 1.39 and T'_1/T_{max} = 1.74.
Thus, from Fig. 3-36 it is seen that a small triangle of the EDC is present, encompassing the

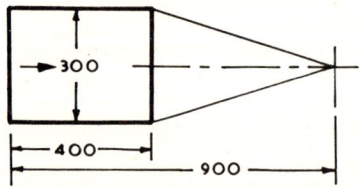

Fig. 3-43. Fetch details for Example 14.

EXAMPLES

band from T_1/T_{max} = 1.39 to 1.62. As seen in the figure the actual EDC indicates that energy is contained in a zone where $T'_1/T_{max} > 1.62$, but for engineering purposes this can be considered negligible. The remaining times are listed in Table 3-XVI.

TABLE 3-XVI

Computations for Example 14

Time	S/t	$\dfrac{S}{t-t_e}$	T_1	T'_1	$\dfrac{T_1}{T_{max}}$	$\dfrac{T'_1}{T_{max}}$	$H^2_{1/3}\%$	$H_{1/3}$	T_{max}
60	20	25	13.2	16.5	1.39	1.62	6– 0 = 6	3.84	14.3 *
80	12.5	14.3	8.2	9.4	0.86	0.99	69–52 = 17	6.45	8.8
100	8.3	9.1	5.45	6.0	0.58	0.63	93–90 = 3	2.72	5.7
120	6.25	6.65	4.11	4.38	0.43	0.46	99–98 = 1	1.58	4.25

* Average of 13.2 and 1.62 × 9.5 = 14.3 sec.

It should be noted in Table 3-XVI that as time progresses the $H_{1/3}$ value decreases, due to the smaller wave components arriving and to the band width present being contracted since from eq. 3-79 and 3-80:

$$T'_1 - T_1 = St_e/t(t-t_e)1.52$$

so that, as t increases the difference becomes smaller.

Values of t when waves exist at this location can be found by dividing 500 NM by the speed in knots of T_U and T_L; these are respectively 21.4 and 99 h, or 56.4 and 134 h after the commencement of the storm, which takes 35 h to reach its optimum wave conditions for the fetch available.

15
A stationary FAS was produced by a 32.2-knot wind at 10 m height in a 212 NM wide fetch (see Fig. 3-44). It is required to know the wave characteristics 3 days after the commencement of the storm at a point which was 500 NM downwind of the fetch and 200 NM to one side of the alignment of the storm, which lasted 2 days.

From Fig. 3-8, U_{10} = 32.2, $U_{19.5}$ = 1.085 × 32.2 = 35 knots.
From Table 3-VII, $(H_{1/3})_{FAS}$ = 22.6 ft., $(T_{max})_{FAS}$ = 11.6 sec, t_{95} = 37.6 h, tan α = 0.364, α = 20°, $D = (500^2 + 182^2)^{1/2}$ = 531, D/W = 531/212 = 2.5.

From Fig. 3-35, θ_1 = 6.42°, θ_2 = 6.99°
$S/t = \dfrac{531}{72-37.6}$ = 15.5 from Fig. 3-37, T_1 = 9.5, T_1/T_{max} = 0.82
$\dfrac{S}{t-t_e} = \dfrac{531}{72-37.6-(48-37.6)}$ = 24 from Fig. 3-37 T'_1 = 16, T'_1/T_{max} = 1.38

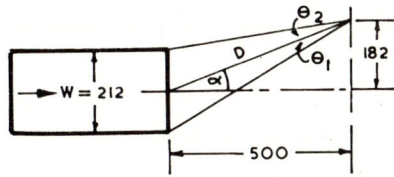

Fig. 3-44. Fetch details for Example 15.

From Fig. 3-36 $H_{1/3}^2\% = 73 - 7 = 66\%$

Since $T_1 = 9.5 > T_{max}/2 = 5.8$, use $\cos^4\theta$ in Fig. 3-34 which for θ values above gives: $H_{1/3}^2\% = 59 - 42 = 17\%$.

Thus $H_{1/3}^2\% = 0.66 \times 0.17 \times 100 = 11.2\%$ so that $H_{1/3} = 22.6 \times 0.334 = 7.5$ ft., and $T_{max} = (9.5 + 16)/2 = 12.8$ sec.

16

Fig. 3-45 provides information recorded at Hongkong during the passage of typhoon Wanda in September 1962. From a chronological plot of the centre it was ascertained that its speed was 10 knots. Compare the heights and periods of the waves generated by assuming the triangular wind distribution as drawn. From these values, determine the equivalent uniform velocity that would produce them, together with the fetch over which it must blow. Compare this with the long-term mean wind velocity in the diagram. (Since the wind velocities were recorded at the

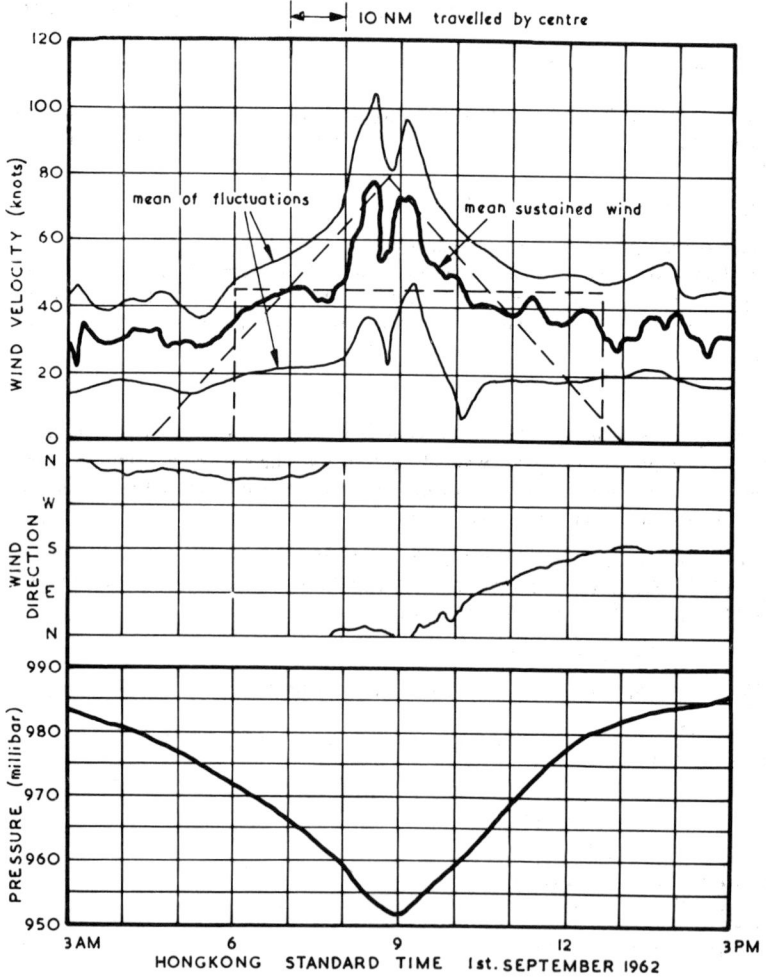

Fig. 3-45. Typhoon data for Example 16.

PROBLEMS

Hongkong Royal Observatory it could be assumed they represent values some 50 m from sea level, thus representing U_{max} in the equations.)

From Fig. 3-45 $\Delta p = (1013 - 952)/34 = 1.8''$.
From eq. 3-86 $U_{max} = 80 \Delta p^{1/2} = 107$ knots (max. at eye edge).
From eq. 3-85 $U_{max} = 60 \Delta p^{1/2} = 80$ knots (mean over fetch width).

This triangular distribution is drawn in Fig. 3-45 and fairly represents the chronological record of wind speed.

From eq. 3-88 $U_R = 0.865 U_{max} + 0.5 V_F = 0.865 \times 80 \times 10/2 = 74$ knots
$V_F/U_R^{1/2} = 10/74^{1/2} = 1.16, R\Delta p = 10 \times 1.8 = 18$
From Fig. 3-40 $H_{1/3} = 25$ ft., $T_{max} = 10.5/0.929 = 11.3$ sec.
From eq. 3-72 $T_{max}/H_{1/3}^{1/2} = 11.3/25^{1/2} = 2.26$.
From Table 3-VIII for $r = 2.26, q = 0.0128$ so that $U_{19.5} = (25/0.0121)^{1/2} = 44$ knots.

This value is shown in the diagram of Fig. 3-45 and fairly represents the mean over 6½ h.

From Table 3-VIII for $r = 2.26, F/F_{FAS} = 22.5\%$.
From eq. 3-56 $F_{FAS} = 3.19 U_{19.5}^{1.5} = 3.19(44)^{3/2} = 942$ NM so that $F = 0.225 \times 940 = 211$ NM.

Since each hour's duration in the figure represents 10 knots travel of the centre it indicates a fetch length of 65 NM. However, the fetch has progressed with the waves so that the generating area is extended and needs only extend for $211 - 65 = 146$ NM back from the final position. This represents 14.6 h of travel for the centre, all of which occurred in this case whilst the centre was on the continental shelf. No measurements of waves are available for verification of the computed values.

PROBLEMS

1
The squares of 96 deviations from a MWL on a 12-min wave record add up to 1584 ft². Find $H_{1/3}$ and $H_{1/10}$ for swell and storm conditions.

2
What are the main differences in approach between that used by Sverdrup and Munk and that used by Pierson, Neumann and James in developing forecasting formulae? What improvements are due to Pierson and Moskowitz?

3
Use the SMB relationships to find $H_{1/3}$ and $T_{1/3}$ when a 28-knot wind at 15 m height is blowing across 500 NM for 30 h. Compare these with values obtained from the PM relationships.

4
A wind at 30 m height from the sea surface is blowing at 19 m/sec. Determining $H_{1/3}$ in metres and T_{max} in seconds when FAS conditions are known to exist.

5
What error is involved in computing $H_{1/3}$ by the PM method when wind velocities accepted as those at 19.5 m height are actually those recorded 7.5 m from the sea surface? Cover the range of wind velocities from 20 to 50 knots for FAS conditions.

6
Determine the height of a 1 sec wide wave band centred on the period of 11 sec when a wind of 40 knots as recorded at 50 m height has developed a FAS.

7

In a certain area of the ocean, measurements taken from synoptic weather charts over a number of years show maximum fetches for various wind speeds as listed below. Determine the $H_{1/3}$ and T_{max} values for optimum storm conditions.

$U_{19.5}$ (knots)	30	35	40	45	50	55
F (NM)	400	350	300	250	200	150

8

The following information is gleaned from wave records taken over a number of days. What can be stated about the FAS producing such a batch of swell?

$H_{1/3}$ (ft.)	2.5	3.5	4.0	3.7	3.1	2.2
T_{max} (sec)	16	14	12	10	8	6

9

Find $H_{1/3}$ and T_{max} by the SMB, PNJ, PM, and Darbyshire relationships for the following three conditions. Indicate the limiting factor in each case.

$U = 30$ knots; $F = 500$ NM; $t = 18$ h
$U = 35$ knots; $F = 200$ NM; $t = 24$ h
$U = 40$ knots; $F = 380$ NM; $t = 27$ h

10

From the PM relationships compute $H_{1/3}$ and T_{max} for the following four fetch conditions:

$U_{10} = 16$ m/sec $F = 660$ km $t = 2$ days
$U_{40} = 52$ ft./sec $F = 600$ miles $t = 29$ h
$U_{7.5} = 41$ mile/h $F = 550$ NM $t = 1.5$ days
$U_{15} = 12$ m/sec $F = 800$ km $t = 48$ h

11

What wind velocity, fetch and duration are required to produce the following waves:

$H_{1/3}$	20	15	70	ft.
T_{max}	10	8	14	sec

12

Draw the equivalent triangular EDC for a FAS when a 30-knot wind ($U_{19.5}$) is blowing. On the same diagram draw the ΔEDC for $F/F_{FAS} = 50\%$ and $t/t_{FAS}, 50\%$. What conclusions can be drawn from this diagram?

13

A 30-knot wind ($U_{19.5}$) blows offshore. Determine the growth pattern at a distance 300 NM from shore of the wave components with periods of 6, 8, 10 and 12 sec from the inception of the wind to the FAS.

14

A 12-sec wave component is built up to 2.5 ft. height in three different fetch conditions in which $U_{19.5} = 30$, 40 and 50 knots. Determine the time for this to occur 500 NM from the upwind end of the fetch in each case.

15

Drilling operations are to be carried out at a distance of 100 NM offshore. These must cease if the significant waves exceed 10 ft. in height. Determine the length of time for which winds of

PROBLEMS

25, 35 and 45 knots can blow offshore before a warning must be issued to the operators. What are the wave periods which the operators should be observing in order to test the approach of this limiting condition themselves? The uniformly sloping bed from shore to the rig has an average depth of 100 ft. Is this sufficient for basing the wave calculations on deep-water conditions?

16
A wind of 35 knots ($U_{19.5}$) raises a FAS. Compute the values of $H_{1/3}$ and T_{max} at the end of the fetch for intervals of 8 h up to 48 h from the start of this wind. Determine also T_U and T_L at each interval. If this wind exists for 48 h, for how long are waves of engineering significance experienced at the downwind end of the fetch? Also, for how long will swell waves from this storm arrive at a point 2,000 NM along the fetch alignment? At this point, how soon after the commencement of swell conditions do the highest waves arrive? How long will this strong swell last?

17
With no meteorological or wave data available, what information could you supply a client wanting to site a port on an oceanic coastline in respect to: (a) peristent swell; (b) storm waves? Consider two cases, both in latitudes 30°–40°, one on the east coast and the other on the west coast of a continent.

18
If you have no other wave data, what height and period would you assume for the persistent swell on an oceanic margin? If a coast borders an enclosed sea, what quick method could you use to derive the height and period of the significant waves for design purposes?

19
Describe, with the aid of sketches, how you can compute waves some distance from a fetch, at a specific time after the commencement of a storm. Note the major factors in the reduction of wave height and give the order of magnitude of these.

20
If the fetch of an extra-tropical cyclone is 1000 NM wide, what is the reduction in wave height at a distance of 1500 NM through angular dispersion alone? Differentiate between the longer and shorter waves and also for points directly down fetch and at 30° to one side of this centre-line.

21
A cyclonic centre contains a fetch with 22.5 knot winds at 10 m height. Its length is 400 NM and its width is 200 NM. The wind blows for 50 h and the waves are directed normal to a coast which is 400, 600 and 800 NM from the end of the fetch. Plot a graph of $H_{1/3}$ and T_{max} against time, starting with the arrival of the first waves, and ending at a time when the T_L band arrives. No allowance need be made for shoaling. For each dispersion distance find the period at which $H_{1/3}$ is optimum and comment on this.

22
A FAS is produced by a 300-NM fetch containing a wind velocity at 19.5 m height of 40 knots. Assess the distribution of wave energy at radii of 300, 600, 900 and 1200 NM from the downwind end of the fetch at angles from its alignment of 0°, 10°, 20°, 30°, 40° and 50°. What conclusions can be drawn from this result?

23

A storm is located 10° from a port and the duration of the 30-knot wind ($U_{19.5}$) is 83.5 h. Determine the maximum $H_{1/3}$ that will arrive at the port if the width of the fetch is 500 NM. What is the T_{max} associated with this maximum wave and at what time after the commencement of the storm will these waves arrive?

24

A FAS is produced by a wind $U_{19.5}$ = 30 knots in a fetch at a distance of 1000 NM from a point directly in line with it at 9.0 a.m. on a certain day and proceeds along this path at 14 knots for 400 NM before being deflected to another direction. Compute the waves experienced at the observation point at 6 hourly intervals after the first waves arrive until the last waves have passed. Omit consideration of waves emanating from the fetch with its second alignment. The width of the fetch is 400 NM. Carry out the same computation for a point offset from the initial fetch centre-line by 350 NM.

25

A ship on the open sea is in direct line and 1000 NM from a fetch whose length is 500 NM and width 300 NM and which has a mean wind velocity at 19.5 m height of 35 knots. It is also 500 NM from another fetch 600 × 400 NM, but is offset from its alignment by 600 NM. The wind in the second fetch has a mean velocity of 40 knots. In both cases the wind has been blowing for 32 h. Determine the average of the highest tenth waves the ship will experience if the winds continue for 12 more hours. What will be the trend if each fetch moves in the direction of its wind during this 12-h period at the speed of the wind?

26

A fetch 600 NM long and 400 NM wide is directed towards the shore some 200 NM distant. The value of $U_{19.5}$ = 30 knots and the fetch itself is travelling parallel to the coast at 20 knots. Calculate the deep-water waves experienced at a point on the coast from the time of arrival of the first waves to that when the fetch has by-passed it.

27

A storm contains a mean wind ($U_{19.5}$) of 30 knots which lasts for several days. It is 400 NM in length and 200 NM wide. A ship is sailing along a path which is parallel to the fetch alignment, but is offset from it by 150 NM. Compute the waves experienced by the ship from the time it is 1,000 NM distance from the downwind end of the fetch until the waves are negligible. The storm is travelling in the direction of its wind at 14 knots whilst the ship maintains a steady 16-knot speed.

28

Cyclonicity data have elucidated the fact that the consistent path of extra-tropical cyclones is at 45° to the coast and cuts it at point A. The optimum characteristics of the fetches are: length 500 NM, width 300 NM, $U_{19.5}$ = 35 knots, and the forward speed of the fetches is 15 knots. It is required to know $H_{1/3}$, T_{max} and mean wave directions throughout the passage of such a cyclone from the open ocean through point A at points B and C which are 300 NM either side of A. (Consider a FAS has been established within the fetch and trace the passage of the storm 1,000 NM from A at 12 hourly intervals. Omit consideration of shoaling and refraction, so that propagation times are based upon deep-water celerities.)

29

What are the main features of a tropical cyclone as they effect the generation of waves? What are the important variables?

30
Calculate $H_{1/3}$ and $T_{1/3}$ at the edge of a hurricane eye whose radius (R) is 20 NM, and also at a radius of $4R$, when the central pressure is 26.4 inches of mercury and the centre is moving at 26 knots.

31
A tropical cyclone contains a central pressure of 2" Hg below normal. Its eye has a diameter of 10 NM and it is approaching the coast at 10 knots. Calculate the maximum deep-water wave height and the period associated with this if the centre strikes the coast in the worst position for a particular port site. What is this critical location for such a cyclone in the Southern Hemisphere?

32
A tropical cyclone is passing along a coast with its centre 200 NM from the shoreline. Compute the highest waves experienced at a port if the following characteristics are known: radius of eye = 5 NM, width of high wind velocity annulus 30 NM, speed of centre 10 knots, central pressure 953 mbar.

What differences occur whether the wind circulation is clockwise or anticlockwise?

REFERENCES

[1] M.S. Longuet-Higgins, 1952. On the statistical distribution of the heights of sea waves. *J. Mar. Res.*, 11: 245–266.
[2] H.V. Sverdrup and W.H. Munk, 1947. Wind, sea and swell theory of relations for forecasting. *U.S. Hydrogr. Office, Publ.*, 601.
[3] Lord Rayleigh, 1880. On the resultant of a large number of vibrations of the same pitch and of arbitrary phase. *Phil. Mag.*, 10: 73–78.
[4] S.O. Rice, 1944–1945. Mathematical analysis of random noise. *Bell System Tech. J.*, 23: 282–332; 24: 46–156.
[5] C. Echart, 1953. The theory of noise in continuous media. *J. Acoust. Soc. Am.*, 25: 195–199.
[6] D.E. Cartwright and M.S. Longuet-Higgins, 1956. The statistical distribution of the maximum of a random function. *Proc. R. Soc., Ser. A*, 230: 212–232.
[7] J. Darbyshire, 1952. The generation of waves by wind. *Proc. R. Soc., Ser. A*, 215: 299–328.
[8] G. Neumann, 1953. On ocean wave spectra and a new method of forecasting wind-generated sea. *Beach Erosion Board, U.S. Army, Tech. Mem.*, 43.
[9] W.J. Pierson Jr., G. Neumann and R.W. James, 1955. Practical methods for observing and forecasting ocean waves. *U.S. Hydrogr. Office, Publ.*, 603:284 pp.
[10] W.J. Pierson, 1964. The interpretation of wave spectra in terms of the wind profile instead of the wind measured at a constant height. *J, Geophys. Res.*, 69 : 5191–5203.
[11] L. Moskowitz, 1964. Estimates of power spectrums for fully developed seas for wind speeds of 20 to 40 knots. *J. Geophys. Res.*, 69: 5161–5179.
[12] W.J. Pierson and L. Moskowitz, 1964. A proposed spectral form for fully developed wind seas based on the similarity theory of S.A. Kitaigorodskii. *J. Geophys. Res.*, 69: 5181–5190.
[13] S.A. Kitaigorodskii, 1961. Application of the theory of similarity to the analysis of wind-generated wave motion as a stochastic process. *Izv. Akad. Nauk S.S.S.R., Ser. Geofiz.*, 1: 105–117. (English transl., 1: 73–80.)
[14] J.W. Johnson, 1950. Relationships between wind and waves, Abbotts Lagoon, California. *Trans. Am. Geophys. Union*, 31: 386–392.

[15] C.L. Bretschneider, 1957. Revisions in wave forecasting deep and shallow water. *Proc. 6th Conf. Coastal Eng., 1957:* 30–67.
[16] M. Hino, 1966. A theory on the fetch graph, the roughness of the sea and the energy transfer between wind and wave. *Coastal Eng. Japan,* 9: 11–26. (Also *Proc. 10th Conf. Coastal Eng.,* 1: 18–37.)
[17] W.H. Munk, 1957. Letter to the Editor. *Trans. Am. Geophys. Union,* 38: 778.
[18] B. Kinsman, 1965. *Wind Waves.* Prentice Hall, Princeton, N.J.
[19] W.J. Pierson, Jr., 1962. *A Unified Mathematical Theory for the Analysis, Propagation, and Refraction of Storm-generated Ocean Surface Waves, 1,2.* College of Eng., Res. Div., N.Y. Univ., New York, N.Y.
[20] A.J. Williams and D.E. Cartwright, 1957. A note on the spectra of wind waves. *Trans. Am. Geophys. Union,* 38: 864–866.
[21] H.V. Roll and G. Fischer, 1956. Eine kritische Bemerkung zum Neumann-Spektrum des Seeganges. *Dtsch. Hydrogr. Z.,* 9: 9–14.
[22] G. Neumann and W.J. Pierson Jr., 1956. *Principles of Physical Oceanography.* Prentice Hall, Princeton, N.J.
[23] L. Moskowitz, W.J. Pierson Jr. and E. Mehr, 1962. Wave spectra estimated from wave records obtained by O.W.S. Weather Explorer and O.W.S. Weather Reporter, 1, 2. *N. Y. Univ., Tech. Rep.,* 3.
[24] P.A. Sheppard, 1958. Transfer across the earth's surface and through the air above. *Q. J. R. Meteorol. Soc.,* 84: 205–224.
[25] J. Wu, 1969. Wind stress and surface roughness at air-sea interface. *J. Geophys. Res.,* 74: 444–455.
[26] M.J. Tucker, 1956. The N.I.O. wave analyser. *Proc. 1st Conf. Coastal Eng. Instr., 1956:* 129–133.
[27] N.F. Barber and F. Ursell, 1948. The generation and propagation of ocean waves and swell, 1. Wave periods and velocities. *Trans. R. Soc., Ser. A,* 240: 527–560.
[28] M.J. Tucker, 1956. A shipborne wave recorder. *Trans. Inst. Nav. Archit.,* 98: 236–250.
[29] J. Darbyshire, 1955. An investigation of storm waves in the North Atlantic Ocean. *Proc. R. Soc., Ser. A,* 230: 299–328.
[30] J. Darbyshire, 1959. The spectra of coastal waves. *Dtsch. Hydrogr. Z.,* 12: 153–167.
[31] J. Darbyshire, 1956. An investigation into the generation of waves when the fetch of the wind is less than 100 miles. *Q. J. R. Meteorol. Soc.,* 82: 461–468.
[32] J. Darbyshire, 1963. The one-dimensional wave spectrum in the Atlantic Ocean and in coastal waters. *Proc. Conf. Ocean Wave Spectra, 1963:* 27–39.
[33] M.S. Longuet-Higgins, 1963. Discussion of Ref. 32. *Proc. Conf. Ocean Wave Spectra, 1963:* 33.
[34] C.L. Bretschneider, 1966. Wave generation by wind, deep and shallow water. In: A.T. Ippen (Editor), *Estuarine and Coastal Hydrodynamics.* McGraw-Hill, New York, N.Y., pp. 133–196.
[35] O.M. Phillips, 1963. Discussion of Ref. 32. *Proc. Conf. Ocean Wave Spectra, 1963:* 33–38.
[36] C.L. Bretschneider, 1963. A one dimensional gravity wave spectrum. *Proc. Conf. Ocean Wave Spectra, 1963:* 41–56.
[37] R. Dorrestein, 1963. Discussion on Ref. 36. *Proc. Conf. Ocean Wave Spectra, 1963:* 57.
[38] T. Inoue, 1967. On the growth of the spectrum of a wind-generated sea according to a modified Miles-Phillips mechanism and its application to wave forecasting. *N. Y. Univ., Phys. Sci. Lab., Rep.,* TR-67-5: 74 pp.
[39] J. Darbyshire, 1959. A further investigation of wind-generated waves. *Dtsch. Hydrogr. Z.,* 12: 1–13.
[40] T.P. Barnett and J.C. Wilkerson, 1967. On the generation of ocean wind waves as inferred from airborne radar measurements of fetch-limited spectra. *J. Mar. Res.,* 25: 292–328.
[41] A.J. Sutherland, 1968. Growth of spectral components in a wind-generated wave train. *J. Fluid Mech.,* 33: 545–560.

REFERENCES

[42] O.M. Phillips, 1957. On the generation of waves by turbulent wind. *J. Fluid Mech.*, 2: 417–445. (See also 4: 426–434; 9: 193–217.)

[43] J.W. Miles, 1957. On the generation of surface waves by shear flows. *J. Fluid Mech.*, 3: 185–204. (See also 6: 568–582; 7: 469–478; 13: 433–448.)

[44] T.P. Barnett and A.J. Sutherland, 1968. A note on an overshoot effect in wind generated waves. *J. Geophys. Res.*, 73: 6879–6885.

[45] P.J. Unna, 1947. Sea waves. *Nature (Lond.)*, 159: 239–242.

[46] M.S. Longuet-Higgins and R.W. Stewart, 1960. Changes on the form of short gravity waves on long waves and tidal currents. *J. Fluid Mech.*, 8: 565–583.

[47] H. Walden, 1963. Comparison of one-dimensional wave spectra recorded in the German Bight with various theoretical spectra. *Proc. Conf. Ocean Wave Spectra, 1963*: 67–81.

[48] R. Silverster, 1963. Design waves for littoral drift models. *Proc. A.S.C.E., Waterways Harbors Div.*, 89 (WW3): 37–47.

[49] J.R. Scott, 1968. Some average sea spectra. *Q. Trans. R. Inst. Nav. Archit.*, 110: 233–239.

[50] M. Darbyshire, 1961. A method of calibration of shipborne wave recorders. *Dtsch. Hydrogr. Z.*, 14: 56–63.

[51] L. Draper and E.M. Squire, 1967. Waves at ocean weather ship station India. *Trans. R. Inst. Nav. Archit.*, 109: 85–93.

[52] M.J. Tucker, 1963. Analysis of records of sea waves. *Proc. I.C.E.*, 26: 305–316.

[53] R. Mayençon, 1969. Etude statistique des observations de vagues. *Cah. Oceanogr.*, 21: 487–501.

[54] M.S. Longuet-Higgins, D.E. Cartwright and N.D. Smith, 1963. Observations of the directional spectrum of sea waves using the motions of a floating buoy. *Proc. Conf. Ocean Wave Spectra, 1963*: 111–132.

[55] F.E. Snodgrass et al., 1966. Propagation of ocean swell across the Pacific. *Phil. Trans. R. Soc. Lond.*, A259: 431–497.

[56] C.L. Bretschneider, 1957. Hurricane design wave practice. *Proc. A.S.C.E., Waterways Harbors Div.*, 83 (WW2): Paper 1238.

[57] C. Holliday, 1969. On the maximum sustained winds occurring in Atlantic hurricanes. *U.S. Dept. Comm., Tech. Mem.*, WBTM-SR-45.

[58] R.H. Kraft, 1961. The hurricanes central pressure and highest wind. *Mariners Weather Log*, 5(5).

[59] H.C.S. Thom, 1971. Asymptotic extreme value distributions of wave heights in the open ocean. *J. Mar. Res.*, 29: 19–27.

[60] T. Ijima, T. Soejima and T. Matsuo, 1968. Ocean wave distribution in typhoon area. *Coastal Eng. Japan*, 11: 29–42.

[61] S. Syono, 1963. Structure of typhoons. *Proc. Int. Regional Seminar Trop. Cyclones, 1962, Tokyo, Japan Meteorol. Agency Tech. Rep.*, 21: 121–131.

[62] R. Silvester and S. Vongvisessomjai, 1970. Energy distribution curves of developing and fully arisen seas. *J. Hydr. Res.*, 8: 493–521.

[63] R. Silvester and S. Vongvisessomjai, 1971. Computation of storm waves and swell. *Proc. Inst. Civil Eng.*, 48: 259–283.

[64] C.L. Bretschneider, 1972. Revisions to hurricane design wave practices. *Proc. 13th Conf. Coast. Eng.* (in press).

Chapter 4

THEORY OF WAVES

In the previous chapters the complexity of sea waves within a fetch have been studied, where many wave trains are travelling in a number of directions. To apply the theories of water-particle motion and other phenomena to this wave propagation, drastically simple models must be used. The following general assumptions are made for this purpose:

(*a*) the wave is a single component of the spectrum;

(*b*) the wave is part of a train that has constant height and period;

(*c*) the height is considered to be very small compared to the length, infinitesimally small in fact;

(*d*) the wave form is sinusoidal, with crest and trough of equal vertical distances from the mean water level (exceptions to this will be noted as they arise);

(*e*) the wave is long-crested and is propagating in a straight line (three-dimensional aspects of wave theory will be considered when appropriate);

(*f*) the depth of water is constant or changes very slowly;

(*g*) the water is considered to be non-viscous, except in special cases to be referred to later.

The main concerns of the coastal engineer in respect to wave theory are the motions of water particles (amplitudes, velocities and accelerations), pressure fluctuations, wave energy and profiles of the water surface. Equations for these take the form of exponential series, the terms of which grow successively smaller as wave steepness to increasing powers occurs. Theory which incorporates only the first term of the series is called *linear,* and hence is the simplest to treat. Fortunately most engineering problems can be handled with sufficient accuracy by the linear theory. Other forms commonly used by the mathematician are second, third- and fifth-order forms, some of which have been shown to be less accurate than the first-order when compared to certain theoretical criteria and flume tests. It is only when waves travel in the shallower zone, and wave steepness increases, that the second- and higher-order theories are required. Whilst the various analyses available for wave phenomena are considered a little imperfect by the mathematician they are well within the accuracy of the wave data presently available, or ever likely to be available, from such complex situations as natural cyclones.

The sinusoidal form assumed in wave theory can be expressed as:

$$a = A \cos t \qquad (4\text{-}1)$$

where a = amplitude of motion (half the height), t = time, A = a constant giving the scale of motion.

This can be simplified as in Fig. 4-1 by using the ratio $a/A = \cos\alpha$ where α is the circular measure of t. As seen in the figure, point P travels around the circle ($x^2 + y^2 = 1$) from which abscissa and ordinate lengths from the axes through the centre are given by $A\sin\alpha$ and $A\cos\alpha$ respectively, where α is the angle turned through by the radius at P from the positive y axis. Also illustrated is the variation in $\cos\alpha$ ($=a/A$) with time measured in circular measure, so that for a complete wave period T the angle $\alpha = 2\pi$ radians. During this period T, $\cos\alpha$ varies from $+1.0$ to zero at $\alpha = \pi/2$, to -1.0 at $\alpha = \pi$, to zero at $\alpha = 3\pi/2$ and $+1.0$ at $\alpha = 2\pi$.

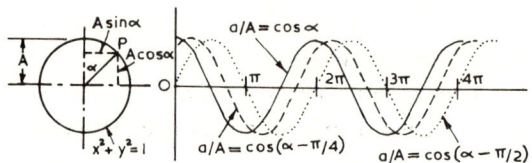

Fig. 4-1. Definition sketch of sinusoidal functions.

As seen in Fig. 4-1 the wave form can be displaced to the right by using the form:

$$a/A = \cos(\alpha \pm \epsilon) \tag{4-2}$$

where ϵ is termed the phase shift of the wave (not to be confused with ϵ expressing the width of the spectrum). Curves for ($\alpha-\pi/4$) and ($\alpha-\pi/2$) are illustrated, the latter phase shift being expressable as:

$$a/A = \cos(\alpha-\pi/2) = \sin\alpha \tag{4-3}$$

Thus, if the position of the wave form in respect to any fixed point is of no consequence, the form $\cos\alpha$ and $\sin\alpha$ can be used indifferently, hence the term *sinusoidal* wave.

Besides the circular functions of $x^2 + y^2 = 1$ in wave theory, others arise such as hyperbolic, parabolic, elliptical and cylindrical functions. The most important of these is that of the rectangular hyperbole $x^2 - y^2 = 1$. A comparison with the circular functions can be made in Fig. 4-2, where it is seen that the sinh, cosh and tanh functions (equivalent to the sin, cos and tan) do not trace out the form of the wave on the water surface.

Two parameters which arise frequently in mathematical treatises on waves are:

$$\sigma = 2\pi/T \tag{4-4}$$

known as the *radian frequency*, and:

$$k = 2\pi/L \tag{4-5}$$

THEORY OF WAVES

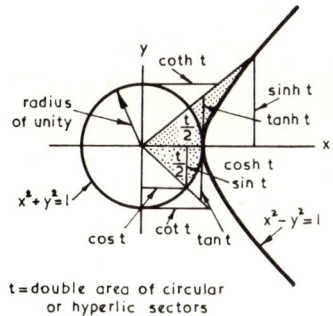

t = double area of circular or hyperlic sectors

FIg. 4-2. Comparison of circular and hyperbolic functions.

known as the *wave number*. Engineers generally prefer to leave their equations in the form of $2\pi/T$ and $2\pi/L$ since both the wave period T and wave length L are dimensions which can be visualized, thus maintaining a scale effect in the mind. They have been employed throughout this tome.

The above ratios do not generally occur alone, rather do they occur as dimensionless fractions such as $2\pi t/T$ (where t is a measure of time throughout a wave cycle), or as $2\pi d/L$ (where d is the depth of water in which the length of the wave is L). The hyperbolic functions previously mentioned arise in all wave phenomena, so that their limiting values as the ratio d/L varies are important to recall. These are listed in Table 4-I in their exponential form.

The ranges and limits of d/L given in Table 4-I denote the wave propagation conditions which are termed; shallow, transitional or deep. It can be envisaged that equations for the transitional depths are likely to be more complex than those for shallow or deep conditions. This is an important zone of ocean wave travel for the coastal engineer, since the bulk of the changes occur in the form of the wave and of its water-particle motions during such passage. Numerical values for the various hyperbolic functions are listed in Appendix I for incremental values of d/L.

TABLE 4-I

Limiting values of hyperbolic functions

d/L	< 0.05	0.05 to 0.5	> 0.5
sinh $2\pi d/L$	$2\pi d/L$	$\frac{1}{2}[\exp(2\pi d/L) - \exp(-2\pi d/L)]$	$\frac{1}{2}\exp(2\pi d/L)$
cosh $2\pi d/L$	1	$\frac{1}{2}[\exp(2\pi d/L) + \exp(-2\pi d/L)]$	$\frac{1}{2}\exp(2\pi d/L)$
tanh $2\pi d/L$	$2\pi d/L$	$\dfrac{\exp(2\pi d/L) - \exp(-2\pi d/L)}{\exp(2\pi d/L) + \exp(-2\pi d/L)}$	1
description	shallow	transitional	deep

PROGRESSIVE WAVE: LINEAR THEORY

A progressive wave, as the name implies, progresses across the ocean, so that successive crests pass a fixed station. This distinguishes it from the *standing wave* whose crests occur at certain fixed points at successive intervals. The theory associated with this simple form, with the restrictions outlined previously, was first developed by Airy and hence the title "Airy wave". It can be shown, by ignoring surface tension effects, that:

$$C \equiv L/T = (gT/2\pi) \tanh 2\pi d/L = [(gL/2\pi) \tanh 2\pi d/L]^{1/2} \qquad (4\text{-}6)$$

or:

$$L = gT^2/2\pi) \tanh 2\pi d/L \qquad (4\text{-}7)$$

where C = wave celerity, L = wave length, T = wave period, d = water depth, g = acceleration due to gravity.

It is seen that, whilst wave celerity C is determined directly by wave period T, it is also influenced by the water depth through the term $\tanh 2\pi d/L$.

As noted already, limits exist for the various hyperbolic functions when $d/L \geqslant 0.5$. The wave characteristics for this deep-water condition will be denoted by a subscript o (o for original, since most waves are generated in deep-water). Symbols not employing this subscript will refer to transitional or shallow-water values.

From Table 4-I it is seen that, as $d/L \geqslant 0.5$, $\tanh 2\pi d/L \to 1.0$, so that, from equation (4-6):

$$C_o = gT/2\pi \quad \text{or} \quad (gL_o/2\pi)^{1/2} \qquad (4\text{-}8)$$

Thus in deep-water the celerity of a wave is dictated solely by its period T. The ratio of $g/2\pi$ varies with the dimensions used, so that:

$$C_o = 5.12\, T \text{ (ft./sec)} = 1.56\, T \text{ (m/sec)} = 3.03\, T \text{ (knots)} \qquad (4\text{-}9)$$

when T is in seconds.

Similarly from eq. 4-7 the deep-water wave length is:

$$L_o = C_o T = gT^2/2\pi = 5.12\, T^2 \text{ (ft.)} = 1.56\, T^2 \text{ (m)} \qquad (4\text{-}10)$$

It is readily seen that:

$$C/C_o = \tanh 2\pi d/L \qquad (4\text{-}11)$$

and also that:

$$L/L_o = \tanh 2\pi d/L \qquad (4\text{-}12)$$

Table 4–II provides some idea of the speeds and lengths of waves in deep water. It can be seen that the longer components of a wave spectrum can move at great

TABLE 4-II

Celerities and lengths of deep-water waves

T (sec)	C_o (ft./sec)	C_o (m.p.h.)	L_o (ft.)	$\tfrac{1}{2}L_o$ (ft.)	$\tfrac{1}{2}L_o$ (fathoms)
4	20	13.6	82	41	6.83
7	36	24.6	251	125	20.9
10	51	34.8	512	256	42.6
13	66.5	45.4	665	332	55.3
16	82	55.9	820	410	68.3

speed, for example, the 10-sec wave at 35 miles/h. The wave length also is large (512 ft. for the same wave train). The normal cargo ship is of this length and the largest bulk carriers are only twice this length. Half of the deep-water wave length is listed $\tfrac{1}{2}L_o$ since this equals approximately the depth at which the wave enters the transitional zone. It is at this depth that waves commence to decelerate from their deep-water speed and suffer commensurate changes in other characteristics. The edges of the continental shelves of the world are around 65 fathoms, so that the longer waves are seen to enter their transition zone well out to sea, some hundreds of miles offshore in some areas.

The deep-water limit of $d/L \doteq d/L_o = 0.5$ is of engineering convenience, but does not infer that celerity and length changes do not occur beyond this point. Observation of the tables in Appendix I indicates that at $d/L_o = 0.5$ (when $d/L = 0.5018$) tanh $2\pi d/L = 0.9964$, and that not until $d/L_o = 0.84$ does tanh $2\pi d/L = 1.000$. The error involved in assuming tanh $2\pi d/L = 1.0$ at $d/L_o = 0.5$ is given by:

$$\frac{C_{o(0.5)}}{C_{o(deep)}} = \frac{0.9964}{1.000} \qquad (4\text{-}13)$$

representing an error of 0.36%, which is negligible compared to the errors involved in forecasting waves.

At the shallow-water limit, as $d/L \to 0$, so tanh $2\pi d/L \to 2\pi d/L$ (Table 4-I), so that:

$$C = (gd)^{1/2} \qquad (4\text{-}14)$$

In this case celerity is dictated only by depth, values of which are listed in Table 4-III.

Comparison with values in Table 4-II indicates how fortunate for mankind that waves are slowed down as they approach the shore. In depths where waves are likely to break, these velocities are in the order of 7 to 10 miles/h.

The shallow-water limit at which the celerity can be approximated by $(gd)^{1/2}$

TABLE 4-III

Wave celerities in shallow water

d (ft.)	C (ft./sec)	C (m.p.h.)	C (m/sec)	d (m)
20	25.3	17.3	7.7	6.1
15	21.9	14.9	6.7	4.6
10	17.9	12.2	5.5	3.0
7	15.0	10.2	4.6	2.1
4	11.3	7.7	3.5	1.2

should be derived on the same basis of accuracy as the deep-water limit, which was found to be 0.36%. In this case:

$$\frac{C_{(\text{shallow})}}{C_{(\text{limit})}} = \frac{[(gL/2\pi)\tanh 2\pi d/L]^{1/2}}{(gd)^{1/2}} = \left(\frac{\tanh 2\pi d/L}{2\pi d/L}\right)^{1/2} = 0.9964$$

or:

$$\frac{\tanh 2\pi d/L}{2\pi d/L} = 0.994 \qquad (4\text{-}15)$$

Relevant ratios involved in eq. 4-15 are listed in Table 4-IV, from which it is seen that the value $d/L_o = 0.004$ should establish this shallow-water limit. However, engineers choose round figures for convenience, which unfortunately do not correspond. For example, Wiegel [1] suggests $d/L = 1/25$ whilst Eagleson and Dean [2] use $d/L = 1/20$. Accepting this, the value of $d/L_o = 1/100$ seems appropriate. It will be seen later that for calculations of wave refraction such simplifying assumptions are not made.

TABLE 4-IV

Determination of d/L for shallow-water limit

1 d/L_o	2 d/L	3 $2\pi d/L$	4 $\tanh 2\pi d/L$	5 4/3
0.015	0.04964	0.3119	0.3022	0.97
0.010	0.04032	0.2533	0.2480	0.98
0.007	0.03362	0.2113	0.2087	0.986
0.004	0.02534	0.1592	0.1579	0.992
0.001	0.01263	0.07935	0.07918	0.999

From eq. 4-12 it can be seen that:

$$\frac{d/L_o}{d/L} = \tanh 2\pi d/L \qquad (4\text{-}16)$$

so that if d/L is known, so also is d/L_o. In any equations, therefore, terms may be expressed in either ratio. It is preferable to work in d/L_o since L_o is readily computed for a wave of specific period. The length L, on the other hand, changes with the depth.

Water-particle motion

There are two methods of viewing the motion of fluid particles. One is to concentrate on a fixed point in space and note the changes that occur with time. Another is to travel with the particle and record the temporal variations. The former gives a picture of the flow field in the form of streamlines and is termed the *Eulerian* presentation, whilst the latter provides the trajectory viewpoint and is termed the *Lagrangian* approach. In the Lagrangian analysis the kinetic variables are considered not to alter with the displacement of the particle from a mean position. Thus velocities and accelerations are obtained by substituting mean depths or horizontal locations of points in the Eulerian equations. At the water surface, where vertical oscillations are optimum, there will be a difference in horizontal velocity at the crest and at the trough in the Eulerian solution, whereas they will be equal when the mean water surface level is substituted to obtain the Lagrangian equation. Hence the Eulerian solution is preferable for velocities and accelerations, whilst the Lagrangian is more readily adapted to amplitudes of orbital motion. As the bed is approached the two sets of equations give very similar results.

The above mentioned concepts are illustrated in Fig. 4-3 for deep, transitional and shallow water-conditions. This figure should be referred to as the following linear equations are examined. The notation to be used is as follows:

u = horizontal velocity, positive in the direction of wave advance
v = vertical velocity, positive in the upward direction
H = crest height, double the amplitude in this assumed sinusoidal wave
T = wave period (sec)
y = depth, positive upwards and zero at the surface (= $-d$ at bed)
d = mean depth of water
L = wave length
x = distance along wave length, origin at wave crest
t = time through wave cycle, origin at wave crest.

Subscripts o are used for wave length in deep water (L_o) and for mean positions of water particles in any depth for the Lagrangian solution (x_o, y_o).

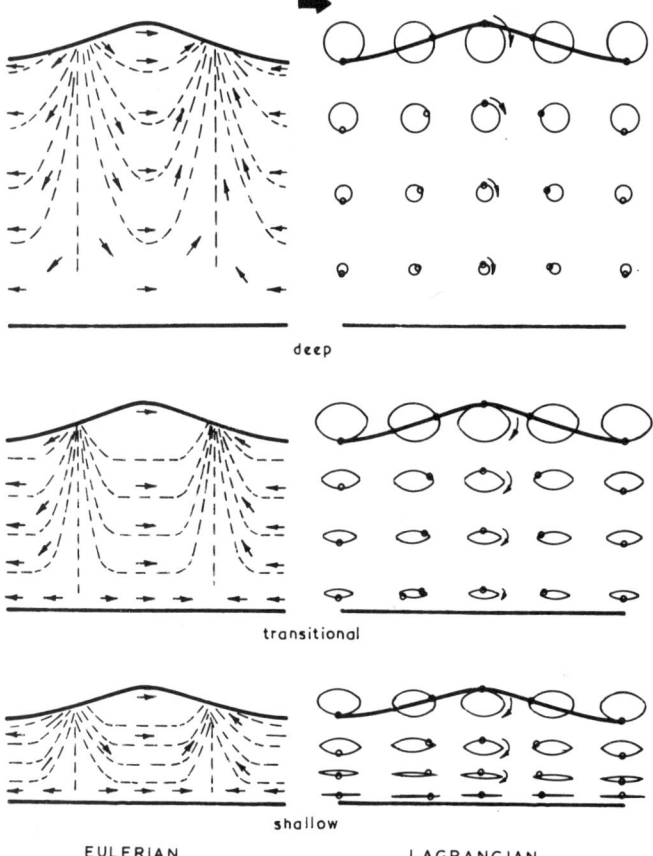

Fig. 4-3. Water-particle motions in progressive waves.

The motions of particles at any point vary with time, or the ratio t/T. These variations are followed by particles at the same depth but located further along the wave length or at various ratios of x/L. Hence in the circular functions of *cos* or *sin* in the equations the ratios x/L and t/T are used alternatively for spatial or temporal variations, respectively. Since Stokes [3] in 1847 put the theory of Airy into its present accepted form and extended it to higher orders, the following first-order equations might be termed Stokes I.

Velocities (Eulerian): transitional depths

$$u = \frac{\pi H}{T} \left[\frac{\cosh 2\pi(y+d)/L}{\sinh 2\pi d/L} \right] \cos 2\pi(x/L - t/T) \tag{4-17}$$

dimension | depth factor | phase

$$v = \frac{\pi H}{T} \left[\frac{\sinh 2\pi(y+d)/L}{\sinh 2\pi d/L} \right] \sin 2\pi(x/L - t/T) \qquad (4\text{-}18)$$

The three elements in eq. 4-17 and 4-18 are to be noted. It is seen that the dimension of the velocities is directly proportional to wave height and inversely proportional to wave period. The depth (y) of the particle from the surface influences magnitude of the velocity as also the depth of water itself. At the bed, where $y = -d$, the vertical velocity becomes zero, as dictated by the physical nature of the motion. The change of velocity with time is given by the phase term which varies between zero and unity. In many problems the coastal engineer will be concerned only with maximum values, when the circular function is assumed as unity. It is advisable to utilize the subscript (max) in such cases in order to highlight the fact that temporal variations have been excluded. It is seen that when the horizontal velocity is maximized the vertical velocity is zero, and vice versa. This is obvious from the Lagrangian presentation in Fig. 4-3 where the orbits are seen to be circular or elliptical.

Velocities (Eulerian): deep-water depths

$$u = \frac{\pi H}{T} \exp(2\pi y/L_o) \left[\cos 2\pi(x/L_o - t/T) \right] \qquad (4\text{-}19)$$

$$v = \frac{\pi H}{T} \exp(2\pi y/L_o) \left[\sin 2\pi(x/L_o - t/T) \right] \qquad (4\text{-}20)$$

Eq. 4-19 and 4-20 are obtained from eq. 4-17 and 4-18 by substituting the limiting values for the hyperbolic functions as in Table 4-I. It is readily seen that:

$$u^2 + v^2 = \frac{\pi^2 H^2}{T^2} \exp(2\pi y/L_o) \qquad (4\text{-}21)$$

which indicates that the tangential velocity of the particle is constant throughout its orbit, which is therefore indicated as circular. The magnitudes of u and v are also similar and are constant at any given depth. They decrease exponentially and when $y/L_o = -1/2$ the resultant velocity $u^2 + v^2 = e^{-\pi} = 0.043$. Thus the velocities at the assumed deep-water limit are about 4% those at the surface.

Velocities (Eulerian): shallow-water depths

$$u = \frac{Hg^{1/2}}{2d^{1/2}} \cos 2\pi(x/L - t/T) \qquad (4\text{-}22)$$

$$v = \frac{H(y+d)}{Td} \sin 2\pi(x/L - t/T) \qquad (4\text{-}23)$$

Here again eq. 4-22 and 4-23 are derived by substituting the limiting values of the hyperbolic functions as in Table 4-I. It is seen that the horizontal velocity (u) is independent of (y) the depth and of period (T), so that u is uniform throughout the depth and determined by the shallow-water celerity. The vertical velocity (v) varies linearly with depth and inversely with T, being zero when $y = -d$ as in the transitional zone.

Displacements (Lagrangian): transitional depths

$$x = -\frac{H}{2}\left[\frac{\cosh 2\pi(y_o+d)/L}{\sinh 2\pi d/L}\right]\sin 2\pi(x_o/L - t/T) \tag{4-24}$$

$$y = \frac{H}{2}\left[\frac{\sinh 2\pi(y_o+d)/L}{\sinh 2\pi d/L}\right]\cos 2\pi(x_o/L - t/T) \tag{4-25}$$

The displacements x and y are either side of a mean position x_o, y_o. Eq. 4-24 and 4-25 can be shown to form an elliptical orbit with the major axis of length $H\left[\dfrac{\cosh 2\pi(y_o+d)/L}{\sinh 2\pi d/L}\right]$ and focal distance $H/\sinh 2\pi d/L$. The values of x and y are influenced by depth y_o and also the depth of the water d.

Displacements (Lagrangian): deep-water depths

$$x = -\frac{H}{2}\exp(2\pi y_o/L_o)\left[\sin 2\pi(x_o/L_o - t/T)\right] \tag{4-26}$$

$$y = \frac{H}{2}\exp(2\pi y_o/L_o)\left[\cos 2\pi(x_o/L_o - t/T)\right] \tag{4-27}$$

The radius of the circular orbit is thus $\frac{1}{2}H\exp(2\pi y_o/L_o)$ which is seen to reduce exponentially with depth as did the uniform velocity.

Displacements (Lagrangian): shallow-water depths

$$x = -\frac{HTg^{1/2}}{4\pi d^{1/2}}\sin 2\pi(x_o/L - t/T) \tag{4-28}$$

$$y = \frac{H(y_o+d)}{2d}\cos 2\pi(x_o/L - t/T) \tag{4-29}$$

In eq. 4-28 it is seen that x varies with T but not y_o, so that amplitude of horizontal motion is uniform throughout the depth and approximates $HL/2\pi d$. The vertical oscillation (y) varies directly with y_o, becoming zero at the bed, thus the elliptical orbits flatten the deeper the water particles.

It is useful to take the hyperbolic terms to one side of the above equations and omit the circular functions. In this way eq. 4-17, for example, can be expressed as:

PROGRESSIVE WAVE: LINEAR THEORY

$$\frac{u_{max}T}{\pi H} = \frac{\cosh 2\pi(y+d)/L}{\sinh 2\pi d/L} = X \tag{4-30}$$

In this way values of X can be graphed against d/L or d/L_o with y/d as an extra variable as in Fig. 4-4. It is then only a minor calculation to derive u_{max} for any given H and T. As indicated in this figure, values of v_{max}, du/dt, dv/dt, x_{max} and y_{max} can be similarly derived from appropriate equations noted in the inset of the figure.

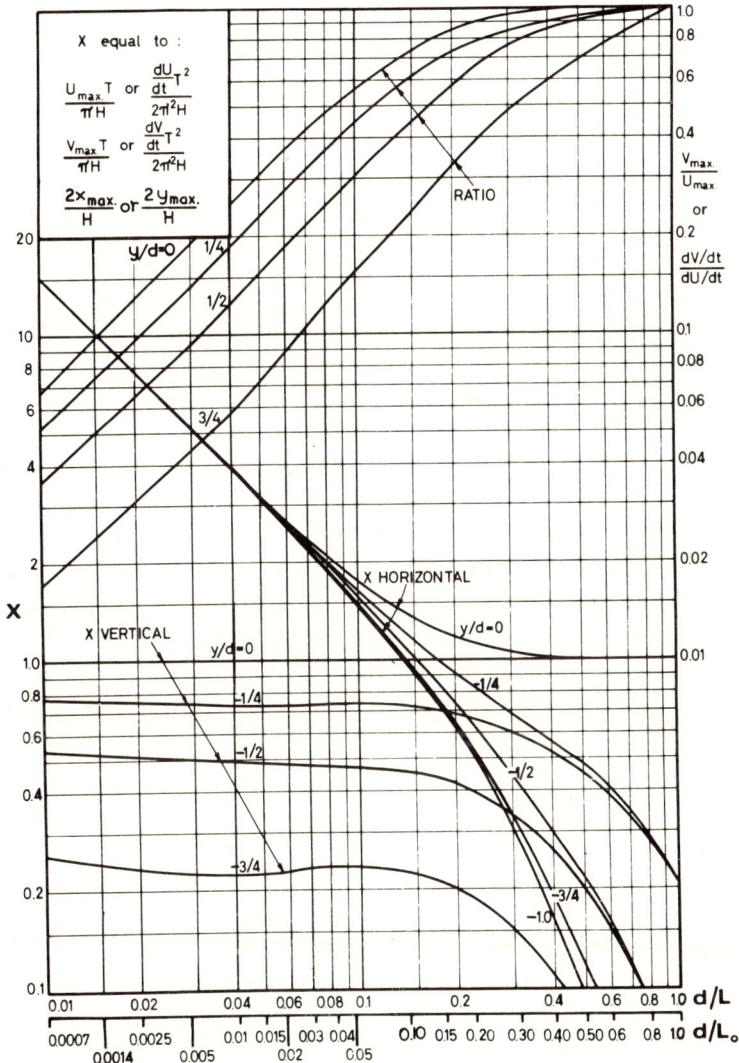

Fig. 4-4. Velocity acceleration and amplitude parameters for water particles in progressive waves — linear theory.

Also included are curves for the ratios v_{max}/u_{max} and $(dv_{max}/dt)/(du_{max}/dt)$ which are useful in the calculation of forces on immersed objects.

By dividing equations for u_{max} by those for x_{max} it can be shown that:

$$u_{max} = 2\pi x_{max}/T \qquad (4\text{-}31)$$

so that when either the amplitude or maximum velocity is known, the other can be computed for any given wave period T.

Wave energy

For the assumption of small-amplitude progressive waves the kinetic and potential energies of this sinusoidal wave form are equal. Integration of this energy over a wave length shows that:

$$E_k = wH^2L/16 \quad \text{and} \quad E_p = wH^2L/16 \qquad (4\text{-}32)$$

so that the total wave energy:

$$E_t = E_k + E_p = wH^2L/8 \qquad (4\text{-}33)$$

per unit length of crest. The energy per unit area of ocean thus becomes:

$$E = E_t/L = wH^2/8 \qquad (4\text{-}34)$$

It will be recalled that in the wave-forecasting formulae energy was computed as (height)^2.

Power in a wave consists of a force times a velocity through any plane normal to wave propagation, which equals (pressure) × (area) × (velocity). The energy transmitted forward is based therefore on the variable pressure and particle velocities throughout the wave cycle. Integration of this power results in:

$$P = \frac{wH^2L}{8T} \cdot \frac{1}{2}\left[1 + \frac{4\pi d/L}{\sinh 4\pi d/L}\right] = \frac{wH^2L}{8T}n \qquad (4\text{-}35)$$

or:

$$P = \frac{wH^2}{8} \cdot \frac{C}{2}\left[1 + \frac{4\pi d/L}{\sinh 4\pi d/L}\right] = \frac{wH^2}{8} C_g \qquad (4\text{-}36)$$

where n is the ratio of group wave velocity to individual wave celerity (i.e., $n = C_g/C$). The RH portion of eq. 4-36, which excludes the energy per unit area of ocean $wH^2/8$, is termed the *wave group velocity* and is an important feature of wave propagation. It is designated by C_g. In deep water the factor $(4\pi d/L)/\sinh 4\pi d/L$ tends to zero since the denominator becomes infinite, so that $C_g = C/2$. Likewise in shallow water this factor tends to unity, making $C_g = C$. Thus, in deep water wave energy is transmitted forwards at half the celerity of the waves, whilst in shallow water it travels at the same speed. The ratio C_g/C is designated as n which varies from 1/2 to 1.

For waves approaching a coast normally, no refraction taking place, the power transmitted forward at one depth may be considered equal to that transmitted further inshore, assuming no losses between. Considering a deep-water and transitional depth for $P_o = P$, then:

$$w H_o^2 C_{go}/8 = w H^2 C_g/8 \tag{4-37}$$

so that:

$$H/H_o = (C_{go}/C_g)^{1/2} = (C_o/2C_g)^{1/2} \tag{4-38}$$

The group wave velocities are a function of d/L, so that the ratio H/H_o can be listed directly with this variable in Appendix I. It is obvious that the wave height increases as the wave comes into shore for this normal approach.

Where two wave trains occur together their energies are additive so that $H = (H_1^2 + H_2^2)^{1/2}$. There is an exception: when two progressive waves of equal period are travelling in the same direction but arrive from different sources. In this case the heights are added directly so that $H = H_1 + H_2$. This can happen on a coast subject to persistent swell from an extensive storm area, for example the "roaring forties". Long waves of similar period could be arriving from almost identical directions even though generated by two different storms. They arrive in the shallow zone travelling at identical speeds and in phase, and are termed "king waves". They are particularly dangerous on a rocky coast where reflection can take place, and have been known to wash many fishermen off rocky ledges. A small calculation will illustrate the problem. Assume two waves each of 1 ft. height in deep water. Addition of the crests, plus amplification due to shoaling, could make the king wave 6 ft. as it approaches a cliff. Reflection will again double the height making a 12 ft. wave at the face. This can readily dislodge the ardent fisherman who perches within a few feet of the water surface.

Consider now a group of waves travelling across calm water. The leading wave must start some motion in the still fluid. This absorbs energy from this wave and in fact absorbs half the energy, the other half proceeding forwards in the deep water. The resulting picture can be gained from Table 4-V where an energy level of 128 is continuously supplied by a mechanical wave generator.

After one wave period ($N = 1$) a wave crest one wave length from the blade contains half the original energy 64. At the end of another period T (i.e., $N = 2$) the crest at one wave length from the blade retains half the original energy ($= 32$) and transmits the other 32 forwards to two wave lengths distance. The energy level at $L = 1$ is therefore 64 received from the blade plus 32 remaining ($= 96$). After $3T$ energy at the same location ($L = 1$) is $64 + 48 = 112$. Observation of the energies after seven wave periods, or the percentage values, shows that although the leading wave has travelled 7 wave lengths the bulk of the energy is contained in the first

TABLE 4-V

Model of wave group velocity

N(T)	L = 0	1	2	3	4	5	6	7	8
1	128	64							
2	128	96	32						
3	128	112	64	16					
4	128	120	88	40	8				
5	128	124	104	64	24	4			
6	128	126	114	84	44	24	2		
7	128	127	120	99	64	29	.8	1	
%	100	99	94	77	50	23	6	1	

3–4 waves. It is at this point that the waves would first be really visible. This is a good illustration of how groups of waves in deep water travel at half the celerity of each wave in the train. If the illustration were taken to 200 wave lengths the front of the wave train would be much steeper and therefore more clearly defined.

In a similar way, energy is left behind the rear of a group of waves. The water particles once in motion cannot stop immediately, so that a new wave continually appears at the trailing end of the group. The picture is one of waves forming, travelling through the train and disappearing at the front. The body of waves thus travel more slowly than the individuals which constitute it, this ratio being 1/2 for deep-water conditions.

Knowledge of wave group velocity is important in tracing waves across the ocean. It is also important in the wave-generating process, for the components of a spectrum advance in the same manner even though they are propagating across an already disturbed sea. For a group of waves to stay under the influence of a wind the fetch must travel at its group-wave velocity. Build-up is therefore optimum in a fetch which travels at the C_g of the T_{max} band in the spectrum.

PROGRESSIVE WAVE: FINITE HEIGHT

The characteristics derived by the linear theory above imply small-amplitude waves and sinusoidal profile. Employment of this simplified form permits the use of strong mathematical tools such as Fourier analysis. More complex wave systems can be considered as summations of several sinusoids. But ocean waves are not infinitely small in amplitude in deep water, and, as they arrive in the shallower depths, steepening introduces distortions in the surface profile and the orbital paths of particles. The linear theory then no longer is applicable.

Gerstner in 1802 [4] was the first mathematician to tackle this problem and his solution applies strictly to deep-water conditions but it was Stokes in 1880 [3] who first put the analysis in the modern form now accepted. Many other workers have been active in this area, deriving slightly different equations depending upon the boundary conditions assumed. Two extremes exist. The one used by Gerstner is to assume vorticity in the liquid and exclude any current or progressive movement of particle orbits [4, 5, 6]. The other used by Stokes is to exclude vorticity and accept a wave current. It has been found that Stokes' irrotational solution is the more correct, so it only will be presented here. A great variety of solutions is possible by assuming varying degrees of vorticity and wave current and is a veritable mathematicians' gold mine [7, 8, 9, 10].

Only the second-order Stokes solutions will be presented herein. Others are available elsewhere [11, 12, 13]. Even these are not applicable throughout the whole transition and shallow-water zone, so that other analyses will have to be introduced for these steeper wave conditions.

Surface profile

The 2nd-order equation for the surface profile is given by:

$$y_s = \frac{H}{2} \cos 2\pi(x/L - t/T)$$
$$+ \left[\frac{\pi H^2}{4L}\left(1 + \frac{3}{2 \sinh^2 2\pi d/L}\right) \cosh 2\pi d/L\right] \cos 4\pi(x/L - t/T) \quad (4\text{-}39)$$

This consists of a first term which is the sinusoidal form, plus a second term the bracketed section of which represents an elevation (ΔH) of the mean water level (MWL) above that of the still-water level (SWL). This is a symmetrical profile with steeper crests and flatter troughs than the sinusoidal wave and is depicted in the inset of Fig. 4-5. The MWL is an imaginary surface midway between the crest and the trough of the wave, so that it is an amplitude $a = H/2$ from each. However, the engineer is more interested in the height to the crest from the SWL as this is the datum he will normally use. This is designated by a_c, as is the depth to the trough from SWL symbolized as a_t; always $a_c + a_t = H$.

As noted earlier, the wave height is amplified from that in deep water, once reasonably shallow water is reached, assuming that no refraction is taking place. This increase is depicted in the lower half of Fig. 4-5, where a wave of given (H_o/L_o) in deep-water increases in steepness as d/L decreases. This occurs until an instability is reached when the wave breaks at a theoretical value [14] of:

$$(H/L)_b = 0.142 \tanh 2\pi d/L \quad (4\text{-}40)$$

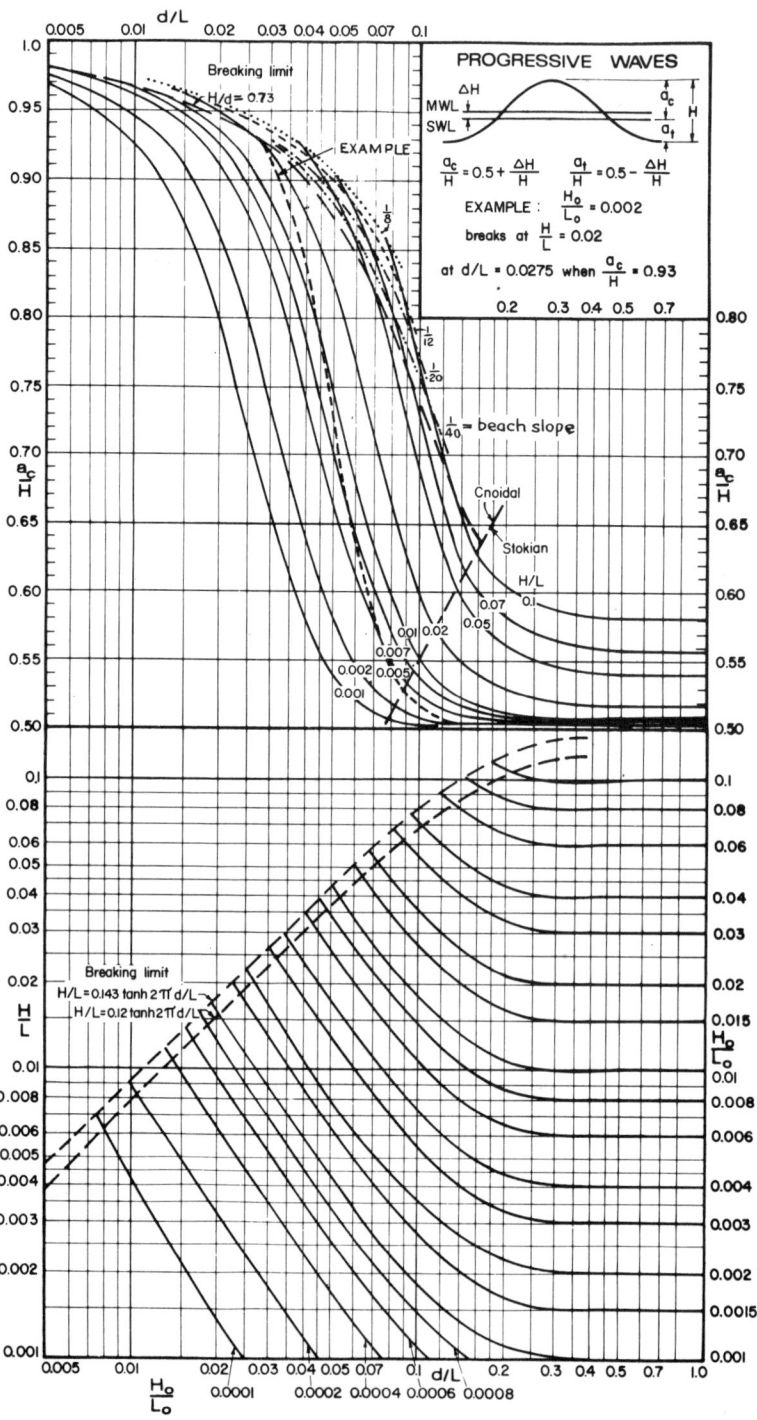

Fig. 4-5. Steepening and vertical asymmetry of progressive waves.

PROGRESSIVE WAVE: FINITE HEIGHT

A more realistic value obtained in tests and field observations [15] is:

$$(H/L)_b = 0.12 \tanh 2\pi d/L \tag{4-41}$$

Both these limits are drawn in the figure as dotted lines.

In the upper half of Fig. 4-5 are curves of a_c/H versus d/L in which the limit of the Stokean theory is demarcated [16]. Beyond this limit the cnoidal theory is more appropriate which will be discussed later. The breaking limit is again indicated by dotted lines for various beach slopes as well as an essentially horizontal bed when $H/d \doteq 0.73$.

Beyond $d/L = 0.5$ the curves become horizontal and the limiting value is determined solely by the steepness H/L, which in this case also represents H_o/L_o. This deep-water condition is given by:

$$a_c/H = \tfrac{1}{2}[1 + 1.57 \, H/L] \tag{4-42}$$

and:

$$a_t/H = \tfrac{1}{2}[1 - 1.57 \, H/L] \tag{4-43}$$

For example, a wave with H_o/L_o (= H/L) of 0.1 has a crest height which approximates 0.59 the wave height.

Before wave breaking occurs the flat trough of the Stokean wave becomes unstable and a secondary crest is formed at its centre [14]. The steepness at which this occurs is given by:

$$H/L = \sinh^2 2\pi d/L \, \tanh(2\pi d/L)/3\pi \tag{4-44}$$

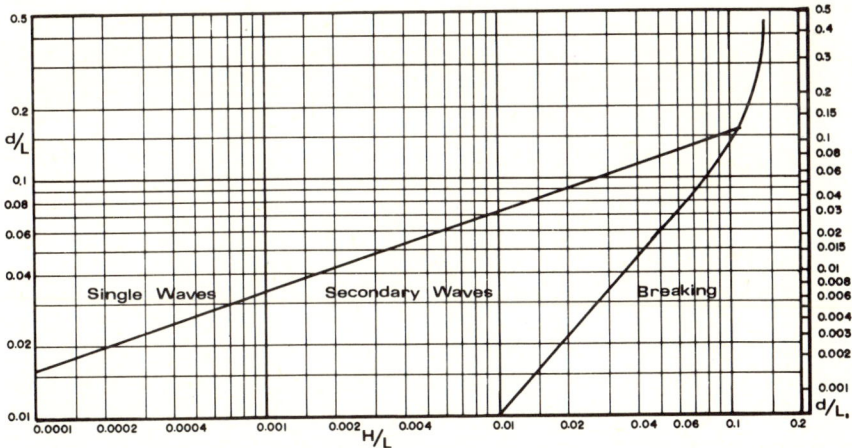

Fig. 4-6. Regions of specific profiles for progressive waves.

which is depicted in Fig. 4-6, where the breaking limit is also shown for an appropriate range of d/L and d/L_o.

The wave celerity and wave length in the second-order theory are similar to those for first-order (see eq. 4-6 and 4-7). For third- or higher-order theory these characteristics depend upon wave steepness, an increase in which gives higher speed and hence longer length:

$$C/C_o = \tanh 2\pi d/L \left[1 + \left(\frac{\pi H}{L}\right)^2 \left(\frac{14 + 4 \cosh^2 4\pi d/L}{16 \sinh^4 2\pi d/L}\right)\right] \qquad (4\text{-}45)$$

which for the limiting steepness of 0.143 increases C_o by only 20%. However, no swell waves would ever approach this deep-water steepness, and storm waves of this order exist only momentarily when the majority of the spectral components combine at one place, so that such increase is only of academic importance.

Water-particle motion

With the same notation as used for linear theory the velocities, accelerations, and amplitudes of orbital motion are presented below [17]. These are seen to consist of a first term which is the same as the first-order theory and a second which includes expressions containing wave steepness. Equations are given only for the transitional depths, the limiting values for deep- or shallow-water conditions can be derived by substitution or by reference to Fig. 4-7 where even the 3rd-order components are incorporated [11].

Velocities (Eulerian): transitional depths

$$u = \frac{\pi H}{T} \left[\frac{\cosh 2\pi (y+d)/L}{\sinh 2\pi d/L}\right] \cos 2\pi (x/L - t/T)$$

$$+ \frac{3}{4} \left(\frac{\pi H}{T}\right) \left(\frac{\pi H}{L}\right) \left[\frac{\cosh 4\pi (y+d)/L}{\sinh^4 2\pi d/L}\right] \cos 4\pi (x/L - t/T) \qquad (4\text{-}46)$$

$$v = \frac{\pi H}{T} \left[\frac{\sinh 2\pi (y+d)/L}{\sinh 2\pi d/L}\right] \sin 2\pi (x/L - t/T)$$

$$+ \frac{3}{4} \left(\frac{\pi H}{T}\right) \left(\frac{\pi H}{L}\right) \left[\frac{\sinh 4\pi (y+d)/L}{\sinh^4 2\pi d/L}\right] \sin 4\pi (x/L - t/T) \qquad (4\text{-}47)$$

The second term in eq. 4-46 fluctuates with twice the frequency of the first, so adding and subtracting values to those of the linear theory. A differential thus results, with a higher velocity forwards occurring at the crest than at the trough. The period $(T/2)$ for each half orbit remains the same, so that the water particle travels further forward due to this higher forward velocity than backward during the reverse velocity. This is confirmed in the amplitude equations.

PROGRESSIVE WAVE: FINITE HEIGHT

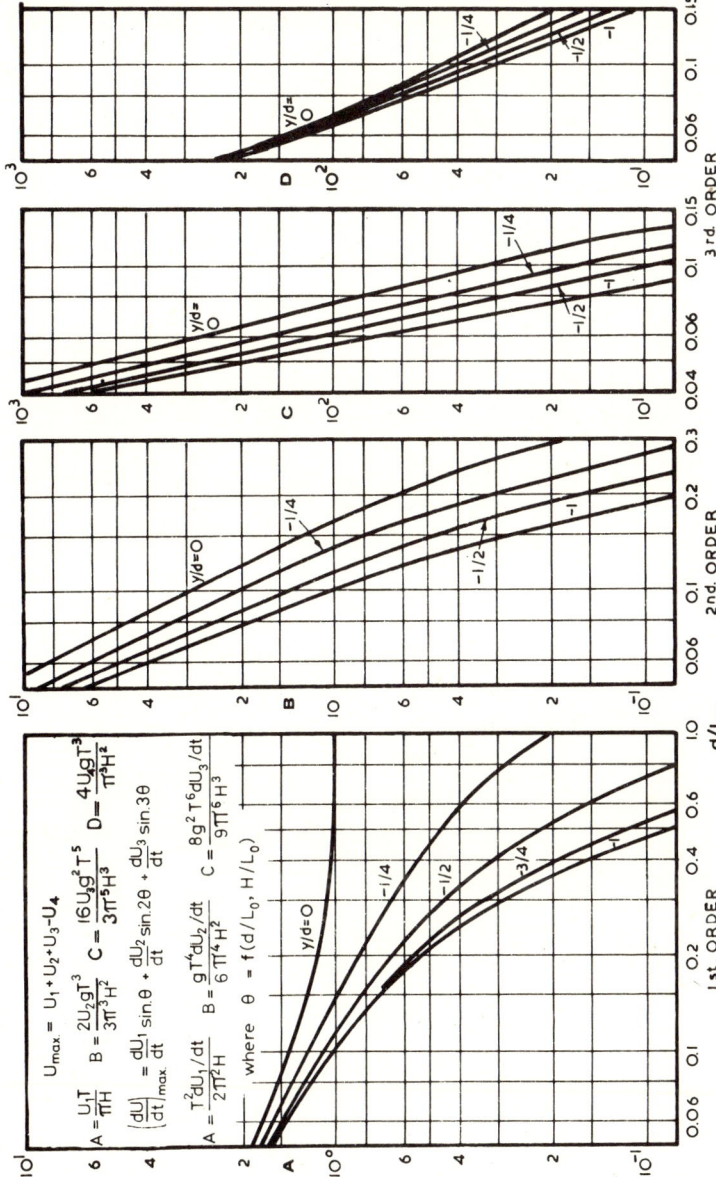

Fig. 4-7. Parameters for maximum horizontal velocities and accelerations for progressive waves — 2nd- and 3rd-order theory.

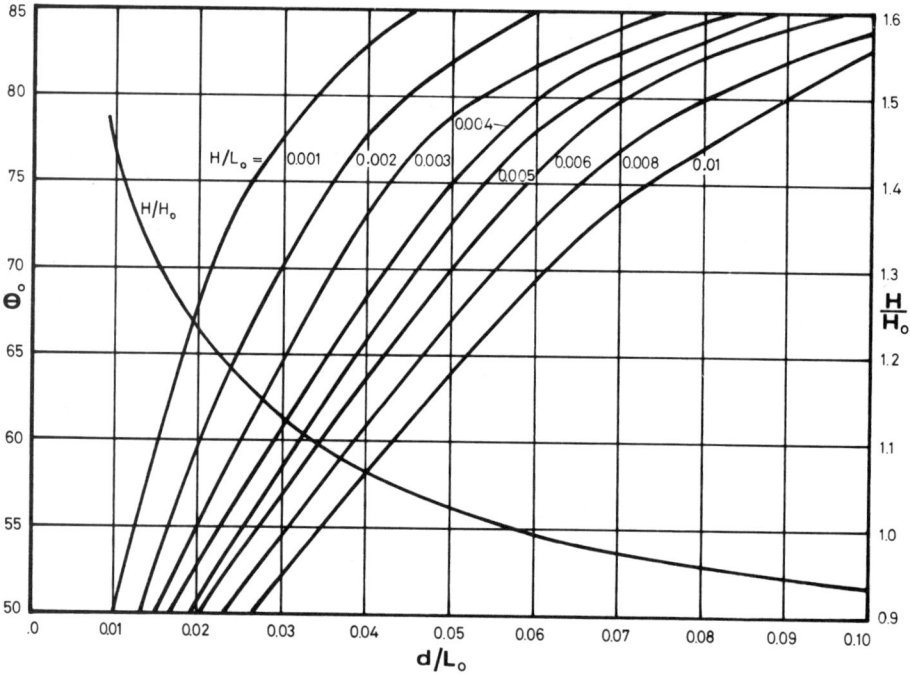

Fig. 4-8. Phase angle θ for substitution in equations of inset in Fig. 4-7.

The 3rd-order Stokes theory for u_{max} and $(du/dt)_{max}$ graphed in Fig. 4-7 has been plotted from tables derived by Skjelbreia [11]. To evaluate the acceleration the additions involve a knowledge of phase angle θ. This has been plotted in Fig. 4-8 utilizing the simplification of $L = L_o \tanh 2\pi d/L$ which is correct until $H/L > 0.02$. It is unlikely that swell waves, for which these theories strictly apply, will approach this steepness. The hyperbolic functions have all been taken to one side and represented as factors A, B, C or D derived from d/L_o and y/d.

Displacements (Lagrangian): transitional depths

$$x = -\frac{H}{2}\left[\frac{\cosh 2\pi(y_o+d)/L}{\sinh 2\pi d/L}\right]\sin 2\pi(x_o/L - t/T)$$

$$+ \frac{\pi H^2}{8L \sinh^2 2\pi d/L}\left[1 - \frac{3\cosh 4\pi(y_o+d)/L}{2\sinh^2 2\pi d/L}\right]\sin 4\pi(x_o/L - t/T)$$

$$+ \frac{\pi H^2}{4L}\left[\frac{\cosh 4\pi(y_o+d)/L}{\sinh^2 2\pi d/L}\right]\frac{2\pi t}{T} \qquad (4\text{-}48)$$

$$y = \frac{H}{2}\left[\frac{\sinh 2\pi(y_o + d)/L}{\sinh 2\pi d/L}\right]\cos 2\pi(x_o/L - t/T)$$

$$+ \frac{3\pi H^2}{16L}\left[\frac{\sinh 4\pi(y_o + d)/L}{\sinh^4 2\pi d/L}\right]\cos 4\pi(x_o/L - t/T)$$

$$+ \frac{\pi H^2}{8L}\left[\frac{\sinh 4\pi(y_o + d)/L}{\sinh^2 2\pi d/L}\right] \tag{4-49}$$

The net movement forward each period T is given by the third term in eq. 4-48 by substituting $t = T$ and then dividing by T. This results in a velocity known as *mass-transport velocity*, which is given by:

$$U = \frac{1}{2}\left(\frac{\pi H}{T}\right)\left(\frac{\pi H}{L}\right)\left[\frac{\cosh 4\pi(y_o + d)/L}{\sinh^2 2\pi d/L}\right] \tag{4-50}$$

In deep-water this simplifies to:

$$U_o = \left(\frac{\pi H}{T}\right)\left(\frac{\pi H}{L_o}\right)\exp(4\pi y_o/L_o) \tag{4-51}$$

Stokes' solution to eq. 4-51 had an additional constant added to the RHS which, computed on the basis that no net motion occurs throughout the complete depth of water, results in the deep-water solution of:

$$U_o = \left(\frac{\pi H}{T}\right)\left(\frac{\pi H}{L_o}\right)\left[\exp(4\pi y_o/L_o) - \frac{L}{4\pi d}\right] \tag{4-52}$$

Although this discussion of mass-transport due to irrotational motion of water particles has arisen from equations pertaining to transitional depths its variation throughout the water depth, and particularly at the bed, are not given correctly by the above solutions. This topic is to be discussed fully in a later section of this chapter. However, values of mass-transport near the water surface are predicted fairly well by eq. 4-50, 4-51, 4-52.

Cnoidal theory

When the water depth decreases to approximately $d/L = 0.1$, the Stokes theory is no longer valid. Another to serve these shallower conditions is called the cnoidal theory, which was derived originally by Korteweg and De Vries in 1895 [18]. Many workers have manipulated its equations to make them suitable for direct application to engineering problems [19, 20, 21]. However, the surface profile and wave celerity are the only characteristics available at this stage, velocities and accelerations of water particles still requiring tedious substitution. This profile, or more particularly the ratio a_c/H will be dealt with here since a more accurate theory is now available for other kinematic quantities.

The steepening of waves in shoaling water is mainly effected by the rise of the crest above the still water and flattening of the trough. This introduces sharp curvatures in the water surface and vertical accelerations must be accounted for. Solutions for such waves of translation must be expressed in terms of elliptic functions designated as cn-functions, so introducing the term cnoidal, analogous to sinusoidal.

The celerity of the cnoidal wave is given by:

$$C = (gd)^{1/2}\left[1 + \frac{H}{d}\frac{1}{k^2}\left(\frac{1}{2} - \frac{E(k)}{K(k)}\right)\right] \qquad (4\text{-}53)$$

where k is an elliptic parameter or modulus, $K(k)$ is the complete elliptic integral of k of the first kind, $E(k)$ is the complete elliptic integral of k of the second kind, and other symbols have their normal meaning. These variables are graphed in Fig. 4-9 against the parameter HL^2/d^3.

The celerity given by eq. 4-53 falls between two limits, that of the sinusoidal wave for very small ratios of H/d and that of the solitary wave for large ratios of H/d. For the former $k^2 = 0$ so that:

$$C = (gd)^{1/2}\left(1 - \frac{2\pi d^2}{3L^2}\right) \qquad (4\text{-}54)$$

which approximates the celerity in linear theory of:

$$C = \left(\frac{gL}{2\pi}\tanh 2\pi d/L\right)^{1/2} = \left[gd\left(1 - \frac{4\pi d^2}{3L^2}\right)\right]^{1/2} \qquad (4\text{-}55)$$

When the solarity wave is approached, $k^2 = 1$ and the celerity is given by:

$$C = (gd)^{1/2}(1 + H/2d) \qquad (4\text{-}56)$$

which is within 2% of that for the solitary wave of:

$$C = [(gd)(1 + H/d)]^{1/2} \qquad (4\text{-}57)$$

In this case it can be shown that $T = \infty$, but reduction of k to 0.9999 reduces T to 7π, indicating that solitary waves can well be considered of finite period or wave length. The period when $k = 0$ can be shown to be 2π, implying that only waves of 6 sec and longer should be considered by cnoidal theory.

PROGRESSIVE WAVE: FINITE HEIGHT

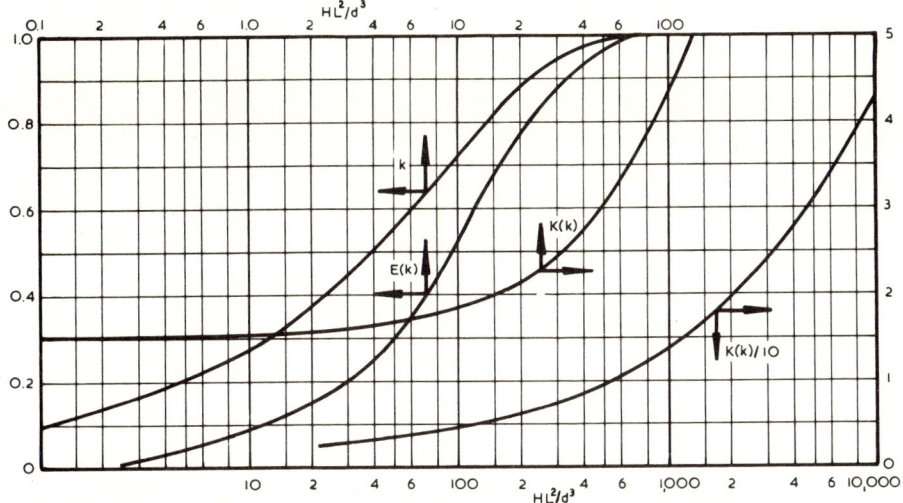

Fig. 4-9. Values of elliptic functions in the cnoidal theory.

The wave length is given by:

$$L = \left(\frac{16d^3}{3H}\right)^{1/2} k\, K(k) \tag{4-58}$$

and the crest to wave height ratio by:

$$\frac{a_c}{H} = \frac{16}{3}\left(\frac{d}{L}\right)^3 \frac{L}{H}\{K(k)[K(k) - E(k)]\} \tag{4-59}$$

This equation has been graphed in the upper half of Fig 4-5, where it is seen to extend beyond the previous Stokes theory. The combined influence of d/L and of H/L is clearly portrayed, together with an example (shown dotted) of a wave approaching the shore with a deep-water steepness of $H_o/L_o = 0.002$. As it traverses the transitional zone it steepens (as given by the lower half of Fig. 4-5) and so cuts across iso-steepness lines in the upper diagram. This continues until breaking occurs at the limit specified as $H/d = 0.73$.

This ratio, as will be seen in Chapter 5, is very much dependent on bed slope, curves for which are included in Fig. 4-5. The RHS of eq. 4-59 contains a parameter L^2H/d^3 known as the Ursell parameter [22], by which changes from Stokean to cnoidal theory are indicated. It is often quoted as a value of 10, but computation from the limit line indicated in Fig. 4-5 results in a range from 2 to 24 over the range of wave steepnesses from 0.001 to 0.1.

Presentation of the cnoidal equations above is based upon the linear theory.

Second-order solutions are available, but as shown below these have been found inapplicable to wave profiles and other measures of validity for a wave theory.

Hyperbolic theory

Iwagaki [23] has simplified the cnoidal wave equations by substituting $k = 1$ and $E(k) = 1$ for $K(k) \geqslant 3$. This does not imply solitary wave occurrence in which $K \to \infty$, but that K is finite and can be related to parameters of T, d and H. The limitation expressed in $K(k) \geqslant 3$ results in the Ursell parameter $L^2 H/d^3 \geqslant 48$ with $d/L \leqslant 1/12$ for application of a pseudo-cnoidal theory. An alternative limit [23] is:

$$d^2 \leqslant \pi H L_o / 24 \tag{4-60}$$

The simplifications permit the equations of wave characteristics to be expressed in primary functions as outlined below. The relationship between $K(k)$ and other wave characteristics for $H/d \leqslant 0.55$ is [23]:

$$[K(k)] d/L = \left(\frac{3H}{16d}\right)^{1/2} \left[1 - 1.3\left(\frac{H}{d}\right)^2\right]^{1/2} \tag{4-61}$$

which is graphed in Fig. 4-10. It is seen that for the bulk of H/d values the term in the square bracket could be omitted, resulting in the expression:

$$[K(k)]^2 = \frac{3}{16} L^2 H/d^3 = \frac{3}{16}\left(\frac{L}{d}\right)^3 \frac{H}{L} = \frac{3}{16}\left(\frac{L}{d}\right)^2 \frac{H}{d} \tag{4-62}$$

The crest height above still-water level is given by:

$$\frac{a_c}{H} = 1 - \frac{1}{K(k)}\left[1 - \frac{H}{12d} - \frac{1}{12K(k)}\left(\frac{H}{d}\right)^2\right] \tag{4-63}$$

which is also represented by Fig. 4-5.

The wave celerity is expressed as:

$$\frac{C}{\sqrt{gd}} = \left(1 - \frac{1}{K(k)}\frac{H}{2d}\right)\left[1 + \left(1 + \frac{1}{K(k)}\frac{H}{d}\right)\frac{H}{d}\left(\frac{1}{2} - \frac{1}{K(k)}\right)\right.$$
$$\left. + \left(1 + \frac{1}{K(k)}\frac{2H}{d}\right)\left(\frac{H}{d}\right)^2 \left\{\frac{1}{K(k)}\left(\frac{1}{K(k)} - \frac{1}{4}\right) - \frac{3}{20}\right\}\right] \tag{4-64}$$

This can be expressed also as a proportion of deep-water celerity by:

$$\frac{C}{C_o} = \left(\frac{2\pi d}{L_o}\right)^{1/2}\left[1 - \frac{\pi^2}{2}\left(\frac{H_o}{L_o}\right)^2\right]\left(1 - \frac{1}{K(k)}\frac{3H}{2d}\right)\left[1 - \frac{5H}{8d}\left(1 + \frac{1}{K(k)}\frac{H}{d}\right)\right]^{-1} \tag{4-65}$$

equating the square bracketed term of eq. 4-61 to unity. This results in an increased speed of the hyperbolic wave over the Airy wave ($H_o/L_o = 0$) at the point of

PROGRESSIVE WAVE: FINITE HEIGHT

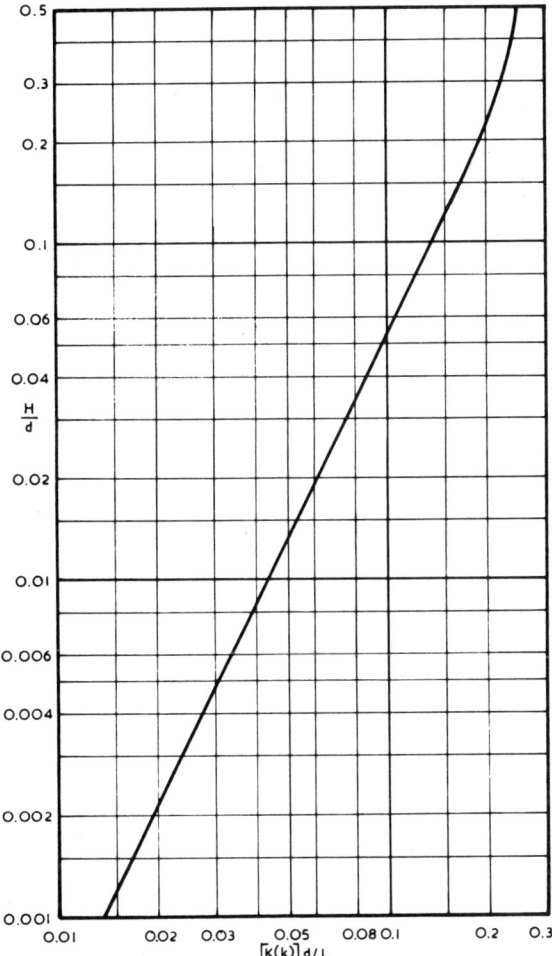

Fig. 4-10. Functions for hyperbolic waves from eq. 4-61.

breaking which is illustrated in Fig. 4-11. This modest increase confirms the statement made previously that acceleration of the wave need not be considered for most engineering problems.

The wave length L in depth d of water is given by:

$$\frac{L}{d} = \left(1 - \frac{1}{K(k)} \frac{3H}{2d}\right) \frac{4K(k)}{\sqrt{3}} \left(\frac{16d}{3H}\right)^{-1/2} \left[1 - \left(1 + \frac{1}{K(k)} \frac{H}{d}\right) \frac{5H}{8d}\right]^{-1} \quad (4\text{-}66)$$

The ratio of L/L_o is given by that for C/C_o namely eq. 4-65.

Iwagaki [23] used relationships derived by Le Méhauté and Webb [24] for deep-water conditions, assuming no wave dissipation, as a wave propagated directly

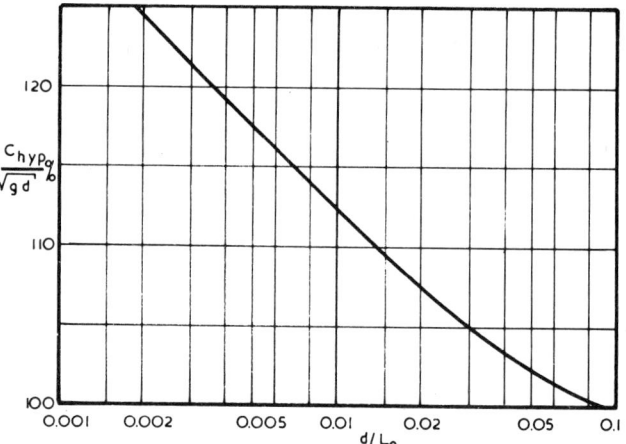

Fig. 4-11. Celerity for waves near breaking point as given by eq. 4-61 and 4-64.

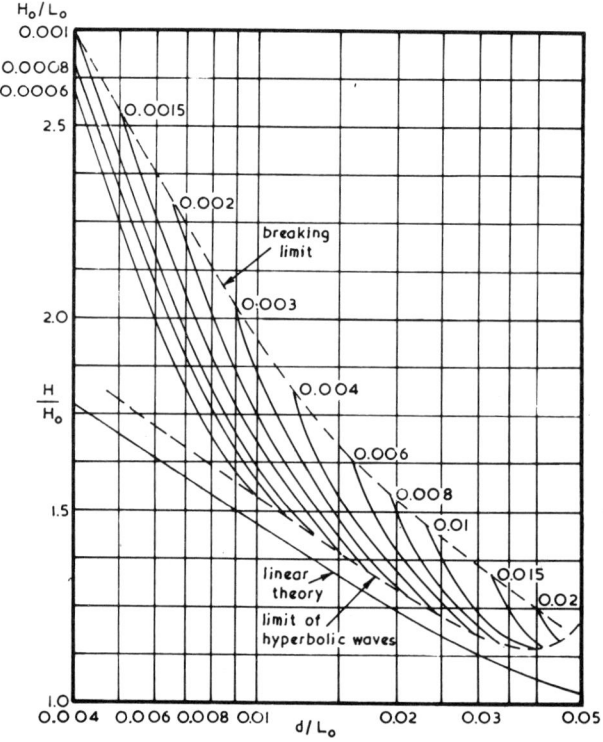

Fig. 4-12. Wave height ratio for progressive waves in shallow water from eq. 4-67.

shorewards without refraction. This gives the following wave height ratio:

$$\frac{H}{H_o} = \frac{3}{16}\frac{L_o}{d}\left(\frac{H_o}{4L_o}\right)^{1/3}\left[1+\left(\frac{\pi H_o}{L_o}\right)^2\right]\left[1-\frac{1}{K(k)}\frac{H}{d}+\frac{1}{12K(k)}\left(\frac{H}{d}\right)^2\right]^{-1/3} \quad (4\text{-}67)$$

which is graphed in Fig. 4-12. Experimental verification of this ratio by Iwagaki was exceptionally good.

Stream function

Dean [25] has proposed for non-linear gravity waves a stream function, whose form is selected so that the solution of the Laplace equation for interior motions satisfies boundary conditions at the bed and at the surface. Parameters in the stream function expression are chosen by a numerical perturbation procedure that results in the greatest correspondence to the kinematic and dynamic conditions at the free surface.

At the bed no vertical motion of the water particle is countenanced, whilst at the surface two criteria must be met. Firstly, the water particles should orbit in accord with the motion of the free surface, and secondly, the pressure predicted at the surface should be satisfied.

Dean applied his analysis to profiles of finite height waves measured by a step-gauge in the ocean and compared his predictions with the recorded surface. Three of the waves were produced by a storm and the fourth by a hurricane, the approximate characteristics of which are listed in the first 3 columns of Table 4-VI.

To check that these waves were within a fetch and hence were comprised of many components of the spectrum, more particularly those centered on T_{max} or the peak of wave energy distribution, calculations as per the remaining columns of Table 4-VI are included. Assuming the listed period and height to represent T_{max}

TABLE 4-VI

Wave and fetch characteristics in Dean's analysis

Wave No.	Height (ft.)	Period (sec)	Depth (ft.)	$\frac{T_{max}}{(H_{1/3})^{1/2}}$	$\frac{F}{F_{FAS}}$ (%)	$\frac{H_{1/3}}{U_{19.5}^2}$ ft/(knots)^2	$U_{19.5}$ (knots)	F_{FAS} (NM)	F (NM)
1	7	7.2	33	2.72	100	0.0185	19.5	280	280
2	15	9.0	33	2.33	40	0.0150	31.6	570	228
3	19	10.8	33	2.48	100	0.0185	32.0	580	580
4	39	14.0	98	2.24	20	0.0121	56.8	1300	260

and $H_{1/3}$, respectively, evaluation of $T_{max}/(H_{1/3})^{1/2}$ permits from Table 3-VIII the derivation of F/F_{FAS}, $H_{1/3}/U_{19.5}^2$, $U_{19.5}$ and from Table 3-VII F_{FAS} and hence F.

Observation of Table 4-VI would indicate that the wind velocities and their related fetches are reasonable and that therefore the waves were in the fetch when recorded. It can be shown that the d/L_o values operating on each occasion would not have affected the wave height significantly, so that the deep-water conditions implicit in the use of Tables 3-VII and 3-VIII are applicable.

Thus the waves measured were part of a spectrum, the orbital velocities and amplitudes being the integrated input from many wave components. The fact that it could be treated as a single wave with a fundamental component and higher harmonics, and be predicted so well by Dean's analysis, indicates that the use of representative monochromatic waves in models and flumes is not attended by so much distortion as is sometimes thought.

Another conclusion to be drawn from Dean's analysis is in regard to pressure distribution on the water surface and mass transport. Because of lack of information Dean omitted quantities respecting both these variables, even though they both could have varied from the zero values assumed. Pressure differentials could have existed over the wave crests, and surface currents in the water would have been in the order of 3% of the wind velocity [26]. However, with the accuracy of ± 0.5 ft. in the wave gauge such influences could have been lost. Whether Dean's analysis would have predicted accurate orbital velocities cannot be verified at this stage.

Dean also applied his stream function to hypothetical waves with specified gross characteristics, such as height, period and water depth. These he termed "theoretical" waves since the profile used in the calculation was derived from theory of a representative natural wave with the given characteristics. A comparison was made [27] with Chappelear's [28] velocity potential method, the two analyses being carried to the 7th order. It was found that the greatest error in the stream function representation occurred at the crest. In the check on a wave near breaking it was found that H_b/d_b was above the normally accepted theoretical value of 0.78, which in turn is above the normally observed value of 0.73. Dean [29] contends that this limit needs further verification.

Von Schwind and Reid [30] attempted to provide workable tables from the stream function. The surface boundary condition was simplified by a conformal mapping procedure and the results compared to Stokes 1st- 3rd- and 5th-order theories. The tables and graphs so produced are not readily applicable by coastal engineers since they contain Fourier coefficients rather than prime functions of waves.

Solitary wave theory

As noted previously, the solitary wave is a special case of the cnoidal wave when k and $E(k)$ are both unity and $K(k)$ is infinite. The wave length becomes infinite and the trough becomes asymptotic to the SWL, the crest rising directly from this level. It is difficult to imagine water for the crest rising directly out of the body of liquid without the surface between being affected. For this reason the solitary wave is considered a hypothetical or mathematical extrapolation of the cnoidal theory. However, its simplification of equations can serve to indicate limits of breaking waves.

As with other theories, expressions of solitary wave characteristics comprise a long series of terms so that first- and higher-order solutions can be computed. Only the first-order as derived by Laitone [31] will be included here, others being readily available elsewhere [1].

The wave celerity has already been presented in eq. 4-57, it depending solely on depth and wave height. Water particle velocities are given by:

$$\frac{u}{\sqrt{gd}} = KN \frac{1 + \cos Mz/d \cosh Mx/d}{(\cos Mz/d + \cosh Mx/d)^2} \quad (4\text{-}68)$$

$$\frac{v}{\sqrt{gd}} = KN \frac{\sin Mz/d \sinh Mx/d}{(\cos Mz/d + \cosh Mx/d)^2} \quad (4\text{-}69)$$

where K is the ratio given by $C = K\sqrt{gd}$; M and N are functions of H/d as graphed in Fig. 4-13; x is horizontal distance from crest; z is vertical distance from bed.

At the crest alignment ($x = 0$) the horizontal velocity is maximum and from eq. 4-67 is given by:

$$\frac{u_{max}}{\sqrt{gd}} = \frac{KN}{1 + \cos Mz/d} \quad (4\text{-}70)$$

which is graphed in Fig. 14-12 for a range of H/d and z/d. It is seen that the maximum velocity at the SWL does not exceed that at the bed by more than 20%, although at the surface of the crest it is almost doubled. The inset of Fig. 4-13 provides a definition sketch which also shows diagrammatically the paths followed by the water particles. As the crest passes from left to right they trace a curved path, this curvature decreasing with depth. During the passage of the trough the particles return to the left in a horizontal plane. There is a net movement in the direction of wave advance which is apparently uniform throughout the depth. Le Méhauté found such uniform mass transport in cnoidal waves [32].

The vertical velocity reaches its maximum value somewhere between the crest and the extremity of the crest, which Munk [33] submits is π/M either side of the

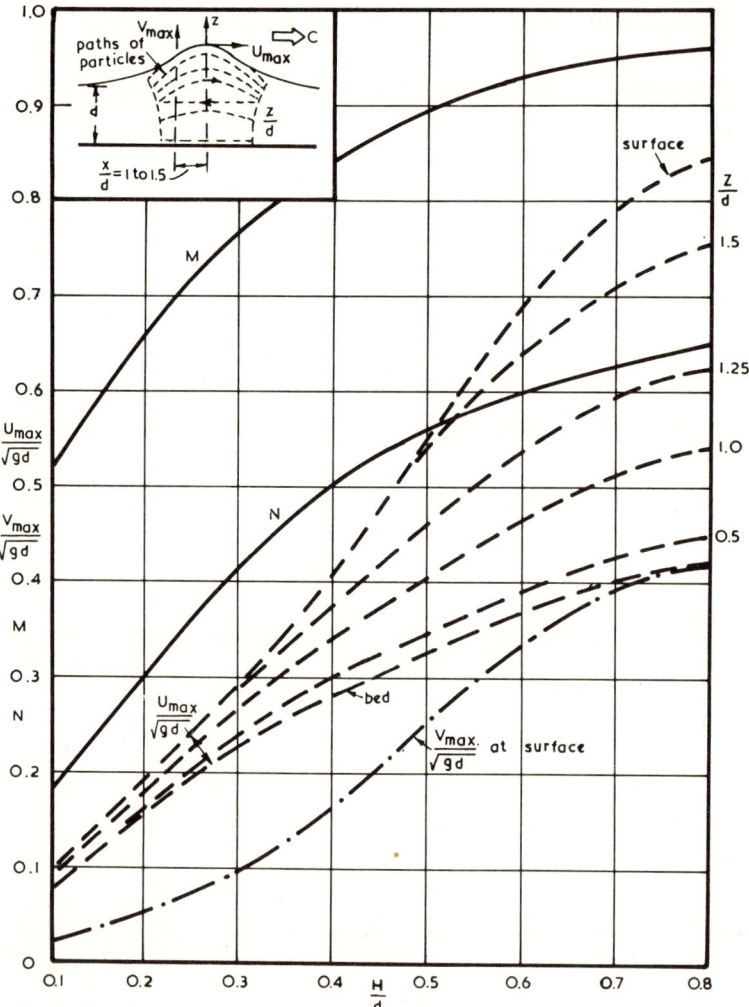

Fig. 4-13. Functions for solitary waves from eq. 4-68, 4-69 and 4-70.

crest, since 92% of the volume and 92% of the solitary wave energy is contained therein. Since the vertical velocity at the water surface is of greatest interest its maximum value is presented in Fig. 4-13. This occurs at distances d to $1.5d$ from the crest depending upon the value of H/d. It is seen that the $v_{max}/u_{max} \doteq 1/2$ to $1/3$ throughout the range of H/d.

The volume of water elevated above the SWL per unit length of crest [33] is:

$$V = \left(\frac{16}{3} d^3 H\right)^{1/2} \tag{4-71}$$

from which the maximum horizontal momentum can be computed by using some average value of u_{max}.

COMPARISON OF PROGRESSIVE WAVE THEORIES

Comparisons of the various wave theories have been conducted on theoretical grounds as well as by laboratory experiments. The former are based upon boundary criteria, whilst the latter compare velocities and amplitudes of water-particle motion against those predicted by the equations. Only the experimental verification is a basic measure of accuracy, since the theoretical test can only be comparative.

Theoretical comparison

Dean [34] has outlined requirements regarding validity of wave theories which assume incompressible fluid, irrotational motion, plus a horizontal and impermeable bed. One condition at the bed is that the equation should predict no vertical movement at the bed, which is met by all theories. At the surface two conditions should be met, firstly, that water particles there should be in accord with the predicted motion of the free surface, and secondly, that uniform pressure should exist at the surface, thus requiring Benoulli's equation to be satisfied. These two surface criteria will be termed "velocity" and "pressure", respectively. To assess the relative validity it is necessary to compare the results as given by the equations.

Dean has derived expressions for the error occurring in the kinematic and dynamic free-surface boundary conditions, which he evaluated for a number of theories over a full-wave profile. This error varied from crest to trough, varying from positive to negative perhaps one or more times. Dean then derived an over-all error which is the r.m.s. of the errors measured at intervals along the wave length. These can be expressed as ratios or percentages.

Dean presented these error data in several diagrams employing d/T^2 in the abscissa with ft./sec units. These have been combined in Fig. 4-14 utilizing percentage error and d/L_o. Although this composite graph appears extremely complex, it serves to highlight the relative accuracies. In the figure the stream function velocity curve does not appear since this theory fits the kinematic criterion exactly (i.e., error = 0). The cnoidal curves for pressure and velocity could almost be represented by a single curve. The Stokes I curves are presented for 4 ratios of H/H_b, namely 0.25, 0.5, 0.75 and 1.0. All other theories are represented only by curves for 0.25 and 1.0. By accepting curves below a specified error the choice of a wave theory can be made. As can be seen in Fig. 4-14, this will depend upon both d/L_o at H/H_b, this latter referring to the breaker height H_b in the relative depth d/L_o at which it occurs. From such a comparison Table 4-VII has been drawn up.

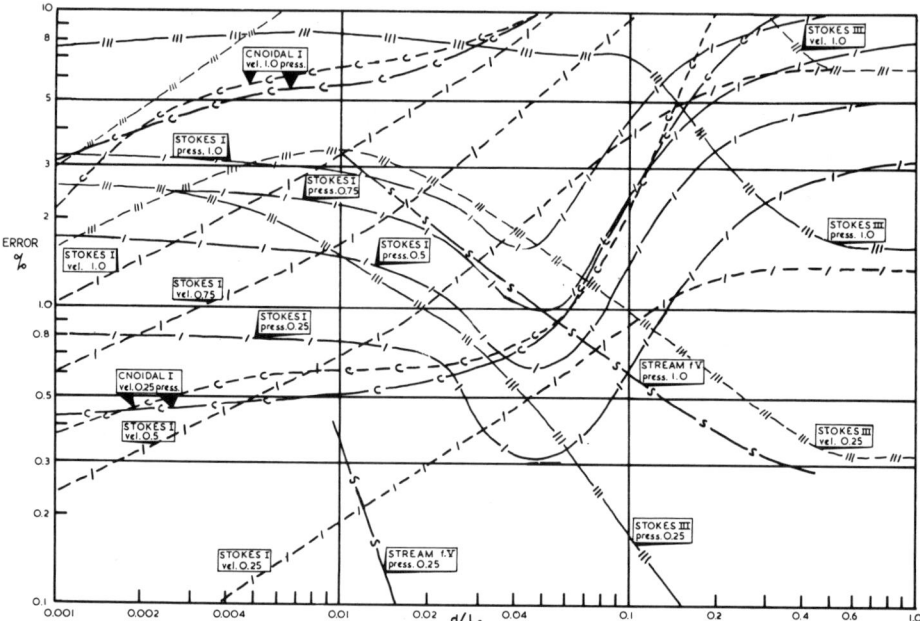

Fig. 4-14. Error of various wave theories as computed by Dean [34]. Velocity comparison is shown dotted and pressure comparison by full lines. Theories included are: Stokes I and III, stream function to fifth order, and cnoidal to 1st order. Figures in the squares represent ratios H/H_b.

TABLE 4-VII

Wave theories selected by Fig. 4-14

Error (%)	H/H_b	Shallow	Transitional	Deep
≤ 1.0	≪ 0.25	Stokes I Cnoidal I	Stokes I Cnoidal I Stream Function V Stokes III	Stokes I Stream Function V Stokes III
≤ 1.0	≤ 0.25	Stokes I Cnoidal I	Stokes I to d/L_o = 0.15 Cnoidal I to d/L_o = 0.08 Stream Function V from d/L_o = 0.05 Stokes III from d/L_o = 0.1	Stream Function V Stokes III
≤ 5.0	≤ 1.0	Stokes I Cnoidal I Solitary I	Stokes I to d/L_o = 0.02 Stream Function V	Stream Function V Stokes III from d/L_o = 0.15

This table clearly indicates the wide application of the Airy or Stokes I theory. For waves very much less than $H_b/4$ it has an error of 1% or less in all depth zones. For the same accuracy and $H/H_b \leqslant 0.25$ the Stokes I theory satisfies the shallow and transitional zones up to $d/L_o = 0.15$. But it should be realized that beyond this limit the wave steepness must be $\geqslant 0.03$ approximately, which it is not likely to be for a single train of waves to which the Stokes I theory strictly applies. For the case of 5% accuracy and $H/H_b = 1.0$ the Stokes I theory applies in the transitional zone up to $d/L_o = 0.02$ in which case $H/L \geqslant 0.05$, again an unlikely situation. The isolines of H/H_b are rather deceptive since H_b varies some 30 fold from shallow to deep-water conditions whilst H decreases some 50%.

Since the linear theory does not predict the asymmetry of the wave in transitional and shallow water it is perhaps preferable to use the cnoidal theory to predict the wave profile as has been done in Fig. 4-5. Recommendations made by Dean are contained in Fig. 4-15, together with the limits as gleaned from Fig. 4-14 and Table 4-VII. These latter are expressed in terms of 1% or 5% accuracy lines. It is seen that for 1% error or less and $H/H_b \leqslant 0.3$ the Stokes I theory serves the shallow-water zone and the greater part of the transitional zone. For slightly less accuracy the deep-water zone can also be treated. For 5% accuracy the linear theory covers

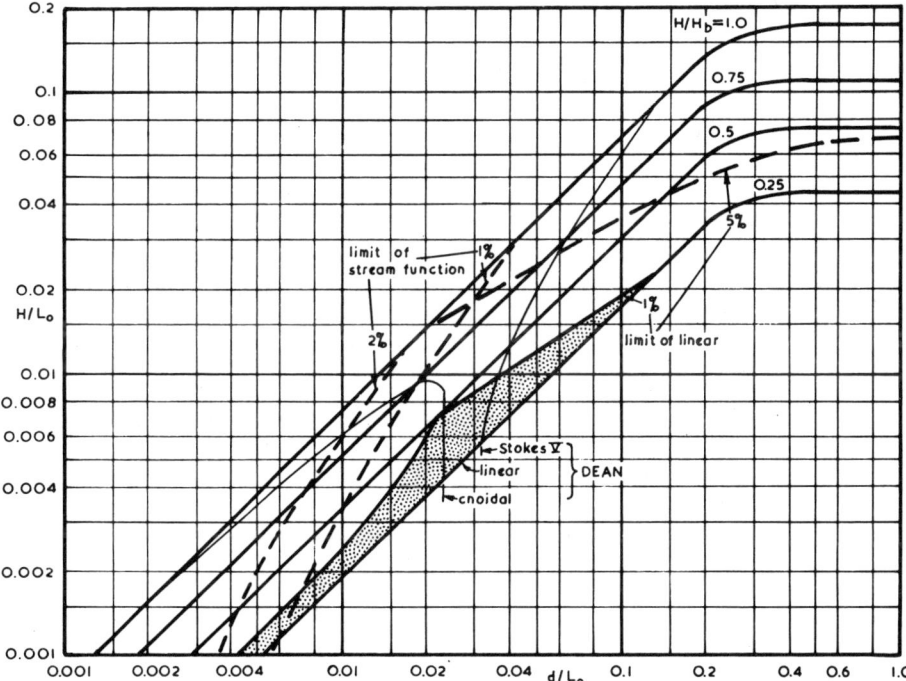

Fig. 4-15. Zones of application of wave theory according to criteria by Dean [34] and Fig. 4-14.

almost all conditions, the only exclusions being transitional and deep-water depths for $H/H_b \geqslant 0.5$.

Experimental verification

Validity checks have normally been made on orbital velocities, particularly those occurring under the wave crest. Correspondence with surface profile is measured concurrently and compared to those predicted by a number of theories. Although the profiles can be predicted quite closely the velocities measured are not so accurate. This may be in part the fault of experimentation, so that validity cannot be definitely proven or disproven at this date. However, comparison of some theories indicates that they are extremely wide of the mark.

The tests carried out by Le Méhauté et al. [35] are listed in Table 4-VIII from which various relevant ratios have been computed. The theories included in his diagrams are listed in Table 4-IX, the symbols (*A,B* etc.) of which are listed in Table 4-VIII in descending order of closeness to the experimental results. Three levels are considered for this purpose, namely, the bed, the SWL and the water surface.

TABLE 4-VIII

Wave characteristics for tests by Le Méhauté et al. [35] with theories listed in Table 4-IX placed in order of accuracy

T (sec)	H (ft.)	d (ft.)	d/L_o	d/L	L	H/L	HL/d^2	Bed	SWL	*Surface*
3.58	0.241	0.556	0.00845	0.037	15.0	0.0161	11.85	*ABK*HFJ IGC	FKIJ *AB*GHC	FJK*IG ABH*C
3.58	0.304	0.555	0.00845	0.037	15.0	0.0203	14.80	B*A*KHF JIGC	K*IFJAB* GHC	JIKGF ABHC
3.06	0.232	0.596	0.0124	0.045	13.25	0.0175	8.64	B*A*KFJI GC	FIJKG ABC	FJKIG *ABC*
3.06	0.293	0.596	0.0124	0.045	13.25	0.0221	10.90	*ABK*FHJ IGC	*FIJK*AB GHC	GJFIK ABHC
2.20	0.323	0.619	0.025	0.065	9.5	0.034	8.05	F*A*KBH JDICG	KAB*FI JD*GHC	KIAB*DG* JFCH
2.20	0.260	0.619	0.025	0.065	9.5	0.0274	6.50	*KFA*BHD JICG	ABKFI DJGCH	IKFDJA GBCH
1.16	0.293	0.587	0.085	0.128	4.6	0.0636	3.90	F*A*KED *BC*JIG	BAKE *DF*CIJG	ABIEK *F*DCJG
1.16	0.255	0.587	0.085	0.128	4.6	0.0555	3.40	*AK*F*E CD*BHJIG	BAEK F*CD*IJGH	ABEI*CF* DKJGH

In the last three columns italics are used where theories give similar values of velocity, in which case the numerical values for accuracy assessment were identical.

TABLE 4-IX

Theories tested by Le Méhauté et al. [35] and their order of accuracy at various depths

	B	SWL	S	OA
A. Airy: Wiegel [1]	1	3	4	3
B. Linear long wave: Wiegel [1]	3	5	6	5
C. Stokes II: Wiegel [1]	7	11	9	11
D. Stokes III; Skjelbreia [11]	6	8	7	8
E. Stokes V: Skjelbreia and Hendrickson [13]	4	4	3	4
F. Cnoidal: Keulegan and Patterson [21, 36]	2	2	2	2
G. Cnoidal I: Laitone [31]	10	10	7	10
H. Cnoidal II; Laitone [16]	5	9	8	9
I. Solitary: Boussinesq [33]	9	6	1	6
J. Solitary: McCowan [33, 37]	8	7	5	7
K. Empirical modification of Airy: Goda [38]	1	1	3	1

B = bed; SWL = still-water level; S = surface; OA = overall.

By giving numerical values to each theory according to its order of proximity to the experimental velocity distribution (as in Table 4-VIII), a total can be evaluated which indicates their order of accuracy. These orders for the three particle levels, as well as overall, are listed in Table 4-IX, where it is seen that theories K, F and A take the first three places. Although Stokes V theory (E) is next in line, the Stokes II (C) is last in overall prediction of velocity for these shallow-water conditions, it also being low in accuracy for each individual level.

It is both surprising and comforting that for the shallower-water conditions the Airy theory (A) predicts velocities so well. For bed values it is equal first with Goda [38], but for SWL and surface levels it is bettered by the 1st-order cnoidal theory [21, 36]. However, for waves in water of $d/L_o = 0.128$ (and presumably deep water) the Airy theory is first or second in prescribing velocities at all depths. It could therefore be accepted for the full transitional zone.

Besides the article of Le Méhauté et al. [35] other workers have reported maximum horizontal velocities (u_{max}) from flume tests [38, 39, 40]. These have been presented in various parameters, but have been combined in Fig. 4-16 as the dimensionless ratio $u_{max}/\sqrt{gd}$. The abscissa of HL/d^2 was selected empirically (after the use of Ursell's parameter HL^2/d^2 proved deficient). There appears to be a reasonable relationship between these two parameters. Because of the nature of the measurements some scatter is bound to occur, but of the tests included those of Le Méhauté et al. and Iwagaki and Sakai [39] could be accepted as the most accurate. They used photography of neutrally buoyant particles and hydrogen-bubble techniques respectively, whereas Goda [38] employed miniature propeller meters. The results of the aforementioned two research groups agree closely with the bed and SWL curves adopted. The traces for 1/4, 1/2 and 3/4 depth have been plotted from

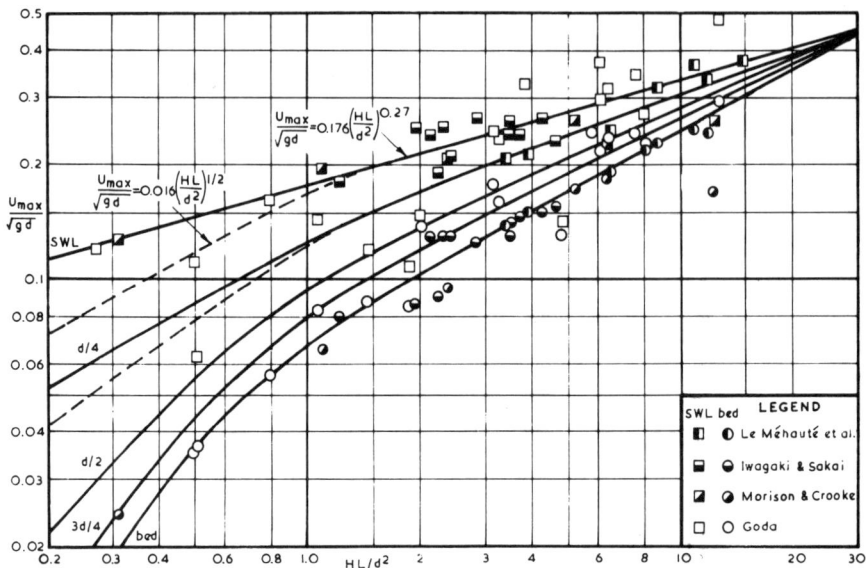

Fig. 4-16. Maximum horizontal orbital velocities at various depths from experimental data.

distributions available in the several reports, for which the plotted points have been omitted for the sake of clarity.

High values of HL/d^2 denote shallow water conditions, since H/d is relatively large and d/L is small. In this region there is an almost linear variation of velocity from SWL to the bed. For completely shallow-water conditions, where theory would indicate uniform velocity throughout the depth, the SWL and bed traces should converge. This appears to be the case at $HL/d^2 = 30$, which represents $d/L \doteq 0.025$ if the breaking criterion of $H_b/d_b = 0.78$ is substituted. Inspection of Fig. 4-5 would confirm that this is the case for an incoming wave for which $H_0/L_0 = 0.002$.

At the other extreme, the LHS of Fig. 4-16, small HL/d^2 implies deep-water conditions, where particle velocities at the bed decrease rapidly. So also do the $3d/4$ and $d/2$ values, even though the SWL velocity parameters and those at $d/4$ are not much reduced from the transitional zone values. The actual maximum velocities u_{max} may indeed be much greater since the ordinates of Fig. 4-16 must be multiplied by $\sqrt{gd}$ to derive them. Because of the dearth and scatter of points in this region for SWL an alternative curve is shown dotted for this and $d/4$ depth. Due to the incompatability of the various theories compared to experiment, in respect to maximum particle velocity distribution (as per Tables 4-VIII and 4-IX), it would appear that Fig. 4-16 could be used for this determination, to be amended as more results become available. Especially is such verification desirable for lower values of HL/d^2 (deeper-water) because of its importance in the design of offshore structures.

STANDING WAVES

Standing waves are produced when two progressive waves are travelling in directly opposite directions. Their characteristics are obtained by applying the theory of the two systems simultaneously. The resulting surface pattern appears to be a transfer of crests from locations one wave length apart to locations similarly spaced but displaced one half wave length from the first. A zone which is a crest at one instant of the cycle becomes a trough a half of the wave period later, forming into a crest $T/2$ later still. This produces pressure and velocity fluctuations with half the period of the wave trains forming the standing wave system.

For a true standing wave to form the two opposing progressive waves should be equal both in height and in period. If either of these two characteristics should differ in the two trains then a partial standing wave is formed, which will be discussed later. As for progressive waves, the analysis can be carried out to first- or second-order terms with Stokes' theory. Standing waves are not amenable to other wave theories even if these were warranted, so that only the Stokes equations will be presented.

Standing waves: linear theory

The wave profile is given for all depths of water by:

$$y_s = \frac{H}{2} \sin \frac{2\pi t}{T} \sin \frac{2\pi x}{L} \tag{4-72}$$

where the usual notation is used, except that in this case H is the crest to trough height of the standing wave, which is double the height of each train forming it. This interpretation of H is used throughout the discussion on standing waves.

The velocities from the Eulerian presentation are as follows:

$$u = \frac{\pi H}{T} \left[\frac{\cosh 2\pi(y+d)/L}{\sinh 2\pi d/L} \right] \sin \frac{2\pi x}{L} \cos \frac{2\pi t}{T} \tag{4-73}$$

$$v = -\frac{\pi H}{T} \left[\frac{\sinh 2\pi(y+d)/L}{\sinh 2\pi d/L} \right] \cos \frac{2\pi x}{L} \cos \frac{2\pi t}{T} \tag{4-74}$$

using the same notation as previously. It is seen that nodes occur (only horizontal motion) when $\cos 2\pi x/L = 0$, i.e., $x/L = 1/4$ and antinodes (only vertical motion) when $x/L = 1/2$. These motions are depicted in Fig. 4-17 for various depth ratios.

Velocities: deep water

Substitution of the limiting values for the hyperbolic functions results in deep-

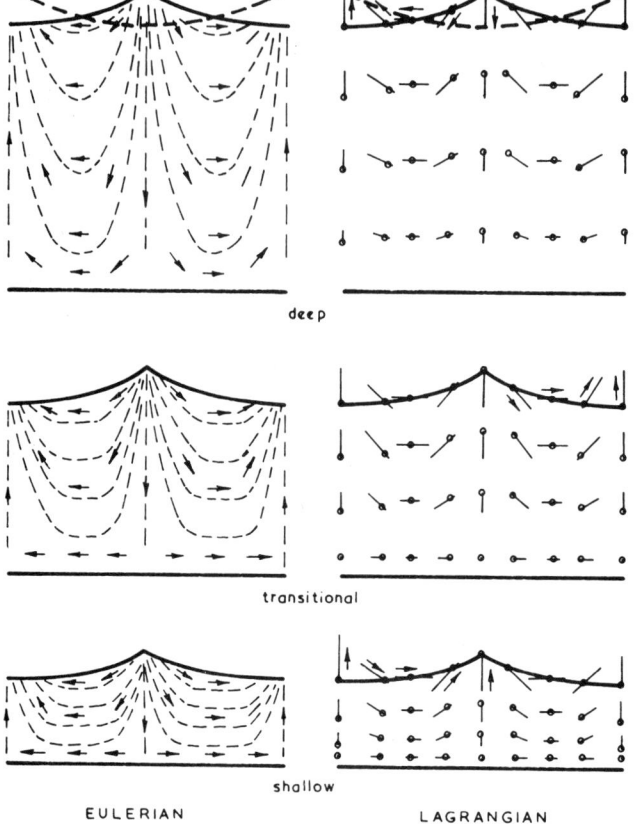

Fig. 4-17. Water-particle motions in standing waves.

water values of:

$$u = \frac{\pi H}{T} \exp(2\pi y/L_o) \left[\sin \frac{2\pi x}{L_o} \cos \frac{2\pi t}{T} \right] \qquad (4\text{-}75)$$

$$v = -\frac{\pi H}{T} \exp(2\pi y/L_o) \left[\cos \frac{2\pi x}{L_o} \cos \frac{2\pi t}{T} \right] \qquad (4\text{-}76)$$

At any depth the horizontal and vertical velocity magnitudes are equal. As noted below the paths followed are rectilinear, so that the resultant velocities ($= \sqrt{u^2+v^2}$) will differ throughout the length of the wave.

STANDING WAVES

Velocities: shallow water

$$u = \frac{HL}{2Td} \sin \frac{2\pi x}{L} \cos \frac{2\pi t}{T} \tag{4-77}$$

$$v = -\frac{H}{Td}(y+d) \cos \frac{2\pi x}{L} \cos \frac{2\pi t}{T} \tag{4-78}$$

In shallow-water u is independent of depth y, whilst v varies linearly with y, being zero at the bed.

The water-particle displacements from their initial locations (x_o, y_o), according to the Lagrangian presentation, are as follows:

Displacements: transitional

$$x = -\frac{H}{2}\left[\frac{\cosh 2\pi(y_o+d)/L}{\sinh 2\pi d/L}\right] \sin \frac{2\pi x_o}{L} \cos \frac{2\pi t}{T} \tag{4-79}$$

$$y = \frac{H}{2}\left[\frac{\sinh 2\pi(y_o+d)/L}{\sinh 2\pi d/L}\right] \cos \frac{2\pi x_o}{L} \cos \frac{2\pi t}{T} \tag{4-80}$$

The particle motions are rectilinear and simple harmonic with amplitude:

$$(x^2+y^2)^{1/2} = \frac{H}{2\sqrt{2}} \frac{(\cos 4\pi x_o/L + \cosh 4\pi y_o/L)^{1/2}}{\sinh 2\pi d/L} \tag{4-81}$$

and inclination θ such that:

$$\tan \theta = \tan 2\pi x_o/L - \tanh 2\pi(y_o+d)/L \tag{4-82}$$

Displacements: deep water

$$x = -\frac{H}{2} \exp(2\pi y_o/L_o)\left[\sin \frac{2\pi x_o}{L_o} \cos \frac{2\pi t}{T}\right] \tag{4-83}$$

$$y = \frac{H}{2} \exp(2\pi y_o/L_o)\left[\cos \frac{2\pi x_o}{L_o} \cos \frac{2\pi t}{T}\right] \tag{4-84}$$

Amplitudes decrease with depth so that when $y_o/L = 0.5$ both x and y are 4% of values at the still-water level.

Displacements: shallow water

$$x = -\frac{HL}{4\pi d} \sin \frac{2\pi x_o}{L} \cos \frac{2\pi t}{T} \tag{4-85}$$

$$y = \frac{H}{2d}(y_o+d) \cos \frac{2\pi x_o}{L} \cos \frac{2\pi t}{T} \tag{4-86}$$

The horizontal displacement x is independent of depth y_o, whilst y varies linearly with it, being zero at the bed.

The Eulerian and Lagrangian picture of particle movement is depicted in Fig. 4-17, which represents conditions for a true clapotis in which opposing progressive waves are of equal height and period.

Standing waves: second-order theory

Miche [14] has derived equations for the standing wave with finite height. The water surface profile is given by:

$$y_s = \frac{H}{2} \sin \frac{2\pi x}{L} \sin \frac{2\pi t}{T}$$

$$- \frac{\pi H^2}{4L} \coth \frac{2\pi d}{L} \cos \frac{4\pi x}{L} \left[\sin^2 \frac{2\pi t}{T} - \frac{3 \cos 4\pi t/T + \tanh^2 2\pi d/L}{4 \sinh^2 2\pi d/L} \right] \quad (4\text{-}87)$$

The first term of this expression is the same as the linear theory, whilst the second represents an incremental depth increase above the SWL. However, unlike the progressive wave, this increase ΔH only occurs under the crests as they appear alternatively at points $L/2$ apart, as exhibited by the factor $\cos 4\pi x/L$. The magnitude of this rise in mean water level is given by:

$$\Delta H = \frac{\pi H^2}{4L} \coth \frac{2\pi d}{L} \left[1 + \frac{3}{4 \sinh^2 2\pi d/L} - \frac{1}{4 \cosh^2 2\pi d/L} \right] \quad (4\text{-}88)$$

As ΔH increases so does the wave height until a maximum is reached which can be approximated [14, 15] by:

$$(H/L)_{max} = 0.22 \tanh 2\pi d/L \quad (4\text{-}89)$$

In deep water the wave profile is given by:

$$y_s = \frac{H}{2} \sin \frac{2\pi x}{L} \sin \frac{2\pi t}{T} - \frac{\pi H^2}{4L} \cos \frac{4\pi x}{L} \sin^2 \frac{4\pi t}{T} \quad (4\text{-}90)$$

and the maximum steepness becomes 0.22 of which 65% of the height is above SWL and 35% below. The above relationships are presented in Fig. 4-18.

Velocities: transitional

$$u = \frac{\pi H}{T} \left[\frac{\cosh 2\pi(y+d)/L}{\sinh 2\pi d/L} \right] \sin \frac{2\pi x}{L} \cos \frac{2\pi t}{T}$$

$$+ \frac{3}{8} \left(\frac{\pi H}{T} \right) \left(\frac{\pi H}{L} \right) \left[\frac{\cosh 4\pi(y+d)/L}{\sinh^4 2\pi d/L} \right] \cos \frac{4\pi x}{L} \sin \frac{4\pi t}{T} \quad (4\text{-}91)$$

$$v = -\frac{\pi H}{T}\left[\frac{\sinh 2\pi(y+d)/L}{\sinh 2\pi d/L}\right]\cos\frac{2\pi x}{L}\cos\frac{2\pi t}{T}$$

$$+\frac{3}{8}\left(\frac{\pi H}{T}\right)\left(\frac{\pi H}{L}\right)\left[\frac{\sinh 4\pi(y+d)/L}{\sinh^4 2\pi d/L}\right]\sin\frac{4\pi x}{L}\sin\frac{4\pi t}{T} \qquad (4\text{-}92)$$

The first terms of the above equations are the same as the linear velocity, whilst the second terms contain wave steepness to the first power. The overall expressions are not dissimilar to those for the progressive wave, but it should be remembered that the height H is double the height of the progressive waves producing the clapotis.

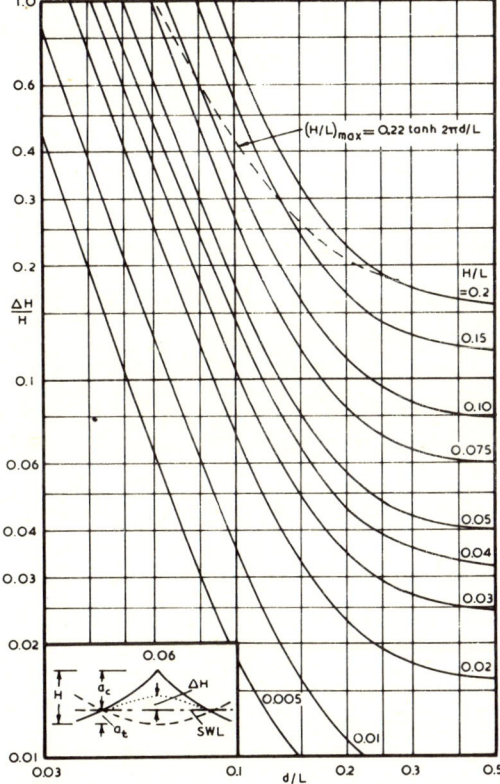

Fig. 4-18. Vertical asymmetry of standing waves where $a_c/H = 0.5 + \Delta H/H$.

Displacements: transitional

The linear displacements of particles from still-water locations $x_o y_o$ are given by:

$$x = \frac{H}{2}\left[\frac{\cosh 2\pi(y_o+d)/L}{\sinh 2\pi d/L}\right]\sin\frac{2\pi x_o}{L}\sin\frac{2\pi t}{T}$$

$$-\frac{\pi H^2}{8L}\frac{\cos 4\pi x_o/L}{\sinh^2 2\pi d/L}\left[\sin^2\frac{2\pi t}{T} + \frac{\cosh 2\pi(y_o+d)/L}{4\sinh^2 2\pi d/L}\left(3\cos\frac{4\pi t}{T} + \tanh^2\frac{2\pi d}{L}\right)\right]$$
(4-93)

$$y = \frac{H}{2}\left[\frac{\sinh 2\pi(y_o+d)/L}{\sinh 2\pi d/L}\right]\cos\frac{2\pi x_o}{L}\sin\frac{2\pi t}{T}$$

$$+\frac{\pi H^2}{8L}\frac{\sinh 4\pi(y_o+d)/L}{\sinh^2 2\pi d/L}\left[\sin^2\frac{2\pi t}{T} + \frac{\sin 4\pi x_o/L}{4\sinh^2 2\pi d/L}\left(3\cos\frac{4\pi t}{T} + \tanh^2\frac{2\pi d}{L}\right)\right]$$
(4-94)

These oscillatory motions will be similar to those depicted in Fig. 4-17, the wave steepness providing an increment in magnitude.

Partial clapotis

In the case of the progressive waves travelling in opposite directions not being of equal height, but of equal period or length, the resulting standing wave will not be a complete clapotis. This can occur when a wave is reflected normally from a vertical wall, although even in this case there can be a slight change in period.

The addition of the two waves of differing height means that nowhere in the surface is there a point of no vertical motion, as was the case, theoretically at least, at the nodes of the complete clapotis. There will, however, be a wavy wetted surface along the sides of the flume in which a partial clapotis exists, constituting the envelope of the crest reach. This will have high points at the antinodes and low points at the nodes as depicted in Fig. 4-19.

Not only do the particles have vertical displacements at all points along the surface, but the orbits also are not rectilinear as for the complete clapotis. This is illustrated in Fig. 4-19, where the interaction of an incoming wave with certain percentages of its reflected component are traced from photographs taken by Wallet and Ruellan [41]. It is seen that the elliptical orbits change the orientation of their axes progressively as the degree of reflection increases. The ellipses also flatten so that for the complete clapotis they become rectilinear.

In the case of equal periods the envelope so described by the surface will remain stationary. However, when the periods of the opposing wave trains differ, the envelope itself will progress slowly along the flume in the direction of propagation of the longer wave.

STANDING WAVES

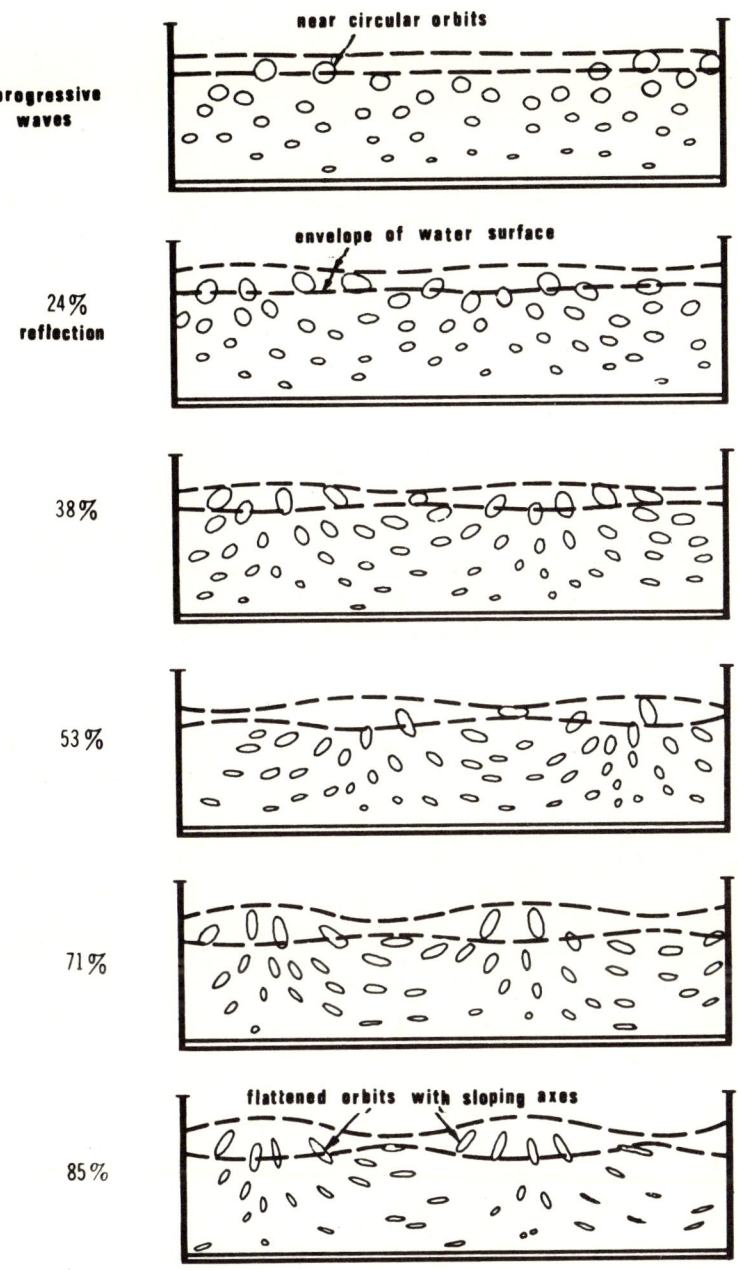

Fig. 4-19. Profiles and water-particle orbits in partial standing waves.

The vertical amplitudes of motion in respect to SWL locations are difficult to define, as they depend upon wave steepness and depth-to-length ratios, but in general it can be stated that for waves of equal period the heights of waves will be additive at antinodes and substractive at the nodes, whilst for waves of differing period the height will be the square root of the addition or subtraction of the individual wave heights squared.

Goda and Abe [42] have carried out an analysis on partial standing waves to the third order in a study of wave reflection. They conclude that the standing wave height is greater than twice the incident height by 10 to 15%. This discussion, although qualitative, serves to emphasize the complexity of motion plus macro-turbulence that is experienced in this simple case of two waves travelling in opposite directions. The implications of this in respect to sediment movement is discussed later.

Clapotis gaufré

When waves approach and are reflected from a structure obliquely they create a surface undulation of a diamond pattern, termed by French engineers as "clapotis gaufré". Although by such reflection the heights and periods of the two progressive waves making up the system are likely to be different, for purposes of discussion it is convenient to consider an example where these are the same.

As seen in the inset of Fig. 4-20 a new variable in the equations of motion is the crest length L', measured transversally to the celerity vector of the combined wave, which bisects the angle between the straight crests of the trains of equal period

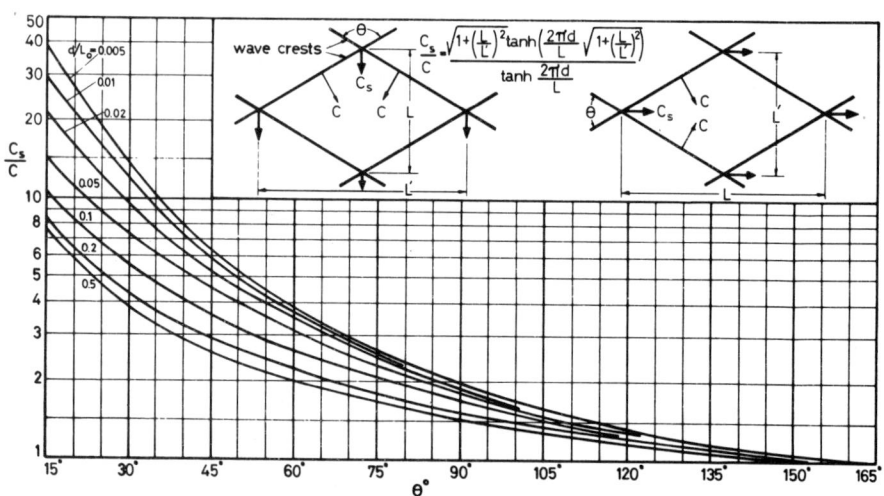

Fig. 4-20. Celerity of combined crests in a short-crested wave system.

STANDING WAVES

making up the system. The celerity of the combined crest (C_s) is given [43] by:

$$C_s = \frac{gT}{2\pi}\sqrt{1+(L/L')^2}\tanh\left[\frac{2\pi d}{L}\sqrt{1+(L/L')^2}\right] \qquad (4\text{-}95)$$

which is obviously greater than C for each wave, since in the same period T it must traverse the diagonal of the diamond pattern, whilst each wave traverses only the normal distance from one side to the other. The ratio C_s/C is presented in Fig. 4-20, where it is seen to vary with the angle between the crests and the depth ratio.

The orbital motions of the water particles are rather complex as seen by the following equations for amplitudes x, y and z for the simplified conditions of two trains of equal period T and equal height H as given by linear theory [43]:

$$\frac{x}{H/2} = \frac{\cosh(2\pi/L)(y+d)\sqrt{1+(L/L')^2}}{\sqrt{1+(L/L')^2}\sinh(2\pi d/L)\sqrt{1+(L/L')^2}}\cos 2\pi Z/L' \sin\gamma \qquad (4\text{-}96)$$

$$\frac{y}{H/2} = \frac{\sinh(2\pi/L)(y+d)\sqrt{1+(L/L')^2}}{\sinh(2\pi d/L)\sqrt{1+(L/L')^2}}\cos 2\pi Z/L' \cos\gamma \qquad (4\text{-}97)$$

$$\frac{z}{H/2} = \frac{(L/L')\cosh(2\pi/L)(y+d)\sqrt{1+(L/L')^2}}{\sqrt{1+(L/L')^2}\sinh(2\pi d/L)\sqrt{1+(L/L')^2}}\sin 2\pi Z/L' \cos\gamma \qquad (4\text{-}98)$$

where $\gamma = 2\pi(x/L - t/T)$, $H = \sqrt{H_1^2 + H_1^2} = \sqrt{2}H_1$ and Z is a measure along the wave crest length (L').

In Table 4-X maximum values of x, y, z (expressed as ratios of $H/2$) and β in degrees are listed for time $t = 0$ at locations specified by X/L and Z/L' for specific values of d/L_o when the two wave trains are normal to each other (i.e., $\theta = 90°$). The expression for orbital inclination (β) is:

$$\beta = \tan^{-1}\left[\frac{\tan 2\pi Z/L}{\sqrt{1+(L/L')^2}\tanh(2\pi/L)(y+d)\sqrt{1+(L/L')^2}}\right] \qquad (4\text{-}99)$$

These results are also displayed in Fig. 4-21, where it is seen that the orbits followed by water particles vary spatially both in the horizontal and vertical planes. Along the paths followed by the combined crests ($Z/L' = 0, 1/2, 1$, etc.) the elliptical orbits are aligned continuously with the direction of crest advance. As seen in the section on $Z/L' = 0$ the long axis of the ellipse may be vertical at the surface and horizontal at mid-depth, the oscillation being rectilinear at the bed.

Midway between the above alignments ($Z/L' = 1/4, 3/4, 5/4$, etc.) the orbit is again rectilinear, but with the motion transverse to the direction of combined crest advance. The position of the particle at different points in the plan is indicated in Fig. 4-21 by the dots.

TABLE 4-X

Characteristics of water-particle orbital motion in angled wave trains ($\theta = 90°$) of equal period and equal height (x, y and z expressed as ratios of $H/2$); see Fig. 4.21

$d/L_0 = 0.4$

Y/d	X/L	$\vec{Z/L'}$	0	1/8	1/4	3/8	1/2	5/8	3/4	7/8	1
0	1/4	x	−0.708	−0.500	0	+0.500	+0.708	+0.500	0	−0.500	−0.708
	0	y	−1.000	−0.707	0	+0.707	+1.000	+0.707	0	−0.707	−1.000
	0	z	0	+0.500	+0.708	+0.500	0	−0.500	−0.708	−0.500	0
		$\beta°$	0	+35.3	90	−35.3	0	+35.3	90	−35.3	0
−1/4	1/4	x	−0.289	−0.204	0	+0.204	+0.289	+0.204	0	−0.204	−0.289
	0	y	−0.405	−0.286	0	+0.286	+0.405	+0.286	0	−0.286	−0.405
	0	z	0	+0.204	+0.289	+0.204	0	−0.204	−0.289	−0.204	0
		$\beta°$	0	+35.5	90	−35.5	0	+35.5	90	−35.5	0
−1/2	1/4	x	−0.120	−0.085	0	+0.085	+0.120	+0.085	0	−0.085	−0.120
	0	y	−0.161	−0.114	0	+0.114	+0.161	+0.114	0	−0.114	−0.161
	0	z	0	+0.085	+0.120	+0.085	0	−0.085	−0.120	−0.085	0
		$\beta°$	0	+36.4	90	−36.4	0	+36.4	90	−36.4	0
−3/4	1/4	x	−0.055	−0.039	0	+0.039	+0.055	+0.039	0	−0.039	−0.055
	0	y	−0.056	−0.040	0	+0.040	+0.056	+0.040	0	−0.040	−0.056
	0	z	0	+0.039	+0.055	+0.039	0	−0.039	−0.055	−0.039	0
		$\beta°$	0	+44.6	90	−44.6	0	+44.6	90	−44.6	0
−1	1/4	x	−0.039	−0.027	0	+0.027	+0.039	+0.027	0	−0.027	−0.039
	0	y	0	0	0	0	0	0	0	0	0
	0	z	0	+0.027	+0.039	+0.027	0	−0.027	−0.039	−0.027	0
		$\beta°$	0	90	90	90	0	90	90	90	0

$d/L_0 = 0.2$

Y/d	X/L	$\vec{Z/L'}$	0	1/8	1/4	3/8	1/2	5/8	3/4	7/8	1
0	1/4	x	−0.733	−0.519	0	+0.519	+0.733	+0.519	0	−0.519	−0.733
	0	y	−1.000	−0.707	0	+0.707	+1.000	+0.707	0	−0.707	−1.000
	0	z	0	+0.519	+0.733	+0.519	0	−0.519	−0.733	−0.519	0
		$\beta°$	0	+36.3	90	−36.3	0	+36.3	90	−36.3	0
−1/4	1/4	x	−0.459	−0.324	0	+0.324	+0.459	+0.324	0	−0.324	−0.459
	0	y	−0.587	−0.415	0	+0.415	+0.587	+0.415	0	−0.415	−0.587
	0	z	0	+0.324	+0.459	+0.324	0	−0.324	−0.459	−0.324	0
		$\beta°$	0	+38.0	90	−38.0	0	+38.0	90	−38.0	0
−1/2	1/4	x	−0.300	−0.213	0	+0.213	+0.300	+0.213	0	−0.213	−0.300
	0	y	−0.324	−0.229	0	+0.229	+0.324	+0.229	0	−0.229	−0.324
	0	z	0	+0.213	+0.300	+0.213	0	−0.213	−0.300	−0.213	0
		$\beta°$	0	+42.9	90	−42.9	0	+42.9	90	−42.9	0
−3/4	1/4	x	−0.220	−0.155	0	+0.155	+0.220	+0.155	0	−0.155	−0.220
	0	y	−0.144	−0.101	0	+0.101	+0.144	+0.101	0	−0.101	−0.144
	0	z	0	+0.155	+0.220	+0.155	0	−0.155	−0.220	−0.155	0
		$\beta°$	0	+56.8	90	−56.8	0	+56.8	90	−56.8	0
−1	1/4	x	−0.195	−0.138	0	+0.138	+0.195	+0.138	0	−0.138	−0.195
	0	y	0	0	0	0	0	0	0	0	0
	0	z	0	+0.138	+0.195	+0.138	0	−0.138	−0.195	−0.138	0
		$\beta°$	0	90	90	90	0	90	90	90	0

STANDING WAVES

TABLE 4-X (*continued*)

$d/L_0 = 0.1$

Y/d	X/L	$\vec{Z}/L'$	0	1/8	1/4	3/8	1/2	5/8	3/4	7/8	1
0	1/4	x	−0.833	−0.589	0	+0.589	+0.833	+0.589	0	−0.589	−0.833
	0	y	−1.000	−0.707	0	+0.707	+1.000	+0.707	0	−0.707	−1.000
	0	z	0	+0.589	+0.833	+0.589	0	−0.589	−0.833	−0.589	0
		β°	0	+39.8	90	−39.8	0	+39.8	90	−39.8	0
−1/4	1/4	x	−0.650	−0.460	0	+0.460	+0.650	+0.460	0	−0.460	−0.650
	0	y	−0.675	−0.477	0	+0.477	+0.675	+0.477	0	−0.477	−0.675
	0	z	0	+0.460	+0.650	+0.460	0	−0.460	−0.650	−0.460	0
		β°	0	+43.9	90	−43.9	0	+43.9	90	−43.9	0
−1/2	1/4	x	−0.530	−0.375	0	+0.375	+0.530	+0.375	0	−0.375	−0.530
	0	y	−0.416	−0.294	0	+0.294	+0.416	+0.294	0	−0.294	−0.416
	0	z	0	+0.375	+0.530	+0.375	0	−0.375	−0.530	−0.375	0
		β°	0	+51.9	90	−51.9	0	+51.9	90	−51.9	0
−3/4	1/4	x	−0.463	−0.327	0	+0.327	+0.463	+0.327	0	−0.327	−0.463
	0	y	−0.198	−0.140	0	+0.140	+0.198	+0.140	0	−0.140	−0.198
	0	z	0	+0.327	+0.463	+0.327	0	−0.327	−0.463	−0.327	0
		β°	0	+61.8	90	−61.8	0	+61.8	90	−61.8	0
−1	1/4	x	−0.441	−0.312	0	+0.312	+0.441	+0.312	0	−0.312	−0.441
	0	y	0	0	0	0	0	0	0	0	0
	0	z	0	+0.312	+0.441	+0.312	0	−0.312	−0.441	−0.312	0
		β°	0	90	90	90	0	90	90	90	0

$d/L_0 = 0.05$

Y/d	X/L	$\vec{Z}/L'$	0	1/8	1/4	3/8	1/2	5/8	3/4	7/8	1
0	1/4	x	−1.034	−0.731	0	+0.731	+1.034	+0.731	0	−0.731	−1.034
	0	y	−1.000	−0.707	0	+0.707	+1.000	+0.707	0	−0.707	−1.000
	0	z	0	+0.731	+1.034	+0.731	0	−0.731	−1.034	−0.731	0
		β°	0	+46.0	90	−46.0	0	+46.0	90	−46.0	0
−1/4	1/4	x	−0.908	−0.642	0	+0.642	+0.908	+0.642	0	−0.642	−0.908
	0	y	−0.714	−0.505	0	+0.505	+0.714	+0.505	0	−0.505	−0.714
	0	z	0	+0.642	+0.908	+0.642	0	−0.642	−0.908	−0.642	0
		β°	0	+51.9	90	−51.9	0	+51.9	90	−51.9	0
−1/2	1/4	x	−0.822	−0.581	0	+0.581	+0.822	+0.581	0	−0.581	−0.822
	0	y	−0.459	−0.325	0	+0.325	+0.459	+0.325	0	−0.325	−0.459
	0	z	0	+0.581	+0.822	+0.581	0	−0.581	−0.822	−0.581	0
		β°	0	+60.9	90	−60.9	0	+60.9	90	−60.9	0
−3/4	1/4	x	−0.771	−0.545	0	+0.545	+0.771	+0.545	0	−0.545	−0.771
	0	y	−0.225	−0.159	0	+0.159	+0.225	+0.159	0	−0.159	−0.225
	0	z	0	+0.545	+0.771	+0.545	0	−0.545	−0.771	−0.545	0
		β°	0	+73.8	90	−73.8	0	+73.8	90	−73.8	0
−1	1/4	x	−0.755	−0.534	0	+0.534	+0.755	+0.534	0	−0.534	−0.755
	0	y	0	0	0	0	0	0	0	0	0
	0	z	0	+0.534	+0.755	+0.534	0	−0.534	−0.755	−0.534	0
		β°	0	90	90	90	0	90	90	90	0

TABLE 4-X (continued)

$d/L_o = 0.02$

Y/d	X/L	$\vec{Z/L'}$	0	1/8	1/4	3/8	1/2	5/8	3/4	7/8	1
0	1/4	x	−1.502	−1.062	0	+1.062	+1.502	+1.062	0	−1.062	−1.502
	0	y	−1.000	−0.707	0	+0.707	+1.000	+0.707	0	−0.707	−1.000
	0	z	0	+1.062	+1.502	+1.062	0	−1.062	−1.502	−1.062	0
		β°	0	+56.3	90	−56.3	0	+56.3	90	−56.3	0
−1/4	1/4	x	−1.424	−1.007	0	+1.007	+1.424	+1.007	0	−1.007	−1.424
	0	y	−0.736	−0.520	0	+0.520	+0.736	+0.520	0	−0.520	−0.736
	0	z	0	+1.007	+1.424	+1.007	0	−1.007	−1.424	−1.007	0
		β°	0	+62.6	90	−62.6	0	+62.6	90	−62.6	0
−1/2	1/4	x	−1.369	−0.968	0	+0.968	+1.369	+0.968	0	−0.968	−1.369
	0	y	−0.484	−0.342	0	+0.342	+0.484	+0.342	0	−0.342	−0.484
	0	z	0	+0.968	+1.369	+0.968	0	−0.968	−1.369	−0.968	0
		β°	0	+70.4	90	−70.4	0	+70.4	90	−70.4	0
−3/4	1/4	x	−1.336	−0.945	0	+0.945	+1.336	+0.945	0	−0.945	−1.336
	0	y	−0.240	−0.170	0	+0.170	+0.240	+0.170	0	−0.170	−0.240
	0	z	0	+0.945	+1.336	+0.945	0	−0.945	−1.336	−0.945	0
		β°	0	+79.8	90	−79.8	0	+79.8	90	−79.8	0
−1	1/4	x	−1.325	−0.937	0	+0.937	+1.325	+0.937	0	−0.937	−1.325
	0	y	0	0	0	0	0	0	0	0	0
	0	z	0	+0.937	+1.325	+0.937	0	−0.937	−1.325	−0.937	0
		β°	0	90	90	90	0	90	90	90	0

At intermediate alignments ($Z/L' = 1/8, 3/8, 5/8, 7/8$, ect.) the orbital motions are elliptical, in planes with angles β to the vertical which vary throughout the depth, being horizontal at the bed. By inspection of Table 4-X it is seen that in plan these orbits are circular, the total orbit being made up from this and the vertical amplitude required at the given depth. The direction of circulation of the particles is dependent upon their position across the crest length L'. In each case the side nearest the crest advance alignment ($Z/L' = 0, 1/4, 1$, etc.) has the particle moving in the direction of advance. Arrows in Fig. 4-21 indicate these rotations, which are continuous. Thus, even at the bed, vortex type motion is experienced with a radius of reasonable magnitude. For example, two waves 2 ft. high in water where $d/L_o = 0.1$ will generate circular orbits of water particles at the bed with radii = $(2\sqrt{2}/2)0.441 = 0.624$ ft.

The velocities of water particles in the direction of advance of the combined crest (u) and transverse to it (crest length alignment, w) have maximum values at the bed given by:

$$\frac{2u_{max} C_s}{gH \cos 2\pi Z/L'} = f(d/L_o, \theta) = X \tag{4-100}$$

$$\frac{2w_{max} C_s \tan \theta/2}{gH \sin 2\pi Z/L'} = f(d/L_o, \theta) = X \tag{4-101}$$

EFFECTS OF VISCOSITY

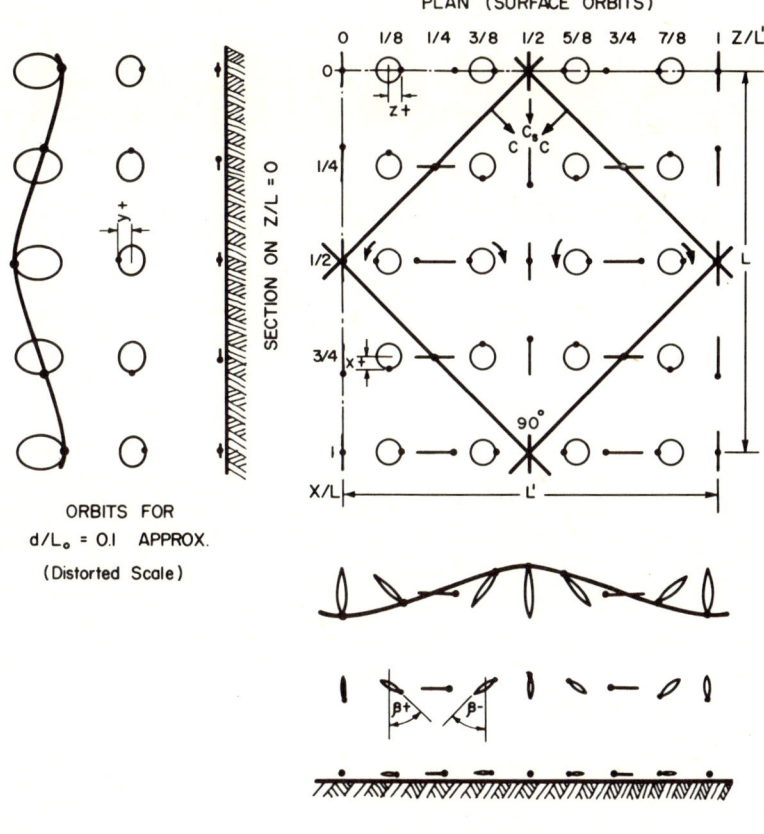

Fig. 4-21. Water-particle orbits in a short-crested wave system.

These relationships are graphed in Fig. 4-22 from which values can be substituted into eq. 4-100 and 4-101 once X is determined from a knowledge of d/L_o and θ. When $\theta = 180°$ the system reverts to a progressive wave and $w_{max} = 0$.

EFFECTS OF VISCOSITY

Although the viscosity of water does not play any vital role in the propagation or particle motions of waves in general, it has a significant influence at the boundaries of bed and surface. The boundary layers developed from the oscillatory flow, in conjunction with the differential forward and backward velocities throughout the orbits, cause a migration of water in the direction of wave advance. This net movement or *mass transport* at surface and bottom is presumably counterbalanced

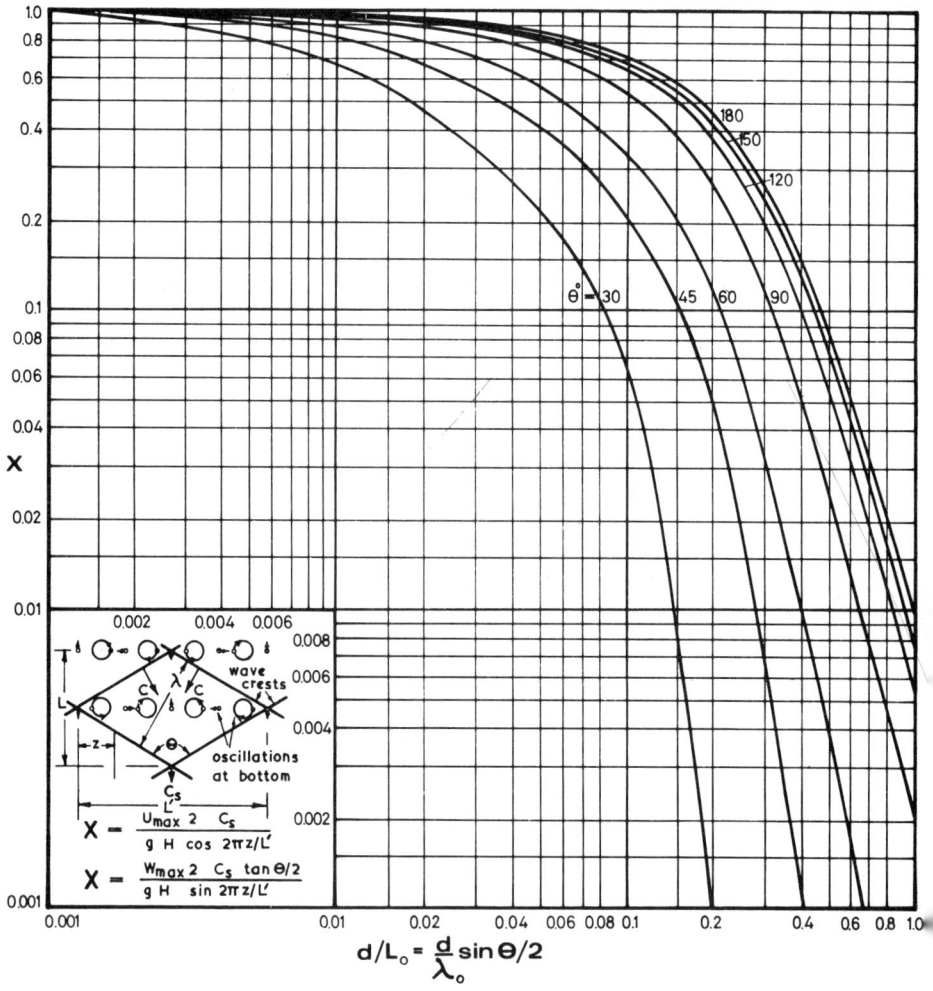

Fig. 4-22. Maximum horizontal velocities of water particles at the bed in a short-crested wave system.

by a backward flow at intermediate depths in order that there be no net drift across the complete vertical plane transverse to the wave orthogonal, although this is difficult to prove in prototype situations.

Besides the above major effect of viscosity there is an attenuation of wave energy through the shear stresses developed in the boundary layer. This influence is greatest on waves in shallow water. Whilst on the subject of wave attenuation, that due to bottom permeability and mud suspensions can also be considered since viscosity and apparent viscous changes enter the calculations.

Viscous damping

It has been accepted previously that for deep-water waves $t_d = L^2/8\nu\pi^2$ were t_d was the time to reduce the wave amplitude to 0.37 of its original value. The expression for transitional and shallow-water conditions is:

$$t_d \cdot 3600 = \frac{L^2}{8\pi^2 \nu} \bigg/ \left[1 + \frac{L}{(T\nu\pi)^{1/2}\, 4 \sinh 4\pi d/L}\right] \qquad (4\text{-}102)$$

were t_d is expressed in hours.

Eq. 4-102 has been graphed in Fig. 4-23 for a 3-sec wave and $\nu = 12 \cdot 10^{-6}$ ft.^{2}/sec. The assumptions implicit in this solution are that water motions at a plane impermeable bed are attenuated by laminar boundary-layer conditions. Comparisons with solutions by Lamb [44], Stokes [44] and Biésel [45] are also illustrated in Fig. 4-23.

Iwagaki and Tsuchiya [46] have derived attenuation relationships for both laminar and turbulent boundary-layer conditions. The former can be converted to the form:

$$t_d/T^{5/2} = 0.01845 \tanh 2\pi d/L\, (\sinh 4\pi d/L + 4\pi d/L) \qquad (4\text{-}103)$$

where t_d is time in hours to reduce a wave to 0.37 of its original height. This equation has been graphed in Fig. 4-24 and can be shown to be identical with the curve for a 3-sec wave train in Fig. 4-23. The relationship for the turbulent boundary layer gives values of t_d which are 0.63 of those for the laminar condition. This also is graphed in Fig. 4-24.

The above energy dissipation occurs within the boundary layer formed by the oscillatory water-particle motion at the bed. This thickness of water is given by:

$$\delta = 2(\pi T\nu)^{1/2} \qquad (4\text{-}104)$$

which is very small even for the longer components in the spectrum of wind-generated waves, as can be noted from Table 4-XI.

It will be seen in the chapter dealing with tidal motion and other long waves that such viscous dissipation is more pronounced and also has the effect of shortening

TABLE 4-XI

Boundary-layer thickness for laminar conditions

T (sec)	4	7	10	13	16
δ (ft.)	0.025	0.035	0.039	0.044	0.049
δ (cm)	0.76	1.0	1.2	1.35	1.5

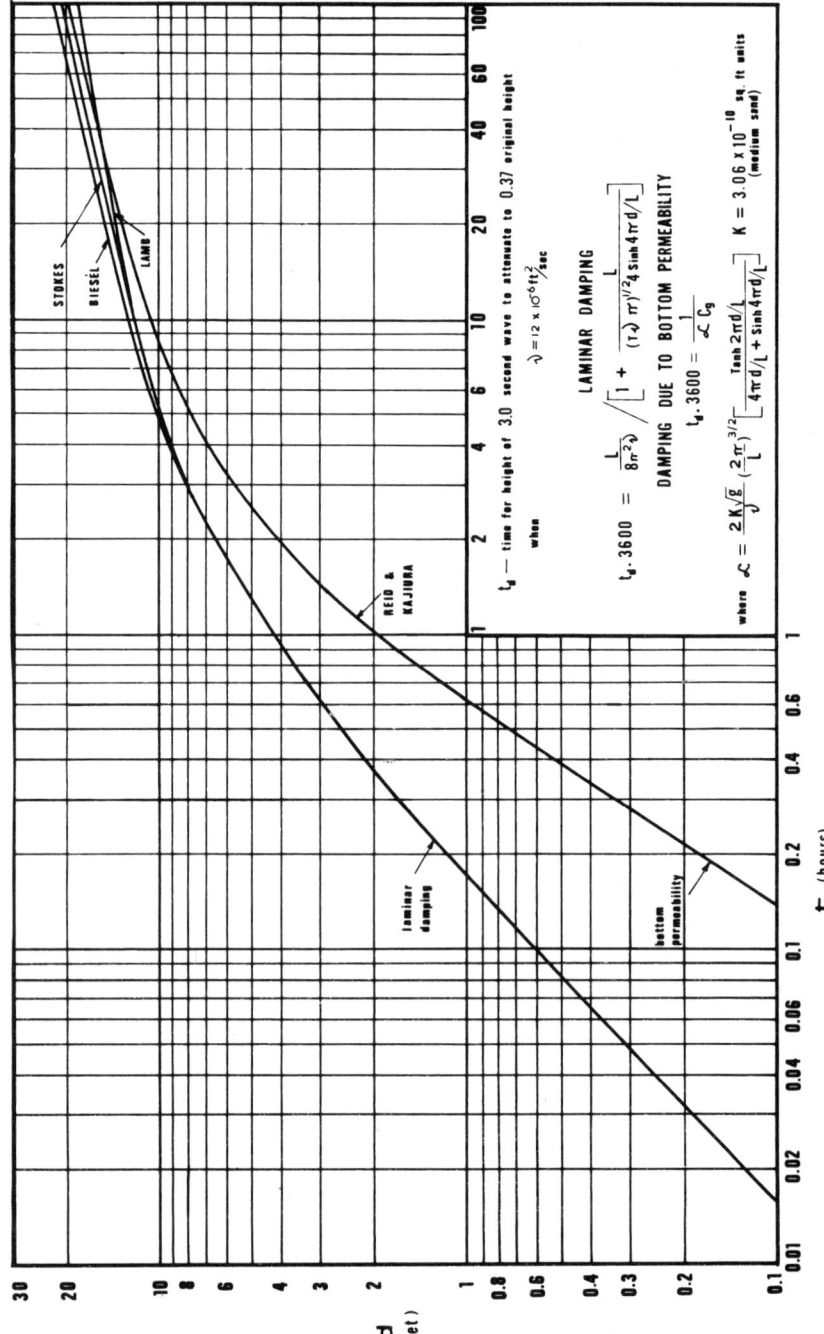

Fig. 4-23. Attenuation of waves by viscous damping and bottom permeability.

EFFECTS OF VISCOSITY 195

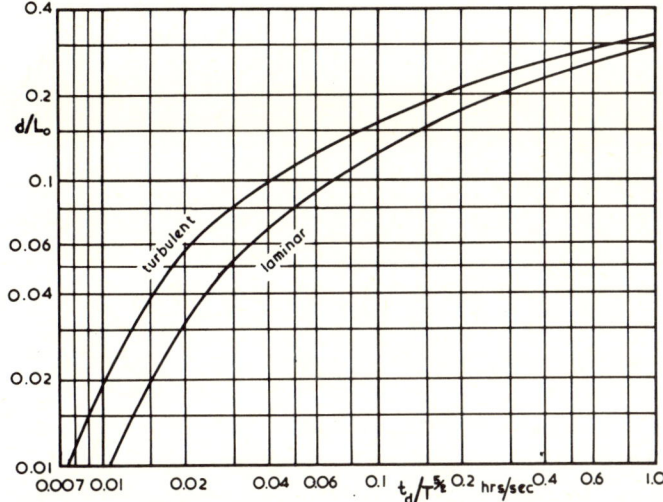

Fig. 4-24. Attenuation of waves by viscous friction in laminar or turbulent boundary layer.

the wave length. However, the boundary layer in these cases will be turbulent so that an exact solution as above cannot be used.

It has been shown [46] that average energy dissipation at the water surface through viscosity is no more than 1% of that at the bed. However, partial breaking in the upper region is bound to consume energy, so that the present analysis applies strictly to swell-type waves.

Laboratory verification of attenuation [47, 48] has shown that model waves suffer a reduction some 2 to 3 times that indicated above. This could be due to an excessive boundary-layer thickness compared to the wave length, as occurs in most hydraulic models. Although this rate of attenuation is still insignificant for most model tests, its importance in sedimentation processes should be stressed, pointing to the need for full-scale reproduction when wave influence on a sandy bed is being studied.

Damping due to bottom permeability

As a wave proceeds shorewards the pressure on the bed fluctuates. This temporal and spatial pressure gradient at the sedimentary face causes water to flow through the top grains. This action absorbs energy and so helps in the attenuation of the wave. On the assumption that pressure reduces exponentially with depth or distance into the bed, an attenuation factor [49] α in the equation:

$$t_d \cdot 3600 = 1/\alpha C_g \qquad (4\text{-}105)$$

where t_d is the time (hours) to reduce the wave height to 0.37 of its original value, is given by:

$$\alpha = \frac{2Kg^{1/2}}{\nu}\left(\frac{2\pi}{L}\right)^{3/2}\left[\frac{\tanh 2\pi d/L}{4\pi d/L + \sinh 4\pi d/L}\right] \qquad (4\text{-}106)$$

where K is the permeability of sand.

For medium sand $K = 3.06 \cdot 10^{-10}$ (ft.^{2}) and substituting $\nu = 12 \cdot 10^{-6}$ (ft.^{2}/sec) a presentation is made in Fig. 4-23 of the attenuation of a 3-sec wave. It is seen that it is of the same order as that for laminar damping.

Since energy dissipation in this case is due to the rate of change of pressure on the bed, it is influenced by wave height and wave period. It has been shown [49] that this gradient is optimum at $d/L_o = 0.13$. This implies that a spectrum of waves will be attenuated differentially, with the central peak components being more greatly affected than the shortest and longest bands at its extremities.

Damping due to non-rigid bottom

Where a band of viscous fluid, with density greater than the water in which waves are propagating, overlays a bed, it will dissipate energy at a much greater rate than where a relatively impermeable and rigid bed exists. Gade [50] has derived a theoretical solution for $d/L < 0.05$ which shows that wave decay is optimum when the thickness of this viscous layer (for example, mud) is about 130 times the normal boundary layer thickness for the wave. Even half this thickness will strongly attenuate waves. Inspection of Table 4-XI will indicate that this non-rigid boundary does not have to be large at all to be an effective dissipator. For example, a 10-sec wave will quickly disappear in water in which the mud at the bed is 8 ft. thick, a 4-sec wave in water with a 3-ft. mud layer. The author has observed waves of around 4 sec arriving in mud flats and in a matter of a few wave lengths are completely attenuated. Gade [50] mentions a mud hole off the Louisiana coast which fishermen use as an emergency harbour in times of storm.

Price et al. [51, 52] have used this concept of a non-rigid bottom to reduce wave heights at a beach where accretion of sediment was desired. They used clumps of stranded polypropylene over a wide area of the beach profile, both in laboratory and in prototype tests, which were inconclusive. Such a solution could bring in its wake other problems of contamination of cooling water intakes and of beaches generally.

Mass transport

It will already have been gathered that water-particle motion due to wave propagation is not simply circular, elliptical, or oscillatory. Each particle does not

TABLE 4-XII

Time to steady state of mass transport through viscous conduction

Depth d (ft.)	10	30	50	100	200	400
Time t (days)	100	900	2500	10,000	40,000	160,000

return to its original position after the passage of a wave; it could be in advance of, or to the rear of this location, depending upon its depth.

This net movement, or mass transport of the water, arises from the viscosity of the water and the differential forward and backward velocities of the particles in their orbits. A boundary layer exists at the bed within which strong velocity gradients occur. These in turn produce large vorticities. Although the viscosity of sea water is extremely small, and hence the boundary layer very thin (see values in Table 4-XI), the vorticity that is generated is finite. This vorticity is diffused from the bed and the surface to other depths by means of viscous conduction and convection of the pseudo-current generated at these boundaries.

The transfer of vorticity to the interior of the fluid by viscous conduction [53] takes time, in fact the order of this time to a steady state is d^2/ν. Assuming the normal value of $\nu = 12 \cdot 10^{-6}$ ft.^{2}/sec, time for a range of depths is given in Table 4-XII.

The long time periods specified in this table may be of little consequence since swell is continually propagating across continental shelves (considered 400 ft. depth at the edge). However, the heights, periods and direction are continually changing so that the possibility of a steady mass transport at intermediate depths is rather remote.

Transfer by convection, which will exist concurrently with that by viscous conduction, will also take time to reach a steady state. In this case conditions must be specified at either end of the propagation zone, length of channel in a laboratory or a proportion of the width of continental shelf in an oceanic situation. Longuet-Higgins [53] gives the order of time as:

$$t = WTL/\pi^2 H^2 \; 3600 \times 24 \tag{4-107}$$

for t = time in days to steady state; W = width of shelf to deep-water limit of wave in ft. (assume slope of 0.002); T = period of wave arriving normal to coast; L = wave length at mid-depth across width W; H = wave height in ft.

Assuming $H = 3$ ft., the time is given in Table 4-XIII for a steady mass-transport distribution to exist through convection from the upper and lower boundaries.

It seems from the rough assessments above that mass-transport "velocity" at mid-depths may not exist for much of the time. The mathematical solution available [53] is difficult to verify in nature, although flume tests [54, 55] have shown

TABLE 4-XIII

Time to steady state of mass transport through convection

Period T (sec)	4	7	10	13	16
Time t (days)	0.82	13.5	80	300	840

the equations to be valid. In the laboratory, lengths for wave propagation are limited enough for steady-state transport to be effected reasonably quickly.

Of the two processes mentioned above that for viscous conduction is the only one for which a solution is available, only then by the omission of inertia terms of the particle motion. This implies that the ratio $H/2\delta \ll 1$, where the boundary layer thickness $\delta = 2(T\nu\pi)^{1/2}$. This severe restriction applies to the interior of the liquid where the convection solution can provide some assistance. However, fortunately for the engineer, the above restriction does not apply to the solution within the boundary layer at the bed.

Progressive wave

As soon as wave action commences the net motion forward at the bed occurs. The distribution of this mass transport within the boundary layer for a progressive wave is graphed in Fig. 4-25A. It is seen that for the majority of the layer the velocity is reasonably uniform. However, a maximum is experienced 0.3 δ from the bed which is 1.1 times the value at the upper boundary. The mass-transport velocity

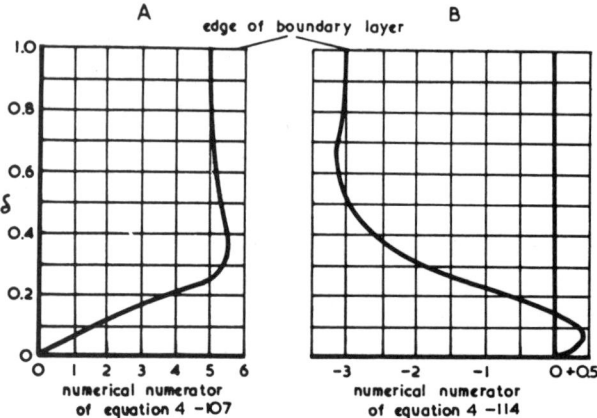

Fig. 4-25. Distribution of mass transport throughout boundary layer. A. of progressive wave; B. of standing wave.

U_{BP} within the boundary layer of a progressive wave is given by:

$$U_{BP} = \frac{5.0\pi^2 H^2}{4TL \sinh^2 2\pi d/L} \qquad (4\text{-}108)$$

or:

$$xT^2/H^2 = \frac{2.41}{\sinh^2 2\pi d/L \tanh 2\pi d/L} \qquad (4\text{-}109)$$

where x is the progress (in ft.) per wave period T (sec). For the peak velocity indicated in Fig. 4-25A the constants in the numerators of eq. 4-108 and 4-109 should be 5.5 and 2.65, respectively. Both these equations are graphed in Fig. 4-26 where they are seen to be quite close.

The distribution of mass transport through the whole depth of water d, where locations are given by depth y measured positively upwards from the surface (i.e., at bed $y = -d$), is given by:

$$U_P = \frac{\pi^2 H^2}{4TL \sinh^2 2\pi d/L} \left[2 \cosh\{4\pi(y+d)/L\} + \frac{2\pi}{dL}(3y^2 + 4yd + d^2) \sinh 4\pi d/L \right.$$

$$\left. + 3 + \frac{3}{d^2}(y^2 - d^2)\left(\frac{\sinh 4\pi d/L}{4\pi d/L} + \frac{3}{2}\right) \right] \qquad (4\text{-}110)$$

It is seen that when $y = -d$, eq. 4-110 reverts to the value of U_B given by eq. 4-108.

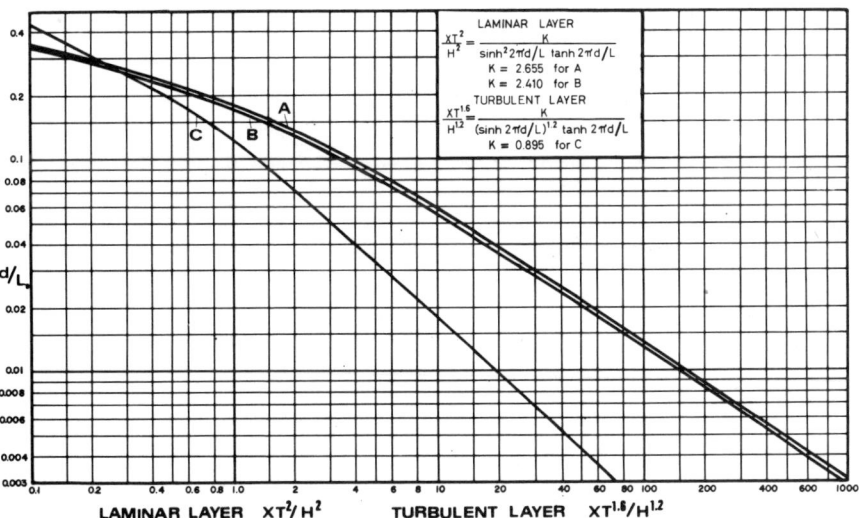

Fig. 4-26. Mass transport at bed produced by progressive waves.

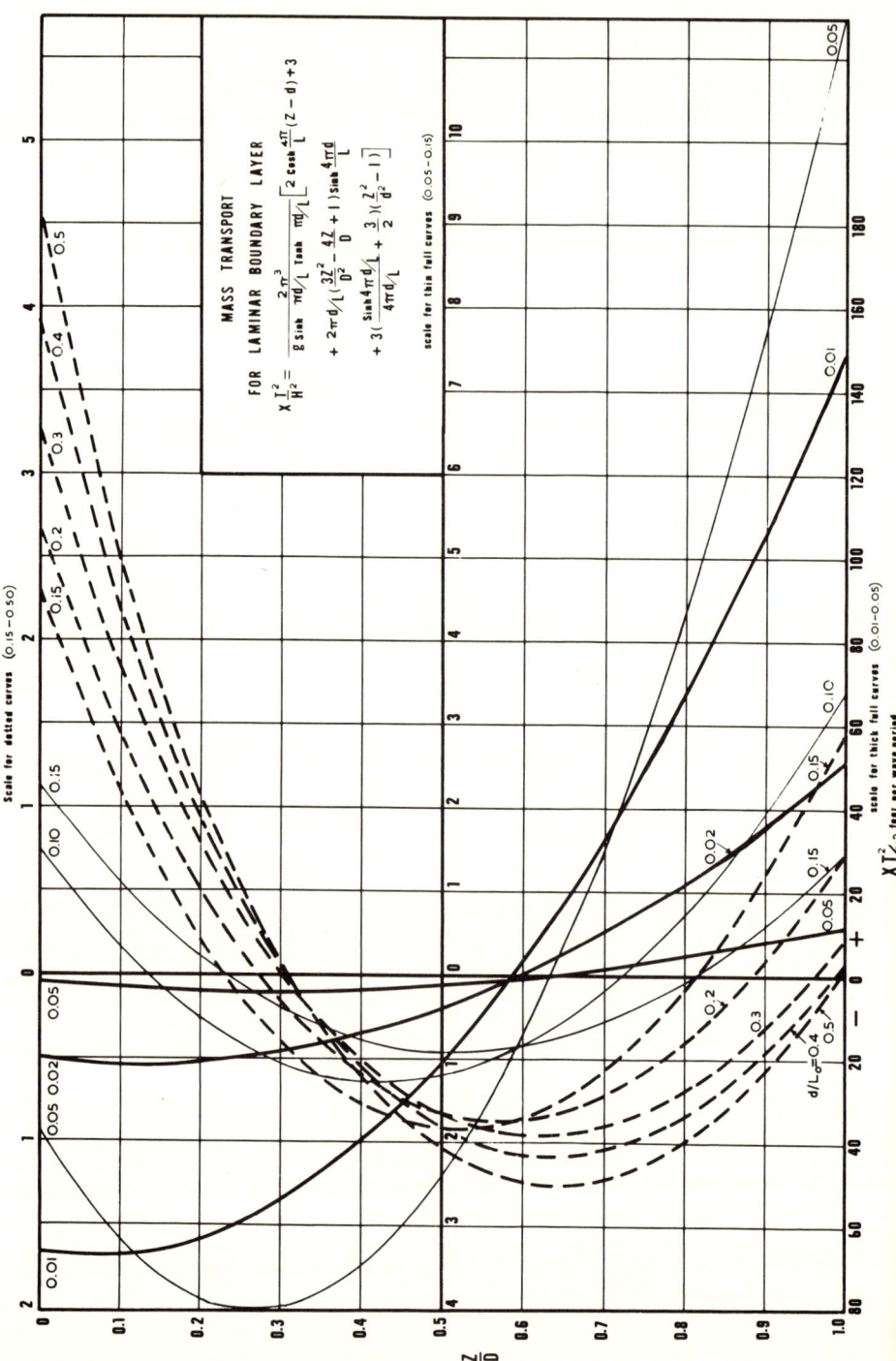

EFFECTS OF VISCOSITY

The initial factor times the first term in the bracket is the non-periodic term in the equation of Stokes [3] for mass transport.

The distribution according to eq. 4-110 is depicted in Fig. 4-27, where three horizontal scales are utilized in order to present the large variety of values of xT^2/H^2. The various curves denote gradations of d/L_o of which limiting values can be determined from the equation by substituting particular values of $2\pi d/L$. For example, for small $2\pi d/L$ eq. 4-110 becomes:

$$U_P = \frac{5H^2(gd)^{1/2}}{32d^4}(3y^2 - d^2) \tag{4-111}$$

making U_P zero when $y = -d/\sqrt{3}$, with a normal approach to the water surface. It is implicit in the analysis [53] that $H/D \ll 1$, but for comparative studies the presentation should be acceptable. When $2\pi d/L$ is very large the mass transport becomes:

$$U_P = \frac{2\pi^4 H^2}{gd\,T^3}(3y^2 + 4yd + d^2) \tag{4-112}$$

making U_P zero at $y = -d$ and $-d/3$, with the tangent to the curve vertical at $y = -2d/3$.

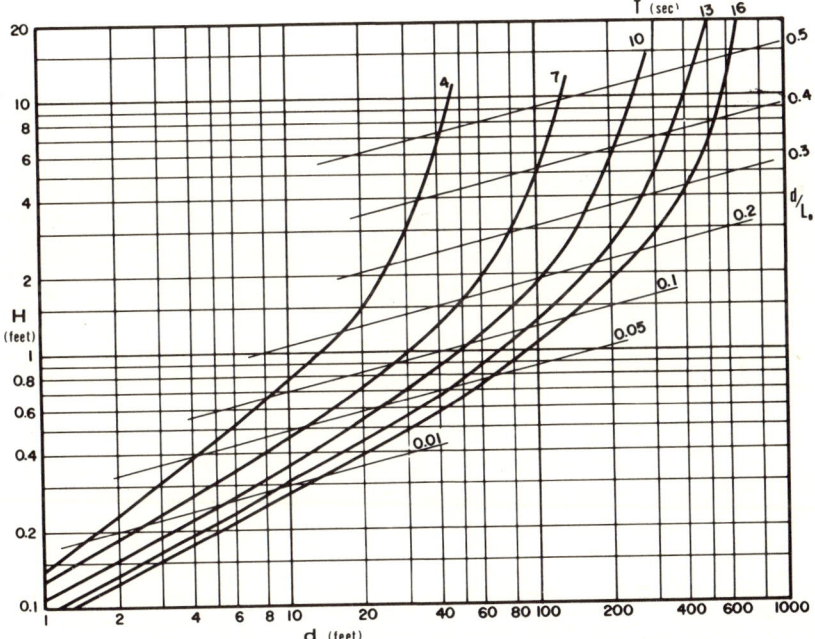

Fig. 4-28. Wave characteristics to produce a turbulent boundary layer on a smooth bed.

The analysis discussed so far is based upon a laminar boundary layer on a smooth bed. Beyond a value of 1000 for the wave Reynolds number (= $U_{max}\delta/\nu$) the boundary layer becomes turbulent. This limit can be expressed in terms of H, T and d as in Fig. 4-28, which depicts the limits at which this critical condition is reached. For example, a 13-sec wave in 70 ft. of water cannot exceed 1 ft. if the boundary layer is to remain laminar.

Very little work has been carried out on mass transport with turbulent boundary layers, which can exist with both smooth and rough beds. In nature the rough bed would be the normal condition, since any sedimentary surface will contain ripples or dunes besides having the particle roughness. Brebner et al. [56] conducted a pilot study with beds of various roughnesses. They found that with a laminar boundary layer over a rough bed the mass transport was in excess of that for the smooth bed predicted by eq. 4-107. This was probably due to distortion of the velocity distribution as in Fig. 4-25A, where the maximum peak was extended at the expense of velocity in the remainder of the boundary layer.

It was also found [57] that when the Reynolds number defined by $U_{max}\delta/\nu = 1000$, the boundary layer was fully turbulent and mass transport velocity at the bed varied as follows:

$$U_{BP} = \frac{K}{L}\left(\frac{H}{\sqrt{T}\,\sinh 2\pi d/L}\right)^{1.2} \tag{4-113}$$

which from the data presented [56] resulted in the equation:

$$\frac{X\,T^{1.6}}{H^{1.2}} = \frac{0.895}{(\sinh 2\pi d/L)^{1.2}\,\tanh 2\pi d/L} \tag{4-114}$$

which has been graphed in Fig. 4-25. Eq. 4-113 needs further experimental verification, but until this is to hand it should serve for qualitative studies of mass transport in prototype situations where a fully turbulent boundary layer will normally exist.

Standing wave

The mass transport in a standing wave, either complete clapotis or a partial one, is determined from the equations for two progressive waves travelling in opposite directions [53]. The profile of this transport within the boundary layer is exhibited in Fig. 4-25B. It is seen to have a negative value at the outer boundary and for the bulk of the width of the layer. At 0.15 of the thickness from the bed this velocity becomes zero and at 0.078 δ it has a maximum positive value which is 0.13 of the maximum negative. The magnitude of upper velocity is given by:

$$U_{BS} = \frac{3\pi^2 H^2}{8TL\,\sinh^2 2\pi d/L}\sin 4\pi x/L \tag{4-115}$$

where x is measured along the wave length (origin at nodal point).

Noda [58] has developed eq. 4-115 by a perturbation method used previously [59] for wave damping due to bottom friction. He has also verified it experimentally for waves of reasonable steepness. Waves of very small height in fact showed much greater mass transport at the edge of the boundary layer.

Eq. 4-115 gives the mass-transport velocity at the upper edge of the boundary layer and establishes cells of circulation in compartments $L/4$ in length as depicted

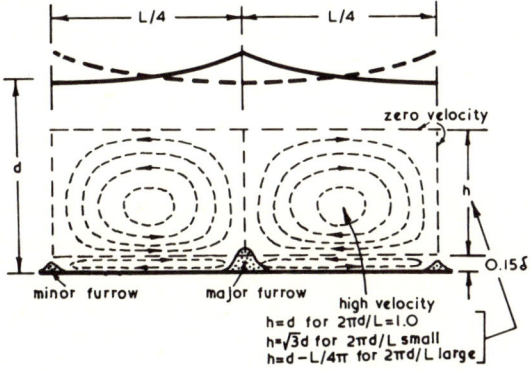

Fig. 4-29. Mass transport cells within standing waves.

in Fig. 4-29. Directions in these alternate each quarter wave length and their vertical extent varies with the ratio d/L; for small d/L the centre is at $d/\sqrt{3}$ from the bed with the circulation extending to the surface, for large d/L the centre is at $L/4\pi$ from the bed. The convection solution indicates that instead of single cells each quarter wave length there are several, each one rotating oppositely to its neighbour above and below, the bottom one being in the same direction as indicated in Fig. 4-29 for the conduction solution.

For purposes of understanding sediment transport due to standing waves the circulation in the bottom cells is extremely important. As noted previously there will be a reverse motion within the boundary layer (see Fig. 4-25B) constituting a cell $L/4$ in length and approximately $0.25\,\delta$ in vertical magnitude. These minor cells are considered unimportant where sediment movement is concerned, since sand grains will be quickly thrown into the larger cell above and so build furrows of sediment at the antinodes of the wave oscillation. Noda [58] refers to records of furrows in harbour basins mainly at the antinodes, but some smaller ones occur at the nodes due to the transport within the lower quarter of the boundary layer. He exhibited these two contrary trends in sediment movement in the laboratory, and also changes when a turbulent boundary layer was established.

Short-crested wave

Theory presented previously for the angled wave trains or short-crested system has indicated a complex motion of water particles which varies in depth and position horizontally. For this purpose axes were established along the line of propagation of the combined crest (diagonal of the diamond pattern) and the crest length measured normal to this line (other diagonal). Wherever orbital motion takes place, particularly near the bed, viscous effects will cause a net advance in the direction in which transient high velocities occur. Because of the changing pattern of orbital motion along alignments parallel to crest advance, so will the mass-transport velocity vary both in magnitude and direction.

Fuchs [43] has presented a formula for this mass-transport velocity which was published prior to Longuet-Higgins' paper [53] on the progressive and standing wave. His relationship indicates the variability across the wave length L', but the magnitude at the bed should agree with the latter paper when $L' \to \infty$ and the two angled wave trains become a single progressive wave. With this modification, the mass-transport at the bed is given by:

$$\frac{xL}{H^2} = \frac{5\pi^2}{8\tanh^2(2\pi d/L)\sqrt{1+(L/L')^2}} \left[(1 - \tanh^2 \frac{2\pi d}{L} \sqrt{1+(L/L')^2} \right.$$

$$\left. + \cos\frac{4\pi z}{L'} \left(\frac{L'^2 - L^2}{L'^2 + L^2} - \tanh^2 \frac{2\pi d}{L} \sqrt{1+(L/L')^2} \right) \right] \qquad (4\text{-}116)$$

where x is the net movement at the bed per wave period for a laminar boundary layer and smooth bed. The dimension of x is dictated by the dimensions of H and L used, which should be consistent.

Along the crest alignments ($Z/L' = 0, 1/2, 1, ...$) mass-transport will be a maximum since $\cos 4\pi Z/L' = +1$. Midway between these crest passages ($Z/L' = 1/4, 3/4, 5/4, ...$) the transport will be least since $\cos 4\pi Z/L' = -1$. At intermediate half-way alignments ($Z/L' = 1/8, 3/8, 5/8, ...$) the value of net movement is between the above two values since $\cos 4\pi Z/L' = 0$. What is more important to note, perhaps, is the direction of mass transport at these various locations. Along the crest alignments, as would be expected it is in the direction of advance, or the bisector of the angle between the two wave trains. At the $L'/4$ points net movement is transverse to crest propagation since the rectilinear oscillation of the water particles is in this direction. At the $L'/8$ alignments water particles rotate in circles but their velocities are not constant. They therefore have a net movement towards the crest paths and partly in the direction of wave advance. Thus the resultant effect on a sedimentary bed would be to remove material from the $L'/4$ zones and force it into and along the paths of the wave crests. This would result in furrows being constructed along the crest paths. When the two wave trains are of unequal height the influence of one wave will be greater than the other, with a concomitant larger mass-transport velocity. The direction of this net movement will not then be along the crest alignments but angled to it. This may be the reason for elongated dunes near the beach where waves are being partially reflected from it [60].

Comparison of mass-transport velocity for a given progressive wave, in a given depth of water for similar bed conditions, with that of the system produced by its reflection from a wall, should take into account the magnified height to be used in the equations. In the case of a short-crested system the height $h = \sqrt{(h_1^2 + h_2^2)}$, which is $\sqrt{2}h_1$ if the reflected wave is equal to the incident wave. The net movement per wave period (x) will then be doubled. When this is considered, in conjunction with the macroturbulence possible from a short-crested wave system, the influence on a sedimentary bed can be appreciated.

This capacity for erosion in front of sea walls is not restricted to sand since Horikawa and Sunamura [61] have reported a three-fold magnification in rock bed removal in such locations over normal cliffed zones where a partial beach exists. This erosive characteristic of short-crested wave systems in shallow water is to be discussed in greater detail later, but the point is so important that it is stressed whilst the theory and physical processes are being considered.

Influence of suspended sediment

The magnitudes of mass transport as given by formulae presently available are based upon certain bed roughnesses and types of boundary layer. The influence of either viscosity or density changes in the fluid near the bed has not to date been evaluated. However, from the knowledge of height attenuation that is effected by a non-rigid bottom it can be visualized that mass transport within a layer of mud can be many times greater than for water of constant viscosity or density throughout. On such muddy coasts it can be observed that local beaches of shell debris can be constructed by waves transporting this material on top of the denser mud layer.

Dore [62] has carried out a theoretical analysis of mass transport in a two-layered fluid system, consisting of two homogeneous fluids of differing density and viscosity. He found that for the progressive wave "the mass-transport velocity can formally be an order of magnitude larger than obtained by Longuet-Higgins [53] for a single homogeneous fluid". The distribution in either layer is independent of the wave length, that in the upper layer being the same as for the single layered system when d/L is negligible [53]. The profile of transport velocity is illustrated in Fig. 4-30 where it is seen that the interfacial values are equal and in the direction of wave advance. The parabolic distributions provide for negative velocities, necessitated by the need for no net drift over the complete vertical plane for either fluid.

The application of this knowledge to river and estuary mouths where fine sedi-

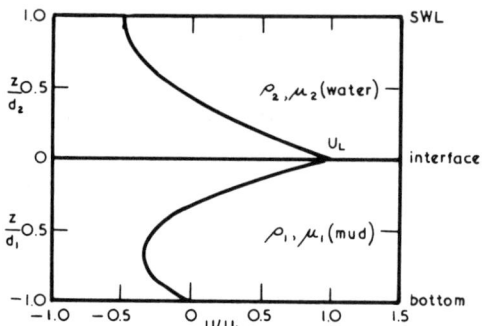

Fig. 4-30. Mass transport velocity distribution in two-layered fluid system (Ref. 62).

ment is predominant and swell waves can enter, explains the ready return upstream of mud and contaminants it may contain. This problem should be considered in conjunction with the penetration of the saline wedge and transport of waves by a flooding tidal current. Such aspects are discussed in the chapter on shoreline processes.

EXAMPLES

1
Aerial photographs of a coastline display the presence of two wave systems, one with crests 200 ft. apart and another more complex with crests at 40 ft. spacing. Timing of major breaking on the beach in the same period indicated a wave period for the longer waves of 10 sec. What was the depth of water in the zone of wave observation and what was the period of the minor wave system?

From eq. 4-7, $L = (gT^2/2\pi)$ tanh $2\pi d/L$, so that $200 = (g\ 10^2/2\pi)$ tanh $2\pi d/L$ or tanh $2\pi d/L = 0.39$.
From Appendix I $d/L = 0.062$, $d = 12.4$ ft.
For the second wave train $40 = (gT^2/2\pi) \times$ tanh $2\pi 12.4/40$ so that $T = 6.4$ sec.

2
At about what depth do all wind-generated waves assume the same speed?

From Table 4-IV, for an accuracy of 0.36% in use of $(gd)^{1/2}$, the respective $d/L_o = 0.004$. If waves of engineering significance are accepted as those between 7 and 16 sec, the associated depths are given by $d = 0.004 \times 5.12\ T^2 = 1.0$ and 5.25 ft., respectively. If the normally accepted value of $d/L_o = 0.01$ is used these depths become 2.5 and 13.1 ft. All other periods of wind-generated waves will lie between these limits, with the greatest energy being concentrated around the 12-sec period, which has respective shallow-water limits of 3 and 7.5 ft. Thus, for normal swell waves similar speeds, dependent solely upon depth, do not occur until waves are near breaking point (see Chapter 5).

3
A 12-sec swell wave is approaching a coast normally. In deep water it has a height of 3 ft. At

EXAMPLES

locations d/L_o = 0.5, 0.1 and 0.01, find the values of u_{max} at the surface, mid-depth and at the bed by 1st-, 2nd- and 3rd-order Stokes theory. Compare the results with values from eq. 4-70 or Fig. 4-13 for the d/L_o = 0.01 case and with readings from Fig. 4-15 for all depths ratios.

H_o = 3.0, from eq. 4-38 and 4-36 or Appendix I, H/H_o gives heights of 2.8 and 5.07 at d/L_o = 0.1 and 0.01, respectively. For Stokes theory u_{max} is given by eq. 4-17, 4-19, 4-22 or Fig. 4-4 and eq. 4-30. Results and intermediate values are given in Table 4-XIV.

To utilize Fig. 4-13, H/d must be determined from vaules of d computed at the head of Table 4-XIV. For parameter HL/d^2 in Fig. 4-16, H/L from Fig. 4-5 is determined for the d/L ratio operating.

Observations of the results in Table 4-XV would prompt the following conclusions:

(a) Stokes-III theory is applicable only in transitional depths.

(b) In transitional and shallow water the solitary wave theory gives higher u_{max} values. If actual surface velocities were used these would be much higher still.

(c) The empirical data from Fig. 4-16 are results between the solidarity wave values for shallow-water conditions (i.e., high values of HL/d^2). At d/L_o = 0.1 they give velocities similar to the solitary for y/d = −0.5 and −1.0 but much greater values at SWL. In deep-water conditions this is also the case, even when the alternative dotted SWL line is used for evaluation. Inspection of results in the references cited [35, 38, 39, 40] would indicate that linear theory underestimates SWL and surface horizontal velocities.

4

A 13-sec wave with height of 3.5 ft. in deep water approaches the coast normally. Compare the values of C, H/H_o and a_c/H at depth ratios d/L_o = 0.5, 0.2, 0.1, 0.05, 0.02, 0.01 and 0.005 by Stokes-I, Stokes-III, Cnoidal-I and Hyperbolic wave theories.

For H_o = 3.5 ft. and T = 13 sec, H_o/L_o = 0.004, the relative values of d/L and d are listed in Table 4-XV.

Stokes-I theory for C/C_o is from eq. 4-11, H/H_o from eq. 4-38, $a_c/H \equiv 0.5$.

Stokes-III theory for C/C_o is from eq. 4-45. This makes little alteration to the linear value except in shallow water, where for d/L_o = 0.01 a multiplying factor of 2.44 results, (0.603/0.248), which is obviously erroneous. Values of H/H_o are taken from Fig. 4-5 by tracing up curve H_o/L_o = 0.004 and taking the ratio $[(H/L)/(H_o/L_o)]L_o/L$. It is seen that this differs so slightly from the linear values that the latter's use is recommended. The a_c/H values are taken from Stokes-II theory in eq. 4-39 and it is seen again that little influence is made until shallow water is reached when it exaggerates the value.

The cnoidal values for C/C_o depend upon values of H/d and HL^2/d^3 for their evaluation, for which purpose the linear theory for H/H_o should be used. These are listed in Table 4-XV, together with values of $k, E(k)$ and $K(k)$ from Fig. 4-9. Three lines of C/C_o values are listed successively from eq. 4-53, 4-54 and 4-56. Those from eq. 4-53 are seen to be excessive in deeper water, where the cnoidal theory should not be used in any case. In such water the alternative eq. 4-54 also distorts the picture for d/L_o = 0.5, and gives an apparently high figure compared to the linear theory. Eq. 4-56, to be used in the shoaler water, gives values commensurate with those from eq. 4-53.

The hyperbolic equations should be used according to eq. 4-60 only for $d^2 \leq \pi HL_o/24$, which for $H \doteq 4$ ft. means depths less than 20 ft., or columns d/L_o = 0.02 and 0.01 only. Values of C/C_o come from eq. 4-62 and 4-65 as well as Fig. 4-11 for waves at breaking point, no matter what H_o/L_o; H/H_o from eq. 4-67 or Fig. 4-12; and a_c/H from eq. 4-63 which is very close to values from Fig. 4-5.

Thus, the conclusions from this problem are that:

(a) values of C/C_o from linear theory can be accepted at all depths;

(b) values of H/H_o can also be accepted from linear theory, except perhaps in water shallower than d/L_o = 0.05 when the hyperbolic equations in Fig. 4-10 may be more accurate;

(c) values of a_c/H based on the cnoidal theory, as provided in Fig. 4-5, should be used for all depths and H/L ratios.

TABLE 4-XIV

Results of calculations for Example 3

	$d/L_o = 0.5$ $d/L = 0.502$ d (ft.) = 369 H (ft.) = 3.0			$d/L_o = 0.1$ $d/L = 0.141$ d (ft.) = 73.7 H (ft.) = 2.8			$d/L_o = 0.01$ $d/L = 0.04$ d (ft.) = 7.37 H (ft.) = 5.07		
$y/d =$	0	−0.5	−1.0	0	−0.5	−1.0	0	−0.5	−1.0
Linear									
X: Fig. 4-4	1.0	0.22	0.085	1.4	1.1	0.95	3.9	3.9	3.9
u_{max}: from $X(u_1)$	0.785	0.173	0.067	1.03	0.806	0.697	5.17	5.17	5.17
Stokes III									
B: Fig. 4-7	—	—	—	3.0	1.4	1.0	—	—	—
C: Fig. 4-7	—	—	—	32.5	9.5	4.6	—	—	—
D: Fig. 4-7	—	—	—	45	35	31	—	—	—
u_2 from B	—	—	—	0.02	0.0093	0.0067	—	—	—
u_3 from C	—	—	—	0.00016	0.000047	0.000027	—	—	—
u_4 from D	—	—	—	0.05	0.039	0.0345	—	—	—
$u_{max} = u_1 + u_2$	—	—	—	1.05	0.815	0.704	—	—	—
$u_{max} = u_1 + u_2 + u_3 - u_4$	—	—	—	1.0	0.776	0.67	—	—	—
Solitary									
H/d: Fig. 4-13	—	—	—	0.038	0.038	0.038	0.69	0.69	0.69
$u_{max}/(gd)^{1/2}$: Fig. 4-13	—	—	—	0.03	0.03	0.03	0.505	0.42	0.4
u_{max}: Fig. 4-13	—	—	—	1.46	1.46	1.46	7.78	6.46	6.16
Empirical									
H/L: Fig. 4-5	0.004	0.004	0.004	0.0062	0.0062	0.0062	0.024	0.024	0.024
HL/d^2	0.0159	0.0159	0.0159	0.312	0.312	0.312	15	15	15
$u_{max}/(gd)^{1/2}$: Fig. 4-16	0.0575 [1]	—	—	0.13 [1]	0.034	0.02	0.37	0.34	0.30
u_{max}: Fig. 4-16	6.25	1.48 [3]	0.535 [3]	6.33	1.51	0.975	5.70	5.24	4.62
$u_{max}/(gd)^{1/2}$: Fig. 4-16	0.0186 [2]	—	—	0.092 [2]	—	—	—	—	—
u_{max}: Fig. 4-16	2.025 [2]	0.48 [3]	0.174 [3]	4.5	—	—	—	—	—

EXAMPLES

TABLE 4-XV

Results of calculations for Example 4

Theory	Equations	Variable						
		d/L_o	0.5	0.2	0.1	0.05	0.02	0.01
		d/L	0.502	0.225	0.140	0.094	0.058	0.04
		d	433	173	86.5	43.3	17.3	8.7
		L	862	770	618	461	299	218
Stokes I	4-11	C/C_o	0.996	0.888	0.709	0.531	0.347	0.248
	4-38	H/H_o	0.990	0.918	0.933	1.023	1.227	1.435
		H	3.46	3.21	3.26	3.58	4.29	5.02
		a_c/H	0.5	0.5	0.5	0.5	0.5	0.5
Stokes II	4-45	C/C_o	0.996	0.888	0.709	0.534	0.391	0.603
or III	Fig. 4-5	H/H_o	1.0	0.943	0.957	1.035	1.215	1.457
	4-39	a_c/H	0.5	0.5	0.5.	0.506	0.536	2.245
Cnoidal I		H/d	0.0081	0.0191	0.0385	0.0835	0.246	0.586
		HL^2/d^3	0.032	0.375	1.97	9.45	73.0	365
		k	0	0.18	0.37	0.71	1.0	1.0
		$E(k)$	0	0.02	0.15	0.51	1.0	1.0
		$K(k)$	1.5	1.5	1.57	1.82	3.65	8.0
	4-53	C/C_o	–	1.4	0.896	0.581	0.374	0.307
	4-54	C/C_o	0.842	1.0	0.759	–	–	–
	4-56	C/C_o	–	–	0.809	0.583	0.398	0.326
	Fig. 4-5	a_c/H	0.5	0.505	0.51	0.545	0.725	0.88
Hyper.	4-65	C/C_o	–	–	–	–	0.36	0.28*
	Fig. 4-11	C/C_o	–	–	–	0.57	0.38	0.282
	4-67	H/H_o	–	–	–	–	1.34	2.20*
	4-63	a_c/H	–	–	–	–	0.735	0.86*

* These waves indicated as broken in Fig. 4-11.

5
A 10-sec wave approaches the coast normally and at 10.25 ft. depth strikes a vertical jetty wall. If its deep-water height was 3 ft., find the maximum horizontal and vertical orbital velocities and displacements at the surface, mid-depth and the bed. Assume that the wave is completely reflected. Is this likely? If the bed near the wall is sedimentary, where would you expect to find furrows formed?

$T = 10$ sec, $H_o = 3.0$ ft., in 10 ft. depth $d/L_o = 0.02$, $d/L = 0.0576$, $L = 173.5$ ft., $H/H_o = 1.226$, $H = 3.67$ ft. Because clapotis is complete $\bar{H}$ for standing wave = 7.34 ft. u and v are given by eq. 4-73 and 4-74 which for maximum values are similar to eq. 4-17 and 4-18, which in turn are represented by Fig. 4-4. Similarly, eq. 4-79 and 4-80 for particle displacement can be equated by eq. 4-24 and 4-25 for maximum values. Results are as in Table 4-XVI. Such optimum values are unlikely, since some energy is lost in the reflection process. Furrows will form at antinodes which are at the wall face and quarter wave length intervals from it = 43.4 ft.

TABLE 4-XVI

Results of calculations for Example 5

y/d	0	$-1/2$	-1
X_{hor}	2.6	2.6	2.6
X_{vert}	1.0	0.48	0
u_{max}	6.0	6.0	6.0
v_{max}	2.3	1.11	0
x_{max}	9.5	9.5	9.5
y_{max}	3.67	1.76	0

It should be noted that if u_{max} or v_{max} were computed from the 2nd-order equations (4-91 and 4-92) these would not exceed the linear values, although intermediate values are changed. The reach of the wave from linear theory = 3.67 ft. above SWL but from Fig. 4-18 2nd-order theory shows $\Delta H/H = 0.52$ or $a_c = 1.02 H \doteq 7.5$ ft.

6

A wave train 5.5 sec in period and 2 ft. height strikes a vertical wall at an angle of 45° in 22 ft. of water. Accepting that no energy is dissipated in the process, determine the height and celerity of the combined crests, the maximum horizontal velocities at one eighth crest length intervals across the wave pattern, and the amplitude of motion of the water particle at these locations.

$T = 5.5$ sec, $H = 2$ ft., $H_c = (\sqrt{2}).2 = 2.82$ ft., $\theta = 90°$, $d = 22$ ft.,
$d/\lambda_o = 22(5.12 \times 5.5^2) = 0.142$, $d/\lambda = 0.1766$, $d/L_o = (d/\lambda_o) \sin 45° = 0.1$.
  Fig. 4-22: $X = 0.53$; Fig. 4-20: $C_s/C = 1.7$.
$C_s = 1.7 \times 5.12 \ T \tanh 2\pi d/\lambda = 1.7 \times 5.12 \times 0.804 = 38.4$ ft./sec.

Eq. 4-100: $X = \dfrac{u_{max} \ 2 C_s}{g H \cos 2\pi Z/L'}$ so that $0.53 = \dfrac{u_{max} \ 2 \times 38.4}{g \ 2.82 \cos 2\pi Z/L'}$

$u_{max} = 0.627 \cos 2\pi Z/L' = 0.627$ at $Z/L' = 0, 1/2$, etc.
$\phantom{u_{max} = 0.627 \cos 2\pi Z/L' } = 0.444$ at $Z/L' = 1/8, 3/8$, etc.
$\phantom{u_{max} = 0.627 \cos 2\pi Z/L' } = 0$ at $Z/L' = 1/4, 3/4$, etc.

Eq. 4-101: $X = \dfrac{w_{max} \ 2 C_s \tan \theta/2}{g H \sin 2\pi Z/L'}$

$w_{max} = 0.67 \sin 2\pi Z/L' = 0.627$ at $Z/L' = 1/4, 3/4$, etc.
$\phantom{w_{max} = 0.67 \sin 2\pi Z/L' } = 0.444$ at $Z/L' = 1/8, 3/8$, etc.
$\phantom{w_{max} = 0.67 \sin 2\pi Z/L' } = 0$ at $Z/L' = 0, 1/2$, etc.

Table 4-X for $d/L_o = 0.1$,
  $x = 0.441 \ H/2$ at $Z/L' = 0, 1/2$, etc.
  $ = 0.312 \ H/2$ at $Z/L' = 1/8, 3/8$, etc.
  $ = 0$ at $Z/L' = 1/4, 3/4$, etc.
  $z = 0.441 \ H/2$ at $Z/L' = 1/4, 3/4$, etc.
  $ = 0.312 \ H/2$ at $Z/L' = 1/8, 3/8$, etc.
  $ = 0$ at $Z/L' = 0, 1/2$, etc.

EXAMPLES   211

At $Z/L' = 1/8, 3/8$, etc., the motion is circular with diameter of $0.312 \times 2.82 = 0.88$ ft. and average circumferential velocity of 0.444 ft./sec, particles rotating a complete circle in 5.5 sec.

7

A bay 20 NM in length is fully open to receive swell of 14-sec period. It may be considered of uniform depth equal to 10 fathoms. Find the reduction in height through damping in a turbulent boundary layer over this distance. Compute similarly for bottom permeability using K and ν values as in Fig. 4-23.

$T = 14$ sec, $d = 6 \times 10 = 60$ ft., $d/L_o = 60 (5.12 \times 14^2) = 0.06$. From Fig. 4-24, $t_d/T^{5/2} = 0.021$, $t_d = 15.6$ h to reduce the wave to 0.37 of its original height. The travel time along the 20 NM long bay is dictated by the group wave velocity so that actual $t_D = x/(C_o/2) \tanh 2\pi d/L = 20 \times 6080/(5.12 \times 14/2)0.575 \times 3600 = 1.64$ h.

Since eq. 4-102 and 4-103 are derived from $H/H_o = e^{-\infty C_g t_d}$ (which gives $0.37 = e^{-1}$), the exponential value now becomes $e^{-(1.64/15.6)} = e^{-0.105} = 0.895 = H/H_o$, so that the wave is reduced in height by 10.5%.

Attenuation due to bottom permeability is given by eq. 4-106 so that:

$$C_g t_d \cdot 3600 = \frac{2K\sqrt{g}}{\nu}\left(\frac{2\pi}{L}\right)^{3/2}\left(\frac{\tanh 2\pi d/L}{4\pi d/L + \sinh 4\pi d/L}\right)C_g t_d \cdot 3600$$

$$= \frac{2 \times 3.06 \times 10^6 (32.2)^{1/2}}{10^{10}\, 12}\left(\frac{2\pi}{5.12 \times 14^2 \times 0.575}\right)^{3/2}\left(\frac{0.575}{1.31 + 1.72}\right)$$

$$\times \frac{5.12 \times 14 \times 0.575}{2} 1.64 \times 3600$$

$$= 0.0132$$

so that $H/H_o = e^{-0.0132} = 0.987$, or the wave height is reduced 1.3% due to bottom permeability. It can be concluded that for prototype situations reduction of wave height can be neglected.

8

A progressive 12-sec wave, 5 ft. high, propagates in 15 ft. of water towards the shore. Compute the mass-transport velocity at the bed for laminar and turbulent boundary layer conditions. What is the limiting wave height at which transition from one layer to the other occurs? What is the mass transport at the surface and at mid-depth for the same wave, based upon a turbulent boundary layer?

$T = 12$ sec, $H = 5.0$ ft., $d = 15$ ft., $d/L_o = 15/5.12 \times 12^2 = 0.02$. From Fig. 4-26, $XT^2/H^2 = 51$ for laminar BL, $XT^{1.6}/H^{1.2} = 9$ for turbulent BL, so that

$$U_B \text{ (lam.)} = \frac{X}{T} = \frac{51\, H^2}{T^3} = \frac{51 \times 5^2}{12^3} = 0.74 \text{ ft./sec.}$$

$$U_B \text{ (turb)} = \frac{X}{T} = \frac{9\, H^{1.2}}{T^{2.6}} = \frac{9(5)^{1.2}}{(12)^{2.6}} = 0.097 \text{ ft./sec.}$$

The progress of particles within the boundary layer per wave period X is 8.9 ft. and 1.16 ft. respectively. This should be compared to the amplitude of water-particle motion, which from Fig. 4-4 is given by $2x_{max}/H = 2.6$ or $2x_{max} = 13$ ft.

Limiting wave height for laminar conditions as given by Fig. 4-28 is 0.6 ft., so that $X = 1.16$ ft. per wave cycle.

The distribution of mass transport throughout the depth should be in proportion to the value at the bed, whether this is produced under laminar or turbulent conditions, so that from

Fig. 4-27 the proportions at mid-depth and surface for $d/L_o = 0.02$, are $-8/50$ and $-20/50$, respectively, of the bed values, i.e., 0.016 and 0.039 ft./sec opposite to the direction of wave advance. It has been shown [53] that it takes a long time for mass transport to be diffused or convected into the interior of the water mass. In the present case it would be a matter of days, by which time the wave height and period would have changed, even if the direction remained substantially the same in this relatively shallow water.

PROBLEMS

1
Waves arrive normal to a coast from a fetch 500 NM from shore, across a continental shelf 50 NM wide, which can be considered of uniform slope to its edge at 60 fathoms depth. Compute the proportion of time that 7-and 13-sec waves take in travelling across the transition and shallow-water zones to that across the deep-water zone for the particular wave. Assume a mean speed based upon the mean depth for the transitional and shallow-water sections.

2
A swell of 12 sec period and 4 ft. height traverses a 30 ft. deep zone offshore. Determine the maximum horizontal and vertical orbital velocities and accelerations of water particles at the surface, mid-depth and bottom, for linear theory only. Find also the maximum displacements of the water particles in these locations. Compare the u_{max} values with those given by Fig. 4-15. Is this wave a single train or has it secondary crests in the main troughs?

3
Compute the height of a 14-sec progressive wave required to give a maximum orbital velocity of 20 cm/sec at the edge of the continental shelf which is 60 fathoms deep. If this were considered HT_{max} for a FAS what wind velocity ($U_{19.5}$) has produced it, and for how long at least has it blown?

4
A 10-sec wave approaches the coast normally. In a beach area, where energy is completely dissipated, find the crest height above SWL when the wave, whose height was 3 ft. in deep water, is traversing a 12 ft. deep zone of water. Find the reach of the same wave in the same depth if the wave is fully reflected from a wall normally.

5
Two similar wave trains of 10 sec period and 3 ft. height are angled at 60° to each other when in 10 ft. of water. What is the celerity of the combined crests? Determine also the water-particle velocities and amplitudes (maximum values) at one-eighth intervals of the crest length across the line of propagation of the combined crests.

If this line of propagation is directly along the beach, compare the alongshore component of mass transport at the bed on the alignment of the combined crests with that of the incoming train angled 30° to the beach.

6
A 12-sec wave, 2.5 ft. high, is travelling in 20 ft. of water. Compute the mass-transport velocity at the bed for a laminar and turbulent boundary layer, assuming a smooth bed. What is the limiting height for such a wave for transition in the layer to occur? Determine the mass transport at the surface and at mid-depth for the same wave, based upon a laminar boundary layer. How long would it take the full mass-transport velocity distribution to be established throughout the depth if the continental shelf is 25 NM wide?

PROBLEMS

7
A 35-NM wind ($U_{19.5}$) develops a FAS at the edge of the continental shelf whose edge is 400 ft. deep. Treating the $H_{1/3}$ wave as a monochromatic wave with period T_{max}, find the maximum horizontal velocity exerted at the bed. Discuss the implications of this in terms of the frequency of occurrence, the extent of the floor so influenced, and other factors present that could promote sediment disturbance.

8
A 10-sec wave of 3 ft. height reflects at an angle of 45° from a wall in 18 ft. of water. Considering that no energy loss is encountered, determine the diameter of the water-particle orbits when these are circular at the bed. From this compute the average peripheral velocity of these vortices. What influence should these have on a sedimentary bed?

9
An offshore oil drilling rig is to contain a platform which must be set above the reach of the highest tenth of the waves likely to arrive. The worst storm conditions in the area can be considered to be a FAS from a 30-knot wind ($U_{19.5}$). If the rig is in 100 ft. of water, compute the height required for the underside of the platform above the highest tide and surge level. If the rig is not far distant from a cliffed coast where 30% of the wave energy is reflected, what should be this safe elevation above SWL?

10
A 12-sec wave, 2 ft. high, arrives normal to a jetty wall where the bed has been dredged to 30 ft. Find the mass transport velocity near the bed and the maximum horizontal orbital velocity that can occur. Determine the distances of major and minor furrows in the bed from the wall alignment.

11
A section of continental shelf can be considered of a uniform depth of 30 fathoms where it is 120 miles wide. Compute the attenuation of waves with 7, 10, 13 and 16-sec period due to viscous effects in a turbulent boundary layer. What are the thicknesses of these layers? Besides energy loss in this zone, what other possible sources of loss can occur to swell waves as they approach the shore?

12
What changes in mass-transport distribution throughout the depth would you expect in shallow water by considering cnoidal or other theory as against the Stokes theory normally used for this analysis. Why should the Longuet-Higgins solution not be used under shoal conditions?

13
A 13-sec wave which is 3.5 ft. high in deep water approaches the coast normally. Determine the mass transport velocity at depth ratios d/L_o = 0.5, 0.2, 0.1 and 0.05 based upon the turbulent boundary layer. If such a calculation were to be used to assess the alongshore transportability of sediment for an oblique approach by this wave, what other considerations would be necessary?

14
A 35-knot ($U_{19.5}$) wind at sea generates a FAS. A large tanker is proceeding through this storm area and encounters an $H_{1/3}$ wave which has smaller period components breaking at its crest. At what speed does such a body of water strike the ship if the hull is in close proximity to the crest? Are such occurrences predictable from observation of waves from the vessel? Will such high waves always approach from the wind direction?

15
Give the formulae for and discuss the importance of wave group velocity.

16
What produces the net movement of water particles known as mass transport? How is the whole body of water affected? Discuss the times involved to establish a stable condition in this respect. How important is mass transport to the engineer?

17
Draw the distribution of mass-transport velocity throughout the depth of water for various depth to wave length ratios. Discuss the possibility of their occurrence in oceanic margins and in enclosed bodies of water such as the Meditteranean Sea or the U.S.A. Great Lakes.

18
Sewage effluent issuing into the ocean from a pipe may be lighter or heavier than sea water, causing it to rise to the surface or remain at or near the bed, after a certain degree of mixing. Discuss the possibility of beach pollution for both cases when storm conditions or long swell waves are present.

19
Describe the action taking place within a short-crested wave system. Is it of engineering significance? Cite as many instances as you can where such patterns may occur.

20
Discuss the major changes in the profile of a progressive wave as given by the second-order Stokes theory compared to the first-order. What engineering significance is there in such changes?

21
As a wave propagates from deep water to the shore, its profile is best determined by Stokes-I, Stokes-II, cnoidal, hyperbolic and solitary wave theory in that order. The Ursell parameter $L^2 H/d^3$ has been used to predict conditions where each theory is most appropriate. Discuss this parameter in terms of the ratios it contains.

REFERENCES

[1] R.L. Wiegel, 1964. *Oceanographical Engineering.* Prentice Hall, Princeton, N.J., 532 pp.
[2] P.S. Eagleson and R.G. Dean, 1966. Small amplitude wave theory. In: A.T. Ippen (Editor), *Estuary and Coastline Hydrodynamics.* McGraw Hill, New York, N.Y., pp. 1–92.
[3] G.G. Stokes, 1847. On the theory of oscillatory waves. *Trans. Camb. Phil. Soc.,* 8: 441.
[4] F.A. Gerstner, 1809. Theorie der Wellen. *Gilbert Ann. Phys.,* 32: 412–440.
[5] J. Boussinesq, 1872. Théorie des ondes et de remais qui se propagent le long d'un canal rectangulaire horizontal, en communiquant au liquide contenu dans ce canal des vitesses sensiblement paralleles de la surface au fond. *J. Math. Lionvilles,* 17:55.
[6] J. Kravtchenko and A. Daubert, 1957. La houle à trajectoires fermés en profondeur finie. *Houille Blanche,* 12: 408–429.
[7] T. Levi-Civita, 1925. Détermination rigoureuse des ondes permanentes d'ampleur finie. *Math. Ann.,* 93.
[8] D.J. Struik, 1926. Détermination rigoureuse des ondes irrotationelles périodiques dans un canal à profondeur finie. *Math. Ann.,* 95.
[9] A.I. Nekrassov, 1951. *The Exact Theory of Steady Waves on the Surface of a Heavy Fluid.* Akad. Nauk S.S.S.R., Moscow (in Russian).

REFERENCES

[10] J.V. Wehausen and E.V. Laitone, 1960. Surface waves. In: *Handbuch der Physik*.
[11] L. Skjelbreia, 1959. Gravity waves, Stokes third-order approximation: tables and functions. Counc. Wave Res., ASCE.
[12] L. Skjelbreia and J.A. Hendrickson, 1961. Fifth-order gravity wave theory. *Proc. 7th Conf. Coastal Eng.*, 1: 184–196.
[13] L. Skjelbreia and J.A. Hendrickson, 1962. Fifth-order gravity wave theory and tables and functions, *Natl. Eng. Sci. Co.*
[14] R. Miché, 1944. Mouvements undulatoires des mers en profondeur constante ou décroissante. *Ann. Ponts Chaussées*, 114: 25–78; 131–164; 270–292; 369–406.
[15] P. Danel, 1952. On the limiting clapotis. In: *Gravity Waves – Natl. Bur. Std., Circ., 521:* 35–38.
[16] E.V. Laitone, 1962. Limiting conditions from cnoidal and Stokes waves. *J. Geophys. Res.*, 67 : 1555–1564.
[17] F. Biesel, 1952. General second-order equations of irregular waves. *Houille Blanche*, 7: 372–376.
[18] D.J. Korteweg and G. De Vries, 1895. On the change of form of long waves advancing in a rectangular canal, and on a new type of long stationary wave. *Phil. Mag., 5 Ser.*, 39: 422–443.
[19] J.B. Keller, 1948. The solitary wave and periodic waves in shallow water. *Comm. Appl. Math.*, 1 : 323–339.
[20] J.E. Chappelear, 1962. Shallow water waves. *J. Geophys. Res.*, 67 : 4693–4704.
[21] F.D. Masch and R.L. Wiegel, 1961. *Cnoidal Waves, Tables of Functions*. Univ. Calif.,
[22] F. Ursell, 1953. The long-wave paradox in the theory of gravity waves. *Proc. Camb. Phil. Soc.*, 49 : 685–694.
[23] Y. Iwagaki, 1968. Hyperbolic waves and their shoaling. *Proc. 11th Conf. Coastal Eng.*, 1; 125–144.
[24] B. Le Méhauté and L.M. Webb, 1964. Periodic gravity waves over a gentle slope at a third-order approximation. *Proc. 9th Conf. Coastal Eng.*, 23–40.
[25] R.G. Dean, 1965. Stream function representation of nonlinear ocean waves. *J. Geophys. Res.*, 70 : 4561–4572.
[26] K.E. Kenyon, 1969. Stokes drift for random gravity waves. *J. Geophys. Res.*, 74 : 6991–6994.
[27] R.G. Dean, 1965. Stream function wave theory: validity and application. *Proc. Santa Barbara Conf. Coastal Eng.*, 269–299.
[28] J.H. Chappelear, 1961. Direct numerical calculation of wave properties. *J. Geophys. Res.*, 66; 501–508.
[29] R.G. Dean, 1968. Breaking wave criteria, a study employing a numerical wave theory. *Proc. 11th Conf. Coastal Eng.*, 1: 108–123.
[30] J.J. Von Schwind and R.O. Reid, 1968. Characteristics of gravity waves of permanent form. *Texas A M Univ., Rep.*, 68–7T.
[31] E.V. Laitone, 1959. Water waves, 4. Shallow water waves. *Univ. Calif., I.E.R. Rep.*, 82–11.
[32] B. Le Méhauté, 1968. Mass transport in cnoidal waves. *J. Geophys. Res.*, 73; 5973–5979.
[33] W.H. Munk, 1949. The solitary wave theory and its application to surf problems. *Ann. N.Y. Acad. Sci*, 51: 376–424.
[34] R.G. Dean, 1970. Relative validities of water wave theories. *Proc. ASCE*, 96 (WW1): 105–119.
[35] B. Le Méhauté, D. Divoky and A. Lin, 1968. Shallow-water waves: a comparison of theories and experiments. *Proc. 11th Conf. Coastal Eng.*, 1: 86–107.
[36] G.H. Keulegan and G.W. Patterson, 1940. Mathematical theory of irrotational translation waves *J. Res. Natl. Bur. Std.*, 24: 47–101.
[37] J. McCowan, 1891. On the solitary wave. *Phil. Mag.*, 32; 45–58.

[38] Y. Goda, 1964. Wave forces on a vertical circular cylinder: experiments and a proposed method of wave force computation. *Port Harbour Tech. Res. Inst., Rep.*, 8.
[39] Y. Iwagaki and T. Sakai, 1970. Horizontal water particle velocity of finite amplitude waves. *Proc. 12th Conf. Coastal Eng.*, 1 : 309–326.
[40] J.R. Morison and R.C. Crooke, 1953. The mechanics of deep water, shallow water, and breaking waves. *Beach Erosion Board, Tech. Mem.*, 40.
[41] A. Wallet and F. Ruellan, 1950. Trajectories of particles within a partial clapotis. *Houille Blanche*, 5: 483–489.
[42] Y. Goda and Y. Abe, 1968. Apparent coefficient of partial reflection of finite amplitude waves. *Rep. Port Harbour Res. Inst.*, 7 (3).
[43] R.A. Fuchs, 1952. On the theory of short-crested oscillatory waves. In: *Gravity Waves – Natl. Bur. Std., Circ.*, 521 : 187–200.
[44] H. Lamb, 1932. *Hydrodynamics*. Macmillan, N.Y., 6th ed.
[45] F. Biésel, 1949. Calcul de l'amortissement d'une houle dans un liquide visqueux de profondeur finie. *Houille Blanche*, 4:630–634.
[46] Y. Iwagaki and Y. Tsuchiya, 1967. On the mechanism of laminar damping of oscillatory waves due to bottom friction. *Bull. Disaster Prev. Res. Inst., Kyoto Univ.*, 16; 49–75.
[47] C.E. Grosch and S.J. Lukasik, 1961. Attenuation of shallow water waves. *Bull. Am. Phys. Soc.*, 6; 210–311.
[48] P.S. Eagleson, 1962. Laminar damping of oscillatory waves. *Proc. ASCE*, 88; (HY3): 155–181.
[49] R.O. Reid and K. Kajiura, 1957. On the damping of gravity waves over a permeable sea bed. *Trans. Am. Geophys. Union*, 38; 662–666.
[50] H.G. Gade, 1952. Effects of a non-rigid, impermeable bottom on plane surface waves in shallow water. *J. Mar. Res.*, 16 :61–82.
[51] W.A. Price, K.W. Tomlinson and J.N. Hunt, 1968. The effect of artificial seaweed in promoting the build-up of beaches. *Proc. 11th Conf. Coastal Eng.*, 1 : 570–578.
[52] W.A. Price, K.W. Tomlinson and D.H. Willis, 1970. Laboratory tests on artificial seaweed. *Proc. 12th Conf. Coastal Eng.*, 2; 995–1000.
[53] M.S. Longuet-Higgins, 1953. Mass transport in water waves. *Phil. Trans. R. Soc.*, 245A: 535–581.
[54] R.C.H. Russell and J.D.C. Osorio, 1958. An experimental investigation of drift profiles in a closed channel. *Proc. 6th Conf. Coastal Eng.*, 1958; 171–193.
[55] R.A. Bagnold, 1947. Sand movement by waves: some small scale experiments with sand of very low density. *Proc. Inst. Civil Eng.*, 27; 447–469.
[56] A. Brebner, J.A. Askew and S.W. Law, 1966. The effect of roughness on the mass-transport of progressive gravity waves. *Proc. 10th Conf. Coastal Eng.*, 1; 175–184.
[57] J.I. Collins, 1963. Inception of turbulence at the bed under periodic gravity waves. *J. Geophys. Res.*, 68; 6007–6014.
[58] H. Noda, 1968. A study on mass transport in boundary layers in standing waves. *Proc. 11th Conf. Coastal Eng.*, 1; 227–247.
[59] Y. Iwagaki and Y. Tsuchiya, 1966. Laminar damping of oscillatory waves due to bottom friction. *Proc. 10th Conf. Coastal Eng.*, 1: 149–174.
[60] C.J. Sonu, J.M. McCloy and D.S. Arthur, 1967. Longshore currents and nearshore topography. *Proc. 10th Conf. Coastal Eng.*, 1; 525–549.
[61] K. Horikawa and T. Sunamura, 1970. A study on erosion of coastal cliff and of submarine bedrocks. *Coastal Eng. Japan*, 13; 127–139.
[62] B.D. Dore, 1970. Mass transport in layered fluid systems. *J. Fluid Mech.*, 40: 113–126.
[63] I.G. Jonsson, C. Skougaard and J.D. Wang, 1970. Interaction between waves and currents. *Proc. 12th Conf. Coastal Eng.*, 1: 489–508.

Chapter 5

EFFECTS OF SHOALING WATER

Many of the effects of decreasing depth on a wave have already been noted, such as reduction of its celerity and increase in its height. In this section the propagation of waves in shoaling water will be discussed in more detail, together with their ultimate breaking or reflection near the beach. One of the most important facets of depth variability is that of wave *refraction*, a knowledge of which is necessary in determining wave characteristics in transitional and shallow-water depths from those computed for deep-water conditions offshore.

WAVE REFRACTION

Waves in deep water advance at a rate dictated by their wave period (T) only since $C_o = gT/2\pi$. At a depth of around $L_o/2$ the wave starts to reduce in speed, the speed then being given by $C_d = gT/2\pi \tanh 2\pi d/L$, so that:

$$C_d/C_o = \tanh 2\pi d/L = L/L_o \tag{5-1}$$

A wave travelling obliquely to a series of bed contours, or obliquely to the general slope of the bed, will have one part of its crest in deeper water than another. Differential celerities will ensue so that the crest will become twisted, as also the orthogonal, which is the path followed by an imaginary point on the crest line. This can best be illustrated by considering a wave passing over a step (reflec-

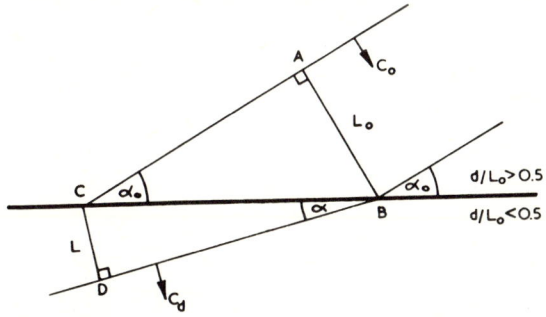

Fig. 5-1. Refraction of wave over a step.

tion and friction effects being ignored). As illustrated in Fig. 5-1, a point on a wave crest moving in deep water at celerity C_o moves from A to B in say time T (= wave period) a distance L_o. Over the step another point on the crest will travel CD (= L) in the same time T, which is less than AD because here the velocity is C_d ($< C_o$). In each case the orthogonal distances AB, CD are measured normal to the wave crest AD. If the hypotenuses of these right-angled triangles are equal (i.e., BC) it is seen that:

$$\frac{\sin \alpha}{\sin \alpha_o} = \frac{L}{L_o} = \frac{C_d}{C_o} = \tanh 2\pi d/L \qquad (5\text{-}2)$$

Eq. 5-2 provides the angular change of direction of the wave due to its passage into the shoaler water. The same relationship applies to light waves refracted through substances of differing refractive index, and is known as Snell's law.

Where the change in depth is gradual (i.e., not stepped), the wave crest will be curved instead of angled as in Fig. 5-1. Consider the simple case of a beach with a uniformly sloping offshore zone, that is, the bed contours are straight and parallel to the water line. These depths can be expressed as ratios of the wave length of a given wave train with period $T(d/L$ or $d/L_o)$. Initially it will be assumed that beyond a depth of $d/L_o = 1/2$ the wave celerity is constant irrespective of depth. Waves approaching such a coast are depicted in Fig. 5-2 as successive positions of the one wave crest, or the simultaneous positions of successive waves. One orthogonal is drawn which is normal to all these wave crests along its path. The wave train has arrived at the $d/L_o = 0.5$ contour at an angle α_o, as measured between the contour

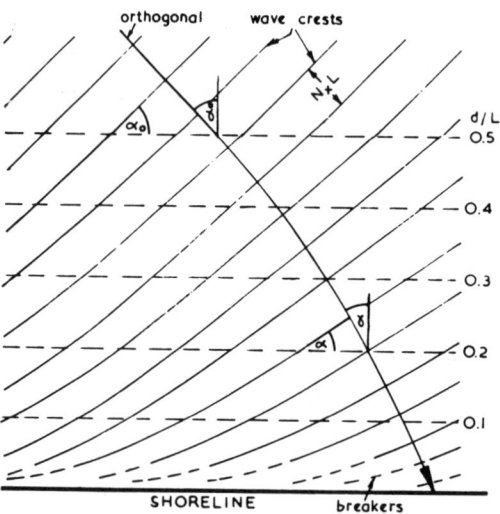

Fig. 5-2. Oblique waves traversing a uniformly sloped shelf.

WAVE REFRACTION

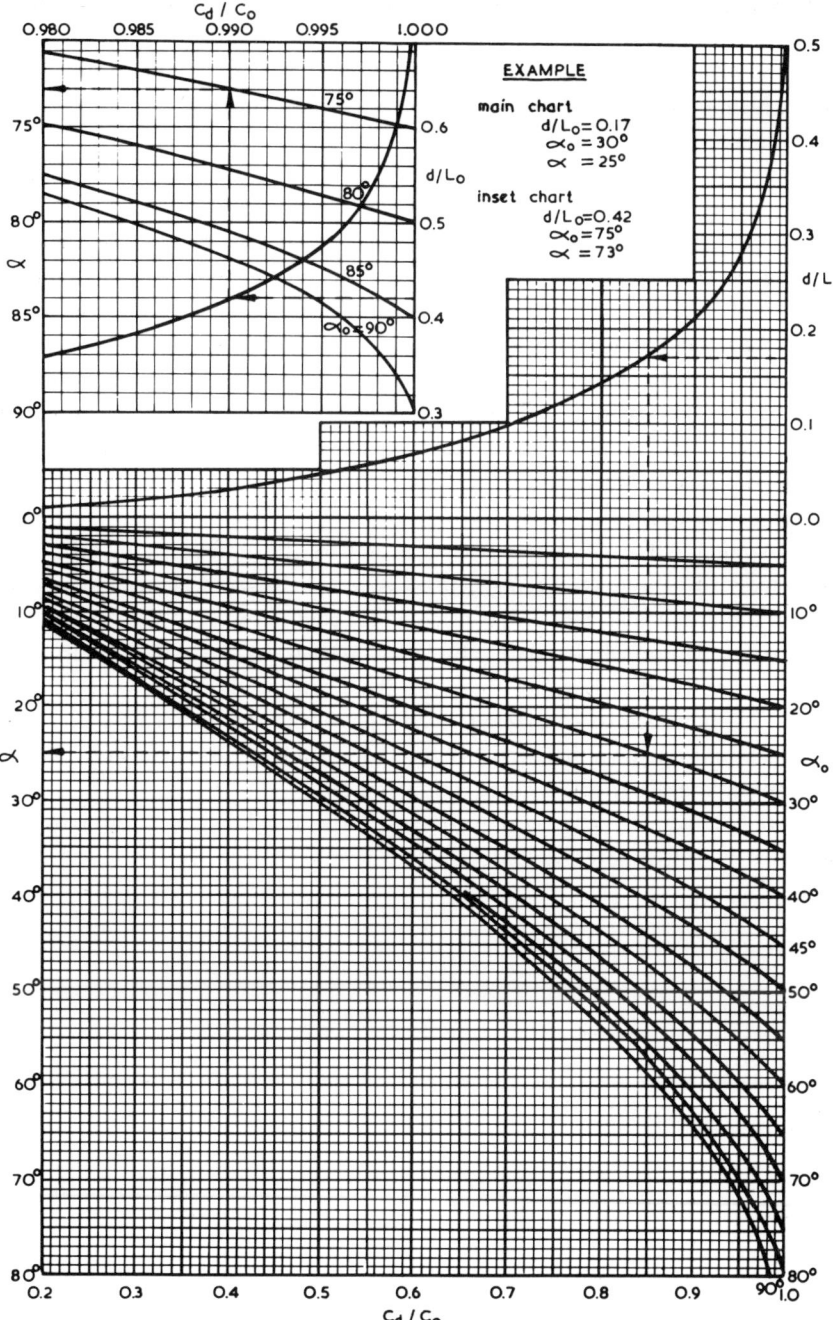

Fig. 5-3. Relationship between d/L_0, C_d/C_0, α_0 and α.

and the crest line, or between the orthogonal and the normal to the bed contour. At some intermediate depth this angle has reduced to α.

The variables contained in eq. 5-2 are related as in Fig. 5-3 where d/L_o gives C_d/C_o, which, in turn, determines angle α for a specific value of α_o.

Parallel contours

Continuing the simplified picture of Fig. 5-2, the angle α at any intermediate depth can be determined directly from Fig. 5-3 for any known α_o. For example, when $\alpha_o = 30°$ the angle α at $d/L_o = 0.17$ is $25°$. Table 5-I has been constructed for a range of d/L_o and of α_o for ready application in the case of parallel contours. It is to be noted from this table that at $d/L_o = 0.5$ the angles are less than those in the deepest water, some twisting has already occurred. This arises from the fact that at this d/L_o ratio $\tanh 2\pi d/L \neq 1.0$ (in fact it equals 0.9964), and it is not unity until $d/L_o = 0.84$. This is illustrated in the inset of Fig. 5-3 where C_d/C_o approaches unity as d/L_o exceeds 0.5.

As illustrated in Fig. 5-4, the bottom contours can be represented by d/L_o or C with suffixes to denote succeeding depths. A wave represented by an orthogonal intersects the deep-water contour (considered for convenience at $d/L_o = 0.5$ or C_o) at angle α_o to its normal. At some subsequent contour d_1/L_o (or C_1) the orthogonal must be angled at α. For most purposes the change in direction, which can be assumed as abrupt, can be considered to take place midway between the contours (implying a uniform slope between). If the zone of mid-depth is known from the hydrographic data this should be used in preference.

As noted already the angle α can be obtained from Fig. 5-3 directly, but a second approach is worthy of note. At contour C_2 we have:

$$\frac{\sin \alpha_2}{\sin \alpha_1} = \frac{C_2}{C_1}$$

and since:

$$\frac{\sin \alpha_1}{\sin \alpha_o} = \frac{C_1}{C_o}$$

then:

$$\frac{\sin \alpha_2}{\sin \alpha_o (C_1/C_o)} = \frac{C_2}{C_1}$$

thus giving:

$$\frac{\sin \alpha_2}{\sin \alpha_o} = \frac{C_2}{C_1} \frac{C_1}{C_o} = \frac{C_2}{C_o} \tag{5-3}$$

TABLE 5-I

Angles between wave crest and shoreline for parallel contours

d/L_o	α_o																	
	5	10	15	20	25	30	35	40	45	50	55	60	65	70	75	80	85	90
0.005	0.88	1.84	2.70	3.59	4.30	5.00	5.84	6.55	7.20	7.80	8.35	8.81	9.25	9.61	10.00	10.10	10.20	10.35
0.01	1.23	2.45	3.70	4.90	6.00	7.11	8.20	9.20	10.15	10.95	11.74	12.41	12.99	13.50	13.90	14.15	14.25	14.35
0.04	2.50	4.90	7.21	9.60	11.75	13.99	16.11	18.05	19.89	21.75	23.30	24.75	25.99	27.08	27.90	28.49	28.80	28.95
0.06	2.89	5.79	8.60	11.40	14.20	16.85	19.42	21.85	24.05	26.30	28.32	30.14	31.25	33.01	34.10	34.81	35.30	35.55
0.08	3.25	6.50	9.69	12.85	15.95	19.00	21.95	24.75	27.40	29.95	32.30	34.39	36.34	37.91	39.20	40.10	40.65	40.95
0.10	3.50	7.10	10.55	14.05	17.45	20.75	24.05	27.11	30.15	32.98	35.60	38.00	40.15	41.95	43.45	44.50	45.10	45.40
0.15	4.00	8.22	12.30	16.21	20.25	24.11	28.00	31.75	35.44	38.91	42.25	45.20	47.90	50.25	52.35	53.76	54.60	55.00
0.20	4.39	8.90	13.30	17.60	22.10	26.31	30.60	34.80	38.98	42.95	46.61	50.25	53.55	56.45	59.00	60.91	62.22	62.70
0.25	4.60	9.30	13.95	18.60	23.20	27.79	32.35	36.89	41.21	45.60	49.70	53.86	57.82	61.35	64.20	66.80	68.30	68.82
0.30	4.77	9.61	14.41	19.20	23.95	28.79	33.55	38.20	42.82	47.45	51.95	56.39	60.70	64.71	68.29	71.35	73.45	74.15
0.35	4.88	9.75	14.65	19.55	24.45	29.30	34.18	38.95	43.80	48.52	53.30	57.89	62.38	66.80	70.89	74.41	77.05	78.05
0.40	4.93	9.87	14.80	19.73	24.70	29.60	34.50	39.40	44.30	49.20	54.00	58.80	63.55	68.20	72.70	76.55	79.80	81.15
0.45	4.97	9.93	14.90	19.89	24.89	29.81	34.72	39.68	44.60	49.55	54.43	59.35	64.20	68.95	73.60	77.97	81.69	83.45
0.50	4.98	9.96	14.94	19.92	24.90	29.88	34.85	39.82	44.79	49.75	54.70	59.63	64.56	69.42	74.22	78.85	82.97	85.07

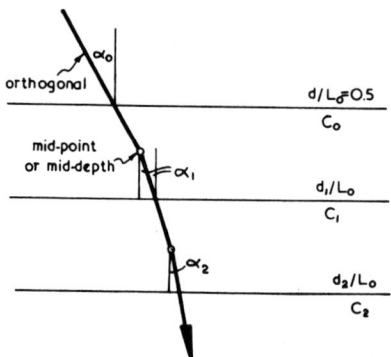

Fig. 5-4. Definition sketch for refraction over parallel contours.

Hence, when the ratio C_2/C_o is known, or any other such celerity ratio to the deep-water value, α_2 can be read directly from Fig. 5-3. But also we have:

$$\frac{\sin \alpha_2/\sin \alpha_o}{\sin \alpha_1/\sin \alpha_o} = \frac{C_2/C_o}{C_1/C_o} = \frac{\sin \alpha_2}{\sin \alpha_1} \tag{5-4}$$

In this case α_2 can be determined from α_1 (instead of α_o) if in Fig. 5-3 the variable denoted α_o is used for α_1 (the approach angle from the deeper water) and the variable $C_d/C_o = (C_2/C_o)/(C_1/C_o)$ is employed. For example, when $\alpha_o = 30°$, then: at $d_1/L_o = 0.17$, $C_d/C_o = 0.85$ so that $\alpha_1 = 25°$; at $d_2/L_o = 0.05$, $C_d/C_o = 0.52$ so that $\alpha_1 = 15°$. By taking $\alpha_1 = \alpha_o = 25°$ and computing $(C_2/C_o)/(C_1/C_o) = 0.52/0.85 = 0.612$ and reading this as C_d/C_o in Fig. 5-3, $\alpha_2 = 15°$ is obtained as before.

Non-parallel straight contours

In Fig. 5-5 are depicted four bed contours that are angled to each other, in this case their obliquity to the deep-water contour C_o is indicated by θ_1, θ_2 etc. for ease in discussion. The incoming orthogonal crosses contour C_o at angle α_o to the normal. Midway between this and the next contour C_1 (i.e., on the bisector of angle θ_1) the orthogonal is considered to change direction, to cross C_1 at angle α_1. The approach angle to this second contour is $\alpha_o - \theta_1$. The addition or subtraction of θ_1 is determined from the geometry of the figure, or a sign notation may be evolved with practice. Should $\theta_1 = \alpha_o$ in Fig. 5-5 the orthogonal will be normal to C_1 and so no refraction would occur. However, if $\alpha_o - \theta_1$ is finite, the angle α_1 is given by:

$$\frac{\sin \alpha_1}{\sin(\alpha_o - \theta_1)} = \frac{C_1}{C_o} \tag{5-5}$$

which can be read from Fig. 5-3 by reading α_o in the graph as $\alpha_o - \theta_1$ and $\alpha = \alpha_1$.

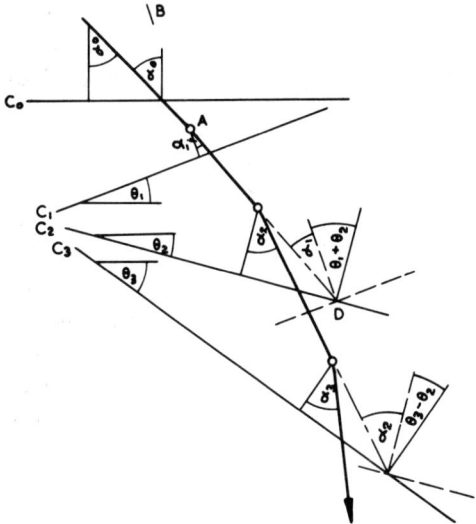

Fig. 5-5. Definition sketch for refraction across straight contours angled to each other.

(Note that α_o represents any approach angle from the deeper water direction.)

The use of angle θ_1 between contours C_o and C_1 is mainly by way of explanation, the actual approach angle to C_1 can be measured directly on a plan of contours. Referring to Fig. 5-5, a protractor can be placed with its base along C_1 and the 90° line through the chosen turning point A, in order to mark the alignment B which is the normal to contour C_1. Then, with the protractor centre on A and the 90° line still on alignment B, the approach angle ($= \alpha_o - \theta_1$) can be measured and also the traverse angle α_1 can be marked. The orthogonal is thus continued to the next turning point, on the bisector of angle $(\theta_1 + \theta_2)$.

The approach angle to contour C_2 is $\alpha_1 + \theta_1 + \theta_2$, as seen by the extension of the orthogonal to D (Fig. 5-5) and the construction of the dotted line parallel to contour C_1. The traverse angle α_2 is given by:

$$\frac{\sin \alpha_2}{\sin (\alpha_1 + \theta_1 + \theta_2)} = \frac{\sin \alpha_2/\sin \alpha_o}{\sin (\alpha_1 + \theta_1 + \theta_2)/\sin \alpha_o} = \frac{C_2/C_o}{C_1/C_o} = \frac{C_d}{C_o} \qquad (5\text{-}6)$$

In Fig. 5-3 the values of d_2/L_o and d_1/L_o give the respective values C_2/C_o and C_1/C_o, the ratio of which gives C_d/C_o to be used with $\alpha_o = \alpha_1 + \theta_1 + \theta_2$ to give α_2. Angle α_2 is then constructed by means of the procedure outlined for contour C_1.

From Fig. 5-5 it is seen that the approach angle to contour C_3 is $\alpha_2 + \theta_3 - \theta_2$ so that:

$$\frac{\sin \alpha_3}{\sin (\alpha_2 + \theta_3 - \theta_2)} = \frac{C_3/C_o}{C_2/C_o} = \frac{C_d}{C_o} \qquad (5\text{-}7)$$

It is possible for the orthogonal to change from the left-hand to the right-hand quadrant. Although the principle remains the same, care must be taken in refracting the orthogonal in the appropriate quadrant. As the approach angle approaches 0° so the angular change of direction tends to zero.

Curved contours

It is assumed that in refraction diagrams any small curvatures in the bed contours will be smoothed out, since these do not affect the overall wave pattern greatly. However, if the radius of curvature is 5 wave lengths or greater and the distances between contours warrants it, curved contours must be considered. In this case it is necessary to determine an appropriate tangent line at the likely intersection with the deviated orthogonal. When this tangent has been established the angle of approach can be measured.

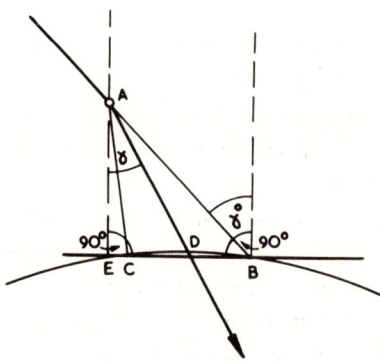

Fig. 5-6. Method of measuring α_o and constructing α for orthogonals approaching curved contours.

Referring to Fig. 5-6, the incoming orthogonal is extended from A to intersect the contour at B. A normal to the contour from A meets it at C. The deviated orthogonal will cross the contour somewhere between B and C, say at D, the tangent at which can be considered to be parallel to BC. From α_o measured at B the traverse angle α is determined from Fig. 5-3, from a knowledge of the relevant C ratios of this contour and the previous one. To draw in orthogonal AD construct the normal AE to the pseudo-tangent BC and mark angle $EAD = \alpha$.

Inspection of Fig. 5-3 will show that AD will bisect the angle EAB (= α_o), that is $\alpha = \alpha_o/2$, when C_d/C_o values are around 0.5, in which case the construction outlined is the most accurate. Should C_d/C_o approach unity then D will be closer to B. Thus, with a knowledge of C_d/C_o, the orientation of the tangent line BC can be varied, from which both α_o and α will vary.

WAVE REFRACTION

The construction lines in Fig. 5-6 have been included in order to help the description, in practice most of them are omitted. To determine α_o point B is marked by a ruler aligned with the incoming orthogonal through A. Point C can be found by eye, so that the protractor base line (0–180°) is placed along BC with the normal (90°) through B, in order to measure α_o. For this purpose the orthogonal may have to be extended backwards depending upon the radius of the protractor in respect to the length of the orthogonal legs. When α is assessed, $\alpha_o - \alpha$ can be measured with the protractor along AB and centered on A. Whilst drawing AD it is advisable to extend or mark the alignment backwards a little for the reason given previously.

Method when α_o approaches 90°

Where the incident angle α_o is near 90° (orthogonal almost parallel to the contour) and $d/L_o \geqslant 0.4$, the main graph of Fig. 5-3 is difficult to read. This condition is served by the inset to the figure. It is in the deeper water near the edge of the continental shelf that such obliquity is likely to be encountered. To retain the same accuracy of the refraction diagram as in the shallower zone, smaller intervals of d/L_o or of C_d/C_o should be used. If no other detail is available a uniform slope should be assumed between the $d/L_o = 0.2$ and $d/L_o = 0.5$ contours so that curves for $d/L_o = 0.3$ and 0.4 can be drawn in. The orthogonal is then traced from mid-depth to mid-depth as before, cognisance being taken of curvature and obliquity.

It is possible with curved contours that the deviated orthogonal will have an approach angle to the next curved contour which is greater than 90°. In this event an orthogonal a little closer inshore must be chosen in order that the bed can "attract" the new orthogonal to the beach.

Tracing orthogonals seawards

Since Fig. 5-3 is based solely upon Snell's Law and does not incorporate any geometrical approximations, it can be used to trace wave rays both shorewards and seawards. When tracing seawards Fig. 5-3 is entered from angle α (since the orthogonal approaches a contour from the shallow side), but the C_d/C_o value is computed as before from the ratio $(C_2/C_o)/(C_1/C_o)$ (as though entering from deeper water). The value of α_o is thus determined and the orthogonal continued seawards. If the respective contours are parallel the angle α_o is drawn directly, but if they are oblique the angle between them is added to or subtracted from α_o to obtain the correct alignment.

The above procedure can best be illustrated by an example, depicted in Fig. 5-7. Consider first the orthogonal approaching contour $d_1/L_o = 0.2$ from the deeper water at angle $\alpha_1 = 40°$ to the normal. The approach angle to contour $d_2/L_o = 0.1$ is then $\alpha_o = \alpha_1 + 30° = 70°$. Then $C_d/C_o = (C_2/C_o)/(C_1/C_o) = 0.71/0.89 = 0.8$, from which $\alpha_2 = 48.5°$.

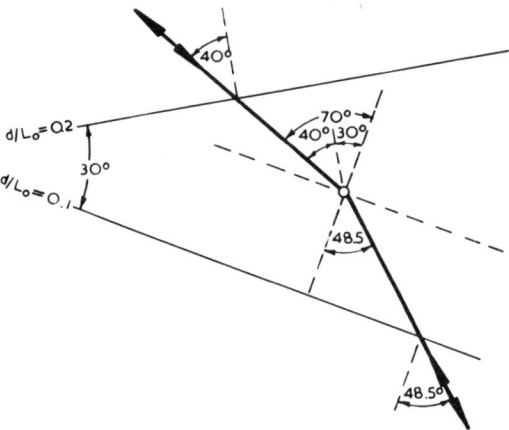

Fig. 5-7. Example of tracing orthogonals seawards or shorewards.

In tracing this orthogonal seawards the approach angle to $d_2/L_o = \alpha_2 = 48.5°$. The C_d/C_o ratio is the same as before, namely 0.8, resulting in $\alpha_o = 70°$ from Fig. 5-3. But since the contours are angled 30° to each other the angle to the normal of contour d_1/L_o is obtained by subtracting 30°, giving the original deeper-water approach angle $\alpha_1 = 40°$.

Choice of depth increments

The smaller the depth increments (d/L_o) chosen the more accurate is the wave ray trace, but economy in time must be considered. When contours are parallel the angular change at any depth contour from that across the deep-water contour will be the same, no matter how many intervals are considered between. However, the actual path of the orthogonal will alter with the depth interval chosen and the location of the mid-depth point, where the actual change in direction is considered to take place. In the case of contours radiating from a centre both the path and final ray direction will diverge.

Where contours are sensibly parallel, intervals of d/L_o should be employed which produce equal angular changes. Reference to Fig. 5-3 indicates that the $\alpha_o - C_d/C_o$ curves are nearly linear for $C_d/C_o < 0.9$ and $\alpha_o < 70°$. For equal changes in direction ($\alpha_o - \alpha$), requiring equal changes in C_d/C_o, variable changes in d/L_o are necessary. This trend is exhibited in Table 5-II where d/L_o values have been chosen to give more-or-less equal differences in C_d/C_o. It is seen that successive values of d/L_o should be approximately 0.6 of the preceding value. In this particular zone of the figure it should suffice to use the interval $d/L_o = 0.5$ to 0.2. Where $C_d/C_o > 0.9$ and $\alpha_o > 70°$ it is found that equal differences in α result from nearly equal differences in d/L_o (these occurring only between 0.2 and 0.5 or greater).

TABLE 5-II

Choice of d/L_o for equal angular change

d/L_o	0.2	0.13	0.08	0.05	0.03
C_d/C_o	0.885	0.775	0.650	0.520	0.410
$\Delta(C_d/C_o)$		0.110	0.125	0.130	0.110
$(d_2/L_o)/(d_1/L_o)$		0.650	0.615	0.625	0.600

The actual increments chosen will depend upon the accuracy of the data available and the nature of the problem being studied. If the distribution of wave energy is to be computed the refraction diagrams need to be more accurate in terms of orthogonal position than if only approach angles to a beach or structure are required. Intervals of 0.1 are suggested in the region $d/L_o = 0.5$ to 0.2, with even smaller decrements if the contours are curved or the incident angle approaches 90°. As can be noted in Fig.5-3, an approach angle $\alpha_o = 90°$ gives $\alpha = 63°$ at $d/L_o = 0.2$ (for parallel contours), so that the previously suggested intervals for $d/L_o < 0.2$, of a ratio of 0.6, is still applicable. Thus, for near equal changes in direction at each contour the d/L_o values as in Table 5-III are suggested.

TABLE 5-III

Values of d/L_o suggested for refraction diagrams

$\alpha_o < 70°$	0.5	0.2	0.12	0.07	0.04	0.025	0.015
$\alpha_o > 70°$	0.5	0.4	0.3	0.2	0.04	0.025	0.015

It should be remembered that, to stipulate a value of d/L_o, not only is a depth of water involved, but also a wave length or wave period. So that each refraction diagram refers to a specific wave train. In order to test the pattern for a variety of wave periods new bed contours would have to be drawn if specific d/L_o values were to be employed. This sophistication may not be warranted, at least until a preliminary study has been carried out to determine the wave components most important to the problem.

Alternative methods

The author published the above procedure in 1966 [1]. Orthogonal methods used prior to this time were based upon changes in wave length [2, 3] or velocity [4, 5, 6] as waves proceeded shorewards or seawards. The differential length methods involved the use of graphs from which the orthogonal direction was computed and then plotted. The second series of methods employed templates based upon velocity ratios, which could be applied to any scale of hydrographic chart without the need for reading angles. Both systems require ratios of length or velocity to be computed for consecutive contours. These methods are based upon trigonometrical and geometrical qualities of the construction diagrams, with certain approximations that are correct only for values of α_o between 0° and 80°. Modifications are provided when α_o approaches 90°.

The analysis above and the publications reported utilize linear theory and deal only with monochromatic waves. Some workers have questioned the accuracy of this procedure, particularly for shallow-water conditions where steepening of the wave through shoaling may influence its celerity. Suquet in 1949 [7] quotes a wave-speed formula that indicates an error of 20% in water $d/L_o = 0.10$ when the wave is approaching its maximum steepness for that depth ($\doteq 1/10$). Such a situation could only occur inside a fetch in which case a procedure for monochromatic waves is no longer valid. Battjes [8] has also computed errors arising from shoaling and bed slope, but even for such a shallow condition of $d/L_o = 0.015$ and a steep slope of 10% the error in wave celerity was only 1.5%.

Wave refraction lends itself readily to computer solution. The parametric representation of wave rays was first presented by Munk and Arthur [9], when they derived an equation for wave intensity, from which wave energy could be predicted given only the ray path and wave velocity on and near the ray. Griswold [10] calculated orthogonal spread by a numerical method. Harrison and Wilson [11] expanded this work into a workable computer programme, which utilized incremental distance along the orthogonal as a variable which minimizes errors due to bed undulations close inshore. Wilson [12] later devised a means of using a digital computer and plotter for determining and drawing orthogonals. Le Petit [13, 14] has combined the concepts of Griswold, and Munk and Arthur to develop a computer programme which Orr and Herbich [15] have applied to some complex hydrography with success. Jen [16] has also employed the computer to draw wave patterns and determine height distributions over 200 square miles of ocean margin at San Pedro Bay, California. Computer application is not restricted to progressive wind waves. Keulegan and Harrison [17] have reported a study of tsunami propagation across the Pacific Ocean and into the embayment of Crescent City, California, the waves in this case being necessarily a shallow-water phenomenon.

It is deemed unnecessary to outline the procedure for programming wave and hydrographic data for computer solution in this treatise. The references included can be used readily for this. The manual method described gives a picture of the variables involved and provides tools for quickly drawing the path of an orthogonal. This simple approach should serve many coastal engineering problems where the inexactness of the wave information warrants mainly a qualitative solution. Where the wave incidence is both complex and well documented, and detail is required over a large expanse of shoreline, use of the computer is recommended. It should be

WAVE REFRACTION

remembered, however, that the output is only as accurate as the input; so that heights, periods and directions should be known for the important wave groups influencing the problem. When this is concerned with forces on structures storm conditions are implicated; when long-term sediment movement is under study it is the persistent swell that should be included.

Wave energy

As noted already, the celerity of waves is not changed significantly in the shoaling process to warrant the use of high orders of theory. When considering energy, which is involved with wave height, it might be conceded that wave profile changes may warrant greater theoretical sophistication. However, Koh and Le Méhauté [18] have shown that wave heights calculated by third- and fifth-order accuracy differ by only 3% from those derived by linear theory. This is well within the accuracy of the wave data ever likely to be available, so that linear theory will be employed in the following discussion of wave energy.

Consider firstly a wave approaching normally to the shore so that no refraction takes place. Assume also that the power transmitted forward by the wave remains constant over time and space. This implies that there is no loss of energy through viscous or eddy sources and that there is no reflection seawards. This last restriction limits the application to bed slopes of less than 1/20.

As noted previously, the power transmitted forward per unit length of wave crest is $P = wH^2 C_g/8$ so that:

$$w H_o^2 C_{g_o} = w H^2 C_g \tag{5-8}$$

where C_g is the group wave velocity and suffix o denotes deep-water values. Thus:

$$H/H_o = (C_{g_o}/C_g)^{1/2} = D_d \tag{5-9}$$

where D_d is termed the *shoaling coefficient*, which can be directly related to d/L or d/L_o.

Consider now a wave train approaching a uniform shoreline (straight parallel bed contours) obliquely. Two orthogonals, spaced a distance b_o apart in deep water, will trace out exactly the same curve towards the shoreline (see Fig. 5-8). This means that they will have the constant displacement b_s along the waterline direction at all depths. However, due to the changing direction of each orthogonal (and the wave crest), the distance between them will increase as the water becomes shoaler, so that:

$$b_s = b/\cos \alpha = b_o/\cos \alpha_o \tag{5-10}$$

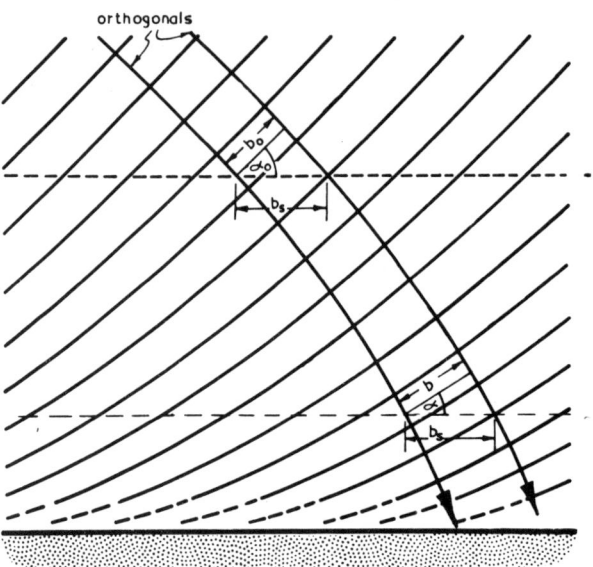

Fig. 5-8. Orthogonal spacing over a uniformly sloped shelf.

or:

$$\frac{b}{b_o} = \frac{\cos \alpha_o}{\cos \alpha} \tag{5-11}$$

The power transmitted between the two orthogonals is assumed to be constant so that:

$$wH_o^2 b_o C_{g_o}/8 = wH^2 b C_g/8 \tag{5-12}$$

or:

$$\frac{H}{H_o} = \left(\frac{C_{g_o}}{C_g}\right)^{1/2} \left(\frac{b_o}{b}\right)^{1/2} = \left(\frac{C_{g_o}}{C_g}\right)^{1/2} \left(\frac{\cos \alpha_o}{\cos \alpha}\right)^{1/2} = D_d K_d \tag{5-13}$$

where K_d is known as the *refraction coefficient*. From the previous relationship of Snell's Law, α can be determined at any d/L ratio once α_o is known, so that both D_d and K_d are functions only of d/L or d/L_o. Thus the wave height can be predicted for oblique wave approach to a uniform shoreline as in Table 5-IV from a knowledge of H_o, α_o and T. It will be noticed in the table that for normal approach ($\alpha_o = 0$) the height ratio decreases from its deep-water value of 1.0 to a minimum of 0.913 at $d/L_o = 0.15$ before rising to 1.692 at $d/L_o = 0.005$. This reduction in the transition zone is due to the combined effect of n (= C_g/C) and C_o/C which is depicted in Fig. 5-9.

TABLE 5-IV

Ratio of wave height at specific depth ratio to deep-water height for waves refracting across a uniformly sloped shelf

d/L_o	α_o																	
	0	5	10	15	20	25	30	35	40	45	50	55	60	65	70	75	80	85
0.005	1.692	1.689	1.680	1.664	1.642	1.614	1.578	1.536	1.486	1.429	1.363	1.307	1.223	1.124	1.013	0.877	0.722	0.504
0.01	1.439	1.435	1.427	1.413	1.395	1.372	1.342	1.307	1.267	1.218	1.160	1.096	1.026	0.944	0.850	0.740	0.606	0.430
0.04	1.063	1.062	1.057	1.048	1.038	1.023	1.004	0.980	0.953	0.923	0.883	0.840	0.789	0.753	0.656	0.576	0.473	0.335
0.06	0.993	0.992	0.987	0.982	0.973	0.959	0.944	0.926	0.903	0.874	0.841	0.801	0.753	0.699	0.633	0.553	0.457	0.326
0.08	0.953	0.953	0.950	0.946	0.938	0.927	0.919	0.898	0.878	0.852	0.823	0.787	0.742	0.690	0.622	0.548	0.453	0.326
0.10	0.931	0.931	0.929	0.925	0.918	0.910	0.898	0.883	0.866	0.843	0.816	0.783	0.743	0.693	0.630	0.555	0.459	0.329
0.15	0.913	0.913	0.912	0.909	0.905	0.898	0.892	0.882	0.869	0.853	0.832	0.805	0.769	0.726	0.669	0.595	0.498	0.353
0.20	0.917	0.917	0.916	0.913	0.910	0.907	0.901	0.886	0.894	0.873	0.857	0.840	0.813	0.773	0.724	0.650	0.548	0.397
0.25	0.932	0.932	0.931	0.931	0.929	0.926	0.923	0.918	0.912	0.904	0.894	0.879	0.858	0.829	0.787	0.713	0.620	0.453
0.30	0.949	0.949	0.949	0.948	0.947	0.945	0.943	0.941	0.937	0.932	0.925	0.915	0.902	0.882	0.850	0.796	0.699	0.524
0.35	0.964	0.963	0.963	0.963	0.962	0.961	0.960	0.959	0.956	0.954	0.950	0.942	0.934	0.922	0.897	0.855	0.773	0.600
0.40	0.976	0.976	0.976	0.976	0.975	0.975	0.974	0.974	0.972	0.971	0.968	0.965	0.960	0.951	0.937	0.910	0.846	0.684
0.45	0.985	0.985	0.985	0.985	0.985	0.984	0.984	0.984	0.983	0.982	0.980	0.978	0.975	0.971	0.963	0.945	0.902	0.769
0.50	0.991	0.991	0.991	0.990	0.990	0.990	0.990	0.990	0.989	0.989	0.988	0.987	0.985	0.983	0.977	0.967	0.940	0.841

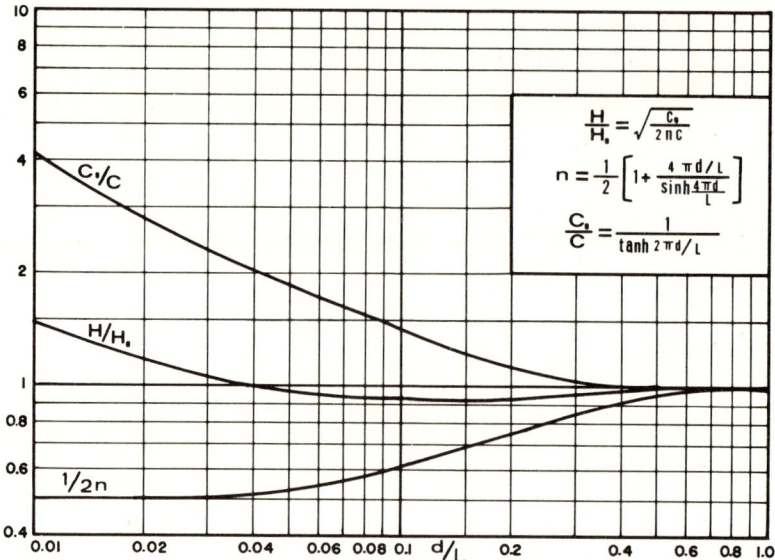

Fig. 5-9. Variation of n, C_0/C and H/H_0 with depth (d/L_0).

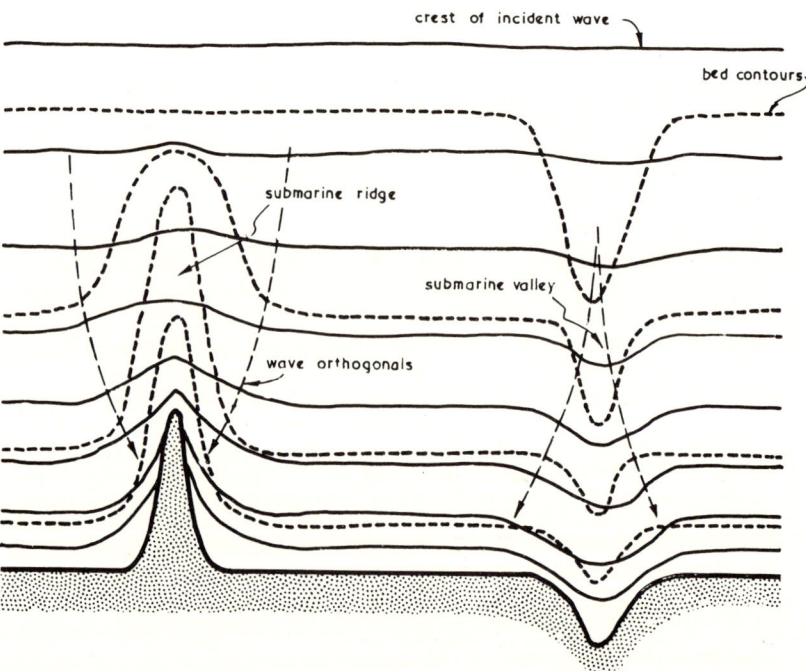

Fig. 5-10. Wave refraction over a headland shoal and submarine valley.

WAVE REFRACTION

Where bed contours are not parallel and suffer curvature the orthogonals of waves passing over them will change their spacing, so altering the energy per unit length of crest and consequently the wave height. Thus relative wave heights can be gauged from a wave refraction diagram as well as changes in wave direction. Two major features can be encountered in this regard, namely, a shoal protruding seawards and a submarine depression protruding landwards, the latter being termed a submarine canyon when on a large scale. These are depicted in Fig. 5-10, where it is seen that the former concentrates the orthogonals on the promontory generally associated with the submarine shoal and the latter deflects wave energy to either side of the embayment normally accompanying the submarine valley. These examples have profound significance in locating harbours or in planning troop landings in time of war.

Refraction of short-crested waves

When two wave trains are approaching the shore there are two possibilities for short-crestedness. There is the case of two trains of equal period angled to each other in deep water, and the occurrence of two aligned trains of differing period. Longuet-Higgins [19] has analyzed these two conditions, the conclusions from which will be illustrated from some simplified examples.

Consider two wave trains of similar height H but of differing period angled to each other in any depth of water as in Fig. 5-11A. The energy brought into the area by one train is added to that of the other train so that the energy contained in crest lengths b_1 and b_2 respectively will now be contained in the crest length b_c of the short-crested system. Thus:

$$w\overline{H}_c^2 b_c/8 = wH^2 b_1/8 + wH^2 b_2/8 \qquad (5\text{-}14)$$

or:

$$\overline{H}_c/H = [(b_1 + b_2)/b_c]^{1/2} \qquad (5\text{-}15)$$

Since $b_1 + b_2 > b_c$ the mean height H_c of the short-crested system is greater than the height of each separate train making it up. This wave height as measured along b_c varies, with a maximum at the intersections of the straight crests of the individual trains of $\sqrt{2}H$. With such variation in energy along the direction of the crest it could be thought that spreading or diffraction would occur, but such distribution exists instantaneously and when the two trains advance a new peak is immediately formed into a new section of each crest, as indicated by arrows in Fig. 5-11A.

Two trains angled to each other in deep water will be refracted differing amounts in shoaler water, so changing the angle between them. Consider two trains of the same period with $\alpha_0 = 60°$ and $40°$, respectively, giving an included angle $20°$. In water of $d/L_0 = 0.1$ Table 5-I shows $\alpha = 38°$ and $27°$, respectively, or $11°$

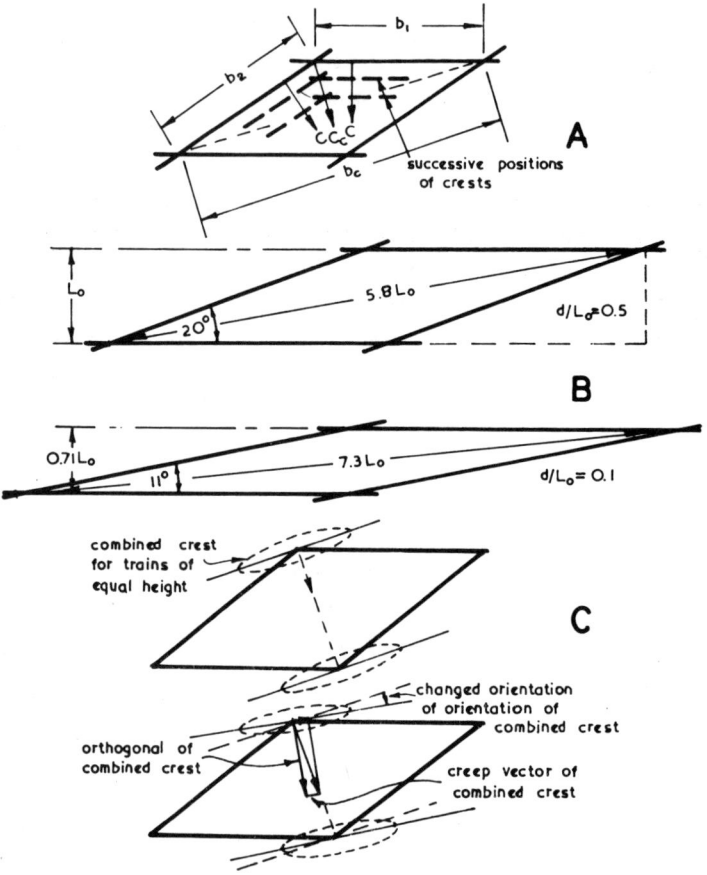

Fig. 5-11. A. Energy of wave trains angled to each other. B. Short-crested wave patterns in deep and transitional depths. C. Alignment of crest and wave advance for angled trains of differing height.

between them (see Fig. 5-11B). The wave lengths will have been reduced to 0.71 of their deep-water values. The crest lengths in each case are $L'_o = L_o/\sin 10° = 5.8 L_o$ and $L' = 0.71 L_o/\sin 5.5° = 7.3 L_o$, so that, in spite of the decreased lengths of the individual trains, the crest length of the combined system has increased. This will be exhibited by a more elongated island of water above the SWL.

Besides the above effect the train with least obliquity in deep water will suffer less refraction and therefore its height at any shoreward depth will be greater than for a similar train refracted more. Thus the elongated crest of the combined system in the shoaler depth will be concentrated more on the crest line of the train least refracted. This results in a crest alignment which is not normal to the direction of

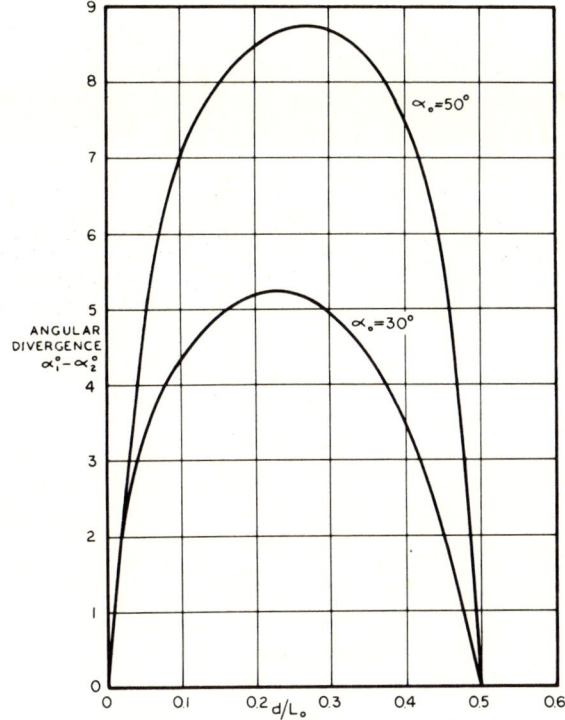

Fig. 5-12. Directional divergence through refraction of wave trains aligned in a water depth d/L_o = 0.5 with respective periods of 7 and 10, 10 and 14, 14 and 20 sec approximately.

advance of the waves, so giving the waves a staggered appearance, whereas in deep water they followed each other along the diagonals of the diamond pattern. In the shoaler conditions the crests still propagate along the diagonal path, but they appear to creep sideways in the maintenance of their staggered appearance. This is illustrated in Fig. 5-11C.

Two wave trains of differing period that are aligned in deep water will undergo different rates of refraction as they propagate into shallower water. Take the example of 8- and 14-sec waves approaching a coast at α_o = 40°. In 50 ft. of water their respective values of d/L_o = 0.15 and 0.05 and α = 31° and 19°. The longer wave has become more normal to the beach. The divergence in direction of trains of slightly different period is depicted in Fig. 5-12, where it is seen that an optimum is reached at d/L_o around 0.25. This will be shown to be significant when recording waves, because it is in such depth ratios that recording is normally carried out.

Refraction of wave spectrum

The refraction of a wave spectrum is an extension of the short-crested system above. Not only are many components of differing period involved, but the bulk of energy in each period band is concentrated at a specific angle to the fetch alignment, along which the longest waves propagate.

Longuet-Higgins [20] has shown that the total energy in a fetch in deep water is transferred to the shallower zone on the basis that each part of the spectrum is transformed independently. Other assumptions are that the area of ocean is wide enough for components not to spread outside it and that no energy is reflected from adjoining sections of coast.

Karlsson [21] has carried out a computer analysis on the Pierson-Moskowitz [22] spectrum, accounting for the directional distribution as given in the PNJ method [23]. This Longuet-Higgins [20] did not do, as he assumed all components travelled in the one direction. Assumptions in Karlsson's analysis are that no wind is present during the refraction process (i.e., the fully arisen sea has the downwind extremity of the fetch at the edge of the continental shelf), that no currents are present and friction is negligible.

In the computer study [21] a fully arisen sea was assumed with a significant wave height of 30 ft. (i.e., a FAS for $U_{19.5}$ = 40 knots and T_{max} = 13.2 sec from Table 3-VII). This was transformed normally to a straight coast across a continental shelf sloping uniformly to a depth of 1000 ft. at 40 miles from shore. This is an atypical profile for such a feature, which should have a mean slope of 0.002 to a depth of around 400 ft. However, this does not detract from the study made since the refraction would take place at similar depths. Karlsson [21] determined the significant wave height along the profile as reproduced in Fig.5-13 where a comparison is shown between it and initial reduction and subsequent amplification of a monochromatic wave of 13.2-sec period (from Table 5-IV). It would appear that the refraction of waves either side of T_{max} causes a slightly greater attenuation than normally is experienced in transitional depths. In this case it approximates 0.6 ft. in a 30 ft. height, which for engineering purposes in such depths of water is inconsequential. The lowest height reached is similar for both wave systems as is the subsequent amplification in the shallow depths, since most of the larger waves in the spectrum are now travelling almost normal to the beach.

Another result from Karlsson's study was the greater concentration of wave energy around the fetch alignment or the direction of the longest waves in the spectrum. Spectral bands with around half the energy of the T_{max} band were initially spread 45° either side of the normal to the coast in deep water but 1 mile from the coast, where the water was 25 ft. deep, these bands were contained within a fan 25° either side of this alignment. Those of smallest period are uninfluenced by the depth until in such shallow water that the linear theory in the analysis no longer

WAVE REFRACTION

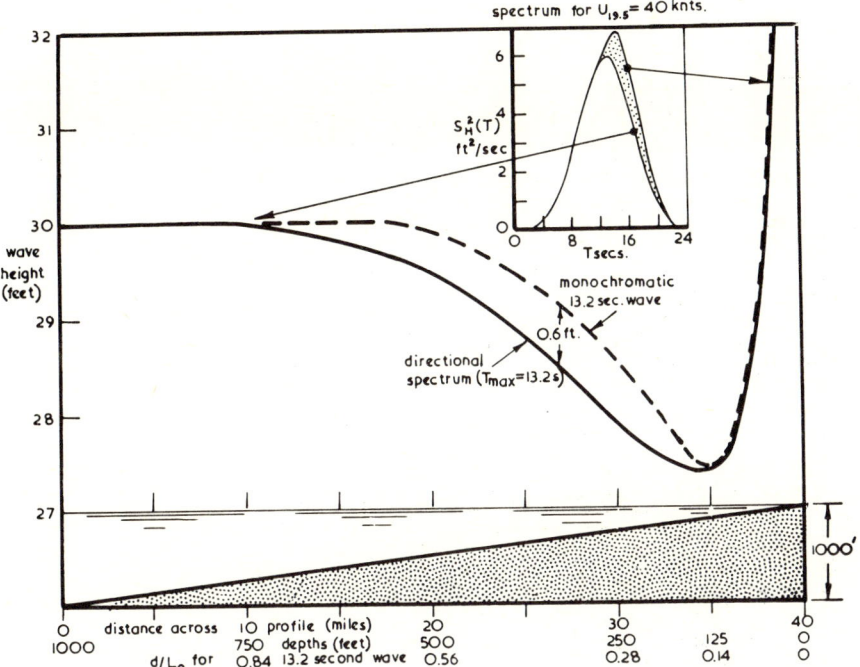

Fig. 5-13. Wave height variation across shelf for a spectrum and monochromatic wave system.

applies. Their directional spread of energy therefore remains the same as for the deep water.

The spectrum occurring in the deep and shallow water is also illustrated in Fig. 5-13, where it is seen that the longer-period waves are amplified more than the shorter ones. This distorts the spectrum, forcing the peak (T_{max}) to a slightly higher value. In this case it moved from 13.2 to 13.8 sec, adding area to the spectrum which increased the significant wave height, mainly due to the shoaling coefficient previously discussed. Longuet-Higgins [20] had concluded that the peak energy of the spectrum would remain at the same frequency band since his analysis was based on all waves being co-linear. Because the waves centred around the peak of the spectrum are more aligned and will receive the greatest amplification, the storm waves near shore will be longer crested than those in the deep-water.

Refraction by currents

In this discussion currents will be assumed uniform throughout the depth. This is reasonable since any velocity distribution will be near uniform in the upper layers where most of the orbital motion due to the wave is taking place. When a current is

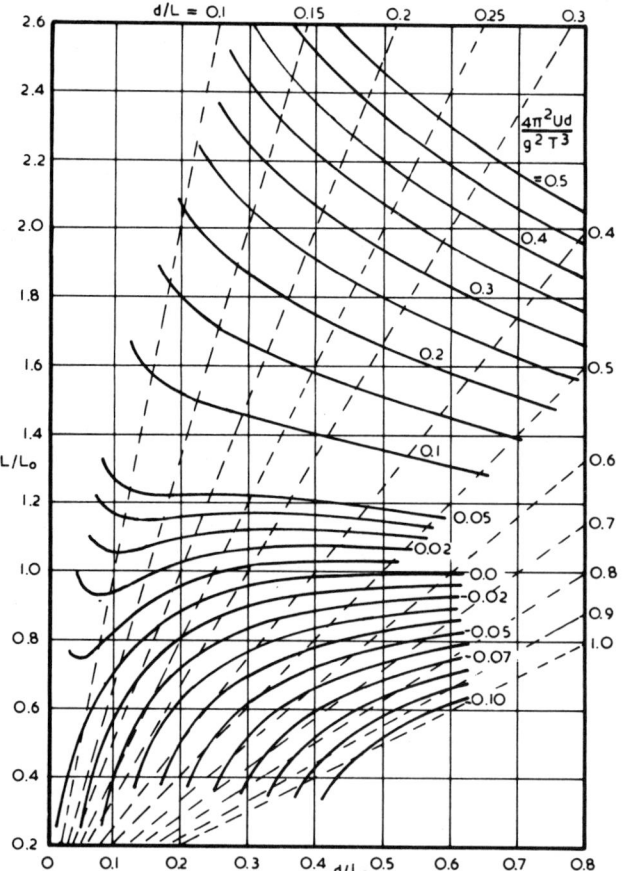

Fig. 5-14. Variation of wave length in a current.

aligned with an orthogonal the wave celerity and length will increase or decrease as the current runs with or against the waves. Using second-order Stokes theory, Jonsson et al. [24] have derived curves and tables for H/H_o and L/L_o in various depths for a dimensionless current parameter reducible to $4\pi^2 Ud/g^2 T^3$ where a positive value of U applies to a following current and a negative one to a contra current. These have been reproduced in Fig. 5-14 and 5-15 where it is seen that for certain positive flows maxima and minima appear for L/L_o. This is due to the opposing influences of an elongating current and a reducing depth ratio. Thus such aligned currents tend to steepen and break the shorter waves present in a spectrum. The clearing of chop from waves viewed from the air can indicate the presence of a current.

When waves propagate obliquely into a zone containing a current their velocity

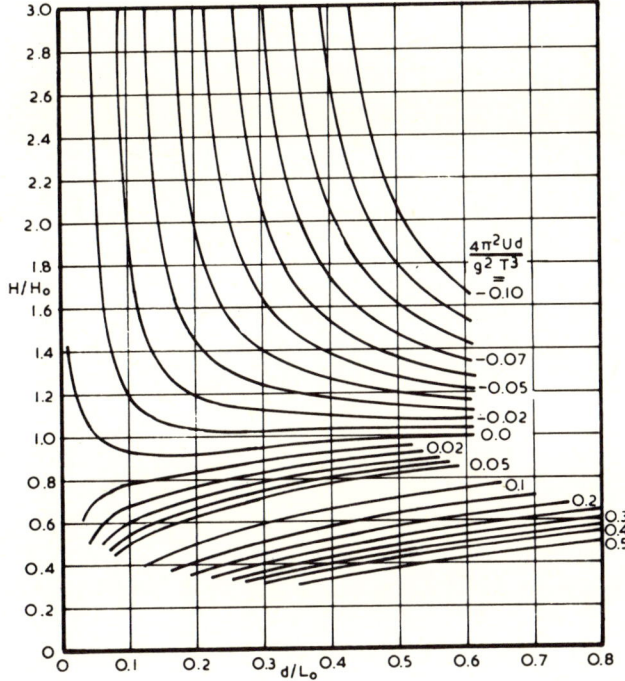

Fig. 5-15. Variation of wave height in a current.

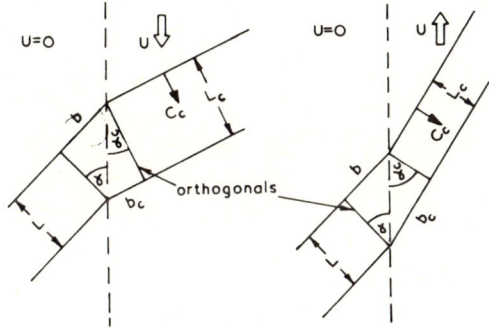

Fig. 5-16. Oblique wave propagation into a current.

will not only change, but also their direction, as depicted in Fig. 5-16. The resulting wave height is dictated by the new wave length L_c and the change of crest width from b to b_c such that:

$$\frac{H_c}{H} = \left(\frac{b}{b_c}\frac{L}{L_c}\right)^{1/2} = \left(\frac{L^2 \tan \alpha}{L_c^2 \tan \alpha_c}\right)^{1/2} \tag{5-16}$$

where suffix c denotes values in the current. Values of L_c can be determined from Fig. 5-14 by using the component of the current in the orthogonal direction. Johnson [25] has shown that this ratio of H_c/H expressed in other terms is, for most oblique angles and directions of currents, greater than unity.

BREAKERS AND SURF

The previous discussion on wave refraction is based on linear theory which becomes inapplicable when the water becomes extremely shallow and the waves are distorted in profile. At this stage the cnoidal or solitary wave theory may aid in determining crest height, but in the actual breaking process recourse must generally be made to experiment.

Thus, in tracing a wave train into shore from deep water, wave refraction theory can be used to some minimum depth, after which cnoidal theory can predict changes to the point where the wave arrives at a beach slope. Whether the wave breaks prior to reaching this final grade, breaks as it travels up it, or is reflected from it, is dictated by the depths where the final slope commences and the wave steepness at this location or offshore. In any event, the water surface will rise up the beach (or even up a vertical wall), the limit of which is termed the "uprush height". This, in turn, is influenced by the slope, roughness and permeability of the slope. For purposes of the present discussion on beach faces, it will be assumed that these are saturated and are therefore smooth and impervious, as far as uprush is concerned.

Monochromatic waves will be treated first to ease the discussion, followed by comments on the shoaling of a spectrum of waves arriving from offshore. The height attenuation inside the surf zone will be described.

Depending upon the slope of the beach face and the wave characteristics, some reflection will take place. This occurrence is inherent in the uprush formulae to be presented, but the nature of the partial standing waves and the path followed by the seaward moving train will be outlined.

Run-up procedure

A useful procedure for wave enhancement in shallow water and final run-up on a beach has been given by Van Dorn [26, 27], which has been modified slightly in the following discussion because of the different graphs available in this treatise. Le Méhauté et al. [28] have made a state-of-the-art report which contains many useful references on the subject.

The situation to be examined is depicted in Fig. 5-17, where a deep-water wave (H_o, L_o) travels up a very mild continental shelf to some depth d at which the bed

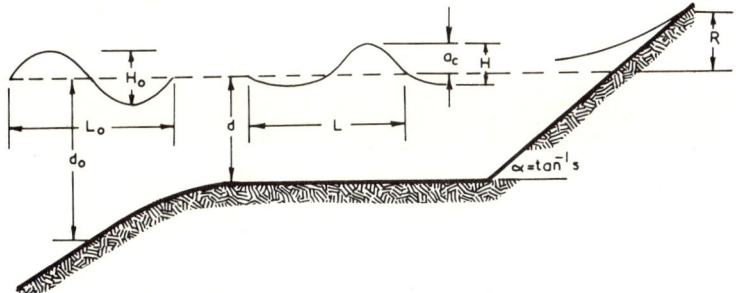

Fig. 5-17. Definition sketch for wave run-up.

slope increases substantially as it passes through the SWL. Although this will be considered a straight uniform slope to the pseudo constant depth d, under natural conditions it is likely to be curved with one or two humps just offshore. However, the beach face slope at the water line should be used in any calculations, which in later chapters will be shown to depend upon the size of sediment constituting the beach.

In Fig. 5-17 it is indicated that the operative depth d should be at least a wave length (L) in width. Even if this near-shore zone is slightly graded with its inshore limit at depth d the relationships to be presented will be acceptable. The run-up (R) is the vertical reach of the wave above SWL along the beach face sloped at $\alpha°$. This slope may be expressed as $s = \tan \alpha$.

By the time a wave has reached the depth d it may have reached instability and be broken. The criterion for this as given theoretically by Miche [29] is:

$$(H/L)_{max} = 0.142 \tanh 2\pi d/L \tag{5-17}$$

although tests by Danel [30] showed that the constant was closer to 0.12. These relationships are displayed in Fig. 4-5, where vertical asymmetry a_c/H is also included. Where waves are refracted across a uniformly sloping continental shelf the refraction and subsequent breaking by the above criterion can be readily computed as in Table 5-V where the ratio H_b/H_o and α_b are given for values of α_o and H_o/L_o. Values in Table 5-V were derived from Groen and Weenink [31] on the basis of linear theory. Le Méhauté and Koh [32] carried out an analysis to third order which in general increased the values of H_b/H_o and α_b by 10%.

If the wave does not break until it has entered the region of the sloping beach face, the slope itself has an influence on the breaking process. It increases both the breaking height and the depth at breaking. Le Méhauté and Koh have analyzed the results of several workers and have suggested the relationship:

$$H_b/H_o = 0.76 \, (s \cos \alpha_b)^{1/7} (H_o/L_o)^{-1/4} \tag{5-18}$$

TABLE 5-V

Angle and height of breaking wave for beach slope 1:30

H_o/L_o	α_o																
	0		10		20		30		40		50		60		70		80
	α_b	H_b/H_o	α_b	H_b/H_o	α_b	H_b/H_o	α_b	H_b/H_o	α_b	H_b/H_o	α_b	H_b/H_o	α_b	H_b/H_o	α_b	H_b/H_o	α_b	H_b/H_o
0.001	0	3.06	1.5	3.03	3.0	2.99	4.3	2.92	5.4	2.80	6.4	2.65	7.0	2.42	7.0	2.12	6.5	1.68
0.002	0	2.42	1.8	2.40	3.7	2.37	5.4	2.30	6.7	2.21	8.0	2.10	8.7	1.93	8.7	1.70	8.2	1.36
0.004	0	1.92	2.3	1.90	4.6	1.88	6.7	1.82	8.5	1.75	9.8	1.66	11.0	1.54	11.0	1.36	10.4	1.08
0.007	0	1.62	2.8	1.61	5.5	1.59	8.1	1.54	10.4	1.48	11.8	1.39	13.0	1.28	13.0	1.13	12.5	0.89
0.01	0	1.46	3.2	1.45	6.4	1.43	9.1	1.39	11.7	1.34	13.3	1.25	14.6	1.15	14.6	1.01	14.0	0.80
0.02	0	1.22	4.2	1.21	8.3	1.19	12.0	1.16	15.0	1.12	17.5	1.05	19.2	0.96	19.2	0.83	18.0	0.65
0.04	0	1.04	5.5	1.02	10.6	1.01	15.5	0.99	19.7	0.95	23.0	0.90	25.2	0.82	25.2	0.71	23.4	0.55
0.07	0	0.94	6.7	0.93	13.2	0.92	19.1	0.91	24.5	0.87	28.7	0.82	31.8	0.75	31.8	0.64	29.3	0.49
0.10	0	0.92	7.6	0.91	15.0	0.90	22.2	0.89	28.5	0.86	34.0	0.81	37.4	0.74	37.4	0.62	34.4	0.46

Note: (1) for other beach slopes and for third-order theory see multiplication factor for α_b and H_b/H_o in Fig. 5-18; (2) depth at breaking is approximately 1.3H_b throughout.

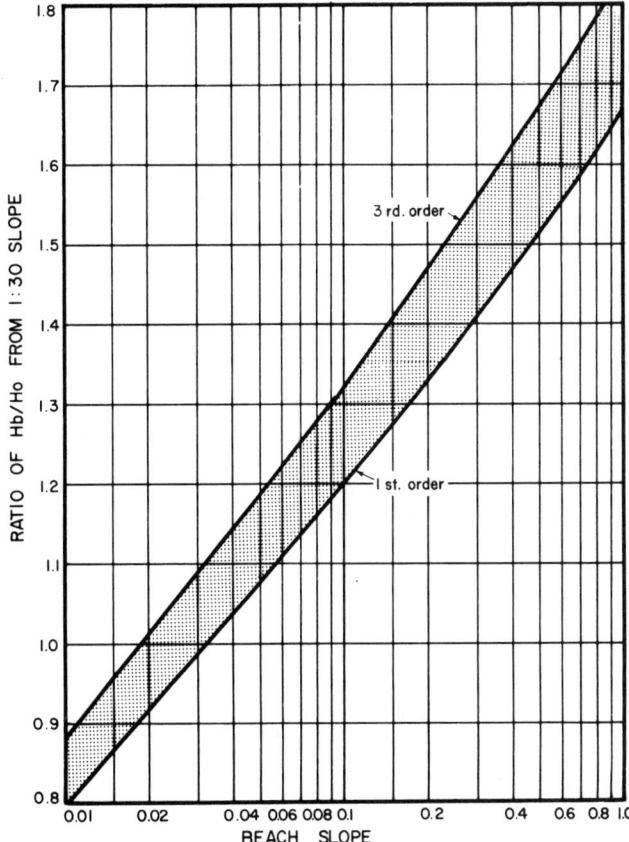

Fig. 5-18. Amplification factor for waves breaking on beach slopes.

The $\cos \alpha_b$ term arises because the beach slope is less for waves breaking at angle α_b to the shore. It can be shown that the values in Table 5-V are equivalent to those for a beach slope of 1 : 30 in eq. 5-18. The respective H_b/H_o ratios for other beach slopes are given in Fig.5-18, together with the 1.1 ratio for the 3rd-order theory.

Collins [33], in an analysis of spectra of breaking waves, has employed an alternative criterion:

$$H_b/d_b = 0.72 + 5.6s \qquad (5\text{-}19)$$

which has been used in Fig.4-5 to predict breaking conditions for slopes up to 1 : 8. Thus a_c/H values can be determined for waves breaking on various mild slopes. Van Dorn [26] has shown this ratio to be useful in computing run-up for non-breaking waves, even though these represent standing waves. This was an empirical approach from an analysis of flume tests by Savage [34] which showed good predictive qualities.

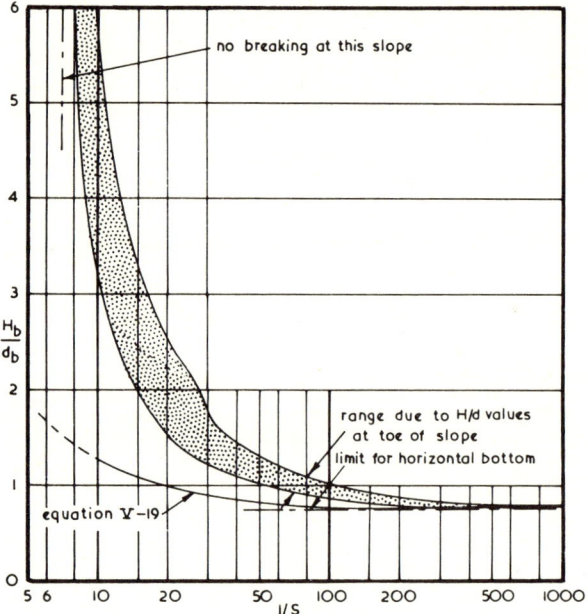

Fig. 5-19. Ratio H_b/d_b for waves breaking on a slope s.

Street and Camfield [35], from tests of their own and others, have derived a generalized diagram of H_b/d_b for solitary waves on a range of bed slopes s as modified in Fig. 5-19. Kishi and Saeki [36] have also carried out similar tests with solitary waves. Eq. 5-19 has been plotted in Fig. 5-19 for comparison of oscillatory waves with the limiting form of solitary wave. There is need for more detailed studies of this phenomenon.

The run-up of a wave on a beach will alter significantly as it breaks or does not break. A criterion for this critical condition must therefore be calculated, which can be taken from the experiments of Hunt [37] in the form:

$$A \gtrless \frac{L_o}{H} \frac{\alpha^2 \pi}{(2a_c/H)(180°)^2} \tag{5-20}$$

where angle α is in degrees and A is given by K in Fig. 5-20 from a further relationship [26]:

$$K = \left(\frac{2\pi d}{L_o}\right)^{1/2} \frac{180°}{\alpha \pi} \tag{5-21}$$

When A in eq. 5-20 is less than RHS the wave is non-breaking, in which case the run-up R is given by the empirical relationship:

$$R = 2a_c A \tag{5-22}$$

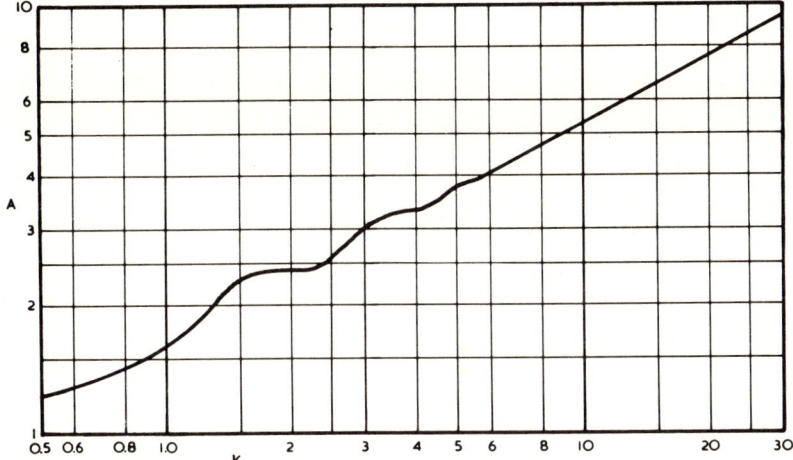

Fig. 5-20. Factor A in eq. 5-20 from factor K in eq. 5-21.

where a_c is the crest height above SWL for the progressive incident wave and factor A is a beach amplification factor.

Should A in eq. 5-20 be greater than RHS the wave will break on the slope, in which case the run-up R is given by [37]:

$$\frac{R}{H_o} = \left(\frac{L_o}{H_o}\right)^{1/2} \frac{\alpha\pi}{180°} \qquad (5\text{-}23)$$

for slopes steeper than 1 : 30. The relationship between deep-water wave height H_o and the height at breaking, or that at the toe of the beach slope, would in this case be based upon a normal approach to the coast. If refraction has taken place over a uniform shelf the local wave height would be smaller and hence R would be some proportion of the former amount as given in Table 5-VI.

Should the wave break on the mild slope prior to reaching the final beach face (according to eq. 5-17) the depth at breaking should be ascertained (eq. 5-19 or Fig. 5-17). During the passage of the surging wave to the toe of the beach face a reduction in wave height to H'_b will occur (see Fig. 5-21B). Tests by Horikawa and

TABLE 5-VI

Percentages of R (eq. 5-23) for various α_o

$\alpha_o°$	20	30	40	50	60	70	80
$\% R$	98	95	92	86	79	69	55

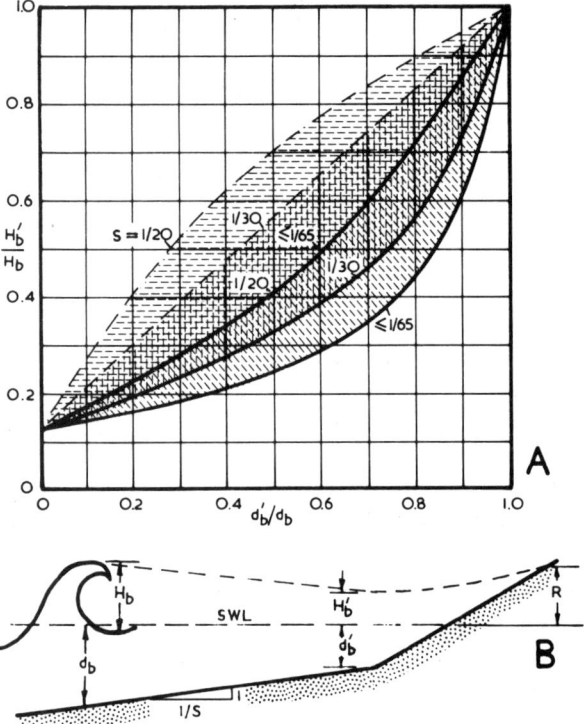

Fig.5-21. A. Height attenuation of breaking waves with depth. Note: hatched zones indicate spread of results due to H/L and experimental variations. B. Definition sketch of wave breaking and ultimate run-up. Note: Convert H'_b to H_o from Table 5-V ($\alpha_o = 0$) and Fig. 5-18 for substitution into eq. 5-21.

Kuo [38] and Nakamura et al. [39] have been summarized in Fig.5-21A by taking the lower and upper envelopes of all test results. The lower extreme is the lowest height possible when no reflection has influenced the measurement and the steepest waves have broken. The upper limit should serve for the waves with least steepness where reflection is likely to be greatest. Hwang and Divoky [40] have attempted to analyze the situation theoretically and have presented design curves compatable with the ranges in Fig. 5-21A.

The slope employed should account for the angle at breaking, so should be $s_{act} \times \cos \alpha_b$. The still-water depth (d'_b) at the toe of the beach face thus gives a wave height (H'_b) which is then transformed into an equivalent H_o for substitution in eq. 5-23 by using column $\alpha_o = 0$ in Table 5-V.

Shoaling of wave spectrum

Where a fetch is formed across the continental shelf a partial or fully arisen sea may result which is normal or angled to shore. It has been shown in the discussion of wave refraction that the spectrum inshore is not greatly different from that in deep water. However, when the many groups of waves enter water shallow enough they will be forced to break. This action is very complex as a number of influences are operating concurrently.

The components of the spectrum are each being steepened at differing rates due to the shoaling. This can cause some of the shorter waves to break. But this group can also be steepened to breaking point by the squeezing effect of the longer waves, acting individually or in combination (see Fig. 3-25). The shortest waves in this respect will be affected by the medium-period waves and they in turn by the longer-period components. The amplification factor could be taken to vary from 1.0 at T_{max} to 2.5 at $T_L = 0.35\, T_{max}$.

Ijima et al. [41] have attempted to trace components into shallow water in order to study the effect on the spectrum. By using a limiting steepness criterion for breaking they submit that a rising curve across the spectrum will cut off a large section of the high-frequency or low-period end. The minimum reduction occurs for the waves greater than T_{max}. If an attempt is made to construct such a cut-off curve it is found that very small d/L_o ratios are required to effect any reduction in spectral area.

Collins [33] has approached the problem from a probability angle, taking waves of a specific period and a distribution of wave heights. This distribution is shown to steepen slightly at the point of breaking, with the peak skewed to higher waves the longer the period considered. The steepness at breaking was virtually unaffected by approach angle α_o, but differed from 1/20 for steepness limitation to 1/12 for depth limited breaking.

It would seem that the interaction of the waves must be considered as they combine to form high waves designated by $H_{1/3}$ and $H_{1/10}$, etc. The formation of these extra high waves will be limited as shallow water is reached through the steepness criterion previously noted. It has already been illustrated from the wave-refraction analysis of Karlson [21] that by the time shallow water is reached the shoaling of a spectrum is essentially the same as that of a monochromatic wave with period T_{max}. Thus the breaking and dissipation of the many wave trains of the spectrum could be based upon such a wave using the height $H_{1/3}$.

Using the breaking criterion of eq. 5-17 which can be expressed as:

$$H_{1/3}/T_{max}^2 = 0.12(5.12 \tanh^2 2\pi d/L) \tag{5-24}$$

for $H_{1/3}$ in feet and L referring to the period T_{max}. This is graphed in Fig. 5-22, where also is traced the $H_{1/3}/T_{max}^2$ ratios from Table 3-VIII for various percentage

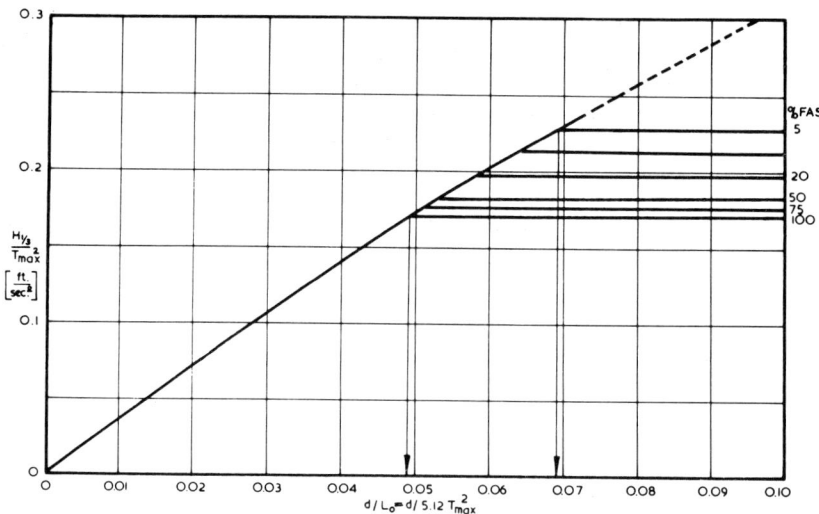

Fig. 5-22. Limiting values of $H_{1/3}/T^2_{max}$ in shoaling depths, with deep-water maxima for various percentages of FAS.

ratios of F/F_{FAS}. The intersection of these lines indicates the d/L_o values at which depth starts to control the significant wave height present. Initially this reduction will be effected through dissipation of the lower-period end of the EDC curve, finally at lower depths these major components of the system suffer reduction. This concentration of energy at the $T_{max} - T_U$ end of the EDC in shallow water confirms the use of monochromatic waves in model tests of marine structures and coastal processes.

Breaking processes

Breaking waves have been classified into four categories by Galvin [42], namely, spilling, plunging, collapsing and surging. The successive profiles of these four breakers are illustrated in Fig. 5-23, altered in position only for clarity. Tests gave the transition values listed in Table 5-VII [42]. Between the limits shown in Table

TABLE 5-VII

Transition values between breaker types

Parameter	Surge-plunge	Plunge-spill
$H_o/L_o\, s^2$	0.09	4.80
$H_b/g\, s\, T^2$	0.003	0.068

BREAKERS AND SURF

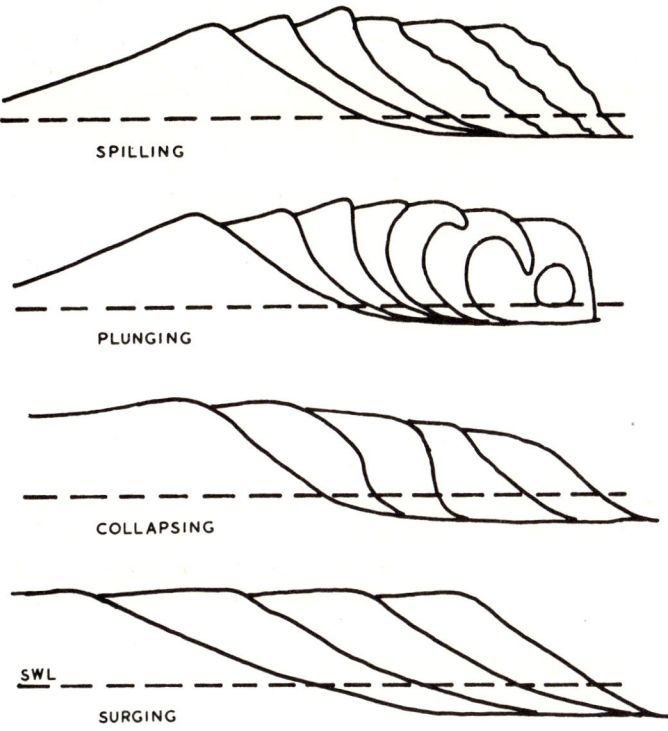

Fig. 5-23. Successive profiles of waves of various types.

5-VII plunging breakers will occur, whilst outside them surging or spilling waves exist. This information could be of engineering significance if a slope were to be provided for complete dissipation of a wave, which takes place when the breaker is truly plunging. Even if some reflection from the beach should influence the breaking mechanism the reflecting wave in such circumstances is unlikely to propagate seawards further than the breaker line. For example, taking a value of $H_o s^2/L_o$ = 1.0 for a reasonable intermediate value and 1 : 200 for a normal swell steepness, the beach slope to produce plunging breakers is 1 : 14. With such a slope waves of steepness from 0.0005 to 0.024 could be expected to break in this manner.

EXAMPLES

1
Trace an orthogonal ($\alpha_o = 60°$) across parallel contours for which d/L_o = 0.5, 0.2, 0.1, 0.05, 0.02, 0.01, by using C_d/C_o and $(C_2/C_o)/(C_1/C_o)$ procedures. What effect is observed if one angular change is made between the 0.5 and 0.01 contours?

The solution is presented in Fig. 5-24. The traverse angles at each contour are listed and the step procedure for finding them indicated by the arrows. When one step is used at the mean depth 0.255 between $d/L_o = 0.5$ and 0.01, the final direction across the 0.01 is the same as previously, but the path is offset from it. The correct orthogonal for this uniformly sloping bed will be curved.

2

Trace an orthogonal across the parallel contours $d/L_o = 0.5$ and 0.40 which is parallel to them in deep-water.

The deep-water approach alignment implies 90° to the contour normal. From Table 5-I it is seen that at $d/L_o = 0.50$, $\alpha = 85.07°$. In Fig. 5-24 this traverse angle is drawn as 85.3°. The subsequent angles for $d/L_o = 0.48, 0.46, 0.44, 0.42$ and 0.40 are drawn in the figure, together with the single angular change. As in Example 1, the orientation of the final orthogonal is the same even though the paths differ slightly.

3

For the five straight-bed contours illustrated in Fig. 5-24 trace the orthogonal approaching the normal to the $d/L_o = 0.5$ contour at 30°.

The values of α_o in terms of α and θ are listed in the figure from which the traverse angle α is found from Fig. 5-3. Since contours $d/L_o = 0.5$ and 0.02 are parallel it would be possible to trace directly in one step and obtain an exit angle of 10° but the path followed by the orthogonal across the intermediate depths would be erroneous to a degree.

4

From a centre construct radii at 10° intervals to represent d/L_o contours as illustrated in Fig. 5-24. Trace an orthogonal with $\alpha = 0°$ at the $d/L_o = 0.20$ contour for five or a single step.

The angles of orthogonals to the contour normals are as depicted in the figure. The single step orthogonal in this case, issuing from the mean depth of $d/L_o = 0.15$ exists at an angle different from that of the 5-step solution.

5

Four curved contours are drawn in Fig. 5-24, for which the approach angles are illustrated by means of normals to pseudo tangents at the point of passage of the orthogonal. Compute the traverse angles.

When α_o is known plus $(C_2/C_o)/(C_1/C_o)$ the value of α is readily determined from Fig. 5-3. The choice of tangent alignment depends somewhat on the C_d/C_o ratio operating, or whether $\alpha_o = 2\alpha$.

6

Construct concentric circular bed contours such that $r/r_o = C/C_o$ as in Fig. 5-24. Trace orthogonals approaching the outer contour at 45° and 60° and compare the result with the logarithmic spirals derived theoretically by Arthur [43].

The trace derived by Arthur is that of the logarithmic spiral $r_o/r = c^{(\theta - \theta_o)} \cot \theta_o$ where θ and θ_o are as defined in Fig. 5-24. It is seen that the step procedure adopted in the present method pulls the orthogonals towards the island centre too sharply. If more contours were used the closer would they match the theoretical solution.

7

Trace an orthogonal which is directed seawards and is 45° to the straight coast when in a depth of $d/L_o = 0.01$, assuming all bed contours are parallel to the water line. What is the greatest d/L_o ratio reached by the orthogonal.

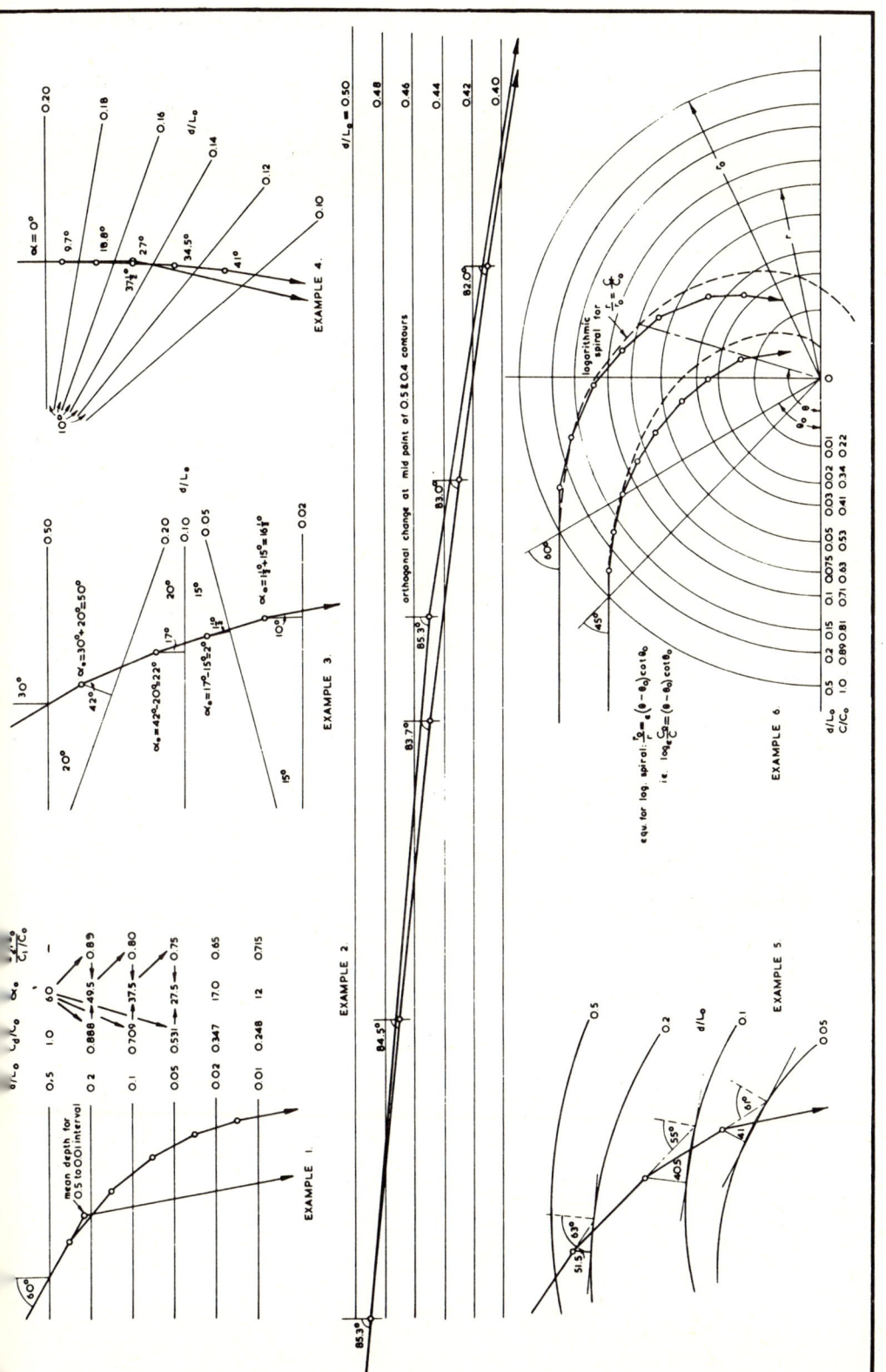

Fig. 5-24. Solutions to Examples 1 to 6.

Now $\sin \alpha_2 / \sin \alpha_1 = (C_2/C_0)/(C_1/C_0)$ where suffix 1 refers to the deeper water. When $\alpha_2 = 45°$, $\sin \alpha_2 = 0.707$ and $C_2/C_0 = \tanh 2\pi d/L = 0.248$. It is more convenient in this case to find the d/L_0 value at which α becomes a specific value, rather than determine α at specific depth values. Thus, to find C_1/C_0 for $\alpha_1 = 50°$ we have $\sin \alpha_1 = 0.766$, so that:

$$\frac{0.707}{0.766} = \frac{0.248}{C_1/C_0} \quad \text{or} \quad C_1/C_0 = 0.2685$$

To find d/L_0 represented by $C_1/C_0 = \tanh 2d/L = 0.2685$, some interpolation is necessary in the tables, for example:

d/L_0	$\tanh 2\pi d/L$			
0.011	0.2598	⎤ 113		
0.012	0.2711	⎦	⎤ 26	$\frac{26}{113} = 0.23$
?	0.2685		⎦	

so that $d/L_0 = 0.01177 = 0.0118$.

With α_2 now equal to $50°$ find C_1/C_0 for $\alpha_1 = 55°$ which gives $d_1/L_0 = 0.0135$ and proceed as in Table 5-VIII. Proceed until $\alpha_1 = 90°$ (*at $d_1/L_0 = 0.01885$*) after which the orthogonal is then traced shorewards (see Fig. 5-25), the first ratios becoming:

$$\frac{\sin \alpha_2}{\sin \alpha_1} = \frac{0.985}{1.0} = \frac{C_2/C_0}{C_1/C_0} = \frac{C_2/C_0}{0.3365}$$

so that $C_2/C_0 = 0.3315$ and $d_2/L_0 = 0.0182$. It will be noted that for parallel contours the angle of the orthogonal is similar for the same depth contour, so that this particular orthogonal is angled $45°$ to the shore at $d/L_0 = 0.01$ on its passage back to the beach.

Such refraction of wave energy back to shore at a point downcoast constitutes what are termed "edge waves". At points along the shoreline where reflection is excessive there are locations elsewhere receiving this energy returned back from the sea, so providing one source of energy variation along the beachline.

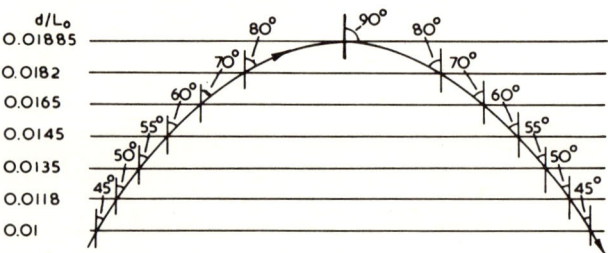

Fig. 5-25. Orthogonal traced seawards and back again as in Example 7.

8

A 30-knot wind ($U_{19.5}$) produces a FAS at the edge of the continental shelf, with the fetch angled at $45°$ to the coast. Assuming all components in the spectrum to be aligned with the wind, find the height and angle of those spaced about 2 sec apart in 50 ft. of water. Reconstruct the spectrum at this point and compare its $H_{1/3}$ value with that of a monochromatic wave with similar $H_{1/3}$ deep-water height and period of T_{max}.

EXAMPLES

TABLE 5-VIII

Refraction computations for Example 7

α_2	d_2/L_0	C_2/C_0	$\sin \alpha_2$	α_1	$\sin \alpha_1$	C_1/C_0	d_1/L_0
Seawards							
45°	0.0100	0.2480	0.707	50°	0.766	0.2685	0.0118
50°	0.0118	0.2685	0.766	55°	0.819	0.2875	0.0135
55°	0.0135	0.2875	0.819	60°	0.866	0.3045	0.0143
60°	0.0143	0.3045	0.866	70°	0.940	0.3165	0.0165
70°	0.0165	0.3165	0.940	80°	0.985	0.3315	0.0182
80°	0.0182	0.3315	0.985	90°	1.000	0.3365	0.01885

α_1	d_1/L_0	C_1/C_0	$\sin \alpha_1$	α_2	$\sin \alpha_2$	C_2/C_0	d_2/L_0
Shorewards							
90°	0.01885	0.3365	1.000	80°	0.985	0.3315	0.0182
80°	0.0182	0.3315	0.985	70°	0.940	0.3165	0.0165
70°	0.0165	0.3165	0.940	60°	0.866	0.3045	0.0143
60°	0.0143	0.3045	0.866	55°	0.819	0.2875	0.0135

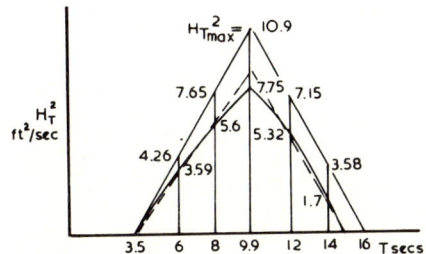

Fig. 5-26. Spectrum of FAS ($U_{19.5} = 30$ knots) and after refraction as in Example 8.

TABLE 5-IX

Wave characteristics in 50 ft. depth for Example 8

T	6	8	9.9	12	14
d/L_0	0.272	0.153	0.100	0.068	0.050
$\alpha°$	42.0	35.5	30.15	25.7	22.0
H/H_0	0.918	0.855	0.843	0.864	0.898
H_T^2	3.59	5.60	7.75	5.32	1.70

FAS for $U_{19.5}$ = 30 knots, gives from Table 3-VII (upper section) $H_{1/3}$ = 16.6 ft., $H'T_{max}$ = 3.3, T_{max} = 9.9 sec, T_L = 3.5 $H'T_{max}$ sec and T_U = 16.0 sec, as in Fig. 5-26. (Values in the lower half of Table 3-VII are more realistic, having been measured in the ocean, see Fig. 3-27). In 50 ft. of water the values of $d/L_0, \alpha$, (from Table 5-I) and H/H_0 (from Table 5-IV) are as listed in Table 5-IX.

The modified spectrum is sketched from which the proportional areas give $H_{1/3}$ in 50 ft. of water as 11.8 ft. A monochromatic wave with period T_{max} = 9.9 sec has a factor H/H_0 = 0.843 giving $H_{1/3}$ = 0.843 × 16.6 = 14.0 ft. In this case the excess of the monochromatic wave over the spectral model is greater than that illustrated in Fig. 5-13, but the conditions differ in that normal approach was assumed in Fig. 5-13 and the components were angled to the fetch alignment.

9

A swell wave of 12-sec period and height of 3 ft. is approaching the coast in deep water at angle of α_0 = 50°. It refracts to a beach whose offshore region slopes at 1 : 65 until the beach-face slope of 1 : 12 intersects it at the SWL depth of 4 ft. Trace the history of this wave to the point where it runs up the beach, assuming the latter to be smooth and impermeable due to its saturated condition.

At 4 ft. depth d/L_0 = 4 (5.12 × 12²) = 0.00542, d/L = 0.0295. For α_0 = 50° from Table 5-IV, H/H_0 = 1.363, so that H = 5.3 ft.

Test for stability $(H/L)_{max}$ = 0.12 tanh $2\pi d/L$ = 0.12 × 0.183 = 0.022. $(H/L)_{act}$ = 5.5/(4/0.0295) = 0.0405.

Therefore, the wave will break prior to reaching the toe of the beach face. The height and angle of breaking from Table 5-V for H_0/L_0 = 3/(5.12 × 12²) = 0.00406 is given by H_b/H_0 = 1.66 and α = 9.8°, so that H_b = 5 ft. for a beach slope of 1:30. The correction factor from Fig. 5-18 for a slope 1/s = 65/cos 9.8° = 66 is 0.91 (average of 1st and 3rd order), so that H_b = 4.55 ft. and α = 8.9°. The depth at breaking from Fig. 5-19 for 1/s = 66 is 0.805 from eq. 5-19 or 0.94 for a solitary wave, say 0.87, so that d_b = 4.55/0.87 = 5.2 ft. The height of the broken wave as it reaches the toe of the beach face (d'_b = 4 ft.) is given by Fig. 5-21 with d'_b/d_b = 4/5.2 ≃ 0.77, so that H'_b/H_b = 0.68 (using top of the range for s ⩽ 1/65 since H_0/L_0 is small) and H'_b = 4.55 × 0.68 = 3.1 ft.

In order to substitute in eq. 5-23 for wave run-up R, the value of H'_b must be converted to H_0 for normal approach, which from Table 5-V gives, for α_0 = 0 and H_0/L_0 = 0.004 H_b/H_0 = 1.92, so that H_0 = 3.1/1.92 = 1.6 ft. For slope of 1:12 α = 4.7° so that eq. 5-23 becomes:

$$\frac{R}{1.6} = \left(\frac{5.12 \times 12^2}{1.6}\right)^{1/2} \frac{4.7\pi}{180} \quad \text{or} \quad R = 2.8 \text{ ft.}$$

The wave thus rushes 2.8/sin 4.7° = 34 ft. up the beach face from the SWL line. In practice this will be reduced somewhat by percolation, especially near the top end of its run where the sand grains will be larger and the face will not be wetted so frequently.

10

If in Example 9 the deep-water wave height were 1 ft. instead of 2 ft., what would be the run-up?

In the depth of 4 ft. H = 1.363 ft. (see above) and H/L = 0.0135, so that the wave is unbroken at the toe of the beach face. It is necessary to test whether it breaks up the slope or just swashes up without breaking, being partially reflected in the process. From eq. 5-21, K = $(2\pi \times 4/5.12 \times 12^2)^{1/2}(180/4.7\pi)$ = 2.25. From Fig. 5-20, A = 2.4. The RHS of eq. 5-20 is $(5.12 \times 12^2/1.363)[4.7^2\pi/(2 \times 0.89)180^2]$ = 0.65 since from Fig. 4-5 a_c/H = 0.89 when H_0/L_0 = 0.0013 and d/L = 0.0295 (H/L = 0.012).

Since A > RHS of eq. 5-20 therefore the wave breaks in the slope and the run-up R is given by eq. 5-23. The equivalent H_0 for normal approach is given in Table 5-IV for d/L_0 = 0.00542

EXAMPLES

by $(H_o)_{50}/(H_o)_0 = 1.363/1.692 = 0.8$. Hence eq. 5-23 becomes:

$$\frac{R}{0.8} = \left(\frac{5.12 \times 12^2}{0.8}\right)^{1/2} \frac{4.7\pi}{180} \quad \text{or} \quad R = 2 \text{ ft.}$$

Thus, in spite of H_o being one third the value in Example 9, the run-up height is only 70% of the former R, due to less energy being dissipated in the breaking process prior to reaching the beach face.

11
What types of breakers are they in examples 9 and 10?

In Example 9 the parameter $H_b/gsT^2 = 4.55 \times 65/32.2 \times 12^2 = 0.064$. From Table 5-VII this wave appears to be a plunging breaker on the verge of spilling.

In Example 10 the parameter $H_o/L_o s^2 = 0.8 \times 12^2/5.12 \times 12^2 = 0.157$. Again the breaker is plunging but closer to a surging type than for Example 9.

12
A FAS is developed by a 35-knot wind ($U_{19.5}$) on the continental shelf. An oil rig is located in 50 ft. of water. What $H_{1/3}$ and T_{max} should it be designed to withstand if such storm conditions are considered to be the worst possible? Compute similarly for a breakwater head in 20 ft. of water.

From Table 3-VII the FAS for $U_{19.5} = 35$ knots gives $H_{1/3} = 22.6$ ft., $T_{max} = 11.6$ sec. From Table 3-VIII, $T_{max}/H_{1/3}^{1/2} = 2.43$.

A 50 ft. depth gives $d/5.12 T_{max}^2 = 50/5.12 \times 11.6^2 = 0.0725$.

Fig. 5-22 shows that FAS conditions exist in 50 ft. of water.

In 20 ft. depth $d/5.12 T_{max}^2 = 0.029$ and Fig. 5-22 gives $H_{1/3}/T_{max}^2 = 0.1$ so that $H_{1/3} = 0.1 \times 11.6^2 = 13.5$ ft.

13
Waves of 9 sec period, and 4 ft. height in deep water, are approaching a river mouth in 30 ft. of water where an outward current of 3 knots exists. Assuming the current to be uniform throughout the depth, compute the length and height of the waves. Do the same for 5-sec waves of similar height. Discuss the results.

$T = 9$ sec, $H_o = 4$ ft., $d = 30$ ft., $U = 3$ knots.
From Fig. 5-14 and 5-15 for

$$\frac{4\pi^2 Ud}{g^2 T^3} = \frac{4\pi^2 (3 \times 6080/3600) 30}{32.2^2 \times 9^3} = -0.008$$

and $d/L_o = 30/5.12 \times 9^2 = 0.072$, $L/L_o = 0.75$ and $H/H_o = 1.4$. Wave steepness $= H/L = 4 \times 1.4/5.12 \times 9^2 \times 0.75 = 0.018$ which is much less than the limiting value of $0.12 \tanh 2\pi d/L$.

$T = 5$ sec, $H_o = 5$ ft., $d = 30$ ft., $U = 3$ knots.
From Fig. 5-14 and 5-15 for $4\pi^2 Ud/g^2 T^3 = 0.008(9^3/5^3) = -0.047$ and $d/L_o = 30/5.12 \times 5^2 = 0.234$, $L/L_o = 0.5$, $H/H_o = 2.0$. Wave steepness $H/L = 4 \times 2/5.12 \times 5^2 \times 0.5 = 0.125$.

Limiting steepness $= 0.12 \tanh 2\pi d/L = 0.12 \times 0.92 = 0.11$ so that these 5-sec waves will break. Thus the 9-sec waves whose deep-water steepness was $4/5.12 \times 9^2 = 0.00965$ now become the predominant wave with double this steepness.

PROBLEMS

1

Draw seven parallel lines 2 cm apart and designate them as bed contours: d/L_o = 0.5, 0.2, 0.1, 0.05, 0.02 and 0.01. Draw an orthogonal at 45° to the d/L_o = 0.5 contour and trace it shorewards using both the C_2/C_0 and $(C_2/C_0)/(C_1/C_0)$ method. Check your angles with those in Table 5-I.

2

Trace a second orthogonal parallel to the one in Problem 1 above and measure the crest length between them at the d/L_o = 0.5 and 0.01 locations. From the ratio of these lengths determine the refraction coefficient and compare with the value given in Table 5-IV and the ratio $(\cos \alpha_o / \cos \alpha)^{1/2}$.

3

Using the contours as in Problem 1 above, trace orthogonals using only contours: (a) 0.5 and 0.01, and (b) 0.5, 0.2, 0.1 and 0.01. Comment on the result.

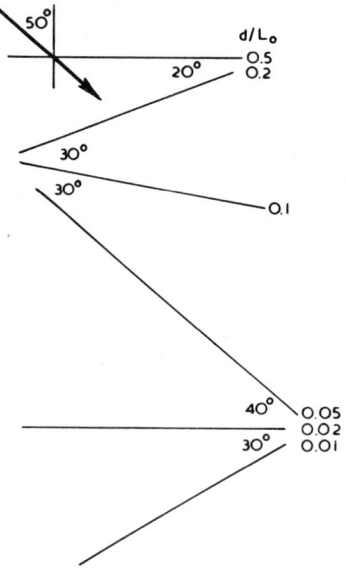

Fig. 5-27. Refraction over straight-angled contours as per Problem 4.

4

For the contours in Fig. 5-27 trace the orthogonal so as to determine the angle over the shallowest contour in respect to the deep-water contour.

5

A corner of a land-mass is depicted in Fig. 5-28 with underwater contours as designated. Trace four orthogonals in approximately the positions indicated so that two arrive on each section of beach. From the orthogonal spacings, compute the relative wave heights in the two regions and comment on the result.

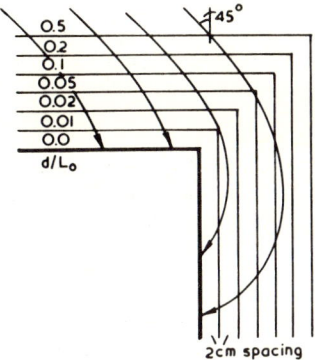

Fig. 5-28. Submarine features at the corner of a landmass as per Problem 5.

6
Draw seven concentric circles with incremental radii of 1 inch designating them as $d/L_o = 0.5$, 0.2, 0.1, 0.05, 0.02, 0.01 and 0.0. Trace the orthogonals of a straight crested wave system approaching this submarine shoal with initial angles between the *wave crests* and the outermost contour of 0°, 30°, 45° and 60°, respectively. Comment on the wave conditions around this minute island.

7
In the complex underwater shoal and canyon depicted in Fig. 5-29 (which should be reproduced to the dimensions shown) trace 6 or more orthogonals to shore in order to determine the respective wave heights along the full length of beach. The straight outer contour is $d/L_o = 0.5$ and the others, in turn, represent 0.2, 0.1, 0.05, 0.02 and 0.01.

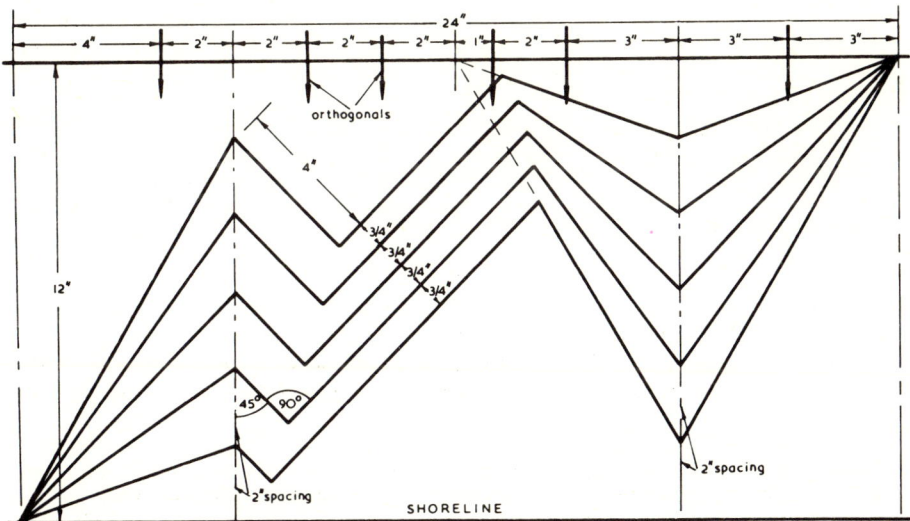

Fig. 5-29. Offshore contours for Problem 7.

8
Trace the orthogonal of a wave moving offshore at an angle of 60° to the beachline when in water $d/L_o = 0.01$, assuming all contours are parallel to the coast. What is the angle of an outgoing wave at this depth which will just not return to the coast?

9
A 35-knot wind ($U_{19.5}$) has a fetch of 80 miles and duration of 24 h when blowing just beyond the continental shelf at an angle of 45°. Compute $H_{1/3}$ in a 40 ft. depth nearshore assuming a uniform submarine slope. Determine the same significant wave height for the purpose of designing a breakwater standing in 20 ft. of water.

10
A 10-sec wave arrives on a uniform continental shelf at 45° to the coast. If its height in deep water is 2.5 ft., what will its breaking height be when it has reached an offshore zone sloping at 1 : 30 that intersects a beach face slope of 1 : 10 in 3 ft. of water. Determine the angle to shore at breaking and the type of breaker.

11
Two wave trains are angled at 30° to each other as they traverse the continental shelf. One train has a height of 3 ft. and a period of 10 sec. The second has a height of 4.5 ft. and a period of 12 sec. Determine the mean height of the short-crested system so formed and the heights at the crests.

12
Two wave trains of 10 sec period are approaching a uniform continental shelf. Train A is angled at 45° to the coast whilst B is angled 25° when in deep water. Compare the crest lengths of this system with that at the 50-ft. contour. Compare the mean crest heights in the two locations.

13
Two trains of 8 and 13 sec period are aligned as they enter the continental shelf zone from deep water at an angle $\alpha_o = 60°$. Draw a graph of the angle between these two trains as they propagate to the beach, assuming a uniform slope for the shelf. If these two trains had a similar deep-water height of 2 ft. what is the breaker height and breaker angle for a beach face slope of 1 : 8 transforming to 1 : 50 at 4 ft. depth. Compute the run-up in each case and the type of breaker.

14
A wave with steepness $H_o/L_o = 0.004$ traverses the continental shelf normally. Draw a graph of its a_c/H ratio for the range of d/L_o before it breaks. On the same graph plot the same ratio for this train arriving at 45° to the coast.

15
A breakwater runs seawards from the coast to a depth of 40 ft. In times of storm it must withstand a broadside attack of the waves. The worst conditions are considered to be a $U_{19.5}$ = 30 knots for 22 h or $U_{19.5}$ = 35 knots for 15 h over respective fetches of 300 NM and 150 NM. Determine $H_{1/3}$ and T_{max} for purposes of designing sections of the breakwater in depths of 10, 20, 30 and 40 ft.

16
The computation of run-up of a wave on a slope is not as simple as the process would indicate when observed at the beach. Outline the procedure, giving all equations and figure references, for tracing waves from an oblique arrival in deep water. What aspects require further research?

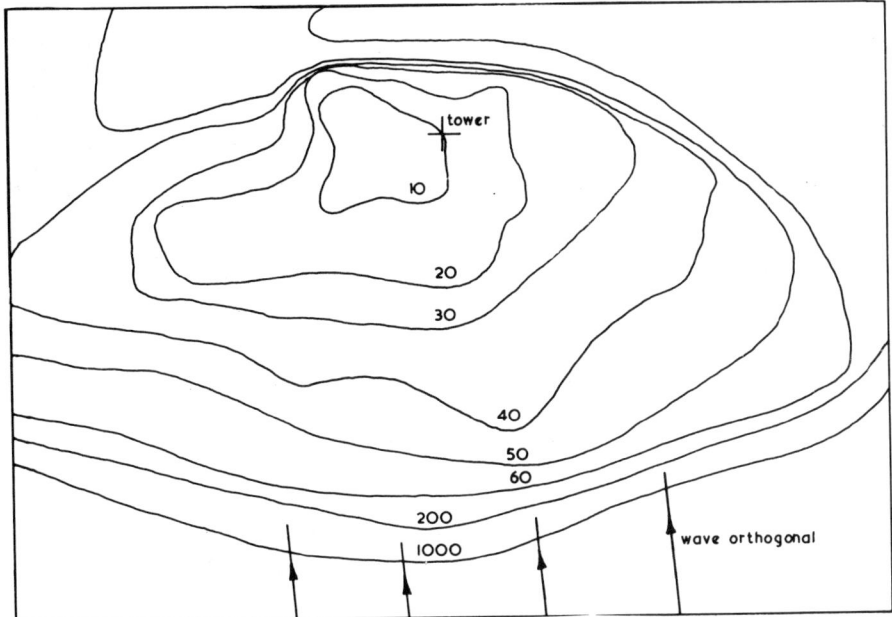

Fig. 5-30. Refraction of waves to offshore tower as per Problem 17.

17
A tower is to be constructed on the continental shelf in submarine conditions indicated in Fig. 5-30. Trace the orthogonals shown across the shoal assuming period values successively of 7, 10 and 13 sec. Comment on the wave conditions at the tower compared to those in deep-water in each case. Depths are in fathoms (1 fathom = 6 ft.)

18
Aerial photographs of a beach zone display a major swell system with crests 230 ft. apart. Simultaneous timing of breakers indicated their period was 10.2 sec. The crests were aligned 42° to the azimuth and the local beach at 28°. The coastline as a whole runs due north and south. Determine the approach direction for these waves in deep water off the edge of the continental shelf.

19
Waves of 12 sec period and 3 ft. height in deep water approach the coast normally (i.e., no refraction). They arrive at an estuary mouth, where the depth is 15 ft., and through which a tidal current of 1.5 knots on the flood and 1.0 knot on the ebb occurs. Determine the wave length and wave height during both these stages of the tide.

20
Swell is arriving at 45° to a coastline which can be considered of uniform slope across the continental shelf. At a depth of 30 ft. this train, whose height is 5 ft. and period 10 sec, encounters a longshore current of 2 knots from a river outlet. Determine the angle of breaking with the shoreline and the height of breaking. Treat for approach angles either side of the normal to the coast.

21

If aerial photographs of the sea indicate a sudden shortening of the wave length, what would you look for in the prints to ascertain whether this was due to shoaling or a contra-current in the area?

REFERENCES

[1] R. Silvester, 1966. An aid to constructing wave-refraction diagrams. *Trans. Inst. Eng. Austr.*, CE 8: 23–127.
[2] J.W. Johnson, M.P. O'Brien and J.D. Isaacs, 1948, Graphical construction of wave refraction diagrams by the wave front method. *U.S.H.O. Publ.*, 605.
[3] T. Saville and K. Kaplan, 1952. A new method for the graphical construction of wave refraction diagrams. *U.S. Beach Erosion Board, Bull.*, 6 (3): 23–34.
[4] R.S. Arthur, W.H. Munk and J.D. Isaacs, 1952. The direct construction of ways rays. *Trans. Am. Geophys. Union*, 33: 885–865.
[5] U.S. Beach Erosion Board, 1961. Shore protection planning and design. *U.S. Beach Erosion Board, Tech. Rep.*, 4.
[6] R.O. Palmer, 1957. Wave-refraction plotter. *U.S. Beach Erosion Board, Bull.*, 11 (1): 13–16.
[7] F. Suquet, 1949. Remarks on graphical computation of wave refraction. *Proc. 3rd Congr. I.A.H.R.*, 1949: 1–8.
[8] J.A. Battjes, 1968. Refraction of water waves. *Proc. ASCE*, 94 (WW4): 437–451.
[9] W.H. Munk and R.S. Arthur, 1952. Wave intensity along a refracted ray. In: *Gravity Waves – U.S. Natl. Bur. Std., Circ.*, 521: 95–108.
[10] G.M. Griswold, 1963. Numerical calculation of wave refraction. *J. Geophys. Res.*, 68: 1715–1723.
[11] W. Harrison and W.S. Wilson, 1964. Development of a method for numerical calculation of wave refraction. *U.S. Army, CERC, Tech. Mem.*, 6.
[12] W.S. Wilson, 1966. A method for calculating and plotting surface wave rays. *U.S. Army, CERC, Tech. Mem.*, 17.
[13] J.W. Le Petit, 1967. Application du calcul numérique à la connaissance de la houle et de ses conditions d'approche. *Proc. 12th Congr. IAHR*, 4: 300–306.
[14] J.W. Le Petit, 1964. Etude de la refraction de la houle monochromatique par la calcul numérique. *Bull. Cent. Rech. Essais Chatou*, 9.
[15] T.E. Orr and J.B. Herbich, 1969. Numerical calculation of wave refraction by digital computer. *Texas A.M. Univ., COE Rep.*, 114.
[16] Y. Jen, 1969. Wave refraction near San Pedro Bay, California. *Proc. ASCE*, 95(WW3)., 379–393.
[17] G.H. Keulegan, and J. Harrison, 1970. Tsunami refraction diagrams by digital computer. *Proc. ASCE*, 96 (WW2): 219–233.
[18] C.Y. Koh and B. Le Méhauté, 1966. Wave shoaling. *J. Geophys. Res.*, 71: 2005–2112.
[19] M.S. Longuet-Higgins, 1956. The refraction of sea waves in shallow water. *J. Fluid Mech.*, 1: 163–177.
[20] M.S. Longuet-Higgins, 1957. On the transformation of a continuous spectrum by refraction. *Proc. Camb. Phil. Soc.*, 53 (I) : 226–229.
[21] T. Karlsson, 1969. Refraction of continuous ocean wave spectra. *Proc. ASCE*, 95 (WW4): 437–448.
[22] W.J. Pierson Jr. and L. Moskowitz, 1964. A proposed spectral form for fully developed wind seas based on the similarity theory of S.A. Kitaigorodskii. *J. Geophys. Res.*, 69: 5181–5190.

REFERENCES

[23] W.J. Pierson Jr., G. Neumann and R.W. James, 1955. Practical methods for observing and forecasting ocean waves. *U.S.H.O. Publ.*, 603: 284 pp.

[24] I.G. Jonsson, C. Skougaard and J.D. Wang, 1970. Interaction between waves and currents. *Proc. 12th Conf. Coastal Eng.*, 1: 489–508.

[25] J.W. Johnson, 1947. The refraction of surface waves by currents. *Trans. Am. Geophys. Union*, 28: 867–874.

[26] W.G. Van Dorn, 1966. Theoretical and experimental study of wave enhancement and run-up on uniformly sloping impermeable beaches. *Univ. Calif., Scripps Inst., Rep.*, S10: 66–111.

[27] W.G. Van Dorn, 1966. Run-up recipe for periodic waves on uniformly sloping beaches. *Proc. 10th Conf. Coastal Eng.*, 1: 349–363.

[28] B. Le Méhauté, R.C.Y. Koh and L.S. Hwang, 1968. A synthesis on wave run-up. *Proc. ASCE*, 94 (WW1): 77–92.

[29] R. Miché, 1900. Mouvements ondulatoires des mers en profondeur constante ou décroissante. *Ann. Ponts Chaussées*, 114: 25–78; 131–164; 270–292; 369–406.

[30] P. Danel, 1952. On the limiting clapotis. In: *Gravity Waves – Natl. Bur. Std., Circ.*, 521: 35–38.

[31] P. Groen and M.P.H. Weenink, 1950. Two diagrams for finding breaker characteristics along a straight coast. *Trans. Am. Geophys. Union*, 31: 398–400.

[32] B. Le Méhauté and R.C.Y. Koh, 1967. On the breaking of waves arriving at an angle to the shore. *J. Hydr. Res.*, 5: 67–80.

[33] I.A. Collins, 1970. Probabilities of breaking wave characteristics. *Proc. 12th Conf. Coastal Eng.*, 1: 399–414.

[34] R.P. Savage, 1959. Laboratory data on wave run-up on roughened and permeable slopes. *U.S. Army, Beach Erosion Board, Tech. Mem.*, 109.

[35] R.L. Street and P.E. Camfield, 1966. Observations and experiments on solitary wave deformation. *Proc. 10th Conf. Coastal Eng.*, 1: 284–301.

[36] T. Kishi and H. Saeki, 1966. The shoaling breaking and run-up of the solitary wave on impermeable rough slopes. *Proc. 10th Conf. Coastal Eng.*, 1: 322–348.

[37] I.A. Hunt, Jr., 1959. Design of seawalls and breakwaters. *Proc. ASCE*, 85 (WW3): 123–151.

[38] K. Horikawa and C.T. Kuo, 1966. A study on wave transformation inside the surf zone. *Proc. 10th Conf. Coastal Eng.*, 1: 217–233.

[39] M. Nakamura, H. Shiraishi and Y. Sasaki, 1966. Wave decaying due to breaking. *Proc. 10th Conf. Coastal Eng.*, 1: 234–253.

[40] L.S. Hwang and D. Divoky, 1970. Breaking wave set-up and decay on gentle slopes. *Proc. 12th Conf. Coastal Eng.*, 1: 377–390.

[41] T. Ijima, T. Matsuo and K. Koga, 1970. Equilibrium range spectra in shoaling water. *Proc. 12th Conf. Coastal Eng.*, 1: 137–150.

[42] C.J. Galvin Jr., 1968. Breaker type classification of three laboratory beaches. *J. Geophys. Res.*, 73: 3651–3659.

[43] R.S. Arthur, 1946. Refraction of water waves by islands and shoals with circular bottom contours. *Trans. Am. Geophys. Union*, 27: 168–177.

Chapter 6

WAVE RECORDING

It is only since the 1939–1945 World War that the measurement of waves by instrument has been possible [1]. As soon as continuous records were available the complexity of the sea-surface undulations became evident. When these records, in turn, were subjected to harmonic analysis the nature of the wave spectrum was displayed, opening the way for a more realistic correlation of waves with the winds that generate them.

It is immediately apparent that scientists need detailed records of waves, particularly in the open ocean, for the purpose of deriving relationships between surface wind conditions and some statistical property of the wave system. However, it might be questioned at this early stage of the discussion whether engineers need to record waves, when confirmed formulae exist for their computation from wind data which is normally available. Admittedly the meteorological information is not always as detailed as might be desired, and in remote locations on a coast may be almost non-existent or be available for only a few years.

But wave recording is not the panacea that some engineers consider it to be. For example, the time available to collect data is usually too short. A client wanting a marine facility built desires it in the shortest time possible. A coastal engineer should consider himself fortunate to have a year's recording on which to base his design wave. Even so, he should examine all meteorological information to see how unique or otherwise was the period of data collection. This will be necessary to determine the general direction of wave approach, which is not amenable at present to instrumental assessment.

Consultants or contractors who feel the need for recording waves, either before, during or subsequent to the construction of a project, should question rigorously the usefulness of this information. Not only are such data difficult and costly to obtain, but there is also the risk of it being mis-interpreted and mis-applied. Unless it is checked systematically a sea wave recorder can accumulate information that can be incorrect by a factor of two without becoming apparent. Even where a true record has been produced, the complexity of the pressure or surface fluctuation requires an experienced person to extract information from it. Such a task is in many cases considered a minor one and passed down the line of authority to a junior who has no respect for the variables involved.

What are some of the reasons for which engineers might invest in wave recording?

(1) A coastal site with complex bed undulations offshore may make the transformation of forecast waves in deep water to shore too difficult or unreliable. The presence of islands, reefs or shoals in the path of incoming waves may make the task of refracting even monochromatic waves dubious.

(2) Meteorological data on a previously undeveloped section of coastline might be minimal and at best unreliable. On many ocean margins the persistent swell arrives from storm zones in the $40°-60°$ latitudes where little specific wind information is available.

(3) Some fetch conditions are not confidently amenable to wave hindcasting procedure, such, for example, as tropical cyclones. In coastal zones situated between $10°$ and $25°$ latitude much more data are required in order to make such predictions reliable.

(4) The refraction of a spectrum of waves across the continental shelf, in the presence perhaps of currents produced by tides or surge, is in its infancy. Components can be traced from several directions in order to identify optimum conditions, but the overall effect near the construction site may require the use of a recorder.

(5) The arrival of very long waves from distant storms, or resonance in nearby bays or basins, is almost unpredictable from meteorological information. Recording of such events requires a slightly different instrument and technique than used for normal wind-generated waves. The problem verges on the tidal-record field, where a continuous trace is kept of surface undulations.

The need for wave measurement having been established from the economic and utility points of view, for how long should records be taken? Since the highest waves possible are required for design purposes, recording over a winter's season is absolutely necessary, but it is preferable to carry the measurement over a full year, or better still two years. Even so, this period should be compared to the previous 10 or 20 years by a study of weather charts or cyclonicity data, as discussed in Chapter 2.

Wave records may be available for a site or installation some miles up or downcoast from the facility being planned. In order to use such information the waves must first be transposed to the deep-water zone, then along the coast in deep water, and finally re-transposed to the new shoreline location. This implies a knowledge of the general wave direction before such refraction or diffraction studies can be carried out. Maxima recorded at one point on the coast may not be possible at another point, even with the same approach of a similar intensity cyclone, due to the presence of islands, reefs or shoals etc. at one and not the other. Great care should be taken in using other people's records, even when these can be considered trustworthy. Full knowledge of the recording conditions should be known for such trust to be well founded.

It is possible to make a continuous record of ocean waves, but this is not necessary nor desirable. In the first place, very little extra accuracy accrues from a 24-h record than a 10-min record. It can be shown [2] by the use of a standard-error curve, that increasing a sample of waves from 60 to 9000 (10 min to 24 h for 10-sec waves) reduces the error by only one half. This is based upon the assumption that the wave conditions remain the same over the two periods, which is most unlikely in the case of the 24-h record. For this reason the 10-min "recording period" is repeated at a "recording interval" of 2 or 3 h, when the error of the interval record is 70% of that of the period record. In order to have sufficient repetitions of the longer waves in a spectrum a recording period of around 12 min is generally used. A recording interval of 3 h conveniently fits the frequency of synoptic weather map preparation.

ANALYSIS OF A SINGLE RECORD

Consider the record depicted in Fig. 6-1. It consists of waves large and small, both in period and height, which interact to give the complex fluctuation shown. The first requirement is to determine a mean line or SWL through the record. This is likely to be parallel to the edges of the recording paper, even for an instrument

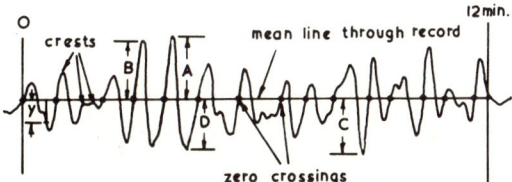

Fig. 6-1. Definition sketch of a wave record.

fixed to a structure or the sea bed. Tidal variations over 12 min are not likely to cause an identifiable slope in this mean curve, although a surge associated with a storm could produce a water-level variation of some inches in such a recording period. In the case of a ship-borne wave recorder [1], the trace should stay in the centre of the paper, unless the ship is sinking, in which case the wave record is of secondary importance, except to the most ardent scientist.

Having located the SWL, it is required to find the deviation of the water surface from it, in fact the root mean square of this deviation y (see Fig. 6-1). This is accomplished by marking off N equal intervals of time along the record (approximately 100 for the 12-min length). The deviation at these intervals, measured either above or below the SWL, is squared (y^2) so that:

$$a_{rms} = \sqrt{\Sigma y^2/N} \qquad (6\text{-}1)$$

This is the same as eq. 3-1, whence it is seen also that:

$$\sqrt{E} = \sqrt{2\Sigma y^2/N} = \sqrt{2}\, a_{rms} \tag{6-2}$$

Values of $H_{1/3}$, $H_{1/10}$ etc. have been related to E as in Chapter 3.

The method described above is tedious and subject to much error in the measuring, squaring and adding procedure. A statistical approach [3,4] with sufficient engineering accuracy is to measure, as in Fig. 6-1:

A = height of highest crest from SWL in the period of record (say 12 min); B = height of second highest crest from SWL in the period of record; C = depth of lowest trough from SWL in the period of record; D = depth of second lowest trough from SWL in the period of record.

From the additions $A + C = H_1$ and $B + D = H_2$ find a_{rms} from Fig. 6-2, representing equations that relate the ratios a_{rms}/H_1 and a_{rms}/H_2 to a function of N_z, the number of zero crossings in the recording period [5,6,7]. As seen in Fig. 6-1, these are the number of times the water surface passes through the SWL, either going upwards *or* going downwards (not both). The results from H_1 and H_2 should be essentially the same, one is a good check against the other. The statistical errors involved in this procedure are of the same order as averaging the highest 1/3 waves in the records. It should be noted that the maximum wave height for the recording period $(A + C)$ is not applicable in a statistical sense.

Another useful measure from a wave record is the width of the spectrum. This

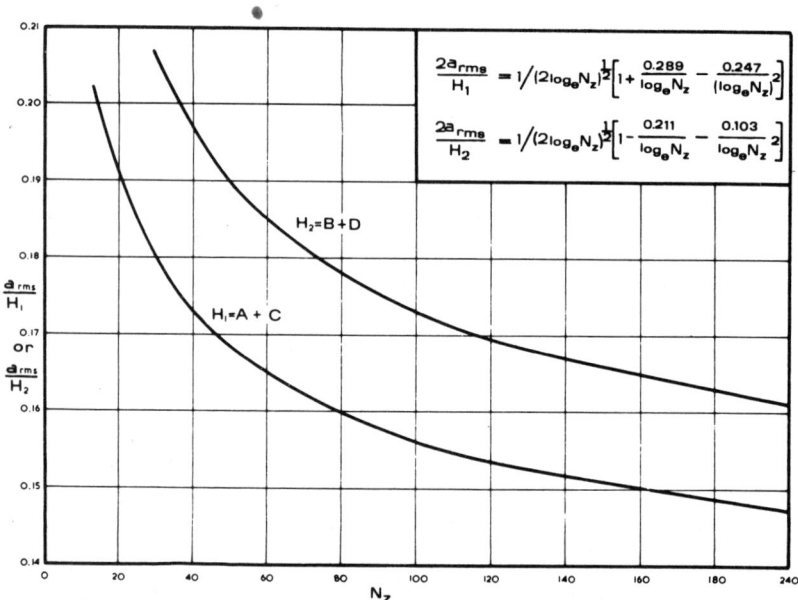

Fig. 6-2. Derivation of a_{rms} from H_1, H_2 and N_z.

ANALYSIS OF A SINGLE RECORD

can indicate whether the waves are within a fetch when measured or are swell from a distant storm. This spectral width ϵ, is given by:

$$\epsilon^2 = 1 - (N_z/N_c)^2 \qquad (6\text{-}3)$$

where N_c is the number of crests or optima in the recording period. These may occur below the SWL as much as above it, as illustrated in Fig. 6-1.

If the recording was of a single wave train, then $N_c = N_z$, so that $\epsilon = 0$. If on this single train was another train of much shorter period $N_c/N_z \to \infty$ and $\epsilon \to 1$. However, this is of no engineering consequence, the limit for ϵ in a FAS being around 0.8. The determination of ϵ helps to assess a more accurate proportion of $H_{1/10}$, $H_{1/3}$ or H_{ave} to $\sqrt{E}$ as in Fig. 3-2. For convenience various wave parameters have been listed in Table 6-I.

TABLE 6-I

Wave height parameters from spectral width (see Fig. 3-2)

ϵ	0	0.2	0.4	0.6	0.8
$H_{1/3}/\sqrt{E}$	2.83	2.85	2.90	2.95	2.95
$H_{1/3}/a_{rms}$	4.00	4.03	4.10	4.17	4.17
$H_{1/10}/\sqrt{E}$	3.60	3.60	3.70	3.85	4.05
$H_{1/10}/a_{rms}$	5.09	5.09	5.24	5.45	5.73
$H_{ave}/\sqrt{E}$	1.77	1.77	1.73	1.60	1.15
H_{ave}/a_{rms}	2.51	2.51	2.45	2.26	1.63

The period of zero crossing (T_z) can be averaged for the 12-min record by:

$$T_z = 12 \times 60/N_z \text{ (sec)} \qquad (6\text{-}4)$$

Scott [8] has examined wave records from the Atlantic Ocean and the Irish Sea and derived the respective relationships:

Atlantic $\qquad T_z = 3.23\, H_{1/3}^{1/3} \qquad (6\text{-}5)$

Irish Sea $\qquad T_z = 2.96\, H_{1/3}^{1/3} \qquad (6\text{-}6)$

It could be expected that eq. 6-5 would relate to the FAS for which eq. 3-43 applies, namely:

$$T_{max} = 2.43\, H_{1/3}^{1/2} \qquad (6\text{-}7)$$

This combined with eq. 6-5 gives:

$$T_{max} = T_z(H_{1/3})^{1/6}/1.33 \qquad (6\text{-}8)$$

For a similar relationship to be derived from eq. 6-6 would require that $T_{max} = 2.22\, H_{1/3}^{1/2}$ or from Table 3-VIII that $F/F_{FAS} = 15\%$. For the 100-NM fetch available in the Irish Sea this would indicate from Table 3-VII that a wind speed ($U_{19.5}$) of 35 knots was present ($F_{FAS} = 660$ NM), which fits the proportion of $(H_{1/3})_{FAS}$ recorded of $H_{1/3} = 13 = 0.59 \times 22.6$ ft. (see Table 3-VIII). Thus, from eq. 6-8, T_{max} can be determined from T_z and a knowledge of $H_{1/3}$.

Also from the single record of some 12 min can be assessed the "probable maximum wave". This wave height is not likely to be measured during the recording period, but is computed to represent the probable maximum over the recording interval, say of three hours. Probability immediately involves the size of the sample from which the estimate is being made. Also, in extrapolating from the shorter to the longer interval it is assumed that the wave conditions remain the same. This necassary error is allowed for in the statistical treatment of random functions [6].

The probable maximum wave height (H'_{max}) is derived from a_{rms} through Fig. 6-3, prepared from Longuet-Higgins' notable paper on wave statistics [9]. Application of this figure can be gleaned from the following example. If T_z averages 10 sec there are 72 major waves during the 12-min recording period, for which $H'_{max}/a_{rms} = 5.95$ from Fig. 6-3. However, this record is to represent an interval of 3 h during which 1080 such waves will arrive (assuming steady conditions). This gives $H'_{max}/a_{rms} = 7.5$, which is 26% greater than for the recording period.

From eq. 3-11 it is seen that $H_{1/3} = 4a_{rms}$, so that the ordinate of the figure can also read as $4H'_{max}/H_{1/3}$.

Such an application of statistics should be based upon an average of a large

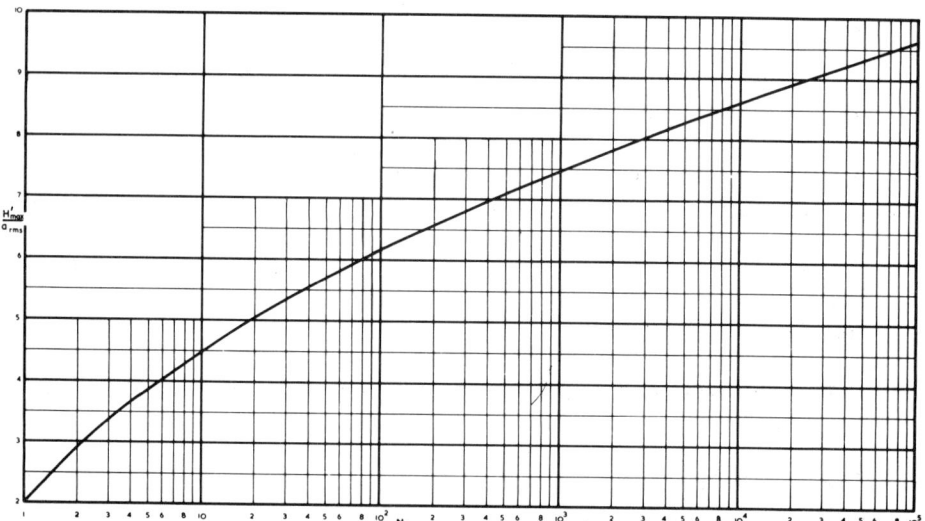

Fig. 6-3. Derivation of probable maximum height from a_{rms} and N_z.

SUMMATION OF RECORDS

number of samples. Although the recording interval is 3 h, it is probably long enough for the wave conditions to have changed sufficiently for successive records to become separate samples. However, this discrepancy must be accepted and the probable maximum height for the 3 h interval used in subsequent averaging processes.

From eq. 3-7 and 3-8 it is seen that $H_{1/10}/a_{rms} = 4.0\sqrt{2} = 5.66$ and $H_{1/3}/a_{rms} = 2.95\sqrt{2} = 4.18$. If $H_{1/3}$ and $H_{1/10}$ are accepted as H'_{max} then N_z is respectively 7 and 50, implying that the third highest wave occurs about 1/7 of the time and the highest 1/10th wave occurs 2% of the time.

SUMMATION OF RECORDS

Each recording interval results in values for $H_{1/3}$, $H_{1/10}$ or H'_{max}, plus T_z or T_{max}, as well as spectral width ϵ. It is now necessary to list these in some manner most manipulative for the engineer. The height and period within a fetch are interrelated, as shown in Chapter 3, but swell waves will exhibit no such correlation, hence the benefit of testing this condition by the assessment of ϵ.

One method of presenting a wave climate for any location is by means of a scatter diagram [10], which is a semi-graphical tabulation of some height measurement against wave period measurement. Increments of each variable are established, into which each interval recording falls and is noted as a single occurrence. Thus, in Fig. 6-4A unit squares of 0.5 ft. and 0.5 sec have been constructed. In these the number of occasions, over 998 × 3 hours = 125 days ≈ 4 months, is recorded when the height and period values have been within these specified limits. After a given time, which might be a season, a quarter, a year, or several years, the concentration of values within the diagram is identifiable. Isopleths can be drawn, to emphasize the trend. As can also be noted in Fig. 6-4A, iso-steepness lines can be drawn since in deep water $H/L = 2\pi H/gT^2$. It is advisable to start with H and T intervals as small as accuracy permits, since additions can always be carried to larger increments later.

Additions of the heights or periods in each increment permit a percentage distribution to be computed. A cumulative sum also leads to a percentage exceedance graph, as for wave height in Fig. 6-4B. The period can be similarly treated, but a histogram of this variable is probably to be preferred, as in Fig. 6-4C.

To provide information on the proximity of the storms generating the waves, a histogram of spectral width ϵ could be drawn as illustrated in Fig. 6-5. If the greater percentage exceeds 0.5 then the bulk of the waves are generated within 200 miles of the recording site, if less than this value swell from more distant storms is predominant.

From the individual 3-hourly records a graph of height or period against time, as

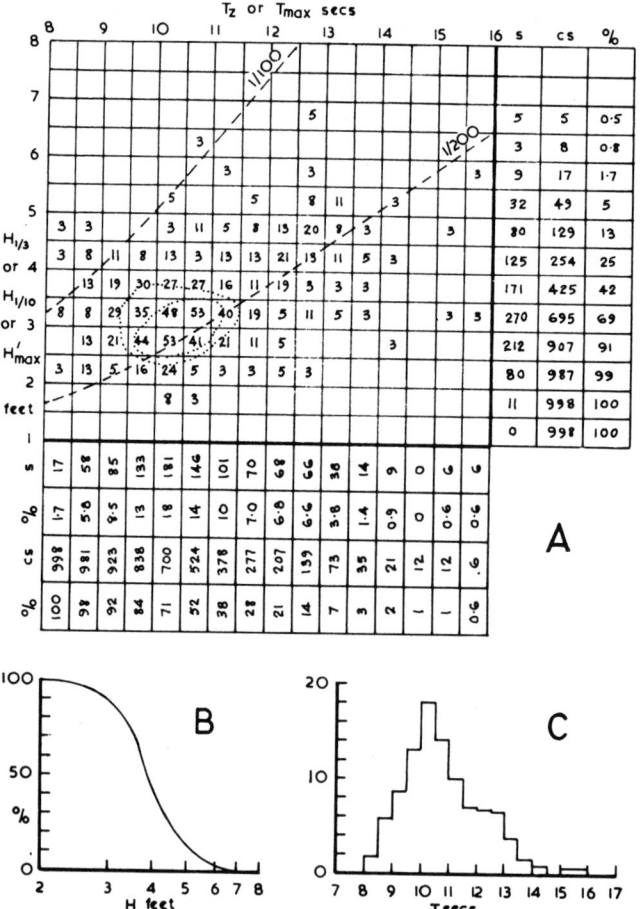

Fig. 6-4. A. Scatter diagram for integrating wave data. B. Percentage exceedance graph for wave height. C. Histrogram of wave period.

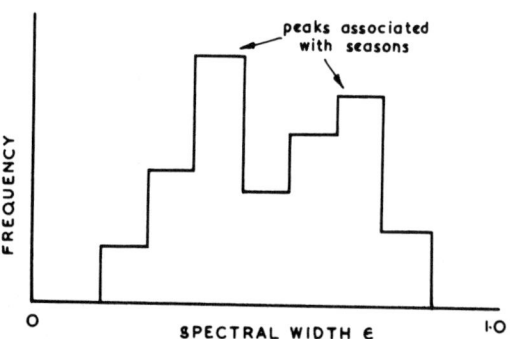

Fig. 6-5. Histogram of spectral width.

EXTREME VALUES

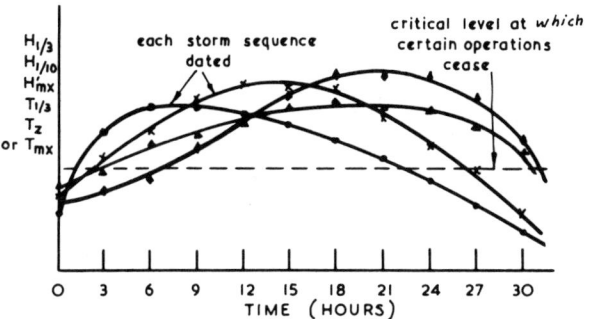

Fig. 6-6. Chronological variations in height or period during storm sequences.

in Fig. 6-6, can supply information on the duration of storm sequences and number of storms per season or year. This can be helpful to planners and contractors alike, in assessing lost time, when certain operations are precluded by either too great a wave height or too long a wave period, because of mooring difficulties, for example.

EXTREME VALUES

Where the highest wave in the lifetime of a structure is required, it is necessary to use statistics of extreme values [3]. The methods described above are not applicable in this case since they deal only with statistics of stationary random processes, or infer that no worse storm can occur than those recorded. There is an element of error in attempting to estimate 100-year conditions (for example) from a 1-year record, but such an engineering risk must be taken. It should certainly be better than no information of the site at all. The point to be made is that the longer the record the more reliable the assessment.

The percentage exceedance, as discussed previously, is plotted on probability paper, to which a straight line can be applied and extended to zones of very infrequent occurrence. This can be adapted to either wave height or wave period. From knowledge of the wave-generating processes and the limiting steepness of waves, it is reasonable to assume that the higher the wave the longer the period associated with it. The graph should not be extended to large heights and related periods without first checking that such waves can exist in the depths existing at the site (see Fig. 5-22, for example).

Each height or period value is assumed to represent waves during the recording interval of 2 or 3 h. The probability density of a specific value implies that it will occur once in $(1/p) \times$ recording interval. For example, a probability density of 0.00033 for a 3-h interval states that this wave characteristic will occur once every $3/0.00033 = 9000$ h, or once per year. Fig. 6-7 is a probability graph sheet with

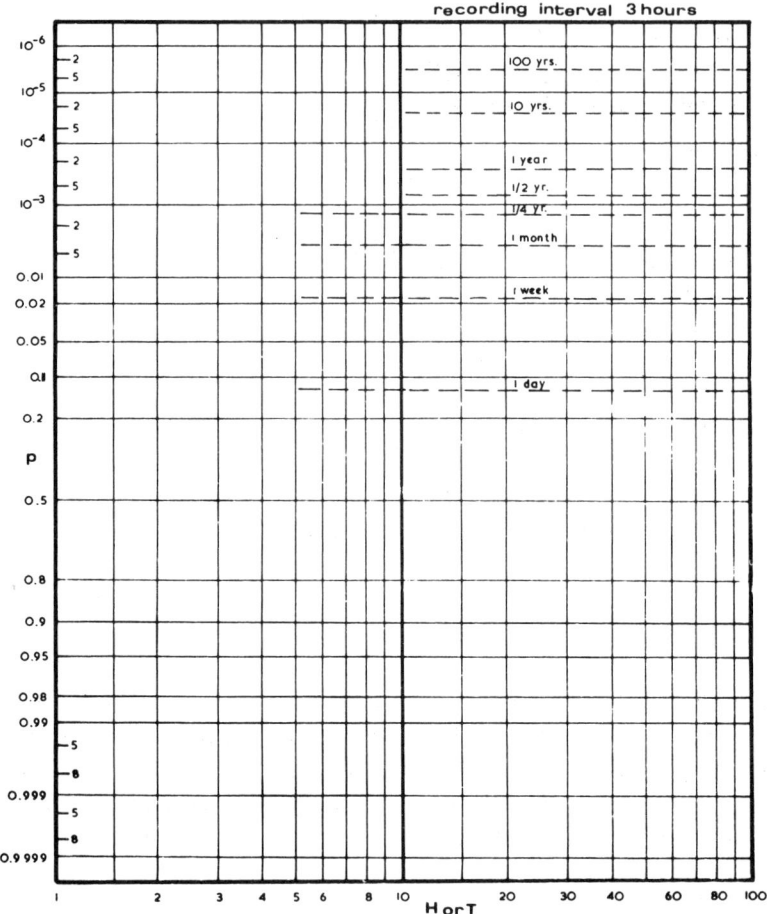

Fig. 6-7. Probability paper adapted to 3-hourly recording intervals.

certain recurrence periods for 3-h recording intervals. The probability for a wave occurring once in 100 years is p = 3 h/100 years = 0.000,0033. Where percentage exceedance is available p = % exceedance/100 if every recording interval provides a wave height even though this may be zero.

DESIGN WAVES

Various wave heights and periods have been suggested above for possible summation and statistical treatment. Which one to employ depends greatly upon the application and the risks involved. The problem is similar to that of the hydraulic

DESIGN WAVES 273

TABLE 6-II

Choice of wave characteristics for design purposes

From 12-min wave record	Average for season	Maximum for year	Maximum over years	Frequency per annum
$H_{1/3}$	A B C	D	G	J K
H_{max}	E	F	H I	L

engineer designing water-transmitting structures. A culvert, for example, can be designed to flood frequently, say every two years. A viaduct or bridge is more costly to replace and the consequences greater, so that a design flood might be permitted a recurrence every twenty years. A dam, on the other hand, must be able to deliver a hundred year flood over its spillway due to the personal and economic risks of failing to do so. In the same way the coastal engineer must choose his probable maximum or most repetitive large wave.

Choice of wave heights and periods for certain design problems listed below are suggested in Table 6-II. The period should be the one most predominant in the scatter diagram, except for harbour seiching and ship ranging problems where the highest periods recorded should be utilized. Designations as in Table 6-II are as follows:

A. Rubble-mound structures such as breakwaters and groynes, in which subsidence of sections is not serious during particularly fierce storms.

B. Rock or soil foundations, supporting concrete structures, which are subject to erosion by standing waves.

C. Study of longshore sediment movement in which duration, period and height, as well as direction, need to be known. Wave records should be checked against the long-term weather cycles for the adjoining ocean.

D. Effect of breakwaters and headlands in giving protection to shipping, mainly small craft.

E. Erodable banks faced with concrete blocks which must remain in place under all circumstances.

F. Heights of jetties and platforms above the waves for all but the worst storm conditions.

G. Overlapping and height of breaking waves which can cause damage to expensive mechanical equipment.

H. Monolithic breakwater structures which must not shift or tilt under any circumstances.

I. Pressures on structural elements of marine structures, the more expensive the element the longer the return period for exceedance of the maximum wave.

J. Seiching in harbours, which requires continuous recording of very long waves, 20–90 sec in period.

K. Ship ranging in enclosed or open harbours, in which period rather than wave height is the criterion.

L. Operation of small craft, hovercraft, sea-planes, or dredgers, in which the frequency of occurrence of waves higher than a specified limit precludes operation.

TYPES OF RECORDER

Draper [11,12] has discussed the various types of recorder used throughout the world, and the National Institute of Oceanography maintains an up-dated information sheet, which should be sought when contemplating such an installation. Recorders may be classified into three groups; (*a*) those measuring above the surface; (*b*) those measuring at the surface; (*c*) those measuring below the surface.

Above surface

(*1*) A movie or lapse photo of the water level against a pole protruding through the surface requires much manpower to read and analyze. When used during specific storm sequences it can be very useful, especially when more sophisticated instruments might be put out of action.

(*2*) A group has used photography of floating buoys anchored to the bed in shallow water off several Japanese beaches and found it useful for comparatively long waves [13].

(*3*) Stereo photography of the sea surface from a plane has been used twice [14,15] for detailed studies of wave growth inside fetches. Not only heights and periods are assessable in such a project, but also the directional spectrum of the waves. This procedure is costly and is not likely to be attempted except by large groups of physical oceanographers. A similar application by a radar altimeter [16] has produced more modest information. A magnetometer and laser have also been tried successfully.

At surface

(*1*) Electrical resistance gauges, either step or continuous, operate on a variation of voltage or of current and hence are always subject to leakage and consequent loss of accuracy. Details of circuitory and possible applications have been documented since 1948 [17,18]. More recent instruments using a change of frequency are more reliable and less subject to corrosion. This type of gauge requires a structure on which to be fitted, which could preclude its use on a virgin site. Continual inspection is another necessity since corrosion, fouling and exposure to floating debris is a constant hazard.

(2) The ship-borne wave recorder [19] has found wide use in vessels of all nations. It is from such instruments, in weather- and light-ships located in the Atlantic Ocean, that records have been used to derive the latest wave generation relationships. Pressure from pickups either side of the vessel is combined electronically with the vertical oscillations and accelerations of the ship. The need to locate the pressure sensors well below the water line (around 10 ft.) causes the smaller-period waves to be attenuated. The ship motions in fact amplify this effect some 2½ times. Draper [12] also reports that waves less than four seconds are virtually omitted. Installation of a ship-borne recorder is expensive enough to warrant long term usage, so does not lend itself readily to civil engineering projects either at the survey or construction stage.

(3) Pressure and accelerometer techniques inside buoys have been used extensively for scientific research [20] and engineering measurements [11,12]. These transmit their recorder output by radio to a vessel or shore station. Any scientific instrument left floating in the sea suffers not only the natural hazards of vertical and horizontal motion, but also the probability of horizontal removal by mariners interested in more than faunal fishing. Especially is this the case on isolated and undeveloped coasts. However, radio transmitting buoy-type recorders have come into wide use due to their ability to record surface fluctuations in deep water offshore.

(4) A float operating within or outside a perforated pipe [21] can be connected to a height integrator by a cord which can also note the number of zero crossings.

Below surface

(1) Measurement of surface waves by the pressure fluctuations produced either at the bed [22] or at some intermediate depth has many advantages in reducing hazards of the sea and of man. Such instruments must either contain their own recording device [23] or be connected by cable to a shore-based one. In both cases an indicator buoy must be used to aid retrieval and in the latter case the submarine cable provides its own problems. As will be noted later, pressure attenuation with depth inserts a natural filter into the measuring system.

(2) To replace the cable, a pressure-sensing device, consisting of a partially inflated tube or bag, can transmit fluctuations through a tube to a land-based recorder [22]. The length of pipe in such an installation is limited to around 300 ft.

(3) An echo sounder has been used on the sea-bed, which was directed upwards to the surface [24]. The beam must be very narrow to prevent reflection from several parts of the wave simultaneously. When waves are breaking and spray is present the image becomes blurred. Perhaps submerged laser beam equipment may provide the answer to the narrow beam requirement.

MAJOR DIFFICULTIES

Some of the problems inherent in using specific types of recorder have already been discussed, but it is pertinent to point out some general difficulties which must be countenanced when embarking on such a programme of data collection. Needless to say, the sea is a treacherous medium when a strong wind has disturbed it. Anyone or anything located in the surface water is subjected to horizontal oscillation and net pull in the direction of wave advance. To stay on location floating objects must therefore be anchored to the bed, the line from which immediately restricts the vertical motion. This could be obviated by anchoring an intermediate float from which a horizontal line holds the recorder [71]. But all floating equipment is subject to impact of waves, ships, and floating debris, to corrosion of the elements and to the improbity of man.

Many safeguards are necessary to prevent even unwitting interruption by man. The sea is everybody's highway, with few no-parking areas demarcated, so that navigation lights and notices in marine manuals are essential. This applies equally to submerged instruments that are subject to damage by anchors and anchor lines.

Because of the hazards of electrical leakage in cables and reduced radio transmission during storms, when the waves really need to be measured, a self-recording unit may appear to be the answer. Such an instrument is necessarily sophisticated electronically and is therefore subject to stoppage. When this occurs in a recorder to be left for some weeks without attention the loss is almost complete. Even a partial breakdown is difficult to detect in a complex of data, when no concurrent visual observation of the waves is available as a check.

As has been noted many times in the literature, there is a dire need at this date for the development of a cheap, simple, robust but reliable wave recorder. Even the cheapness might be disregarded if the reliability is there. The inspection and maintenance of equipment in remote construction sites around the world can exceed by far the initial cost. A time to take advantage of competent personnel on the site during early stages of development is when the hydrographic surveys and bed borings are being made for initial planning. At this stage, of course, it may not be known whether this actual site will be feasible. But if recorders become more economic, in terms of the qualities enumerated above, the rejection of a record would not become a great financial loss. The task, it would appear, should be tackled by the soils consultants.

WAVE DIRECTION

The above discussion has centered on wave-height and wave-period measurement. Another characteristic highly desirable is that of direction of wave propagation. Attempts to measure this by arrays of recorders [25,26], instrumented buoys [27] and ship-borne wave recorders [28] involve complicated analyses [29,30]. To date it is necessary to visually observe wave direction or derive it from synoptic weather maps in which the respective fetches can be identified [31]. A waveclimate study [32,33] should provide the coastal engineer with the general direction of the predominant swell and the location and orientation of the highest storm waves along a given length of coastline. Even if and when wave direction is instrumented, such confirmation from meteorological data is a worthwhile investment, in order to obtain a macroscopic view of the adjacent ocean margin.

SPECTRAL ANALYSIS

It is essential for scientists to know the complete complement of components in the wave systems to be related to characteristics of their fetches. From these relationships and various statistical procedures it is possible for the engineer to glean the information required from a record without the detailed Fourier analysis carried out by scientists.

However, the engineer should be conversant with the methods involved, as well as their errors and limitations. In this way, any spurious results in a record may be picked out and the source of error identified. Digital computers are employed now in analysis of such random functions, but the original work at the National Institute of Oceanography in England was conducted on an analog machine [34,35]. From this, the first knowledge was gained on wave propagation across the ocean [36].

The 12-min record was transformed to a black and white silhouette which was placed around a drum, together with a time-base marking [37]. The drum was rotated to a high RPM and allowed to reduce speed slowly. Photo-electric pick-ups on the wave record and the time base were combined so that at various frequencies the energy contained in each spectral component was recorded as an output. In spite of the errors involved in the mechanical-electrical operation some valuable conclusions were drawn. Further developments at this Institute have expedited the gathering and analyzing of data [22,38].

It would be every engineer's desire to have the output from his wave recorders pre-analyzed in some way, so that some significant parameters were immediately available. This has been attempted [23,39,40] with some degree of success. However, it should be appreciated that with this further stage incorporated into the recording process more difficulty is involved in checking the output.

ERRORS ARISING

These can be divided into two groups, those due to the hydrodynamics of water motion and others specific to location of the recorder. Of the first the major error is that of pressure attenuation, which affects the record taken at some depth. A lesser error arises from the steepness of the waves, which can cause the loss of height in the crest measurement of a wave. In the second group are errors due to refraction, currents, estuarine conditions, non-rigid bottoms and length of record. Finally, errors during the analysis of the record are discussed.

Pressure attenuation

Progressive waves There is little doubt that the easiest method of recording waves is by means of a pressure transducer placed on the sea bed or at some height below the troughs. Such an instrument can be placed in an offshore position where no structure exists to hold a surface instrument. Also it is sheltered somewhat from the excessive water-particle velocities and breaking action which occurs at the surface, especially during storm action.

The pressure variation measured at any depth is associated with velocities of the water particles in that location. Thus the attenuation of this orbital motion with depth is accompanied by a reduction in the pressure fluctuation, so that at $d/L_o = 0.5$ it is practically zero. This limiting depth thus varies with wave period, making the attenuation different for different components of the spectrum. This introduces a natural filter to the surface undulations, which will change as the water depth changes, as for example through tidal action.

Airey derived the first-order theoretical value for pressure fluctuation as:

$$\frac{p}{w} - y = \frac{H}{2} \frac{\cosh 2\pi(y+d)/L}{\cosh 2\pi d/L} \cos 2\pi \left(\frac{x}{L} - \frac{t}{T}\right) \tag{6-9}$$

where p is the pressure at depth y measured positively upwards from SWL (i.e., at the bed $y = -d$) and w is the specific weight of fresh or seawater ($= \rho g$). The variation of the pressure is seen to be sinusoidal and in phase with the profile of the wave, with maximum pressure beneath the crest.

The ratio of the height indicated at depth y (H_y) to that producing it at the surface (H_s) is given by:

$$\frac{H_y}{H_s} = K_p \frac{\cosh 2\pi(y+d)/L}{\cosh 2\pi d/L} \tag{6-10}$$

where K_p is an empirical factor introduced to cater for a discrepancy normally found in the theory. At the bed:

$$\frac{H_b}{H_s} = \frac{K_p}{\cosh 2\pi d/L} \tag{6-11}$$

which has been graphed in Fig. 6-8 against d/L_o for $K_p = 1$.

The reciprocal of K_p is often used in this context and designated n. This will be omitted here, but, in any case, should not be confused with the n associated with wave group velocity. This amplification factor $(1/K_p)$ has been discussed by Homma et al. [41] and shown to vary between research workers from 1.06 to 1.37. This group, from a comprehensive series of laboratory and field data, derived the expression:

$$1/K_p = 1.55 \exp[-5.19 d^{5/2}(1/T - 0.274 d^{-1/2})^5] \tag{6-12}$$

From the combined graph of $1/K_p$ and d/L_o for field and laboratory, a mean curve could be expressed as:

$$1/K_p = 1.5 - 1.25 \, d/L_o \tag{6-13}$$

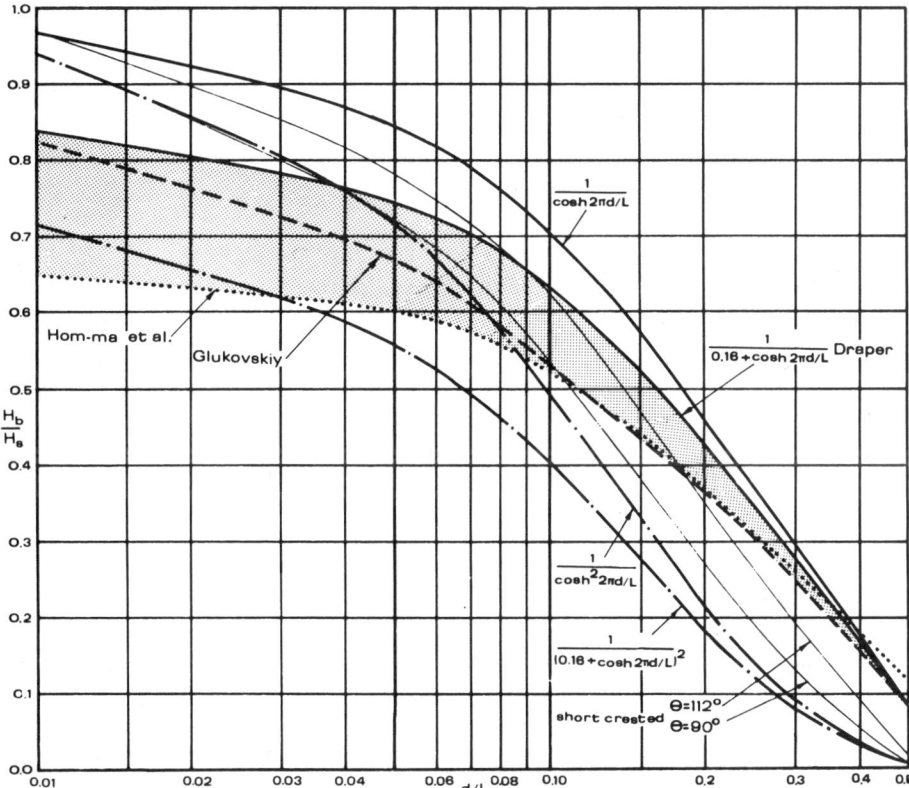

Fig. 6-8. Attenuation of wave pressure head with depth.

which for $d/L_o < 0.3$ has an accuracy ± 0.5 and for $d/L_o > 0.3$ has an accuracy of ± 0.25. This would be the same order of accuracy as for eq. 6-12. Eq. 6-13 has been combined with eq. 6-11 in the plot of Fig. 6-8.

Glukhovskiy [42] had previously derived the expression:

$$1/K_p = \exp[5.5(d/L)^{0-8} - 2\pi d/L] \tag{6-14}$$

which results in the curve indicated in Fig. 6-8.

Draper [43] from field tests has derived the relationship:

$$\frac{H_b}{H_s} = \frac{1}{0.16 + \cosh 2\pi d/L} \tag{6-15}$$

which is also depicted in Fig. 6-8. The total variation in H_b/H_s of these empirical approaches is shown hatched in the figure, which is seen to be greatest for the shallower depths.

The theoretical attenuation is strictly applicable to waves of small amplitude and of sinusoidal form which is unrealistic in deep water, let alone transitional and shoaler depths. For finite amplitude waves the second-order relationship [44] is:

$$\frac{p}{w} - y = \frac{H}{2} \frac{\cosh 2\pi(y+d)/L}{\cosh 2\pi d/L} \cos 2\pi\left(\frac{x}{L} - \frac{t}{T}\right) + \frac{3}{8} \frac{\pi H^2}{L_o} \left(\frac{\cosh 4\pi(y+d)/L}{\sinh^2 2\pi d/L} - \frac{1}{3}\right)$$

$$\times \frac{\cos 4\pi(x/L - t/T)}{\sinh^2 2\pi d/L} - \frac{1}{8} \frac{\pi H^2}{L_o} \frac{\cosh 4\pi(y+d)/L}{\sinh^2 2\pi d/L} \tag{6-16}$$

Of the three terms in the RHS, the first is that of the linear theory, the second contains a wave steepness factor and a double frequency variation due to the $4\pi(x/L-t/T)$ of the cosine term, the third is a non-periodic function produced by the rise in MWL above SWL (ΔH) in the progressive wave of finite height. This last term attenuates very slowly with depth, as does the second term, due to the factor $\cosh 4\pi(y+d)/L$. Thus the double frequency fluctuation will be magnified in respect to the basic wave measurement and be exhibited in any spectral analysis of the record as a spurious bulge.

Bergan et al. [45] carried out flume tests with a surface and submerged gauge and found that correlation for individual waves was not good, either for linear or fifth order theory. They noted the difficulty of applying anything but the first order theory with an attenuation factor.

When considering a spectrum of waves the component amplitudes $S_{H2}(T)$ involve the square of the attenuation factor, which has been graphed for eq. 6-11 in Fig. 6-8. Because of the variation in d/L_o, waves in the lower-period end of the EDC may be omitted altogether, whilst the highest period waves will be attenuated the least. The waves near T_{max}, containing the optimum energy, will still exhibit this characteristic in the pressure response, except at very great depths [46].

When waves are being recorded that have propagated a reasonable distance from a fetch, the heights associated with bands around T_{max} will be predominant, except during intervals before and after the bulk of the waves from the storm zone arrives. From T_z in such a batch of high waves it may be possible to gather information on $H_{1/3}$ inside the fetch some distance away. From the T_z so recorded T_{max} is given from eq. 6-8, which can be combined with eq. 3-72 to give:

$$T_{max} = \frac{H_{1/3}^{1/6} T_z}{1.33} = r H_{1/3}^{1/2} \quad (6\text{-}17)$$

or:

$$H_{1/3} = T_z^3 / 2.35 r^3 \quad (6\text{-}18)$$

From Table 3-VIII values of r for respective percentages of FAS provide ratios of denominators in eq. 6-18 which give parametric values as in Table 6-III.

TABLE 6-III

Ratios of $T_z^3/H_{1/3}$ for various percentages of FAS

	5	10	15	20	30	40	50	70	100
F/F_{FAS} %									
t/t_{FAS} %	14.5	20.5	20.5	31.5	41	50	59	76	100
$T_{max}/H_{1/3}^{1/2}$ (sec/ft.$^{1/2}$)	2.10	2.17	2.22	2.25	2.29	2.32	2.35	2.39	2.43
$T_z^3/H_{1/3}$ (sec^3/ft.)	21.8	24.1	25.7	26.9	28.3	29.4	30.5	32.1	33.7

To employ Table 6-III some assessment of the fetch conditions must be made, which is a worthwhile exercise at any time. Some mean wind velocity would, from Table 3-VII, give F_{FAS} and t_{FAS} from which the percentages in the current case can be gauged, even roughly. Then, from a measure of T_z in one or several records made at 3-h intervals, some idea of $H_{1/3}$ and hence $U_{19.5}$ can be evaluated for this specific storm. Such an estimate might require a trial-and-error approach with corrections determined from the several sources of information.

The ratios listed in Table 6-III will not be those measured (using $H_{1/3}$ from the record) unless the fetch exists over the recording site. So that to use the height values calculated from T_z as in the Table, they must be transferred by appropriate attenuation factors to the downwind end of the fetch. All this procedure is not warranted unless some major storm conditions need to be assessed. This might be necessary when meteorological data for some remote area are not reliable, in which case information traced from wave data can be utilized. Once the fetch conditions are known these can be transferred closer inshore in order to compute the optimum wave heights possible for design purposes.

In respect to component wave heights and $H_{1/3}$ from the spectrum, Bergan et al. [45] found from their laboratory tests that the spectral wave heights were well predicted from a pressure record. For this purpose the power spectra of the pressures were obtained and amplified by linear theory (eq. 6-11) with apparently no empirical factor $1/K_p$, using the T_z for determining d/L. The resulting $H_{1/3}$ values were very close to those measured at the surface.

Standing waves. The pressure in a complete clapotis is given to the 2nd order [47] by:

$$\frac{p}{w} - y = \frac{H}{2} \frac{\cosh 2\pi(y+d)/L}{\cosh 2\pi d/L} \sin \frac{2\pi x}{L} \sin \frac{2\pi t}{T} - \frac{\pi H^2}{8L_o} \frac{\cos^2 2\pi t/T}{\sinh^2 2\pi d/L} [\cosh 4\pi(y+d)/L$$

$$+ \cos \frac{4\pi x}{L} - 1] + \frac{3\pi H^2}{16 L_o} \frac{\cosh 4\pi(y+d)/L}{\sinh^4 2\pi d/L} [\cos \frac{4\pi x}{L} \cos \frac{4\pi t}{T}] + \frac{\pi H^2}{4L_o} \cos \frac{4\pi t}{T} \quad (6\text{-}19)$$

Of the four terms in the RHS the first is as for linear theory, the second and third contain terms with wave steepness, whilst the fourth is a pressure fluctuation, evenly spread across the sea surface (no $\cosh 4\pi x/L$ term), with double the wave frequency that is not attenuated with depth (no $\cosh 4\pi(y+d)/L$ term). In a record obtained at the bed the presence of any standing waves will be indicated by a pronounced batch of energy at probably twice the frequency of the main progressive waves being recorded. However, this fluctuation is not attenuated with depth and is therefore out of proportion in respect to the progressive wave record.

Such variations in pressure over a reasonably wide area of ocean produces pulsations in the earth's core known as microseisms [48], which have been picked up on seismographs and been used to predict the presence of storms at sea [49]. Verification of the wave source for this seismological energy has been gained in the laboratory [50] and the field [51]. In storm centres various fetches exist concurrently and swell from one may meet swell from another head-on and so create clapotis, which produce the double frequency pressures within the water body and on the bed.

It has been suggested that such pulsations measured by a recorder from long waves reflected from a beach or a cliff might be used to forecast the approach of more severe storm waves [52], since the long-period components travel about twice the speed of the T_{max} band.

Reflection from beaches depends upon the beach slope and the wave steepness. Some typical slopes [53] are: pebbles ¼" to 1½", slope 1:5; mixed sand to ¼" pebbles, slope 1:10; medium to fine sands, slope 1:25.

Values of reflection coefficients H_r/H_i measured in the laboratory [54] have

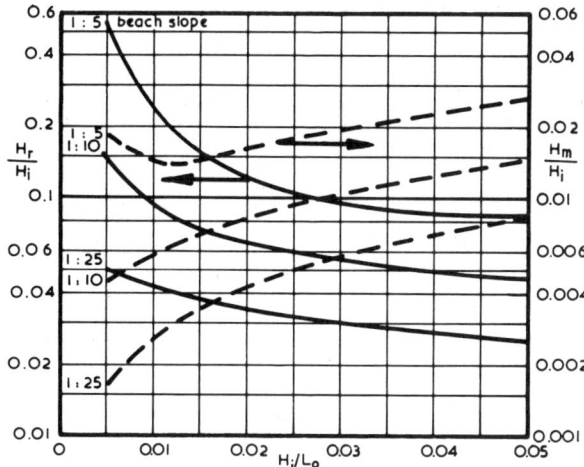

Fig. 6-9. Reflection coefficient H_r/H_i and double frequency response H_m/H_i.

been graphed in Fig. 6-9. Also in the same figure is plotted the non-attenuated double frequency pressure response [48] H_m as given by:

$$H_m/H_i = 2\pi(H_i/L_o)(H_r/H_i) \qquad (6\text{-}20)$$

where H_m is the pressure head recorded, H_i is the height of the incident wave, and H_r is the height of the reflected wave (as given in Fig. 6-9).

It is seen that steeper beaches and longer swell waves produce the greatest reflection and highest H_m/H_i values.

Even the recording of the surface oscillation is influenced by the location of the recorder in the standing wave. At the antinodes a reasonably complete height will be received (up to twice the incident wave), but at the nodes no change in water level occurs, hence no wave recorded. Thus, in conditions of changing periods waves will be devalued sporadically as the nodal point passes over the recorder fixed in position from a beach or wall.

It is possible, from a combination of reflection, attenuation, or positioning, for say a 12-sec wave to produce spurious energy at a 6-sec period, which may add to or replace energy already attenuated with depth. Considering this influence over the spectrum, it can become distorted, with an energy tail added to the short-period end.

Short-crested waves There are few occasions when truly uni-directional waves can be measured. Most natural conditions contain more than one wave train angled to another. These have been discussed in many places throughout this book for their other important influences. The equation for pressure within a short-crested wave system [55] for trains of similar period and height is to the first order:

$$\frac{p}{w} - y = \frac{H}{2} \frac{\cosh 2\pi(y+d)\sqrt{1/L'^2 + 1/L^2}}{\cosh 2\pi d\sqrt{1/L'^2 + 1/L^2}} \sin 2\pi \left(\frac{x}{L} - \frac{t}{T}\right) \cos \frac{2\pi z}{L'} \qquad (6\text{-}21)$$

where L is the distance between the combined crests in the line of propagation, L' is the transverse distance between combined crests normal to propagation, x is measured along L, z is measured along L', y is measured positively upwards from the surface (i.e., at bed $y = -d$), t is time measured through a wave period T.

It is seen that in this case zero pressure fluctuation exists along alignments of $z/L' - 1/4, 3/4, 5/4$, etc., since a trough of one wave occurs when the crest of the other has reached this alignment. It has been seen in Chapter 4 that only transverse horizontal oscillations of water particles occur at these points.

The ratio of wave height recorded at the bed to that at the surface is given by:

$$\frac{H_b}{H_s} = \frac{K_p}{\cosh 2\pi d\sqrt{1/L'^2 + 1/L^2}} \qquad (6\text{-}22)$$

which, compared to eq. 6-11, is less than the similar ratio for a progressive wave. This ratio is plotted in Fig. 6-8 for $L/L' = 2/3$ and 1 or the angle between $\theta = 112°$ and $90°$. Thus, at any location the height of a short-crested wave system will be undervalued by a pressure gauge, whilst at $z/L' = 1/4, 3/4, 5/4$ alignments no fluctuation will be recorded at all. This has to be accepted as an error in measuring waves on the continental shelf, but should serve as a warning not to locate recorders in the vicinity of vertical walls or cliffs where reflection is probable.

Inspection of Fig. 6-8 shows a wide variation in attenuation factors, either through variations of wave steepness in progressive waves or from the many unknowns associated with short-crested systems. This variation can be exhibited in another manner as in Fig. 6-10 where K_p is graphed against d/L_o. Also illustrated is the ratio of the theoretical short-crested H_{sc} to progressive H_p responses at the bed for $\theta = 90°$. Further analyses could be presented of cnoidal, solitary of hyperbolic waves but sufficient has been presented to indicate the large margin of error present in recording through piezometer-type instruments, in spite of their convenience in many other respects.

Another problem exists in that the record made at depth may omit some of the shorter waves altogether. In this case the true surface oscillation cannot be predicted because of such an omission. What is not there cannot be amplified by any factor, even if this were known accurately.

Wave steepness

As already noted in the foregoing the attenuation of pressure is closely allied with wave steepness. The 2nd- and higher-order theories, in fact, contain terms with

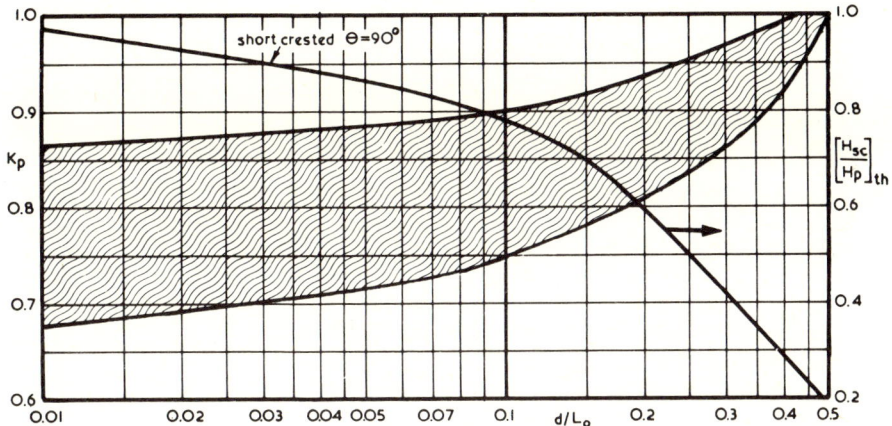

Fig. 6-10. Empirical attenuation factor K_p, and H_{sc}/H_p based on theory.

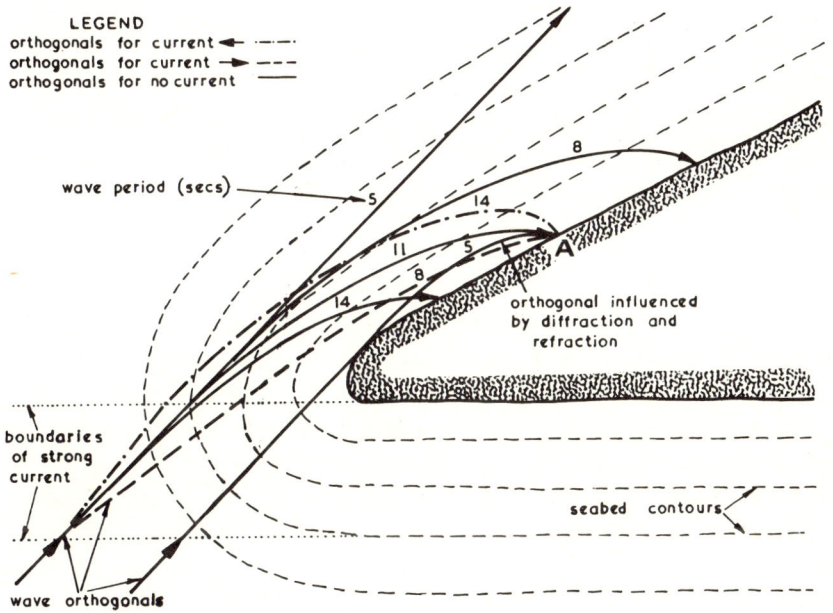

Fig. 6-11. Influence of headland, offshore shoals and currents on wave propagation.

this parameter. Seiwell [56,57] found that even in 75 ft. of water the steepness of the long waves affected his results. Morison [58] in his model experiments obtained reasonable agreement with theory down to $d/L = 0.2$, at which steepening commences. He found that the error arose in steep waves mainly due to the crest pressure being attenuated more than the trough. This is probably due to the concomitant high orbital velocities at the instant of measurement. This can be amplified when water particles are deflected around the body of an instrument.

Refraction and diffraction

As already noted in Chapter 5, waves propagating across the continental shelf, even from the same distant storm centre, will be refracted at different rates and so produce short-crested systems. For this reason two recorders placed in close proximity could give completely different readings, even if they were surface recorders.

Another effect of refraction and diffraction is to filter out certain wave periods, this changing as the water depth varied with tidal frequency. Consider point A in Fig. 6-11 which is in the lee of a headland. The long waves will be refracted into shore prior to reaching the recorder. The middle band of components may refract into the locality of the instrument. The shortest waves may not be refracted at all, but may spread their energy to the beach mainly by diffraction. With a reasonable change in tide the wave period being optimized at point A could fluctuate along the spectrum.

Where waves that are recorded have been influenced by refraction, diffraction, reflection, etc., they should not be used for transferring data to another point of the coast. As noted already, even transforming information elsewhere from an ideal open beach site requires knowledge of wave direction, which at present can only be assessed from observation of synoptic weather maps, or perhaps keen visual observation of wave direction in a known depth of water.

Currents

Currents can clear short waves from a spectrum, a condition that can change throughout the day and throughout the season. Whilst waves so recorded at a site near the coast provide correct information on this aspect, it points to the need for long-term measurement in order to obtain the full annual variations. It emphasizes also the errors arising when such data are transferred down-coast to another site without due consideration being given to the characteristics of either location.

Tidal and other streams can also deflect the orthogonals of waves [59] as displayed diagrammatically in Fig. 6-11. Many tidal currents oscillate back and forth along a coast, so affecting the beach zones where batches of swell will arive. The flow of water to and from a coast can accelerate or retard the arrival of wave energy and so concentrate it in certain periods of the tidal cycle [60]. With 3-h recording intervals four or five records per 12-h tide should be sufficient to average out these variations. However, if particularly high waves occur at the same tidal stage continually, then this is an important fact to note in a wave climate report for the region.

Estuarine conditions

An estuary is a sea zone where fresh water issues and remains separated somewhat from the salt water by density stratification. It is generally associated with an indentation of the coast which is reasonably shallow. Estuaries are extremely important commercially because of the low flat land available for industrial development and the protection for harbours sited thereon. In many cases large tidal ranges occur in estuaries besides the fluctuating discharge of rivers into them.

Because of the complex bed conditions between a port located in an estuary and the open sea, it is difficult to forecast ocean wave incidence for such locations. This is where recording becomes more of a necessity, but due to the greater number of variables involved (tidal streams, floods, etc.) a longer series of data should be gathered. The proximity of one harbour to another in an estuary may prompt the sharing of wave information. This should be done with the utmost care, since the shoals and currents may cause conditions at the two sites to be completely different.

Non-rigid bottom

The attenuation of wave height and the magnification of mass transport due to a mud stratum at the bed has already been discussed. In the present context it will be assumed that the muddy bed is sufficiently compacted to withstand the weight of a wave recorder, or one may be in buoyant suspension above it. A similar situation arises if the recorder rests on prolific seaweed. In either case the pressure fluctuation is attenuated more than if the bed were rigid. It is assumed in the following discussion that the "fluid" bottom is a reasonable thickness compared to the wave length.

In this event the attenuation is given approximately by $\exp(-2\pi d/L_o)$. Ewing and Press [61] have suggested that Seiwell's data [62] fell between this limit and that for a rigid bottom given by eq. 6-11. Abramson and Bretschneider [63] have sub-

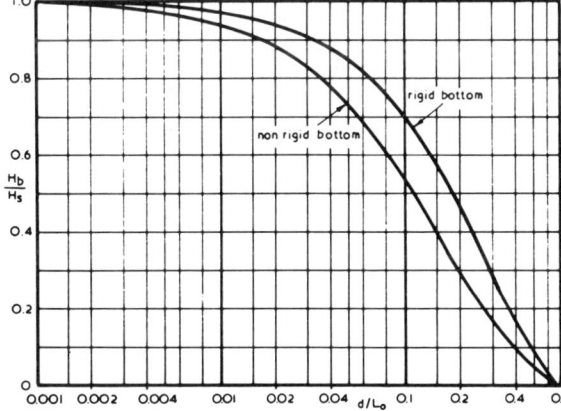

Fig. 6-12. Wave attenuation due to a non-rigid sea bed.

mitted a mean of these two values, or:

$$\frac{H_y}{H_s} = \frac{1}{2}\left[\frac{\cosh 2\pi(y+d)/L}{\cosh 2\pi d/L} + \frac{\exp[-2\pi(y+d)/L_o]}{\exp(-2\pi d/L_o)}\right] \quad (6\text{-}23)$$

where y is the depth of the recorder measured positively upwards from the surface.

The ratios for H_b/H_s on the above basis are illustrated in Fig. 6-12, where the difference between rigid and non-rigid bottom is seen to be maximum in the region of $d/L_o = 0.1$, or the depths most likely used for pressure recording. Actual cases of fluid bed are likely to produce attenuation factors between the two limits in Fig. 6-12. It must be remembered from Fig. 6-8 that even the theoretical rigid bottom relationship is not reliable.

LENGTH OF RECORD

Although a continuous record may appear ideal for giving a complete picture of the waves emanating from a storm centre, economics apart, there are several difficulties which preclude such a step.

In the first place it cannot be so long that the wave conditions change significantly during the recording period. This rate of change will depend upon the proximity of the recorder to the fetch. For swell having travelled long distances the rate of arrival of new bands is slow. For waves inside or just outside a fetch the changes in height and period of the various components of the spectrum can be relatively swift.

In the second place, a record must be long enough to contain sufficient samples of wave-interaction to make the statistical operation accurate. If a 10- and 11-sec wave appeared together it would take the latter approximately 100 sec to pass through a moving wave length of the former. If fifteen repetitions of this action were desired a record 1500 sec in length, or 25 min, would be necessary. For similar combinations of 7- and 8-sec waves this recording period could be halved.

Barber and Ursell [36] have suggested that for swell from storms 1000–2000 miles distant the optimum length of record is 33–47 min. For closer storms 17 min should be sufficient. It has been shown that the random error is reduced the longer the record and the wider the spectrum [3,64].

RECORD ANALYSIS

A frequency analyzer which divides a wave record into its component wave trains assumes that these are sinusoidal. Waves within a fetch are far from this idealized shape so the analyzer reacts as though several sinusoidal trains are present. These harmonics have their energy recorded as first, second or third, with the first

RECOMMENDATIONS

two predominating. The second harmonic is reported with double the frequency of the base wave or half the period.

Pressure recorders omit the steep peaks of the surface undulation and tend to reproduce sinusoidal profiles. Surface recorders will therefore suffer more from the spurious band of unrecorded waves at double the frequency of the major surface waves. Barber and Ursell [36] have shown that the ratio of amplitudes of:

$$\frac{\text{2nd harmonic}}{\text{original wave}} = \frac{a_n}{a_i} \leqslant \left[\frac{3 \cosh 4\pi d/L}{4 \cosh(4\pi d/L) - 4} \operatorname{sech} 2\pi d/L \right] \frac{a_i}{d} \qquad (6\text{-}24)$$

where a_i is the amplitude recorded at the bed, and a_n is the amplitude of the harmonic in the record.

Applying the appropriate pressure response factors in order to compute the equivalent surface fluctuation gives:

$$\frac{\text{2nd harmonic}}{\text{original wave}} = \text{RHS eq. 6-24} \times \frac{\cosh 2\pi d/KL}{\cosh 2\pi d/L} = [\ ] \frac{a_s}{d} \frac{\cosh 2\pi d/KL}{\cosh^2 2\pi d/L} \qquad (6\text{-}25)$$

where K is a fraction determined from the fact that L_o for the harmonic is one quarter of the L_o for the original wave (since $(L_o)_h = g(T/2)^2/2\pi$) and a_s is the amplitude of the original surface wave. The ratio of eq. 6-25 has a minimal value at $d/L_o = 0.08$. Where $d/L_o > 0.08$ the ratio rises due to the pressure response factor (usually those > 2.0 are not used); and where $d/L_o < 0.08$ the increase is due to wave steepness. Thus, for waves of 7- to 14-sec period the depth to minimize error through the analyzer is around 50 ft.

Pierson and Chang [65] have described another analyzer and Pierson [66] has discussed the loss of sharpness in the spectrum due to the use of particular filters. This topic has been further examined by Barber [67] and Tukey [68].

RECOMMENDATIONS

Suggestions for siting and operating recorders depend greatly upon the use to which the records are to be put. If the information is to be solely retained by the data gatherer, for use at a particular point on the coast, then the one recommendation is to take readings in zones away from reflecting walls or cliffs, preferably in bed contours where orthogonals can be easily transferred to other areas of construction.

However, once sets of wave records exist there is strong pressure to disseminate them for the use of other clients. Whilst the widest use of wave-climate information is desirable, its misuse or use in preference to studying overall meteorological conditions, is to be deprecated. To improve the usefulness of such data, a report should accompany them giving full details of the mode of recording and the site where they were taken. In this way users can judge for themselves how to transform the information to any other location nearby or more distant.

Where possible a programme of visual observation of waves should be organized, to operate concurrently with the instrumental programme. Also, synoptic weather charts should be examined for the previous 10 or 20 years to compare the cyclone intensity during the recording series. The dimensions, directions and distances of storms, together with their maximum wind velocities and directions, should be noted during the seasons when records are being taken. In this way an overall picture will be obtained which is of inestimable help in subsequent design and construction tasks.

But before deciding on a wave recording programme count ten, and it will be this figure in terms of thousands of dollars that it is likely to cost. Have a wave-climate study made for the area at the outset to see how amenable the meteorological information for the area is to forecasting storm waves or swell. If this proves inadequate or unreliable then invest in a recorder or several recorders, by obtaining the most up-to-date information on their cost and availability. Finally, place somebody in charge of the installation who has a feel for the important information he is gathering.

EXAMPLES

1

The variables as listed in Table 6-IV have been computed by a number of students from wave records kindly supplied by the National Institute of Oceanography U.K. These were taken aboard the Helwick Light Ship stationed in 22 fathoms of water near the Helwick Bark in the Bristol Channel, where waves have been recorded at 3-h intervals since August 1960 [69]. The dates and times are listed with no attenuation factor taken into account. Values for the same records have been derived by different workers at different times. Comment on the results.

The records analyzed are similar to the one reproduced in Fig. 6-13, where it is seen that 1/8-min intervals are provided. These were used as the 96 points at which the deviation of the water surface from SWL was measured. The columns in Table 6-IV were derived as follows:

(a) y measured at 60/8-sec intervals, squared and added to give $a_{rms} = \sqrt{(\Sigma y^2)}$.

(b) a_{rms} from Fig. 6-2 using $H_1 = A + C$ and $H_2 = B + D$ where A = highest crest from SWL, B = second highest crest, C = lowest trough below SWL, D = second lowest trough.

(c) Number of zero crossings (upwards) N_z and number of maxima (crests) N_c to give spectral width $\epsilon = \sqrt{[1-(N_z/N_c)^2]}$. (The record could be read upside down with little difference — see Fig. 6-13).

(d) Average of a_{rms} in columns a and b to find $H_{1/3}$ and $H_{1/10}$ from Table 6-I using value of ϵ in column c.

(e) Non-statistical maximum wave $H_{max} = A + C$ and probable maximum over recording interval H'_{max} from Fig. 6-3 using a_{rms} and N_z over 3 hours.

(f) Apparent period $T_z = 720/N_z$ sec to give $T_{max} = T_z H_{1/3}^{1/6}/1.33$.

(g) Ratio $T_{max}/H_{1/3}^{1/2}$ to check proximity of fetch.

Comments on results:

(1) The a_{rms} values from the various methods are in most cases quite close. The average from the H_1 and H_2 approach is a reasonable representation of the a_{rms} from the multi-measurement technique of column a. Disconcerting differences have occurred between workers which have been magnified in the determination of $H_{1/3}$ and $H_{1/10}$.

(2) Unexplainable differences, sometimes large in nature, appear in the number of zero crossings and number of crests. Slight variations in either can make large changes in ϵ. The value of N_z enters into the computation of a_{rms} and H'_{max}, as any error of N_z in the latter is multiplied fifteen times. This points to the need for thorough tuition in recognizing zero crossings and of crests. Where a crest just reaches SWL it should be counted as a crossing. Even though a record extends beyond the 12 min, only this length should be read to be consistent with the multiplication factor of 15 for the 3-h interval.

(3) Whilst many values of $H_{1/3}$ and $H_{1/10}$ are close between workers, other cases contain differences of more than 50%. This may have been due to the use of the standard eq. 3-11 and others using the more appropriate ratio of Table 6-I.

(4) The value of $H_{max} = A + C$ is seen to be in general smaller than H'_{max}, as would be expected, but on some records it exceeds it. It should not be used for statistical purposes. There is close agreement between workers for the solitary measurement of $A + C$, but great variation in H'_{max} due to its dependence on a_{rms} and N_z.

(5) T_z, derived directly from N_z, has a similar percentage variation between workers as the latter. Values of T_{max} in turn depend slightly on the value of $H_{1/3}$ already showing differences.

(6) Agreement between workers of $T_{max}/\sqrt{H_{1/3}}$ is reasonable, considering the various computations necessary beforehand.

TABLE 6-IV

Wave analysis by various workers

Helwick Lightship		a	b		c			d		e		f	g	
Date	Time	a_{rms}	H_1 / a_{rms}	H_2 / a_{rms}	N_z	N_c	ϵ	$H_{1/3}$	$H_{1/10}$	$A+C / H_{max}$	$3\,hrs / H'_{max}$	T_z	T_{max}	$T_{max} / H_{1/3}^{1/2}$
2.12.60	0000	2.55	2.33	2.55	72	106	0.73	10.6	14.4	14.4	18.7	10.0	11.2	3.43
		1.90	2.33	2.54	73	102	0.53	9.22	12.1	12.8	18.3	10.0	10.9	3.59
	0300	2.22	1.96	1.93	83	110	0.73	8.85	12.0	13.3	13.0	8.68	9.38	3.15
		2.21	1.90	1.90	94	128	0.68	8.34	11.2	12.1	15.6	8.60	9.22	3.20
		2.18	1.74	1.90	103	107	0.26	7.90	10.0	11.2	15.9	7.85	8.34	2.97
	0600	1.97	1.80	1.82	75	100	0.66	7.75	10.5	11.2	14.0	9.60	10.1	3.65
		1.87	1.94	1.51	98	180	0.70	7.25	9.65	7.74	8.27	8.25	8.63	3.21
	0900	1.79	1.95	1.75	80	93	0.51	7.50	10.2	12.2	13.7	9.00	9.47	3.46
		1.74	1.94	1.96	81	100	0.59	7.73	9.95	12.2	14.1	10.0	10.5	3.80
		1.82	1.94	1.67	87	113	0.64	7.53	10.4	12.3	14.1	9.31	9.80	3.56
	1200	2.48	2.50	2.40	78	100	0.62	10.3	13.9	15.6	18.4	9.24	10.2	3.20
		2.43	2.45	2.34	87	113	0.34	9.80	12.7	15.5	18.2	9.00	9.90	3.17
	1500	1.94	1.95	1.80	83	105	0.61	7.92	10.8	12.2	14.4	8.26	8.80	3.13
		2.21	1.88	1.77	96	119	0.59	7.80	9.90	12.0	15.0	8.13	8.60	3.08
		2.02	1.81	1.94	82	102	0.59	8.00	10.3	12.2	14.8	8.78	9.32	3.30
	1800	1.84	2.06	1.70	92	116	0.61	7.80	10.6	13.1	14.3	7.83	8.30	2.98
		1.89	2.70	1.82	93	132	0.71	9.10	12.2	17.1	17.5	8.10	8.81	2.92
	2100	1.80	1.62	1.87	89	111	0.59	7.38	10.0	10.7	13.5	8.10	8.50	3.13
		1.68	1.70	1.81	100	127	0.48	7.12	9.15	10.9	13.3	7.20	7.50	2.81
		1.78	1.71	1.85	88	102	0.50	7.44	9.70	10.8	13.5	8.20	8.61	3.16

TABLE 6-IV (continued)

Helwick Lightship		a	b		c			d		e		f		g
Date	Time	a_{rms}	$\frac{H_1}{a_{rms}}$	$\frac{H_2}{a_{rms}}$	N_z	N_c	ϵ	$H_{1/3}$	$H_{1/10}$	$\frac{A+C}{H_{max}}$	$\frac{3\ hrs}{H'_{max}}$	T_z	T_{max}	$\frac{T_{max}}{H_{1/3}^{1/2}}$
3.12.60	0000	2.54	2.56	2.56	82	103	0.61	10.1	13.7	15.3	19.3	8.80	9.73	3.16
		2.35	2.51	2.38	94	110	0.52	9.86	12.5	16.0	14.6	8.30	9.13	2.91
	0300	2.90	2.75	2.68	96	114	0.54	11.1	15.0	17.1	19.9	7.50	8.43	2.53
		2.84	2.72	2.74	102	109	0.36	11.4	14.3	17.2	21.0	7.65	8.69	2.57
		2.80	2.76	2.71	91	119	0.64	11.6	15.2	17.6	21.4	7.90	8.96	2.64
	0600	2.34	2.70	2.69	87	112	0.63	10.9	14.7	17.2	15.7	8.27	9.26	2.81
		2.67	2.73	2.70	97	123	0.62	11.5	14.7	17.3	16.7	8.04	8.96	2.64
	0900	2.31	2.25	2.36	87	107	0.60	9.85	13.4	14.3	14.4	8.27	9.12	2.91
		2.32	2.23	2.33	100	116	0.51	9.45	12.2	14.8	17.5	7.80	8.52	2.78
		2.29	2.31	2.40	91	109	0.55	9.74	12.6	14.7	18.1	8.78	9.65	3.09
	1200	2.68	2.71	2.70	85	116	0.68	10.8	14.7	17.0	16.4	8.48	9.47	2.88
		3.04	3.01	3.20	101	120	0.54	12.6	16.1	20.5	20.3	8.00	9.20	2.60
	1500	2.52	2.51	2.64	98	125	0.62	10.4	14.1	16.0	15.9	7.35	8.16	2.54
		2.76	2.97	2.74	96	102	0.36	11.1	14.5	19.0	17.1	7.50	8.46	2.54
	1800	3.14	2.56	2.48	78	102	0.65	11.0	14.9	16.0	20.2	9.21	10.4	3.1
		3.24	3.56	3.37	85	89	0.30	13.6	17.3	22.7	21.1	8.50	9.90	2.68
		—	3.67	3.45	84	100	0.54	14.6	18.8	23.2	27.2	8.60	10.2	2.6
	2100	3.25	3.20	3.36	78	100	0.62	13.9	18.9	20.0	24.6	9.21	10.8	2.8
		3.72	3.54	3.81	86	114	0.66	11.1	14.2	21.4	28.8	9.50	10.7	3.2
4.12.60	0000	5.00	4.20	4.50	84	101	0.32	18.5	23.4	26.3	28.0	9.70	11.9	2.7
		3.75	4.25	4.44	75	84	0.45	17.0	21.6	26.4	32.7	8.60	10.4	2.5
	0300	4.10	3.96	4.10	92	104	0.47	16.5	21.0	24.5	30.6	8.80	10.6	2.6
		4.04	3.88	4.24	84	101	0.56	16.9	21.8	24.7	30.6	8.60	10.4	2.5
	0600	5.70	3.70	3.85	74	90	0.47	16.0	20.4	23.0	28.5	10.5	12.6	3.1
		5.05	3.86	4.24	66	84	0.63	19.3	23.6	23.7	37.1	10.9	13.5	3.0
	0900	3.94	3.96	3.80	77	96	0.36	15.9	20.7	24.7	29.6	10.1	12.1	3.0
		4.08	3.88	3.67	72	94	0.64	16.7	21.7	24.7	29.5	10.0	12.1	2.9
	1200	3.73	3.41	3.70	79	108	0.68	15.1	20.2	21.3	27.0	10.9	13.0	3.3
		4.16	3.27	3.56	68	91	0.67	16.1	20.8	21.2	27.5	10.7	12.8	3.2
	1500	3.34	3.48	3.51	95	124	0.42	13.5	17.3	22.1	25.0	8.20	9.60	2.6
		3.29	3.45	3.33	91	114	0.46	14.0	18.7	22.0	25.0	7.90	9.24	2.4
	1800	3.60	3.70	3.50	88	115	0.65	15.3	19.8	23.5	27.3	8.90	10.6	2.7
	2100	3.35	3.20	3.16	80	107	0.67	13.8	18.3	23.3	23.7	9.60	11.2	3.0
5.12.60	0000	3.00	2.78	2.61	80	110	0.69	11.8	15.7	16.3	18.0	9.40	10.7	3.1
	0300	2.47	2.65	2.70	96	120	0.60	10.7	14.0	17.0	16.1	7.80	8.63	2.6
	0600	2.16	2.26	2.11	98	194	0.70	9.25	12.4	14.4	16.2	8.00	8.71	2.8
	0900	2.70	2.20	2.10	86	106	0.58	9.75	12.7	13.3	13.4	8.35	9.18	2.9
	1200	1.96	1.73	1.75	107	109	0.20	7.77	10.5	12.0	13.5	8.00	8.45	3.0
	1500	1.79	2.08	1.59	103	129	0.75	7.64	9.46	11.9	13.6	7.86	8.28	3.0
	1800	1.52	1.54	1.46	101	140	0.48	6.34	8.60	9.90	12.2	8.00	8.17	3.2

EXAMPLES

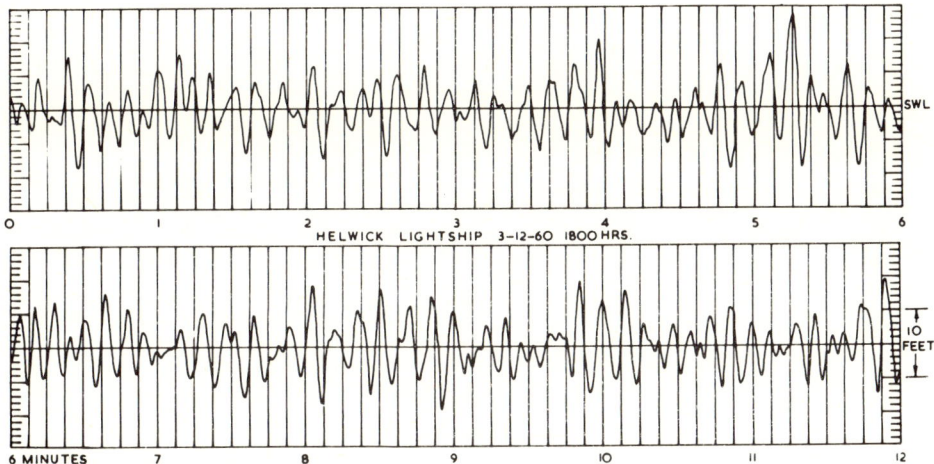

Fig. 6-13. A sample 12-min record from a ship-borne wave recorder.

In order to compare values of this parameter with those in Table 3-VIII for partially and fully arisen seas, it is first necessary to convert the heights in the records to those at the sea surface. For this purpose it will be assumed that the piezometers are set at 10 ft. depth in the ship and that the equivalent depth is 2.5 times this or 25 ft. [46]. The ship was located in 22 fathoms of water [69], so the attenuation factor, assuming $T_z = 10$ sec ($L = 481$), is given by:

$$\frac{H_y}{H_s} = K_b \frac{\cosh 2\pi(y+d)/L}{\cosh 2\pi d/L} = 0.9 \frac{\cosh 2\pi(-25+132)/481}{\cosh 2\pi 132/481} = \frac{1}{1.5} \qquad (6\text{-}26)$$

The range of $T_{max}/\sqrt{H_{1/3}}$ in Table 6-IV is 2.5 to 3.8, which must therefore be divided by $\sqrt{1.5}$ and become 2.04 to 3.11. The value in Table 6-IV equivalent to the FAS value of 2.43 is 2.97.

Consider a fetch, as in Fig. 3-33, which produces a FAS for a lengthy period so that all waves of the spectrum are propagating in the dispersal area. Taking values down the centreline of the fetch, the percentage arriving at distances (D), so many fetch widths (W) away, is given in Fig. 3-33. The value of $H_{1/3}$ at any point provides a value of $T_{max}/\sqrt{H_{1/3}}$ as follows:

D/W	0	1	2	3
$T_{max}/\sqrt{H_{1/3}}$	2.43	2.98	4.23	5.06

At any greater distances than 3 it would be hard to conceive of the whole spectrum being present at the one time.

The range of 2.04 to 3.11 derived above thus indicates that the Helwick Lightship was within one fetch width of storms at all times and inside the fetch on many occasions. The lowest limit is associated with a partially arisen sea where $F/F_{FAS} < 5\%$ or $t/t_{FAS} < 15\%$ (see Table 3-VIII). The location of the ship in the mouth of the Bristol Channel would lend credence to this proximity of storms, especially during the month of December.

(7) Overall, the four-day record represents a fairly rough sea with waves being generated close to the recorder for the whole duration. The wave heights should all be amplified by a factor around 1.5, the actual value being derived for each specific T_z. There are many inconsistencies in the value of ϵ and $T_{max}/\sqrt{H_{1/3}}$ which strongly indicate that swell is combining with more locally generated waves. On the eastern side of the Northern Atlantic Ocean this is very likely, as the majority of the strong winds are from the west.

2

A 12-sec swell wave with 3 ft. height is propagating in 50 ft. of water. Find the magnitude of

TABLE 6-V

Wave heights as recorded in Example 2

y (ft.)	-10	-20	-30	-40	-50
$2\pi(y+d)/L$	0.556	0.417	0.278	0.139	0.
$\cosh 2\pi(y+d)/L$	1.16	1.09	1.04	1.01	1.0
$(H_y)_{th}$	2.75	2.59	2.47	2.40	2.37
$(H_y)_L$	2.00	1.87	1.79	1.74	1.72
$(H_y)_U$	2.45	2.30	2.20	2.14	2.11

the theoretical pressure oscillation (in ft.) that would be measured by a recorder placed at 10-ft. intervals of depth and at the bed. For each location what is the range of readings likely from the many empirical factors available.

$H = 3$ ft., $T = 12$ sec, $d/L_o = 50/5.12 \times 12^2 = 0.0696$, $L = 452$ ft., $\cosh 2\pi d/L = 1.265$, from Fig. 6-10 $K_b = 0.725$ to 0.89.

Thus at the bed, for example, the highest reading is likely to be 11% lower than the theoretical and could be 27.4% lower. With the very mild steepness involved the former result is more probable.

3

If a second swell wave similar to the first in Example 2 above were present, but normal to it, what height would be recorded at the bed along the alignment of the combined crests?

Height of combined crests at surface = $3\sqrt{2} = 4.24$ ft., $d/L_o = 0.0696$, from Fig. 6-10 $H_{sc}/H_p = 0.83$. Thus at the bed:

$(H_y)_{th} = 4.24 \times 0.83/1.265 = 2.79$ ft.
$(H_y)_L = 0.725 \times 2.79 = 2.02$ ft.
$(H_y)_U = 0.89 \times 2.79 = 2.48$ ft.

At locations other than this alignment further reductions would occur; and on planes of $L'/4$ no waves would be recorded. This emphasizes the need to place pressure recorders as far inshore as possible so that waves are in closer alignment, but not so far that steepening accentuates the attenuation.

4

In the wave conditions of Example 2 the floor is covered by 10 ft. of fluid mud so that the recorder is held at 40 ft. depth. Determine the height likely to be received.

The wave length will be the same as for the non-stratified depths of 50 ft., so that $d/L_o = 0.0696$. From Fig. 6-12, the maximum reduction through the theoretical attenuation at this d/L_o due to the mud is 0.645/0.79. With the further reduction due to K_p the range of heights possible at 40 ft. depth is $3 \times (0.645/0.79) \times (1.01/1.265) \times (0.725$ to $0.89) = 1.42$ to 1.74 ft. Under such conditions the waves will be strongly attenuated as they approach the shore, so that the recorded values should be used with care.

PROBLEMS

1

The number of waves recorded within specified height increments for a period of two weeks is listed in Table 6-VI. Draw up a table in order to determine H_{ave}, $H_{1/3}$ and $H_{1/10}$. Compare the ratios of these heights with the statistical values for a wide and narrow spectrum. Process the data in order to plot it on probability graph paper and so determine the probable maximum

TABLE 6-VI

Wave data for Problem 1

H'_{max}	0–10	1.1–2.0	3.0	4.0	5.0	6.0	7.0	8.0
Number of waves	8	36	32	21	9	4	1	1

wave over a year and five years. (This presumes similar generating conditions.) Assume that the waves have resulted from consecutive recording intervals, i.e., no intervals had zero wave incidence.

2

Note the changes in Problem 1 when the number of waves in the increment 0–1.0 ft. becomes 5 and that in the 7.1–8.0 ft. becomes 3.

3

From a particular wave record the following values are extracted: highest crest above SWL = 8.4, 2nd highest crest = 7.9, lowest trough below SWL = 6.3, 2nd lowest trough = 5.5, number of zero crossings = 100, number of crests = 121. Find H'_{max}, T_{max} and ϵ for the recording interval of 3 h, if the recording period is 12 min. Derive ϵ for an erroneous value of 95 zero crossings, and again for an omission of 6 crests.

4

Using the data in the scatter diagram of Fig. 6-4A as $H_{1/3}$ and T_{max} find the probable maximum height (H'_{max}) and T_{max} over a 50-year period, assuming 12-min recording periods and 3-h intervals, assuming the heights are those at the sea surface. Assume also that all recording intervals are successive, or that no intervals contained zero waves.

5

Carry out the same computation as Problem 4 above, assuming that the heights ($H_{1/3}$) were those measured on a recorder at the bed in 40 ft. of water. Use T_{max} to determine d/L_o for the attenuation factor and choose a mean value in the range displayed in Fig. 6-8.

6

A 35-knot wind generates a FAS across the continental shelf. Draw the energy distribution curve (EDC) for the surface oscillation and on the same diagram the spectrum as would be measured by recorders placed on the bed at 50, 100, 200, 300 and 400 ft. depths. Considering Fig. 5.22, what is the greatest depth at which the same FAS as in the deeper water can exist? Draw the EDC's for surface and bed recorders in depths of 25 and 15 ft. Comment on the results.

7

A 25-knot wind blows over a wide expanse of ocean for 24 h. Draw the expected EDC at the surface, at mid-depth and at the bed in 30 ft. of water.

8

A recorder at the bed in 40 ft. of water records an $H_{1/3}$ = 11.5 ft. and T_z = 9.8 sec. An observer at the site notes that the larger and longer waves are arriving from a WSW direction. If the coast runs due N-S, and the continental shelf is of uniform profile along this stretch of it, find the characteristics of the fetch which could produce this, including the direction of the 30-knot wind known to have been blowing in the cyclone.

9

A fetch 400 NM long and 150 NM wide is directed towards a wave-recording site and has reached FAS conditions when still 600 miles away. It travels towards and across the site at an average speed of 30 knots. Draw a graph of $T_{max}/\sqrt{H_{1/3}}$ against time from the commencement of this path until the upwind end of the fetch has traversed the surface recorder.

10

Plot the data of Example 1 on a scatter diagram (Fig. 6-4A) and draw a percentage exceedance curve for wave height (Fig. 6-4B), a histogram for wave period (Fig. 6-4C) and a histogram for spectral width (Fig. 6-5). Can any conclusions be drawn from these? From a graph on probability paper determine the maximum wave height to be expected in 100 years.

11

From the data in Example 1, plot ϵ, $H_{1/3}$, H'_{max}, $A + C$, T_{max} and $T_{max}/\sqrt{H_{1/3}}$ against time for the four days duration. Can any conclusions be drawn from these data?

12

From the 12-min record made on the Helwick lightship, and reproduced in Fig. 6-13, find a_{rms} by vertical measurements each 1/8 min interval and by measuring A, B, C and D (notation as in text). Find also N_z, N_c, ϵ, $H_{1/3}$, $H_{1/10}$, $H_{max} = A + C$, H'_{max}, T_z, T_{max} and $T_{max}/\sqrt{H_{1/3}}$. Compare your result with the three listed in Example 1 for this time and date. (Some deviation will exist due to errors in tracing the original record.) Convert the wave heights to surface values assuming the equivalent submergence of the recorder is 25 ft. and the vessel is in 132 ft. of water. From the $T_{max}/\sqrt{H_{1/3}}$ value derive the state of the fetch and hence state the $U_{19.5}$ wind velocity and the limit of fetch or duration for which it has existed. (See Table 3-VII and 3-VIII.)

13

From the same record (of Problem 12, Fig. 6-13) compute a_{rms} by measuring deviations of the water surface at 1/16-min intervals. Comment on the result.

14

Over a series of days 3-hourly recordings show first long-period low waves and then suddenly higher swell with $H_{1/3}$ = 9.0 ft. and T_z = 10.2 sec, which gradually reduces in height and period over a number of days. From this, what could you infer about the fetch, which is known to be two fetch widths from the recording site?

15

List all the disadvantages of recording ocean waves in an area of coast where major reflections are taking place.

PROBLEMS

TABLE 6-VII

Wave data for Problem 19

Max. wave height (ft.)	Wave period T_z (sec)								Total
	3	4	5	6	7	8	9	10	11	
1	–	–	–	–	–	–	–	–	–	–
2	–	–	1	2	–	–	–	–	–	3
3	–	1	6	3	3	–	–	1	–	14
4	3	13	11	10	8	2	–	1	–	47
5	3	7	14	8	6	3	–	3	–	44
6	–	7	26	17	14	14	2	1	–	81
7	4	17	14	26	13	18	4	1	–	97
8	1	7	22	29	13	11	–	5	–	88
9	7	7	17	24	22	12	1	–	–	90
10	–	4	21	25	11	8	–	2	1	72
11	1	4	8	25	14	11	3	–	2	68
12	1	1	22	35	14	6	4	–	–	83
13	–	1	13	22	15	6	2	–	1	60
14	–	–	11	12	13	8	3	6	–	53
15	–	1	1	14	14	10	4	2	1	47
16	–	–	4	8	12	4	–	3	–	31
17	–	–	1	4	6	3	1	3	–	18
18	–	–	1	4	10	7	2	–	–	24
19	1	–	–	5	6	7	–	–	–	19
20	–	–	1	7	3	2	1	1	–	15
21	–	–	–	1	4	6	–	–	–	11
22	–	–	–	–	6	3	–	–	1	10
23	–	–	–	–	2	–	1	–	–	3
24	–	–	–	–	3	3	–	–	–	6
25	–	–	1	–	–	–	2	1	–	4
26	–	–	–	–	–	4	–	–	–	4
27	–	–	–	–	–	1	–	–	–	1
28	–	–	–	–	2	1	–	–	–	3
29	–	–	–	–	3	–	–	–	–	3
30	–	–	–	–	–	–	–	–	–	0
31	–	–	–	–	–	1	–	–	–	1
Total	21	70	195	281	216	151	30	30	6	1000

16
Discuss the errors that can occur in a wave record obtained from a pressure-actuated instrument placed on the seabed offshore from a beach.

17
Why is it not economical to record waves continuously? What are the factors determining the best length of record for any known wave climate?

18
As a coastal engineering consultant you have been asked by a client whether he should instal a wave recorder in a remote site on the coast. Draft a letter explaining to him the reasons for and

against such an action and the investigations you would first make before answering the question specifically.

19

The data in the Table 6-VII was reduced from records taken on the Helwick Lightship for the months of December 1960 and January, February 1961 [69]. The heights may be taken as $H_1 = A + C$ and the periods as T_z. Convert these respectively to H'_{max} for 3-hourly intervals (12-min recording period) and T_{max}. Construct a percentage exceedance graph for wave height and histogram of wave period. Comment on these results. What is the maximum height to be expected in 100 years? Compare this result with that of Problem 10, which represents four days record at the same site and within this 3-month period. (It is assumed that all intervals recorded waves. It is necessary to convert the repetitions of waves in each group from the per thousand base to actual figures knowing that eight records were available in each of the 90 days.)

20

Mayençon [70] has analyzed visual observations of waves supplied on punched cards by the British Meteorological Surveys for various zones covering the Atlantic Ocean. One of these zones included areas of the Bristol Channel where the Helwick Lightship wave data of Example 1 and Problems 10 and 19 were obtained. He derived a relationship, between $H_{1/3}$ as recorded and the height normally observed on these ships H_{vos}, of $H_{1/3} = 2 + 0.82\,H_{vos}$. This resulted in statistical maxima for $H_{1/3}$ of 10.0 m annually, 12.2 m each decade and 14.4 m each century. Compare these with values derived in Problems 10 and 19, and comment on any differences.

REFERENCES

[1] M.J. Tucker, 1953. Sea wave recording. *Dock Harbour Authority*, 34: 207–210.
[2] L. Draper, 1963. Derivation of a "design wave" from instrumental records of sea waves. *Proc. Inst. Civil Eng.*, 26: 291–304.
[3] M.J. Tucker, 1957. The analysis of finite length records of fluctuating signals. *Brit. J. Appl. Phys.*, 8: 137–142.
[4] M.J. Tucker, 1961. Simple measurement of wave records. *Proc. Conf. Wave Recording Civil Eng., Natl. Inst. Oceanogr., Engl., 1961*: 22–23.
[5] D.E. Cartwright and M.S. Longuet-Higgins, 1956. The statistical distribution of the maxima of a random function. *Proc. R. Soc.*, A237: 212–232.
[6] D.E. Cartwright, 1958. On estimating the mean energy of sea waves from the highest waves in a record. *Proc. R. Soc.*, A247: 22–48.
[7] R.R. Putz, 1954. Statistical analysis of wave records. *Proc. 4th. Conf. Coastal Eng.*, 1: 114–120.
[8] J.R. Scott, 1969. Some average wave lengths on short-crested seas. *Q. J. R. Meteorol. Soc.*, 95: 621–634.
[9] M.S. Longuet-Higgins, 1952. On the statistical distribution of the heights of sea waves. *J. Mar. Res.*, 11: 245–266.
[10] L. Draper, 1966. Waves at Sekondi, Ghana. *Proc. 10th Conf. Coastal Eng.*, 1: 12–17.
[11] L. Draper, 1961. Wave recording instruments for civil engineering use. *Proc. Conf. Wave Recording Civil Eng., Natl. Inst. Oceanogr.*, 1: 7–15.
[12] L. Draper, 1970. Routine sea-wave measurement – a survey. *Underwater Sci. Tech. J.*, 2: 81–86.
[13] H. Higuchi and T. Kakinuma, 1966. Observations of the transformation of ocean wave characteristics near coasts by use of anchored buoys. *Proc. 10th Conf. Coastal Eng.*, 1: 77–98.
[14] L.J. Cote, J.O. Davis, W. Marks, R.J. McGough, E. Mehr, W.J. Pierson Jr., J.F. Ropek, G.

REFERENCES

Stephenson and R.C. Vetter, 1960. The direction spectrum of a wind-generated sea as determined from data obtained by the stereo wave observation project. *Meteorol. Pap. N.Y. Univ.*, 2(6): 88 pp.

[15] T.P. Barnett and J.C. Wilkerson, 1967. On the generation of ocean wind waves as inferred from airborne radar measurements of fetch-limited spectra. *J. Mar. Res.*, 25: 292–328.

[16] J. Darbyshire, 1970. Wave measurements with a radar altimeter over the Irish Sea. *Deep Sea Res.*, 17: 893–901.

[17] J.M. Caldwell, 1948. An ocean wave measuring instrument. *Beach Erosion Board, Tech. Mem.*, 6.

[18] T.L. Russell, 1963. A step-type recording wave gauge. *Proc. Conf. Ocean Wave Spectra, 1963*: 251–258.

[19] M.J. Tucker, 1956. A ship-borne wave recorder. *Trans. Inst. Nav. Archit.*, 98: 236–250.

[20] M.S. Longuet-Higgins, D.E. Cartwright and N.D. Smith, 1963. Observations of the directional spectrum of sea waves using the motions of a floating buoy. *Proc. Conf. Ocean Wave Spectra, 1963*: 111–131.

[21] P.J. Wemelsfelder, 1954. De integrator als golfmeetapparatuur. *Ingenieur*, 7: 1–8.

[22] M.J. Tucker, 1963. Recent measurement and analysis techniques developed at the National Institute of Oceanography. *Proc. Conf. Ocean Wave Spectra, 1963*: 219–226.

[23] R. Bonnefille, P. Cormault and J. Valembois, 1966. Progrès des méthodes de mesure de la houle naturelle en Laboratoire National d'Hydraulique. *Proc. 10th Conf. Coastal Eng.*, 1: 115–126.

[24] J.M. Jordaan Jr., 1963. Experience with recording of storm waves, swell and tide, using an inverted echo-sounder off Durban, South Africa. *Int. Hydrogr. Rev.*, 40: 125–141.

[25] L.E. Borgman and N.N. Panicker, 1970. Design study for a suggested wave gauge array off Point Mugu, California. *Univ. Calif.*, HEL: 1–14.

[26] N.N. Panicker, and L.E. Borgman, 1970. Directional spectrum from arrays of wave gauges. *Proc. 12th Conf. Coastal Eng.*, 1: 117–136.

[27] D.E. Cartwright and N.D. Smith, 1964. Buoy techniques for obtaining directional wave spectra. *Buoy Tech., Wash. Mar. Tech. Soc.*, 1964: 112–121.

[28] D.E. Cartwright, 1963. The use of directional spectra in studying the output of wave recorder on a moving ship. *Proc. Conf. Ocean Wave Spectra, 1963*: 203–218.

[29] N.F. Barber, 1963. The directional resolving power of an array of wave detectors. *Proc. Conf. Ocean Wave Spectra, 1963*: 137–150.

[30] L.E. Borgman, 1969. Directional spectra models for design use for surface waves. *Univ. Calif.*, HEL: 1–12.

[31] W.C. Thompson, 1970. Swell characteristics from coastal wave records. *Proc. 12th Conf. Coastal Eng.*, 1: 33–52.

[32] L. Draper, 1970. The Canadian wave climate study — the formative year. *Proc. 12th Conf. Coastal Eng.*, 1: 1–12.

[33] R. Silvester, 1957. The calculations of climatological ocean waves for specific points on a coastline. *J Inst. Eng. Austr.*, 29: 283–296.

[34] N.F. Barber, F. Ursell, J. Darbyshire and M.J. Tucker, 1946. A frequency analyser used in the study of ocean waves. *Nature*, 158: 329–332.

[35] G.E.R. Deacon, 1952. Analysis of ocean waves. In: *Gravity Waves — Natl. Bur. Std., Circ.*, 521: 209–214.

[36] N.F. Barber and F. Ursell, 1948. The generation and propagation of ocean waves and swell, 1. Wave periods and velocities. *Trans. R. Soc.*, A240: 527–560.

[37] M.J. Tucker, 1956. The N.I.O. wave analyser. *Proc. 1st. Conf. Coastal Eng. Instr., 1956*: 129–133.

[38] M.J. Tucker, 1936. Analysis of records of sea waves. *Proc. Inst. Civil Eng.*, 26: 305–316.

[39] F.E. Snodgrass and R.R. Putz, 1958. A wave height and frequency meter. *Proc. 6th Conf. Coastal Eng., 1958*: 209–224.

[40] H.G. Farmer, 1963. A data acquisition and reduction system for wave measurements. *Proc. Conf. Ocean Wave Spectra, 1963*: 227–233.

[41] M. Hom-ma, K. Horikawa and S. Komori, 1966. Response characteristics of underwater wave gauge. *Proc. 10th Conf. Coastal Eng.*, 1: 99–114.

[42] B.K.H. Glukhovskiy, 1961. Study of wave attenuation with depth on the basis of correlation analysis. *Meteorol. Gidrol.*, 11.

[43] L. Draper, 1957. Attenuation of sea waves with depth. *Houille Blanche*, 12: 926–931.

[44] F. Biésel, 1952. General second-order equations of singular waves. *Houille Blanche*, 7: 372–376.

[45] P.O. Bergan, A. Tørum and A. Traetteberg, 1968. Wave measurements by a pressure-type wave gauge. *Proc. 11th Conf. Coastal Eng.*, 1: 19–29.

[46] L. Draper, 1965. Wave spectra provide best basis for offshore rig design. *Oil Gas Int.*, 5: 58–60.

[47] R. Miché, 1944. Mouvements ondulatoires des mers en profondeur constante ou décroissante. *Ann. Ponts Chaussées*, 1944: 25–78; 131–164; 270–292; 369–406.

[48] M.S. Longuet-Higgins, 1950. A theory of the origin of microseisms. *Phil. Trans. R. Soc.*, A243: 1–35.

[49] J. Darbyshire, 1950. Identification of micro-seismic activity with sea waves. *Proc. R. Soc.*, A202: 439–448.

[50] R.I.B. Cooper and M.S. Longuet-Higgins, 1951. An experimental study of the pressure variation in standing water waves. *Proc. R. Soc.*, A206: 424–435.

[51] G.V. Latham, R.S. Anderson and M. Ewing, 1967. Pressure variations produced at the ocean bottom by hurricanes. *J. Geophys. Res.*, 72: 5693–5704.

[52] G.E.R. Deacon, 1956. Marine physics. *Proc. Inst. Civil Eng.*, 5(I): 661–676.

[53] J.W. Hoyle and G.T. King, 1955. The lateral stability of shingle beaches. *Proc. Inst. Munic. Eng.*, 81: 357.

[54] L. Greslou and Y. Mahé, 1956. Etude du coefficient de réflexion d'une houle sur un obstacle constitué par un plan incliné. *Proc. Conf. Coastal Eng.*, pp. 68–84.

[55] R. Fuchs, 1952. On the theory of short-crested oscillatory waves. In: *Gravity Waves–Natl. Bur. Std., Circ.*, 521: 187–200.

[56] H.R. Seiwell, 1947. Investigations of the underwater pressure records and simultaneous sea surface patterns. *Trans. Am. Geophys. Union*, 28: 722–724.

[57] H.R. Seiwell, 1948. Evaluation of sea-surface roughness from underwater pressure recordings. *Trans. Am. Geophys. Union*, 29: 197–201.

[58] J.R. Morison, 1952. Analysis of subsurface pressure records in constant depths and on sloping beaches. *Univ. Calif., I.E.R. Ser.*, 3: 336.

[59] N.F. Barber, 1949. The behaviour of waves on tidal streams. *Proc. R. Soc.*, A198: 81–93.

[60] P.J.H. Unna, 1942. Waves and tidal streams. *Nature*, 149: 219–220.

[61] M. Ewing and F. Press, 1949. Notes on surface waves. *Ann. N.Y. Acad. Sci.*, 51(3): 453–462.

[62] H.R. Seiwell, 1948. Results of research on surface waves of the western North Atlantic. *MIT Woods Hole Pap. Phys. Oceans Meteorol.*, 10(4).

[63] H.N. Abramson and C.L. Bretschneider, 1954. Some observations concerning the analysis of surface waves when the bottom is non-rigid. *Beach Erosion Board, Tech. Mem.*, 46.

[64] J.W. Tukey, 1949. The sampling theory of power spectrum estimates. *Symp. Appl. Autocorr. Anal. Phys. Probl., Woods Hole, 1949*: p. 47.

[65] W.J. Pierson Jr. and S.S.L. Chang, 1955. A wave spectrum analyser. *Proc. 1st Conf. Ships Waves*: 55–62.

[66] W.J. Pierson Jr., 1954. An electronic wave spectrum analyser and its use in engineering problems. *Beach Erosion Board, Tech. Mem.*, 56.

[67] N.F. Barber, 1963. A plea for the rectangular lag window. *Proc. Conf. Ocean Wave Spectra, 1963*: 151–154.

[68] J.W. Tukey, 1963. What can data analysis and statistics offer today. *Proc. Conf. Ocean Wave Spectra, 1963*: 347–352.

[69] M. Darbyshire, 1963. Wave measurements made by the National Institute of Oceanography. *Proc. Conf. Ocean Wave Spectra, 1963*: 285–291.

[70] R. Mayençon, 1969. Statistical study of wave observations. *Cah. Océanogr.*, 21: 487–501.

[71] J.H. Nath and S. Neshyba, 1971. Two-point mooring system for spar buoy. *Proc. ASCE*, 97(WW2): 295–312.

Chapter 7

EFFECT OF STRUCTURES

When waves arrive at a structure, be it floating or fixed, they will undergo some degree of reflection and dissipation. These processes will involve run-up of the wave over the seaward face of the structure, similar to the wave run-up previously discussed for beaches. Where this run-up exceeds the top limit of the structure overtopping takes place. Such phenomena can be treated on a two-dimensional basis and calculations of quantities made per unit length.

With reflection taking place along the length of a structure, the overall shape of the water body enclosed by this and other structures can influence the resultant wave pattern. If successive reflection back and forth between two nearly parallel walls can occur, as in some harbour basins, standing waves of various degrees of completeness result. Such oscillations are termed harbour seiches, and generally evolve from the longer waves in the spectrum, as these are more readily reflected, even from rubble-mound structures. To impede the entrance of these longer components into wharf areas, resonant basins can be constructed at the entrance which can cancel them by wave interaction.

Where marine structures are backed by water instead of land, there may be a degree of wave transmission on the leeward side, after reflection and dissipation have taken toll of the incident wave. These structures may be transitions within a channel, submerged breakwaters, or arrays of piles. These may be constructed for their expressed influence on waves, or this reaction may be secondary to the main purpose of the structure. This is not to infer that the secondary effect may not nullify the other as, for example, the attenuation of waves by the piles of a jetty causing accretion of sand at the berth it is supposed to serve.

Some installations are designed primarily for wave attenuation, either to produce calmer conditions for marine operations or to effect siltation by reducing wave energy in certain areas of the sea or river bed. Fixed installations for this purpose are the pneumatic and hydraulic breakwaters, both of which employ opposing currents of water to dissipate the incoming waves. Floating breakwaters, on the other hand, rely on reflection, resonance and attenuation for reduction in the transmitted wave.

All structures, even natural features, are finite in length and hence waves will by-pass the ends and propagate into the zone to leeward. This lateral spreading of energy along the wave crests is termed diffraction, and is one of the major consequences to be accounted for by the coastal engineer. Although the energy so distributed into the shadow zone is partly dependent upon the degree of reflection on the structure itself, the main mechanism is completely different from the other influences of structures on waves. It is convenient, therefore, to treat this topic first.

WAVE DIFFRACTION

Initially only long-crested waves are considered and these of swell nature, the influences on directional spectra being discussed finally. Where a single train of waves is intercepted by a breakwater or a headland, the section striking it is dissipated and/or reflected. The remainder continues past the structure, as for example

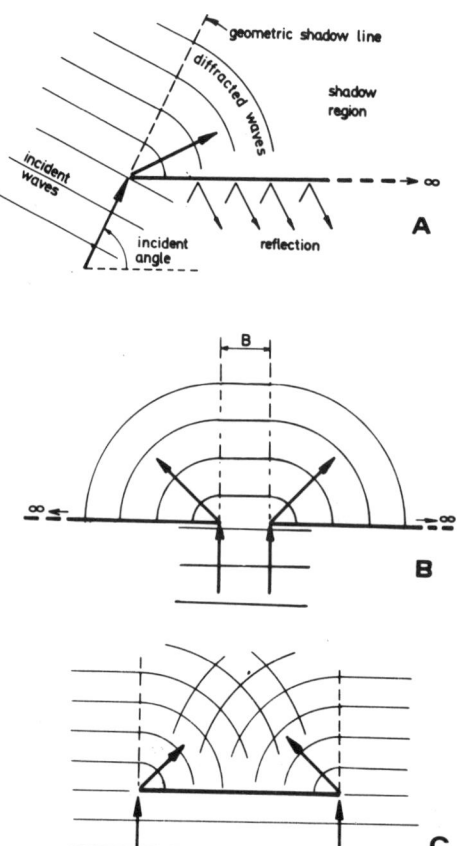

Fig. 7-1.A. Wave diffraction behind a semi-infinite breakwater. B. Diffraction through a breakwater gap. C. Diffraction behind an island or offshore breakwater.

in Fig. 7-1A, where penetration into the shadow zone is by curved crests which are essentially concentric with the tip of the breakwater. It is implied at present that the whole area around the breakwater is of the same depth, so that no concurrent refraction is taking place. As seen in Fig. 7-1A, the line or orthogonal at which curvature commences, necessarily passing through the tip of the breakwater, is termed the *geometric shadow line,* or just the *shadow line*. This does not mean, as will be seen later, that the wave does not undergo changes along this alignment, in fact the height is about half the incident wave height along it.

The situation depicted in Fig. 7-1A is generally designated the semi-infinite breakwater, which is a mis-nomer because the qualification required is that the body of water beyond the breakwater is large or infinite, in order to supply sufficient energy in the diffraction process. The inference of great length of breakwater, of course, should apply equally to the dimension in the opposite direction. Where this is limited the phenomenon is termed the "breakwater gap", as illustrated in Fig. 7-1B, where two shadow lines occur, between which the incident wave must supply the bulk of the energy for diffraction on either side. A third installation worthy of note, but not dealt with in detail herein, is the island or offshore breakwater, around both ends of which waves can diffract (see Fig. 7-1C). This is similar to the semi-infinite water case, but the diffracted waves form clapotis and "clapotis gaufré", the wave energy being supplied from either end. Wada [1] has treated this problem for long waves or surges penetrating behind protective barrages.

Semi-infinite breakwater

Although the diffracting waves depicted in Fig. 7-1A are circular in plan, some slight distortion of this pattern can occur due to the differential celerities of waves of substantially differing heights, but this is not of engineering significance. The heights, of course, do vary along these crests, from full incident magnitude just outside the shadow line to small proportions next to the leeside of the breakwater. Also these heights vary as the waves proceed along their orthogonals, which are radial to the tip of the breakwater. These orthogonals, infinite in number, are straight when no refraction is taking place.

Another distortion to the wave pattern can accrue from the slight phase difference of the incident and diffracted waves, due to the energy spreading process. The theory found applicable to water-wave diffraction [2] is that for optical diffraction studied by Sommerfeld in 1896 [3]. However, it is difficult to visualize the mechanism by which wave height is transferred laterally along the crest. The nearest hydraulic equivalent is the case of the collapsing dam, or more closely the case of the moving dam. In either analogy, as depicted in Fig. 7-2, the water depth at the original dam site remains at 4/9 of the original water depth. This is not unlike the

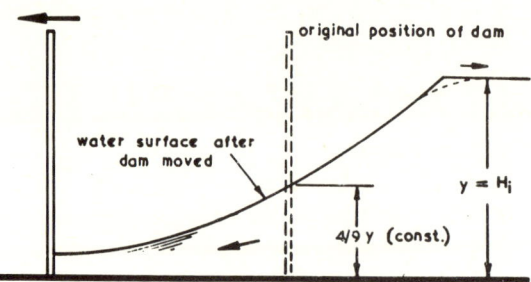

Fig. 7-2. Analogy of wave diffraction to the dam-break problem.

height along the shadow line which remains essentially at one half the incident wave.

It would be imagined that with this comparison there should be a flow of water into the shadow zone. In fact, this does happen within the first 1/5 of a wavelength behind the breakwater [4], but further from the tip no such currents can be observed or measured. At any location the height or head provided is almost instantaneous, the differential water level from trough to crest occurring over a wave period. The time taken to effect this energy transfer may in part explain the theoretical phase lag, although again this is of little engineering consequence.

Other second-order phenomena were derived by Biésel [5], such as wave components shorter in period than the main incident band. Finally, it will also be assumed that the breakwater is of negligible width compared to the length of the wave. In this case the reflected wave, which is also diffracted around into the shadow zone, will be considered to be in phase with the incident wave, so that their amplitudes are additive. Should the structure, such as a headland, be of reasonable width, the diffraction theory should be applied from the inside corner and the reflection component of the total diffracted wave ignored.

The general case for the semi-infinite breakwater is illustrated in Fig. 7-3, in which it is seen that a train of waves is approaching at an angle θ to the breakwater. The location of a point P can be defined by either a polar coordinate system $(\alpha R/L)$, or a circular arc system $(S/L, R/L)$, the origin in either case being the shadow line. This latter system, it will be seen, can be reduced to S/L alone, with little loss of accuracy.

It can also be observed in the figure that waves reflected from the breakwater also diffract whilst propagating seawards. Before extending to the shadow zone they must spread through an angle of $360° - 2\theta$, so that at P the angle of their diffraction is $360° - 2\theta + \alpha$. Outside the shadow zone the interaction of the reflected and incident waves creates a short-crested system, the characteristics of which are available [6]. Immediately outside the shadow zone the two waves are practically aligned and, although a slight phase difference may exist, heights in

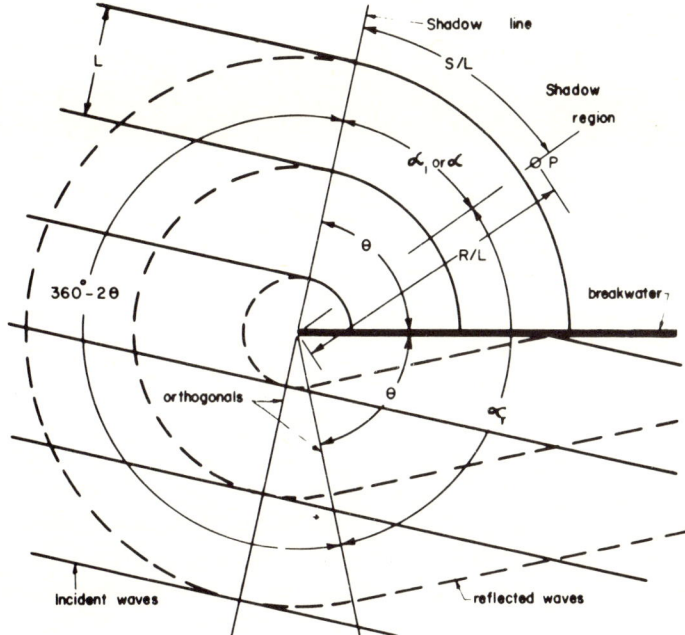

Fig. 7-3. Definition sketch for semi-infinite breakwater.

excess of those of the incident wave are theoretically possible. This has not been exhibited clearly in experiments, but is dependent greatly upon the degree of reflection from the breakwater. This zone is not being treated in the following analysis.

It is readily surmised that the size of the reflected wave in the shadow zone is small, but not insignificant for 100% reflection. In spite of the trend to design breakwaters for the fullest dissipation of waves, diffraction theory is normally based upon 100% reflection. In proximity to the breakwater tip, where the reflection component is greatest, a correct assessment of this additional height could result in worthwhile economies of design.

The Sommerfeld [3] solution of optical diffraction as applied by Penny and Price [2] results in the following equation:

$$F(R_1 \alpha) = f(u_1) \exp(-ikR \cos \alpha_i) + f(u_2) \exp(-ikR \cos \alpha_r) \tag{7-1}$$

where α_i and α_r are as shown in Fig. 7-3, and:

$$u_1 = -\sqrt{8R/L} \sin(\alpha_i/2) \quad u_2 = -\sqrt{8R/L} \sin(\alpha_r/2) \quad k = 2\pi/L$$

$$f(u) = \frac{1+i}{2} \int_{-\infty}^{u} \exp(-i\pi u^2) du \quad f(-u) = \frac{1+i}{2} \int_{-\infty}^{-u} \exp(-i\pi u^2) du$$

$f(u) + f(-u) = 1$

The diffraction coefficient $K = \dfrac{\text{diffracted wave height}}{\text{incident wave height}}$ has a numerical value equal to the modulus of eq. 7-1, so that:

$$K = |F(R, \alpha)| \tag{7-2}$$

In this event the second term of the RHS of eq. 7-1 can be written:

$$f(u_2) \exp\left[-ikR \cos(360° - \alpha_r)\right] \tag{7-3}$$

which represents the diffraction of the reflected wave, from its orthogonal through the breakwater tip around to the polar direction of point P (see Fig. 7-3). The first term of eq. 7-1 represents the fraction of the wave height resulting from diffraction of the incident wave from the shadow line. Thus, in general:

$$K = |f(u) \exp(-ikR \cos \alpha)| \doteq \text{incident} + \text{reflected} \tag{7-4}$$

where α is measured from the shadow line for the incident wave and from the orthogonal of the reflecting wave passing through the breakwater tip. In terms of Fig. 7-3 the diffraction angle for the reflected wave is $360° - 2\theta + \alpha_i$. The values of K are added in the case of 100% reflection and a proportion of the K for the reflection wave is used in the case of partial reflection.

Larras [7] has made a similar approach to the problem by solving the sine and Fresnel functions from the geometry of the point P in terms of orthogonal axes and the use of Cornu spirals. In this case also the diffraction coefficient is the addition of an incident and reflected term, the latter being modified according to the degree of reflection. Separating the components in this manner introduces a slight error for incident angles $\theta \leqslant 45°$, which are not likely to be used for design purposes. The magnitude of this error is illustrated in Fig. 7-4 for $\theta = 30°$ and $R/L = 2.5$. Its maximum is less than 0.5% along the shadow line. For 100% reflection the results agree closely with those given by Wiegel [8].

Eq. 7-1 has been graphed in Fig. 7-5 and tabulated in Table 7-I. The values of K representing incident and reflected components are read from angle α as previously indicated and then added. For example, for $\theta = 60°$, $\alpha_i = 30°$ and $R/L = 10$ in Fig. 7-5 read: $\alpha = 30°$ and $\alpha = 360 - 2(60) + 30 = 270°$, so that $K_i = 0.10$ and $K_r = 0.03$ giving $K_{100\%} = 0.13$ and $K_{50\%} = 0.115$. From Table 7-I the respective values are:

$$K(\alpha = 30°) = 0.096 \; ; \quad K(360° - 2\theta + \alpha = 270°) = K(360° - 270° = 90°) = 0.036 \; ;$$

$$K_{100\%} = 0.132 \, .$$

In reading Table 7-I or Fig. 7-5 it is sufficient for the reflection term to use $2\theta - \alpha$, which in this case $= 120° - 30° = 90°$.

WAVE DIFFRACTION 307

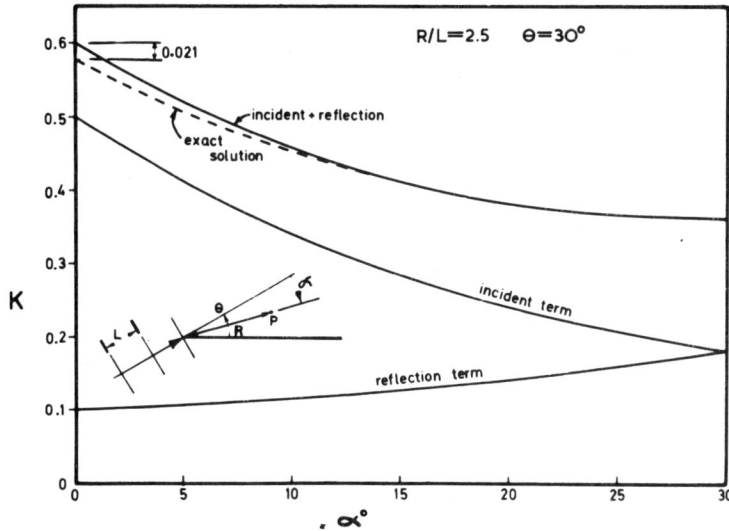

Fig. 7-4. Distribution of incident and reflection components and deviation from exact solution for small θ.

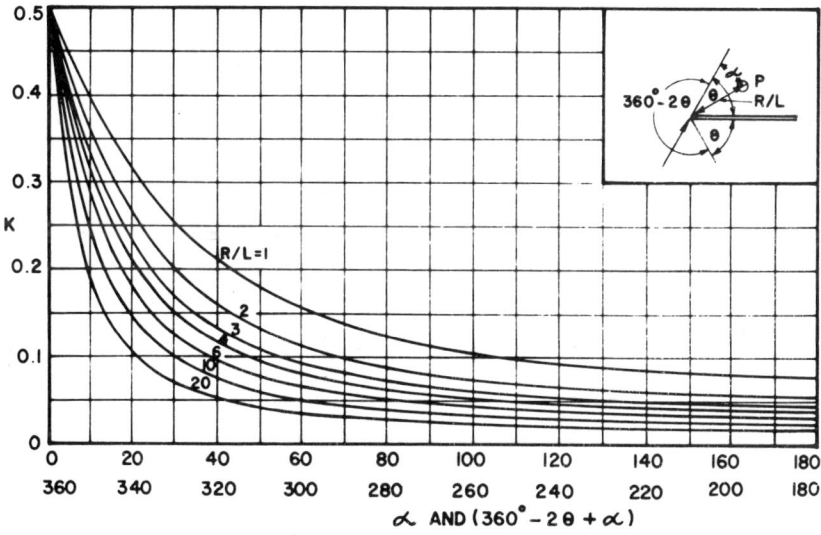

Fig. 7-5. Diffraction coefficient for all incident angles with or without reflection. For $\theta < 60°$ and $R/L < 3$ add 0.1 to K.

TABLE 7-I

Diffraction coefficient for incident wave (α_1) and reflected wave ($360° - 2\theta + \alpha_1$)[1]

$\alpha°$	R/L K × 1000									
	1	2	3	4	5	6	8	10	15	20
0	500	500	500	500	500	500	500	500	500	500
2	476	466	459	453	448	443	435	428	413	402
4	453	435	422	411	402	393	379	367	344	325
6	431	406	388	373	361	350	332	317	288	266
8	411	379	357	340	325	313	292	275	244	222
10	392	355	329	310	294	280	258	241	210	188
12	373	332	304	283	267	253	230	213	182	162
14	356	311	282	260	243	229	207	190	161	141
16	340	292	262	240	222	208	187	170	143	125
18	325	275	244	221	204	191	170	154	128	112
20	310	259	228	205	189	175	155	140	116	101
25	278	225	194	173	157	145	127	115	94	82
30	251	197	168	148	134	123	107	96	79	69
35	228	175	147	129	116	107	93	83	68	59
40	208	157	131	115	103	94	82	73	60	52
45	191	142	118	103	92	84	73	66	54	46
50	176	130	107	93	84	77	66	59	49	42
55	164	120	99	86	77	70	61	54	44	39
60	153	111	91	79	71	65	56	50	41	36
65	143	104	85	74	66	60	52	47	38	33
70	135	97	80	69	62	57	49	44	36	31
75	128	92	75	65	58	53	46	41	34	29
80	122	87	71	62	55	51	44	39	32	28
85	116	83	68	59	53	48	42	37	30	26
90	111	79	65	56	50	46	40	36	29	25
95	107	76	62	54	48	44	38	34	28	24
100	103	73	60	52	46	42	37	33	27	23
105	99	71	58	50	45	41	35	32	26	22
110	96	69	56	49	43	40	34	31	25	22
115	94	67	54	47	42	39	33	30	24	21
120	91	65	53	46	41	38	32	29	24	21
125	89	63	52	45	40	37	32	28	23	20
130	87	62	51	44	39	36	31	28	23	20
135	86	61	50	43	39	35	30	27	22	19
140	84	60	49	42	38	35	30	27	22	19
145	83	59	48	42	37	34	29	26	22	19
150	82	58	48	41	37	34	29	26	21	18
160	80	57	47	40	36	33	29	26	21	18
170	80	56	46	40	36	33	28	25	21	18
180	79	56	46	40	36	32	28	25	21	18

[1] For $\theta < 60°$ and $R/L < 3$, add 0.1 to K.

It is noteworthy that with no reflection the wave height along the shadow line ($\alpha_i = 0°$ in Table 7-I) remains static at 0.5. Also, on the lee-side of the breakwater (where $\alpha_i = \theta$), it is found that the incident and reflection components are each 50% of the total. As already seen above, use of $2\theta - \alpha_i$ in Table 7-I is equivalent to using $2\alpha_i - \alpha_i = \alpha_i$, which is the same value as used for the incident wave. This point is also well illustrated in Fig. 7-4 where $\alpha_i = 30°$. This is significant when the reflected wave might not exist at all due to adequate dissipation on the breakwater.

Fig. 7-5 or Table 7-I is all that is required to determine the height of the diffracted wave at P when the radial distance in terms of wave lengths (R/L) and the deviation from the shadow line (α_i) is known. Where many points, wave periods and wave directions are involved in a study, use of this data may become a little tedious. For this reason some simplifications have been carried out by Silvester and Lim for details of which the reader is directed to the reference [9].

The principle of the concept is the constancy of height reduction along the crest length as measured from the shadow line. Thus, if S/L in Fig. 7-3 were measured around the circular arc for any θ or any R/L, the same height reduction should be found. Using the cumulative procedure outlined above, curves of K versus S/L were drawn for $60° \leq \theta \leq 150°$ and some specific R/L. These curves were very close and so were represented by a single curve. Similar mean traces were obtained for values of $3 \leq R/L \leq 10$, which again could be represented by a single curve. The above simplification can be carried out for full reflection and zero reflection, resulting in

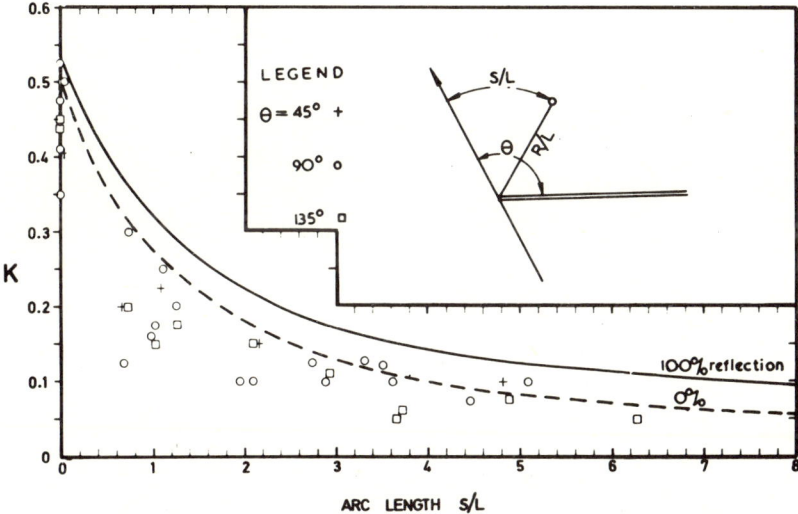

Fig. 7-6. Diffraction coefficient using arc length measurement (S/L) from shadow line. Comparison is made with data from Putman and Arthur [10] where tests excluded reflection. For $\theta < 60°$ and $R/L < 3$ add 0.1 to K.

the curves of Fig. 7-6. The error entering from the averaging of the θ curves and of the R/L curves as above is not optimum at the same S/L values, so they are not strictly cumulative. The resultant error for the ranges of θ and R/L indicated should not exceed 5% and for near-normal wave approach the error should be nearer 3%. These are well within the accuracy of forecasted or recorded values of the incident wave.

To use Fig. 7-6, it is only necessary on the plan of a proposed port to locate the shadow line, and draw an arc from point P to it, centered on the breakwater tip. With a pair of dividers measure this arc length in terms of wave lengths, for the period being considered. The diffraction coefficient is then read directly from Fig. 7-6 for full, zero, or part reflection. When a preliminary study is so carried out, to obtain the general picture of wave energy, more specific layouts can be examined by the use of Fig. 7-5 or Table 7-I.

Putman and Arthur [10] conducted experiments which avoided reflection from the breakwater. As seen in Fig. 7-6, these agreed well with the zero reflection curve and were substantially below the normal theory which implies 100% reflection. Tests by Lim [4] also conveniently omitted the uncontrollable variable of reflection, by the layout in Fig. 7-7, and again displayed the validity of dividing the coefficient into two components. For the case of $\theta < 60°$ and $R/L < 3$ it was found from the experiments that 0.1 should be added to the values of K evaluated by Fig. 7-5 and 7-6 or Table 7-I. The previous comparison of wave diffraction with the dam-burst problem may help explain this deviation from the theory. When θ is

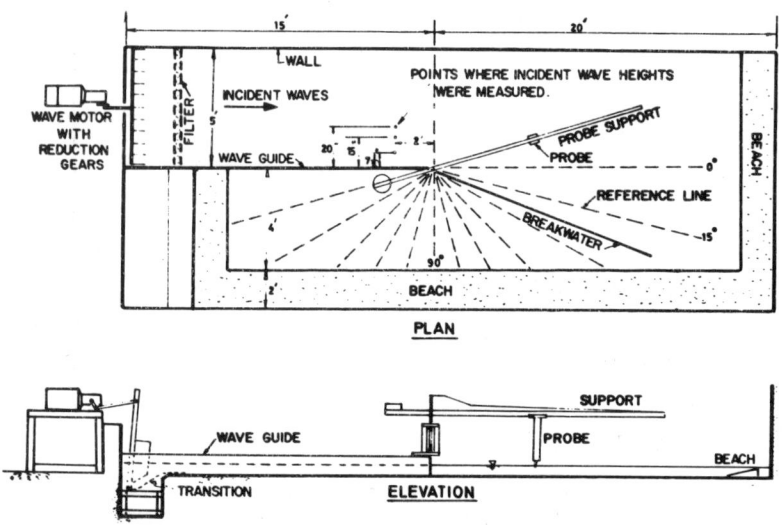

Fig. 7-7. Wave basin used by Lim [4] to eliminate reflection at the breakwater.

small the wave has insufficient room to spread normally, so the crest-length profile does not assume its maximum steepness. This is similar to the dam moving too slowly for the normal dam burst profile to develop in Fig. 7-2.

Breakwater gap

Where two breakwaters are aligned as in Fig. 7-1B, and fairly full reflection is realized from both, the waves in each shadow zone are comprised of the incident wave and the two reflected waves. Since the crest curvature of one of these reflected waves is not centered on the alternate breakwater tip, the resultant wave height measured along the shadow lines, or any other radii α (see inset Fig. 7-8), will fluctuate about the smooth curve of the semi-infinite breakwater solution. These deviations increase as the gap width decreases, as illustrated in Fig. 7-8. The previous solution can be used without great loss of accuracy down to a gap width of $B = 5 L$. Where there is no reflection from the breakwater such undulations as in Fig. 7-8 are not present.

One solution to this problem is the double application of Sommerfeld's analysis for the semi-infinite breakwater [2]. It was found [4, 9] that by graphing values of K against R/L from Table 7-I, that the wave height reduction was proportional to $(R/L)^{1/2}$, which suggested a parameter $K\sqrt{R/L}$ for combining radial and arc influences. It was also convenient to centre the polar coordinate at the mid-point of the breakwater gap. In the knowledge that for $R/B > 5$ the value of $K\sqrt{R/L}$ is essentially constant for any α, a simple series of graphs can be made to represent conditions

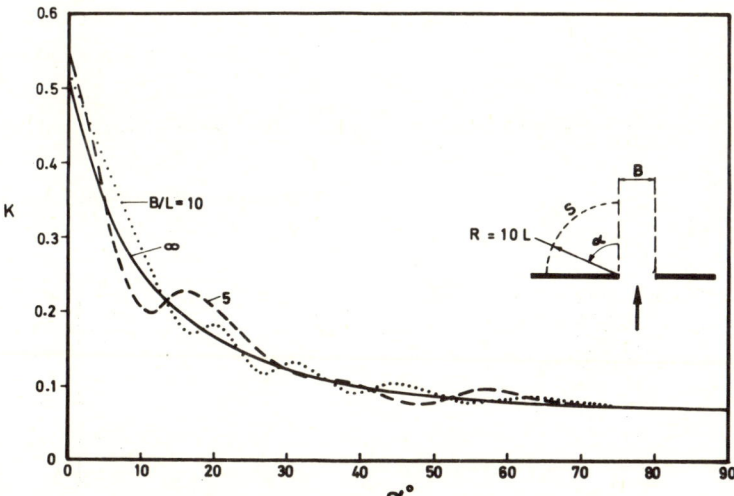

Fig. 7-8. Variation of diffraction coefficient for various breakwater gaps.

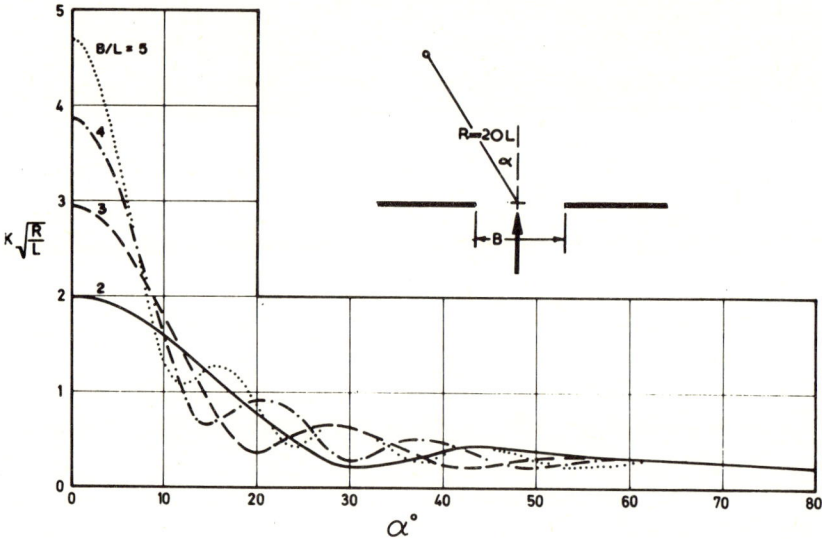

Fig. 7-9. Diffraction coefficient for breakwater-gap widths $B \leq 5L$.

anywhere in the protected basin. An example of this is given in Fig. 7-9 for $R/L = 20$, the largest probable radius to be encompassed. In the absence of any reflection the fluctuations so displayed would not occur, so that averaging them over a range of R/L should not involve undue error in a prototype situation. Fig. 7-10 provides

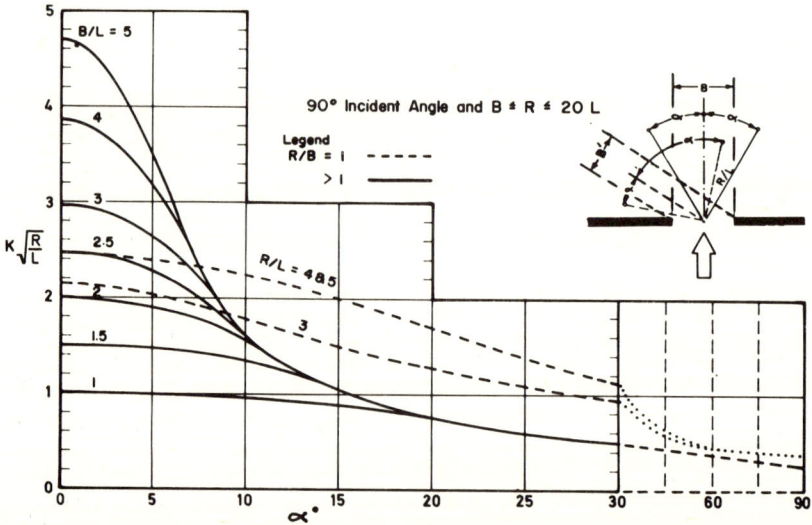

Fig. 7-10. Diffraction coefficient for breakwater gaps from Penney and Price [2] solution.

WAVE DIFFRACTION

the result of the above averaging process for $\theta = 90°$ and $B/L \leq R/L \leq 20$, in which curves are grouped into two categories: $R/B = 1$ and $R/B > 1$. For gaps smaller than $2L$ the single curve (full line) in Fig. 7-10 represents both cases of $R/B \geq 1$.

The above simplifications lead to a maximum error in $K\sqrt{R/L}$ of ± 0.3 at the maxima and minima of the undulations (see Fig. 7-9). The average deviation is in the order of ± 0.2. Since reflection is likely to be much less than 100%, these errors will be greatly reduced. Although Fig. 7-10 applies only to $\theta = 90°$, other incident angles can be treated by the method suggested by Blue and Johnson [11] in which

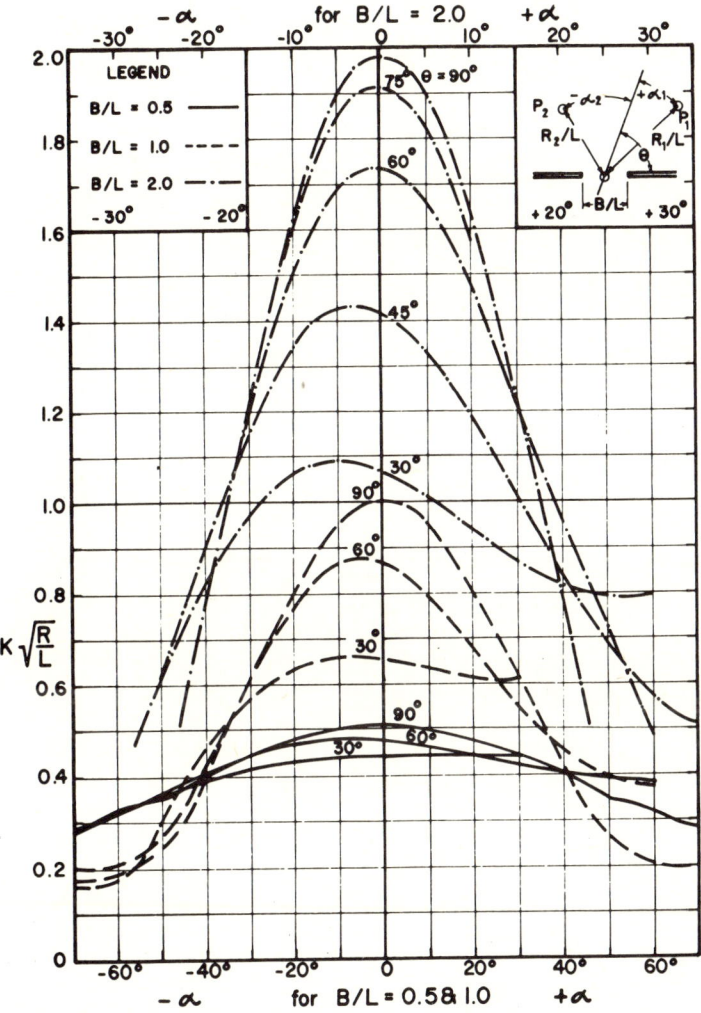

Fig. 7-11. Diffraction coefficient for breakwater gaps from Morse and Rubenstein [12, 13] solution.

the equivalent width B' is used in place of B and arc angles α are measured from the oblique centre-line.

For gap widths $B \leqslant 3L$ an exact solution in optics has been derived by Morse and Rubenstein [12], and applied to water waves by Carr and Stelzriede [13], to

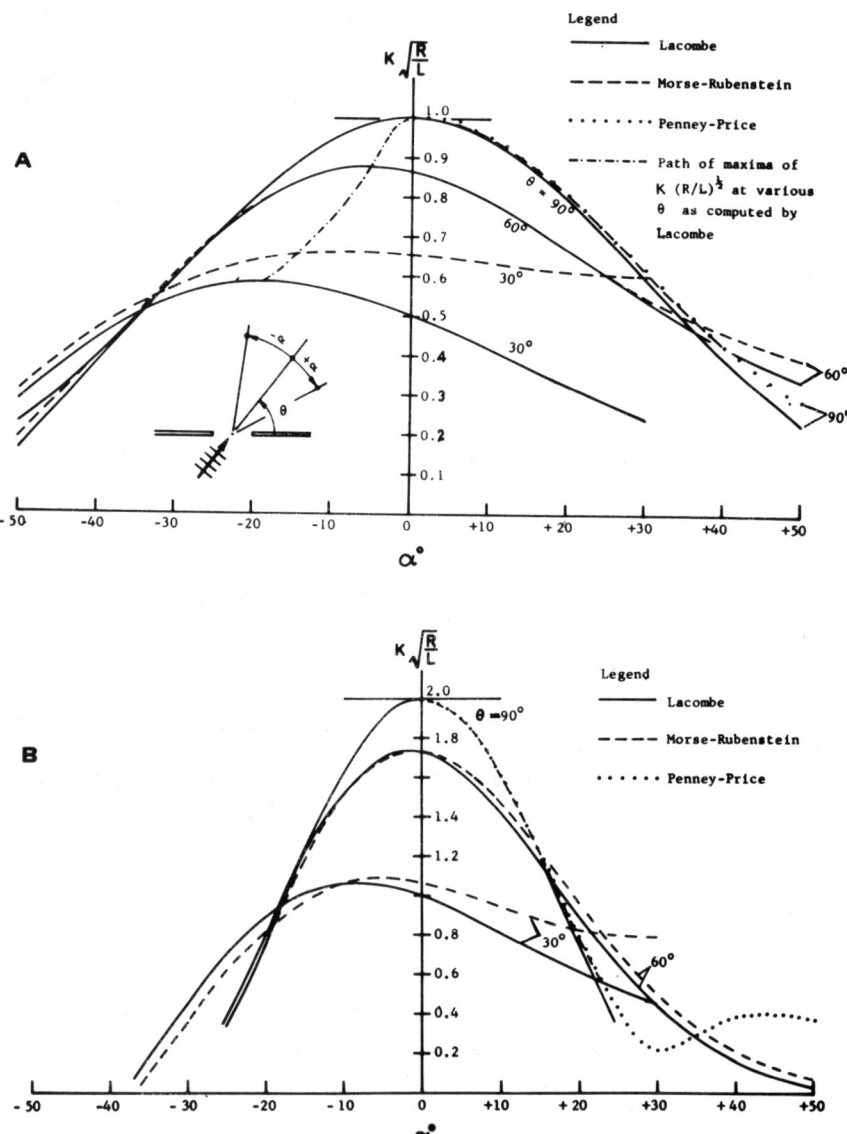

Fig. 7-12. Comparison of theoretical solutions for diffraction through breakwater gaps. A. One wave length wide ($B = L$); B. Two wave lengths wide ($B = 2L$).

which the reader is referred for the relevant equations. Using the parameter $K\sqrt{R/L}$ graphs of $B/L = 0.5$, 1.0 and 2.0 are presented in Fig. 7-11 for $\theta = 30°$, $60°$ and $90°$, in zones where $R > B$.

Lacombe [14] has derived an approximate solution which is based upon a polar coordinate system centered on the mid-gap point. It applies to $R > B$ and gives the maxima values in the fluctuations previously discussed (see Fig. 7-9).

For the smaller gap widths ($B \leqslant 2L$) a direct comparison of the three solutions is possible. As seen in Fig. 7-12A and B, for $B = L$ and $B = 2L$, respectively, all solutions are very close except where $\theta = 30°$, which is an unlikely design condition.

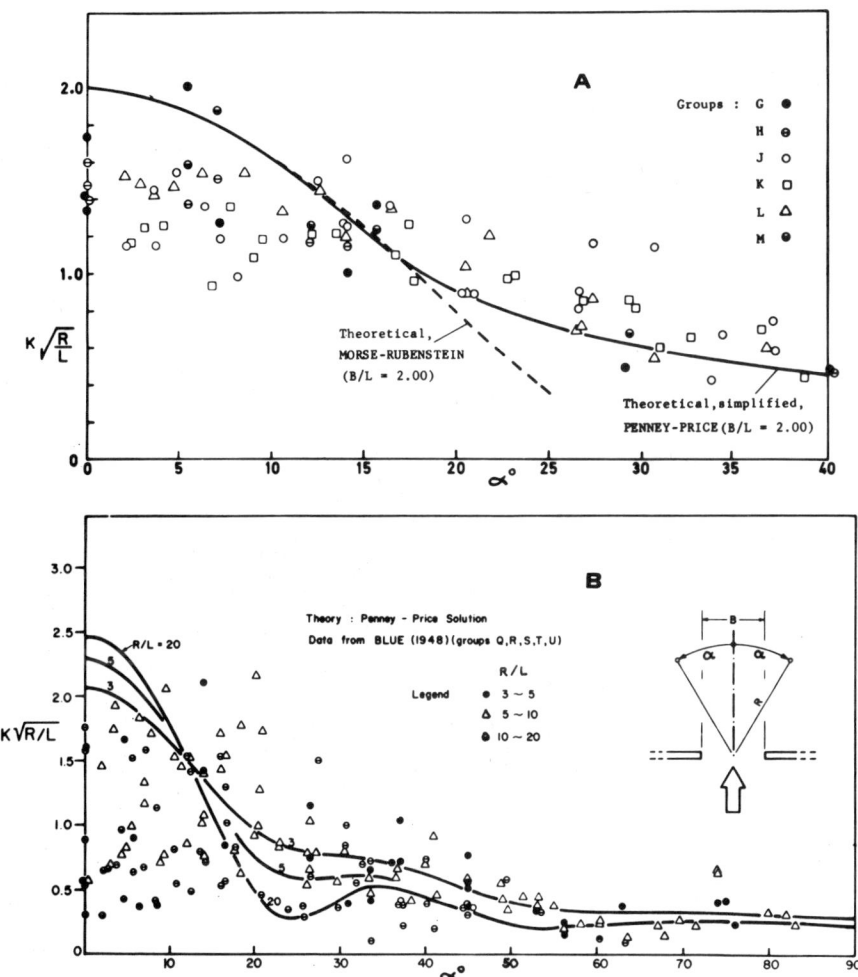

Fig. 7-13. Experimental results from Blue [15] from which group designations are taken. A. For $B/L = 1.95$ and $R/L = 4-10$; B. For $B/L = 2.5$ and $R/L = 3-20$.

The Morse-Rubenstein solution should be used in this case in preference to that of Lacombe because of its conservative tendencies.

Experimental verification of the theory for diffraction through a breakwater gap has not been extensive. The major effort in this regard has been by Blue [15], whose measurements on a square grid system had to be converted to the polar coordinate system for comparison [4]. Only those results could be used, therefore, which approximate the B/L value for the theory. The groups noted in Fig. 7-13A, B are according to the notation by Blue. It is seen that, apart from the averaging that had to be carried out, the scatter precludes any conclusions regarding the validity of the theories, even with or without reflection. The trend is certainly present, but further research is indicated. In order to exclude reflection as a variable, tests could be conducted similar to those by Lim [4] as illustrated in Fig. 7-7, in which waves would essentially pass from a channel into an enlarging basin, around which good absorption beaches are installed. No drastic differences from the semi-infinite break-water tests [9] should be expected, since the only difference would be the limited crest length from which diffraction energy is supplied.

Diffraction of wave spectrum

The analysis above can be applied very effectively to swell waves in which a single period at a time can be assessed. For many large harbours this is the nature of the problem, since attenuation of the most persistent waves is necessary for normal marine activities. When a storm is brewing all large vessels will leave port and ride it out at sea, so that wave disturbance in the harbour area is not a criterion of design at such times. However, with harbours of refuge for fishing and other small craft, protection is required from storm waves. It is necessary therefore to understand the influence of diffraction on either a one-dimensional spectrum or a directional spectrum of waves.

Consider first that all the components are arriving from the same direction, either in the alignment of the wind or at some angle produced by refraction across the continental shelf. This supposition is not far from reality since the largest waves of the spectrum, concentrated around T_{max}, will all be travelling in essentially the same direction. Those at the lower-period end of the spectrum will be the most oblique within the fetch and will be the least affected by refraction.

Thus, with all waves approaching along the same orthogonal through the breakwater tip, the arc S/L to a point P in the shadow zone from this shadow line will have different values for different wave periods. For longer waves it will be smaller and hence K from Fig. 7-6 is greater, whereas shorter waves will be attenuated more. This will distort the spectrum received, in the same manner as that recorded at the bed after selective filtration through pressure attenuation. It is because of such possible distortion that all facts should be published about the siting of recorders in semi-protected places.

WAVE DIFFRACTION

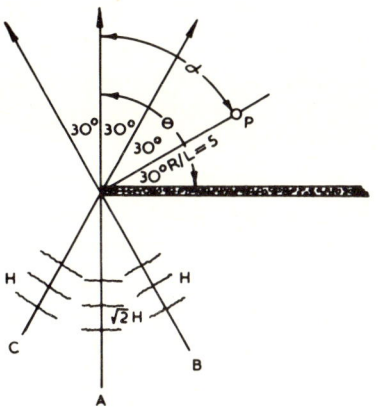

Fig. 7-14. Diffraction of spectral component whose energy is divided evenly each side of resultant orthogonal.

Now consider a single component of the spectrum whose energy, instead of being concentrated on one alignment, is considered to be divided evenly into two major directions slightly oblique to the first. In Fig. 7-14 such a situation is depicted with the two directional elements having height H, so that their resultant with one common approach should be height $\sqrt{2}H$. Using the angles shown in Fig. 7-14, the following values are obtained using Table 7-I:

Wave A: $\theta = 90°, \alpha = 60°, 2\theta - \alpha = 120°$
          $K = 0.071 + 0.041 = 0.112$, height $= \sqrt{2} \times 0.112H = 0.159\,H$

Wave B: $\theta = 120°, \alpha = 90°, 2\theta - \alpha = 150°$
          $K = 0.050 + 0.037 = 0.087$, height $= 0.087\,H$

Wave C: $\theta = 60°, \alpha = 30°, 2\theta - \alpha = 90°$
          $K = 0.134 + 0.050 = 0.184$, height $= 0.184\,H$.

Since the waves from the two directions are not necessarily in phase their energies are cumulative so that the maximum height at $P = (0.087^2 + 0.184^2)^{1/2}H = 0.2\,H$, which is 26% greater than for the single orthogonal A. It should be realized that the deviation angle of 30° is extreme, especially for waves near a coast where most computations will apply, but the example emphasizes the need for considering the directional spread of wave energy. It could indicate the need for some secondary structure, even remote from the harbour itself, to cut off the approach of some portion of this wide-angled energy.

Mobarek and Wiegel [16] have conducted experiments on locally generated waves and measured the spectra at points in the shadow zone. They found a strong skewing tendency of the spectrum to longer-period components. It should be appreciated that the smaller-period waves will be dissipated more readily on the breakwater than the longer ones, so that different percentages of the reflection component should be used in computations.

Combined diffraction and refraction

Many large harbours are dredged to some minimum depth in order that vessels cannot touch bottom under any planned or even unplanned manoeuvre. Thus the assumption of uniform depth within the shadow zone is normally correct. However, it is possible for only a portion of a protected basin to be so treated and wave heights in the remainder might need to be assessed.

Two simple cases are depicted in Fig. 7-15 where the wave crests and orthogonals are shown diagrammatically. The latter are seen to be curved instead of straight as was the case with uniform depth. Any radial from the tip of the breakwater can be traced by refraction procedure across the bed. Studying a point P on the straight radius it is seen that the orthogonal deviates from it, either increasing the S/L distance as in Fig. 7-15A, or decreasing it as in 7-15B. These variations in S/L will decrease or increase K respectively. A further modification to the wave height is due to the shoaling that has taken place. Should the bed contours be circular arcs concentric on the tip of the breakwater, the diffracting orthogonals will be straight, but allowance must be made in the wave height for shoaling.

Blue and Johnson [11] have suggested diffracting waves to five wave lengths and

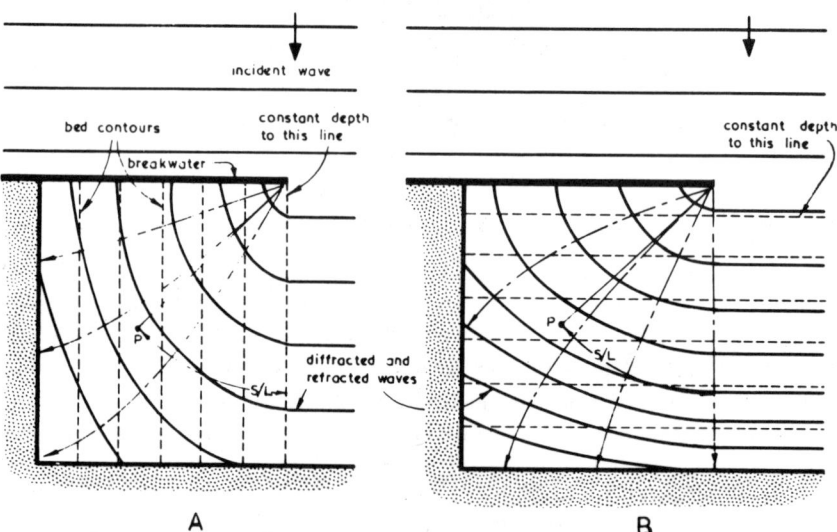

Fig. 7-15. Wave crests and orthogonals subject to diffraction and refraction.

then refracting them onwards into the harbour basin. Mobarek [17] has attempted a combined diffraction and refraction procedure for simple bed contours. Fan and Borgman [18] have constructed a computer model for diffraction which can cope with directional spectra. It can serve for prototype calculations or analysis of experimental data.

A final comment on diffracted waves in harbours is the reminder that if the port basin is large enough wave generation within it can be significant. Build up of waves already present can be much more rapid than with an initially calm sea. Zhukovets [19] has conducted experiments on this topic and presented some relationships, which are difficult to apply since the diffracted wave height on which the wind acts is changing spatially, as is the angle of wind attack on the crest. Since the conditions in which wind blows across the harbour basin will be bringing a wide spectrum of waves to be diffracted, the computation of the initial conditions is extremely tedious to compute. But even so, the energy generated in a reasonably sized enclosed body water should be added to that entering the mouth and a conservative judgment made of the addition of these two effects.

LAND-BACKED STRUCTURES

This class of marine structure may consist of a relatively smooth surface, such as concrete or stone pitching, or be very rough, such as a rock-fill structure in which units are dumped randomly. The latter will necessarily be permeable to the oscillatory flow of water from the wave action and hence dissipate energy strongly, but formed concrete surfaces may also be perforated in order to provide a similar effect, in which water pours into a secondary balancing chamber.

Besides the slope of the seaward face of a structure an important feature is the depth of water at the toe. It will be presumed that the bed from this point is relatively flat, sufficiently so that normal theory can be applied to derive characteristics of the wave prior to its travel up the steeper slope of the structure. This toe depth is extremely important as it dictates the approach celerity of the wave and hence the orbital velocity of the water particles on impact.

The interest here is in the action on the waves by way of reflection or dissipation. These, in turn, will determine the reach of the wave up the slope. If this run-up exceeds the height of the structure above SWL a certain degree of overtopping will take place, which itself influences the degree of overtopping will take place, which itself influences the degree of reflection. The volume of water thus thrown over a structure may be of concern. The forces involved in all these processes will be treated in Chapter 8.

Reflection and dissipation

The distortions occurring as a wave reaches a slope and ultimately breaks and reflects preclude the application of theory to this phenomenon. Resort must be made to flume tests, which involve the task of measuring partial clapotis. As seen previously, these consist of nodal and antinodal zones spaced at odd and even multiples of $L/4$ respectively from the vertical wall or some SWL alignment in the case of a sloping wall (see Fig. 7-16). The method used by Healy [20] is to measure H_{max} at the antinode and H_{min} at the node and derive H'_i, H'_r and K'_r as follows:

$$K'_r = \frac{H'_r}{H'_i} = \frac{H_{max} - H_{min}}{H_{max} + H_{min}} \tag{7-5}$$

where the prime is used to designate the apparent values. Because of the finite height normally existing in these circumstances second-order effects arise, the main one of which is a secondary crest at the nodal point. This will make the measurement of H_{min} larger than it should be, so reducing the reflection coefficient K'_r below the true value K_r.

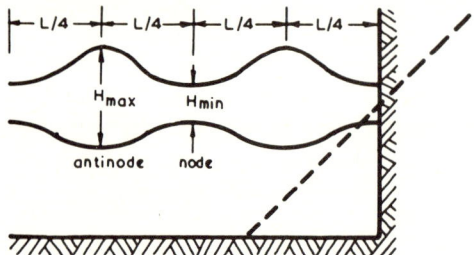

Fig. 7-16. Definition sketch for waves striking a wall.

For this reason Carry [21] applied second-order theory to derive a correction factor to be applied to K'_r, and in a similar way Goda and Abe [22] applied Stokes' third-order theory for the same purpose. This correct coefficient K_r can be obtained from the graphs in Fig. 7-17 which have been derived from those provided by Goda and Abe. By using eq. 7-5, after measuring $H_{max} + H_{min}$ and $H_{max} - H_{min}$, H'_i and K'_r are evaluated, from which K_r is read in Fig. 7-17 for the specific values of d/L operating at the toe of the slope and the wave steepness H'_i/L. For this purpose it is presumed that the wave length is that for linear theory. The correct wave steepness H_i/L can be read from the curved ordinates or from the ratio $H'_i/H_i = (1 + K_r)/(1 + K'_r)$.

It is seen in Fig. 7-17 that, where the apparent reflection coefficient K'_r is less than 1, the correct value K_r could well be unity. Miché [23] derived the theoretical

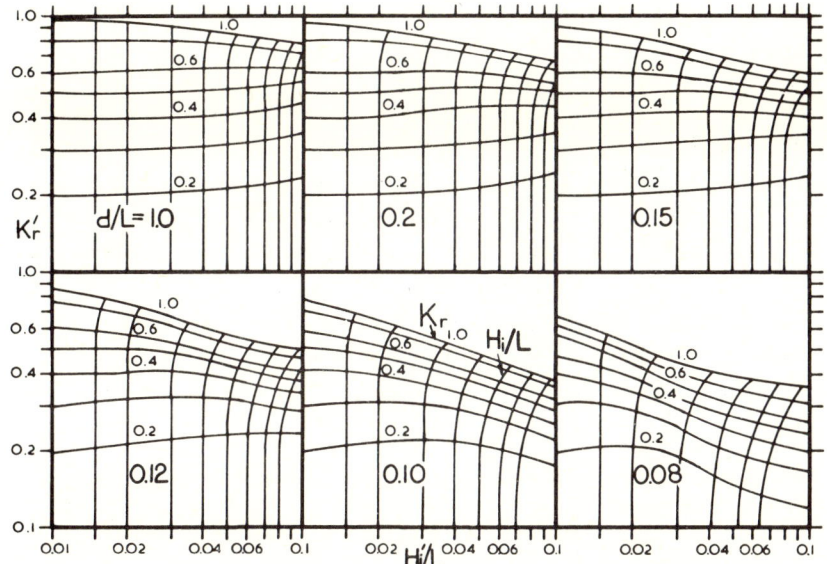

Fig. 7-17. Corrected reflection coefficient K_r from apparent coefficient K_r'.

maximum steepness of a wave which could be completely reflected from a slope as:

$$\left(\frac{H}{L}\right)_{max} = \sqrt{\frac{2\alpha}{\pi}} \frac{\sin^2 \alpha}{\pi} \tag{7-6}$$

the theoretical reflection coefficient being:

$$K_r = (H/L)_{max}/(H_o/L_o) \tag{7-7}$$

when $K_r < 1$. If K_r as calculated exceeds unity the limiting value of 1.0 should be used.

Eq. 7-6 and 7-7 have been graphed in Fig. 7-18, together with corrected experimental coefficients K_r from Moraes [24] and Straub et al. [25] for smooth impermeable slopes. It is seen that the test results are slightly less than the theoretical, especially at small H_o/L_o, although with such waves the percentage accuracy of wave measurements is decreased.

The influence of roughness on reflection has been studied by Moraes who found that the flatter the slope the greater the influence of roughness, measured in terms of stone dimension B to deep-water wave length (B/L_o). Results for the 30% slope have been interpolated in Fig. 7-19 from Moraes' results for three roughnesses, together with curves for determining B/L_o readily. It is seen that the greatest reduction occurs for the flatter waves which have more horizontal orbital motion per unit height than the steeper waves. The vertical wall exhibited no influence of roughness [24].

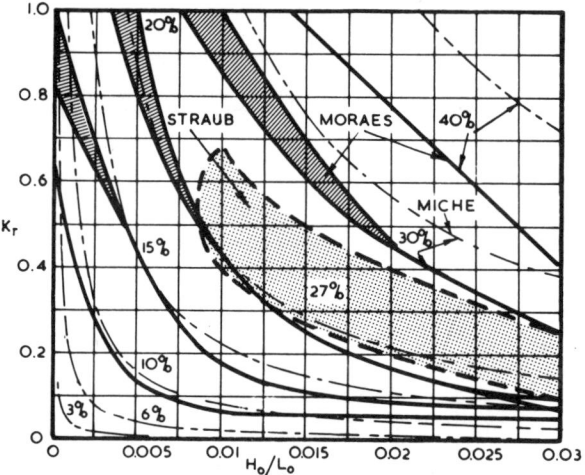

Fig. 7-18. Theoretical and experimental values of K_r for smooth impermable slopes.

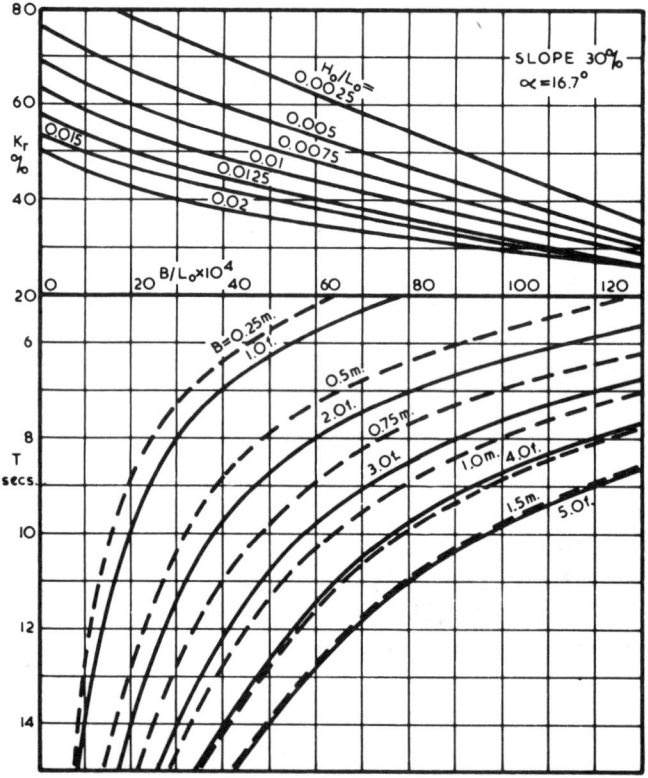

Fig. 7-19. Influence of roughness dimension B on K_r for 30% slope.

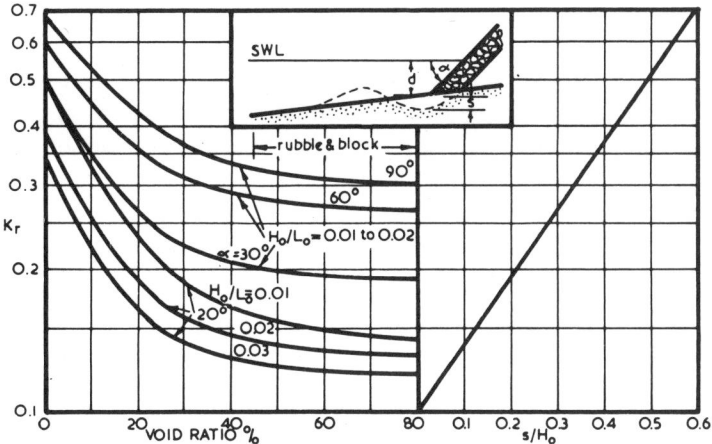

Fig. 7-20. Influence of porosity on K_r and scour resulting from wave reflection.

The effect of permeability has been studied by Sawaragi [26], who used perforated plate for smaller void ratios and model blocks for larger ratios. He also included the results of Straub et al. [25] in his graphs, from which Fig. 7-20 has been prepared by averaging some of the results. Void ratios for rubble-mound structures are around 45%, whilst those for the various fabricated blocks are generally greater than this. Inspection of Fig. 7-20 indicates that for all such ratios (beyond 45%) the minimal reflection has been reached, so that choice of block is not paramount on this score.

Comparison of 20° curves for $H_o/L_o = 0.01$ and 0.02 in Fig. 7-20 with those for Fig. 7-19 where slope 30% is equivalent to $\alpha = 16.8°$, shows that for void ratio of 0% the equivalent roughness B/L_o was $35 \cdot 10^{-4}$. Fig. 7-20 combines the influence of roughness and permeability, whereas Fig. 7-19 applies to concrete slopes or stone pitching where roughness elements are only incorporated to reduce reflection.

Sawaragi [26] also examined the amount of scour at the toe of permeable breakwaters. Although this is a topic of coastal defense it is worthy of note here, since it is closely allied with degree of reflection. The line included in Fig. 7-20 is the resultant curve derived by Sawaragi for an initial sedimentary slope of 1:15. For an impermeable vertical wall it is seen that an allowance of $S/H_o = 0.6$ must be made for scour when waves are reflecting normally to it. In the case of an extremity to a structure, where waves can pass around and behind it, extra allowance must be made for the vortices generated there, which can scour much bigger holes. Also the reflection of oblique waves can expedite removal of material in front of the wall which is a separate issue from the equilibrium scour indicated in Fig. 7-20.

324   EFFECT OF STRUCTURES

Run-up

The reach of waves up the slope of marine structure from the SWL is closely associated with the standing wave developed. This in turn depends upon the depth of water at the toe of the wall or dike; and, when this is small, on the slope of the bed leading to the wall.

Tominaga et al. [27] have conducted flume tests on impermeable walls of various slopes and two different bed profiles. From their series of graphs from experimental data the composite curves of Fig. 7-21 have been compiled. The averaging so carried out is probably well within the scatter of the original data which was not available.

Two major wave conditions exist, that of non-breaking and that of breaking. Of the latter, two subdivisions occur, of waves breaking prior to reaching the dike and those breaking on the slope after having traversed the toe. The breaking on the gently sloped bed seaward of the wall has been dealt with in Chapter 5. The breaker height at various depths within the surf zone is given in Fig. 5-21. However, when the SWL passes through the intersection of the bed and the dike face (i.e., at the toe) there is still a surge height given by the wave set-up. This ΔH has a specific value at the breaker point [28] and decreases as the depth decreases, having a finite value at the toe. Experiments [27] showed the run-up to be 3.5 times the ΔH at the toe, which results in the curves of Fig. 7-22 for the case of vertical and near-vertical walls, and a range of bed slope ($1/m$). As H_o/L_o increases so the ratio R/H_o

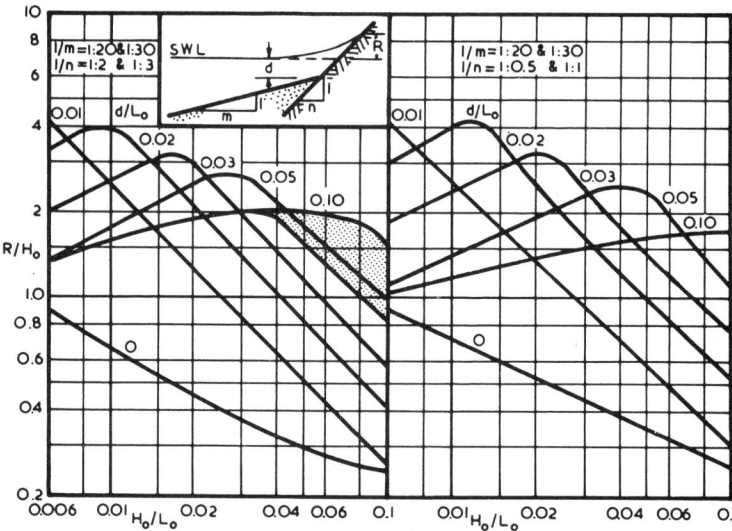

Fig. 7-21. Run-up on smooth sloping walls with differing approach grades.

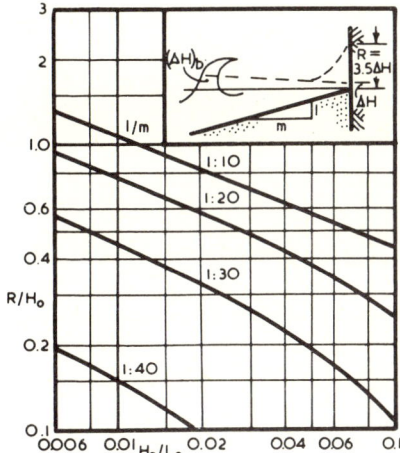

Fig. 7-22. Run-up on a vertical wall when depth is zero at wall.

decreases since the wave breaks further from the wall and loses much of its energy before reaching it. The gentler bed slopes also decrease the run-up height due to the added friction in the wider surf-zone.

Where the dike slope is decreased the possibilities of standing waves and breaking waves produce the maxima as shown in Fig. 7-21. On the LHS of these maxima reflection is producing the run-up, whilst on the RHS breaking is taking place. Where the wave is at maximum steepness and ready to break, maximum run-up is experienced. Further curves are provided for $d/L_o = 0$ similar to Fig. 7-22, but with the smaller $1/n$ ratios the bed slope $1/m$ does not have such a great influence and in fact has been ignored in Fig. 7-21.

The maxima in Fig. 7-21 can be joined by a curve giving $(R/H_o)_{max}$ which is independent of d/L_o. This is shown in Fig. 7-23, in which are also depicted d/L_o values at which these maxima occur for the slope of 1 : 2 as derived from tests by Sato and Kishi [29]. It is seen that longer waves can run-up 2 or 3 times further than shorter waves for the same deep-water wave height. The steeper waves require a greater toe depth to produce maximum run-up. Sato and Kishi note that run-up for a 1 : 2 sloping wall is greater than for a vertical wall.

Where waves arrive at an angle θ to a dike, the run-up is reduced by a factor K_θ as given in Table 7-II, compiled from test results by Hosoi and Shuto [30]. These were obtained for a dike slope of 1 : 2, but in the absence of further data they may be applied to other slopes as given in Fig. 7-21. The rapid transit of sediment in front of such a wall is discussed in *Coastal Engineering II* (Chapter 3).

The influence of slope roughness on wave run-up is presented in a comprehensive report by Machemehl and Herbich [31], to which the reader is referred for an

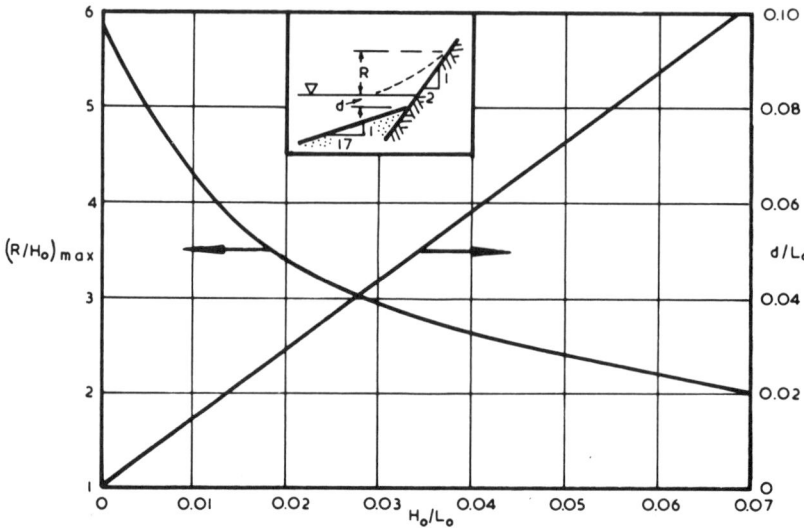

Fig. 7-23. Values of $(R/H_o)_{max}$ at various H_o/L_o and d/L_o for 1 : 2 wall slope.

extensive bibliography. For the slope of 1:3/2 (about 30° or 67%) values of R/H_o were reduced about 25% from the smooth face value. Previously Savage [32] had found that roughness and permeability decreased run-up for increasing wave steepness and flatter slopes.

The composite slopes tested consisted of an approach berm of various lengths and depths. Results were presented in terms of energy density levels for both monochromatic waves and a spectrum provided by generating wind waves in the flume. No significant difference was found [31] between monochromatic and spectral wave conditions for run-up on either the plain or composite slope.

Delft tests [33] with a spectrum of waves indicated a higher run-up for the wider spectra. A relationship based upon that by Hunt [34] for monochromatic

TABLE 7-II

Values of K_θ for run-up of waves angled θ to dike slope of 1:2

H_o/L_o	0.01	0.015	0.02	0.025	0.03	0.04	0.05
$\theta° = 10$	0.97	0.97	0.97	0.97	0.97	0.97	0.97
20	0.94	0.94	0.94	0.94	0.94	0.94	0.94
30	0.90	0.91	0.91	0.90	0.90	0.89	0.88
40	0.88	0.88	0.89	0.87	0.86	0.85	0.84
50	0.80	0.79	0.78	0.75	0.73	0.70	0.65
60	0.77	0.72	0.62	0.54	0.48	0.41	0.34

LAND-BACKED STRUCTURES

waves was found satisfactory in the prediction of run-up on 1 : 4 and 1 : 6 smooth impermeable slopes. For want of further evidence it should be used with caution for other slopes:

$$R = C \tan \alpha \, H_{1/3} \, (gT_{max}^2/2\pi H_{1/3})^{1/2} \tag{7-8}$$

where α is the wall slope, and C is a factor varying with spectral width ϵ, which for a near-shore FAS could be taken as 0.6, giving $C = 0.7$. This factor also is dependent upon the frequency of exceedance and the value quoted is based upon 2% of the waves exceeding the run-up computed by eq. 7-8. For a narrow spectrum it should be reduced by a factor of 0.85. The ratio of $(H_{1/3})^{1/2}/T_{max}$ can be obtained from Fig. 5-22 for a given d/L_o at the shoreline.

Overtopping

When the run-up of a wave exceeds the height of a marine structure overtopping will occur. This can produce serious results by way of flooding facilities landward of the revetment, or destruction of the leeward face of a breakwater where smaller armour blocks are generally used. It is necessary, therefore, to be able to assess the quantity of water passing over the top of a structure per wave period, or averaged over time. In respect to damage of breakwaters the velocity of such water is also significant.

The run-up computed previously is no longer applicable since the overtopping

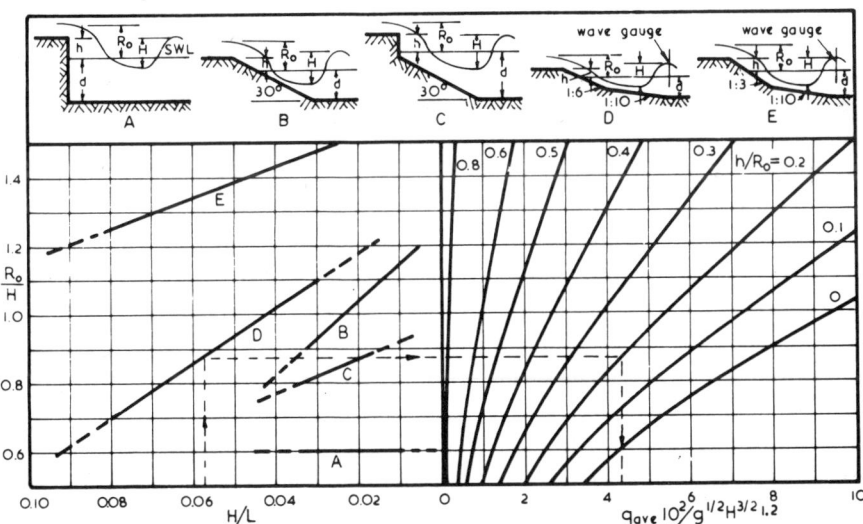

Fig. 7-24. Average overtopping discharge q_{ave} per unit length of walls illustrated.

reduces the reflection and hence the extent of the standing wave. The definition sketch of Fig. 7-24 indicates the variety of variables that can enter the problem. Many workers have carried out flume tests [35, 36, 37, 38] whilst others have endeavoured to derive a theory based upon a weir approach [39, 40]. Many experimental data are expressed in terms of deep-water wave heights and lengths, which involve a d/L ratio at the wall to relate these values to those in its vicinity. It would seem more appropriate to report tests in wave characteristics at the toe of the slope.

The analysis by Kikkawa et al. [39] uses local wave dimensions and treats the overtopping as flow over a broad-crested weir for a changing upstream height. Their analysis appears to agree well with the test data available [39, 40], in spite of the assumption of a triangular shape for the wave crest rather than a sinusoidal or cnoidal profile. By the time the wave reaches the crest of a dike it will either be a standing wave or be breaking. In either case the crest shape should be close to triangular. The equation so derived [40] can be put in the form:

$$\frac{q_{ave}}{g^{1/2}H^{3/2}} = \sqrt{2}\,\frac{2}{15}m\left(\frac{R_o}{H}\right)^{3/2}(1-h/R_o)^{5/2} \qquad (7\text{-}9)$$

where q_{ave} is the average discharge over the weir per unit length of dike, m is the discharge coefficient for flow over the weir, R_o is the maximum reach of the overtopping wave above SWL and h the height of the dike above SWL.

Eq. 7-9 has been plotted in Fig. 7-24 for $m = 0.6$ [40]. Also included are curves for R_o/H from the various dikes illustrated [35–38]. The volume V_T discharging over a length B of the dike in a wave period T is given by:

$$V_T = q_{ave}\,TB \qquad (7\text{-}10)$$

with an average velocity $v = 2q_{ave}/(R_o-h)$, assuming the discharge to take place as a rectangular block for half the wave period. Since the overtopping water body has been considered of triangular cross-section this velocity should be doubled, but verification of such figures should be made in the laboratory.

Tsuruta and Goda [41] have used eq. 7-9 but modified it to obtain other dimensionless parameters, both for monochromatic waves and a spectrum. By using the fourth-order theory for standing waves and normal shoaling relationships they derived the limiting ratio $(H_o/d)_c$ at which no overtopping would take place for a given h/d value. This is sketched in Fig. 7-25.

Another approach by Shiraishi et al. [42] relates the overtopping discharge to the quantity of water moving shorewards in deep water. This involves the problem of relating H_o or $(H_{1/3})_o$ to possible equivalents close to the structure.

Curves by Tsuruta and Goda [41] showed an optimum overtopping when $H_o/d \doteq 1.0$ to 1.3 or $H_{1/3}/d \doteq 0.6$ to 0.8. This will be when the wave is just breaking at the toe of the major slope. At this time the water at the crest is moving forward at about the speed of the wave or $\sqrt{gd}$. This is retarded to about 70% [31]

LAND-BACKED STRUCTURES

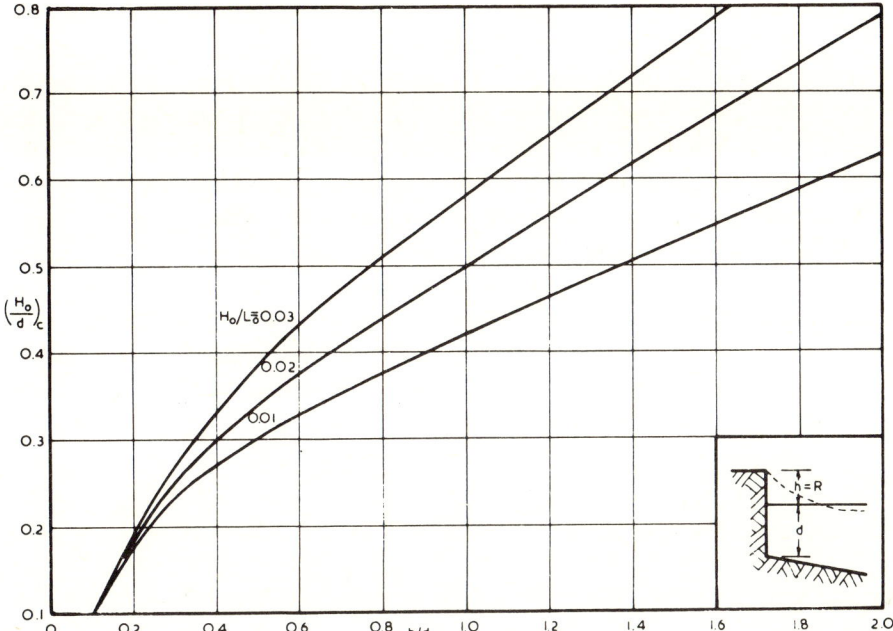

Fig. 7-25. Limiting wave height for no overtopping.

of this velocity due to friction on the slope and the transfer of momentum vertically.

The effect of wind has been studied in wind-wave flumes [43], but the results are in terms of $U/\sqrt{gH_o}$ and therefore difficult to apply. It appears that when d/L_o is large increased overtopping is produced by a following wind, probably due to the larger heights possible just before reaching the dike. With shallower conditions more breaking takes place in the offshore area. Allowance should be made for double the discharge when a wind of 30 knots is blowing shorewards. The maximum $H_{1/3}$ under such storm conditions is given by Fig. 5-22 for the depth in front of the revetment.

Where waves are oblique to a structure the bore from breaking is angled to the dike crest so that overtopping is reduced [35]. Also the longshore component of this velocity will interfere with the breaking of other waves so that the discharge is lessened greatly overall. However, this does not preclude the spilling of large volumes where the wave crest meets the wall, but this discharge traverses the length of the dike.

Seiching

The term "seiching" is normally applied to long-term oscillations in an enclosed body of water, as may be caused by a transient application of wind stress. The standing wave oscillations in a harbour due to the reflection of waves is then called surging, but this can equally be confused with long-period oscillations produced by storm surge or tsunamis. Hence the patterns of standing waves inside port basins will be termed here as harbour seiching.

As noted already, the long-period low waves are readily reflected from revetments, even where these are of rubble mound construction and mild in slope. So that, when waves enter a harbour, perhaps through a breakwater gap, they can reflect from structures or beaches within the harbour basin. If these reflected waves are again reflected from an opposite wall or boundary, standing oscillations can be formed which can generate large horizontal orbital motions of the water particles at the nodes, even though the resultant wave height at the antinodes is negligible. These horizontal oscillations can cause ships at a berth to range back and forth, so straining their moorings dangerously and making loading operations hazardous.

McNown [44] has studied the amplification of waves entering a circular port through a small opening in the circumference. Various standing-wave patterns can be excited by resonance, including a sloshing side to side, crests concentric with the harbour, nodal lines radial at 60°, and concurrent circular and radial nodes. He also observed non-resonant oscillations that can be established and commented on the instability of seiching within a harbour, even for the simplified shape of a circle. In model work the slightest change in wave period (thousandths of a second) can materially change the pattern [45].

It is instructive to consider some simple port plans such as circular, elliptical and rectangular, into which more complex shapes may be equated for an initial attack on a seiche problem. The criteria for employing the theory developed are:

(1) The bed is level throughout, or steps accounted for in the analysis.

(2) Shallow-water waves mainly are considered, so that $L = \sqrt{gd}T$ (or $d/L < 0.04$). It is possible for standing waves to be set up when their celerity is that for transitional water.

(3) All boundaries have 100% reflection capacity, except steps in the bed, where due allowance is made.

In the following discussion no attempt is made to provide relationships for the magnitude of surface oscillations due to resonant or non-resonant conditions. The first concern of the coastal engineer is to avoid port layouts and dimensions which promote seiching. In this respect the ratio of the length of the exciting wave to be size and shape of harbour are paramount.

Circular

Free oscillations in a circular basin of constant depth give an equation for surface amplitude which contains Bessel functions, the roots of which dictate the mode of oscillation. The lowest three occur when:

$R/L = 0.61, 1.12$ and 1.62

The first of these contains a nodal circle at $r/R = 0.628$ and for shallow-water waves has a period of:

$$T = 1.64R/\sqrt{gd} \tag{7-11}$$

Oscillations across one nodal diameter are established when:

$R/L = 0.293, 0.849$ and 1.36

For the first of these nodes:

$$T = 3.41R/\sqrt{gd} \tag{7-12}$$

A similar "sloshing" action can be developed in a square tank when $T = 4R/\sqrt{gd}$ for the square side equal to $2R$.

Elliptical

For an elliptic basin defined by $x^2/a^2 + y^2/b^2 = 1$ where a and b are half the major and minor axes, respectively, the longitudinal sloshing mode has a period:

$$T = \frac{2\pi a}{\sqrt{gd}} [(5 + 2b^2/a^2)/(18 + 6b^2/a^2)]^{1/2} \tag{7-13}$$

Rectangular

A closed rectangular basin $a \times b$ has a free oscillation whose period is given by:

$$T = \frac{2}{\sqrt{gd}} \left[\left(\frac{n}{a}\right)^2 + \left(\frac{m}{b}\right)^2 \right]^{-1/2} \tag{7-14}$$

where n and m are integers representing the number of modes in the a and b directions, respectively.

Where the basin is long and narrow:

$$T = \frac{2a}{n\sqrt{gd}} \tag{7-15}$$

and the first mode has a nodal point at the centre.

Harbours of course are not enclosed basins, unless locks or other structures enclose them from the sea, in which case excitation of resonance by wind-generated wave phenomena or longer-period surges is not possible. Thus harbours must be considered that have a continuous opening to the ocean. For the elongated shape, the effective length is now double the actual length so that the period is given by:

$$T = 4a/n\sqrt{gd} \qquad (7\text{-}16)$$

which indicates the promotion of resonance when $a = L/4, 3L/4, 5L/4$, etc. The first of these ratios is the most important, and as shown by Dorrestein [46], is one factor causing high tides in certain bays. He also shows that the resonant lengths for other shapes of elongated bay are:

(a) triangular in plan and constant depth, $a/L = 0.383$;
(b) constant width, bottom sloping to zero at end, $a/L = 0.192$;
(c) varying width and bottom slope, zero depth and width at end, $a/L = 0.305$.

Ippen and Goda [47] have studied a rectangular harbour connected to the sea by a breakwater gap as illustrated in Fig. 7-26. They examined resonance for vari-

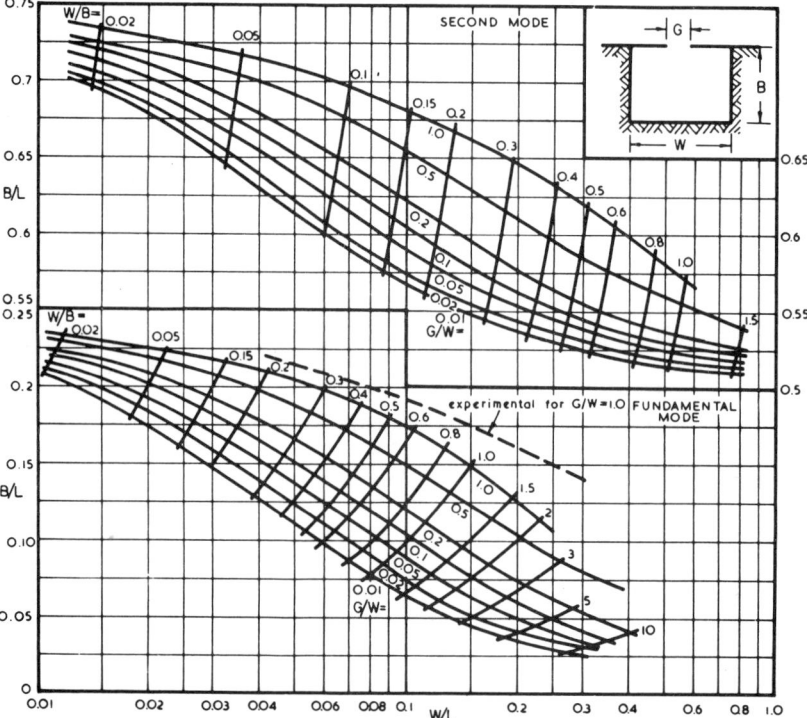

Fig. 7-26. Harbour basin ratios which promote resonance from waves of length L entering the breakwater gap.

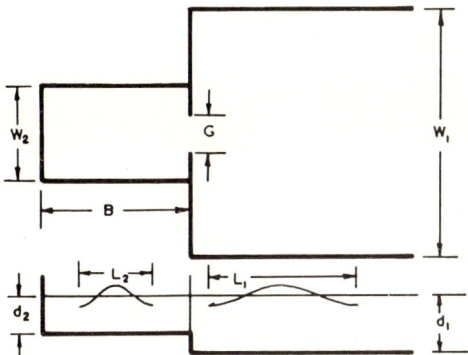

Fig. 7-27. Definition sketch of harbour connected to larger basin of greater depth.

ous offset positions of the gap, but theory for the symmetrical condition only is graphed in the figure. It is seen that the G/W curves for the fundamental mode are asymptotic to $B/L = 0.25$ and for the second mode to $B/L = 0.75$. The experimental results followed the dotted line drawn, indicating that frictional effects of the bottom and at the entrance tend to make the resonant length B a little greater. The curves should be employed with this point in mind.

Although the wave amplification for the second mode of oscillation is less than that of the first, the shorter wave lengths (L) able to generate it, make it a greater possibility from the long-period components of the wave spectrum or from surfbeat on adjacent beaches. In a reasonably sized harbour the wave length to create a seiche is from 5 to 20 times the dimension B and so its period must be in the order of minutes. In small-craft harbours, where the breadth and depth are smaller, even wind-generated waves may create standing waves in badly proportioned basins.

Curves for the amplification factor [47], not reproduced herein, show an increase as the gap G is decreased. This produced the paradox to which Miles and Munk [48] referred, but Le Méhauté and Wilson in discussion of this reference have pointed out that if frictional resistance through the entrance were taken into consideration the infinite amplification predicted for extremely small gaps would not ensue.

Biésel and Le Méhauté [49, 50] have examined the problem of a rectangular basin which has an opening to a larger and deeper basin with finite width, as distinct from the previous case of infinite width to the ocean. With the same notation as in Fig. 7-27 the resonant length ratio is:

$$B/L_2 = (2n\pi + \beta)/4\pi \qquad (7\text{-}17)$$

where n is an integer representing the mode (normally = 1).

$$\beta = \cos^{-1}\left[\frac{2 - \alpha_1^2(1 + L_2 W_2/L_1 W_1)}{2(1 - A\alpha_1^2 L_2 W_2/L_1 W_1)^{1/2}}\right]$$

where α_1 is the proportion of wave height transmitted over the step at entrance into the basin and is given by the empirical relationship:

$$\alpha_1 = \left(\frac{G}{W_2}\right)^{1/2} \left(\frac{W_1}{W_2}\right)^{1/4} \left(\frac{2}{1+AL_2/L_1}\right)$$

and:

$$A = \left(1 - \frac{4\pi d_2/L_2}{\sinh 4\pi d_2/L_2}\right)\left(1 + \frac{4\pi d_1 L_1}{\sinh 4\pi d_1/L_1}\right)^{-1}$$

Eq. 7-17 is based upon the restriction that $W_1 < L_1/2$ and $W_1/W_2 < 10$.
For shallow-water waves the resonant period of the basin is:

$$T_R = \frac{4\pi B}{\sqrt{gd_2}(2n\pi + \beta)} \tag{7-18}$$

Complex harbour shapes

Even for the simplified circular basin the theoretical analysis breaks down if the entrance becomes too large. Complex oscillations result which can only be solved by hydraulic [51, 52], electric analog [53] models, or by numerical methods [54].

For matrix and other methods for analyzing harbours with diverse boundaries and depths the reader is referred to the references given.

When investigating a harbour the waves of concern are the shallow-water variety, in which celerity is determined by depth. Should the harbour experience a reasonable tidal range then the tests should incorporate all limiting and intermediate conditions of any substantial duration. Tide-gauge records for the port site, or those obtained in adjacent areas, should be analyzed for long-period waves to see how prevalent they are in the region. If these are not available, recorders should be installed that are designed to measure these long-period fluctuations.

As can be surmised from the discussion on diffraction, it is not necessary for a wave to approach an entrance directly for it to set up resonance in a harbour basin. Long-period waves can pass across an opening and the rise and fall in water level there will cause a wave to propagate into the harbour and then diffract and refract to the boundaries. Because of the long wave length, total reflection can be expected, so that if the entrance is small the basin will oscillate as an almost closed unit. For many purposes, therefore, it is preferable to keep access to the sea as wide as possible.

Exclusion of resonant waves

It is virtually impossible to prevent long-period waves from entering a harbour. The only method to minimize their effects is to shape the partially enclosed basin so that standing waves cannot form. However, shorter-period waves (those in the wind-generated spectrum) may be attenuated strongly by dissipating their energy at the harbour entrance or through the approach channel, or by reflecting the incident energy seawards by means of wave resonators.

Lee [52] has discussed the provision of a Vee shape to the entrance channel of a port, with spending beaches consisting of rock fill. Battjes [56] has made a semi-empirical study of vertical side strips on the walls of a rectangular approach channel. Wave attenuation in this case is given by:

$$H/H_o = 1/(1 + \beta H_o x/2) \qquad (7\text{-}19)$$

where H_o is wave height at entrance, H is wave height at distance x along channel, and

$$\beta = \frac{8}{3W_c L_o} \frac{C}{C_g} \left(\frac{1}{3} + \frac{1}{\sinh^2 2\pi d/L} \right) C_f$$

where W_c is width of approach channel, C_f is the coefficient of boundary resistance which is a function of x_{max}/k and s/k where x_{max} is the maximum horizontal orbit of water-particle motion (average of bed and surface values), k is the transverse dimension of the strips, and s is the longitudinal spacing of the strips.

The variation of C_f with x_{max}/k for two values of s/k is depicted in Fig. 7-28 where a definition sketch is included. These were derived from tests for which $0.1 < d/L_o < 0.76$. For deep-water waves (x_{max}/k small) the coefficient is very large making for greater attenuation. C_f itself is dependent on x_{max}, which in turn is dictated by the wave height, so that a step procedure should be used for any reasonable length of channel. Battjes found that the strips were effective dampers over a wide range of periods. Only where resonance was set up, for $S/L = 1/2$ and $k/L = 1/4$ did attenuation decrease substantially. Also, to be effective, these ratios could not be too large or too small.

Another method of trapping waves and returning them seawards was suggested first by Valembois [57, 58]. These were rectangular basins open to the approach channel in which reflection of waves in resonance caused a degree of annulment of waves propagating past the basin and a partial standing wave seawards of this region. Such stuctures would be located at the seaward end of any navigation channel. James [59] has carried out extensive model tests on a single resonator from which design curves emerged. These have been interpolated and modified in Fig. 7-29, where it is seen that the main variables are resonator width W_r, resonator length B, channel width W_c and wave length L. In the figure any combinations so chosen provide resonance and therefore significant attenuation. Among the conclusions reached

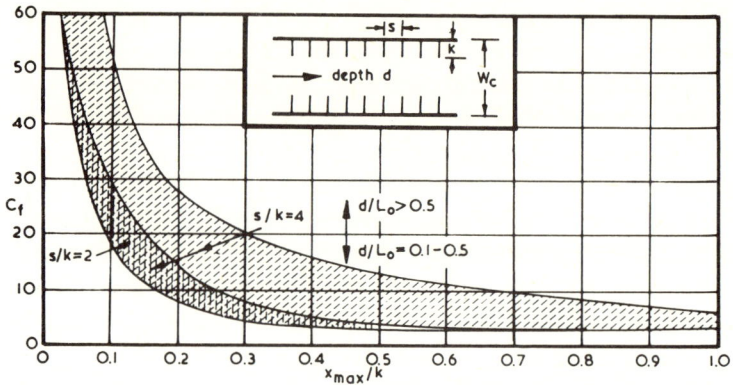

Fig. 7-28. Coefficient of boundary resistance to be used in eq. 7-19.

[59] were the following:

(a) beyond the resonator protection is afforded for only a width $W_c < L$;

(b) wave height has no influence on resonator geometry for resonance but finite height reduces attenuation somewhat;

(c) decreased efficiency is experienced with small W_r/L, with accompanying high velocities which could be a navigational hazard;

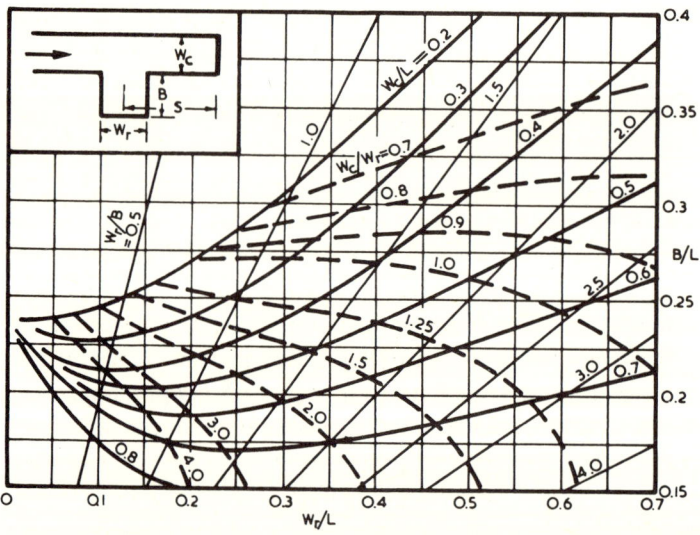

Fig. 7-29. Proportions of entrance channel and resonators to produce optimum attenuation.

(d) where there is partial reflection beyond the resonator the resonator should be located at a distance S (see inset Fig. 7-31) of $NL/2$ where N = 1,3,5, etc.

The single resonator is only effective in attenuating a narrow band of wave periods. To obviate this disability workers have suggested a series of basins with dimensions to influence a wide range of wave trains [57]. James [60, 61] has examined the proximity of such resonators to each other and concluded that they should not be contiguous. In this event the influence of each basin can be predicted from data from the single resonator as in Fig. 7-29. For example, in that figure, if W_r/B is chosen as 1.0 then $0.7 < W_c/W_r < 4.0$ for which $0.2 < W_c/L < 0.7$ and $0.175 < B/L < 0.30$. A series of basins would be chosen with complementary widths and lengths to attenuate wave lengths in this range. It is seen that a range of wave lengths of 1 : 3.5 is possible which will correspond to a certain range of periods as determined by the depth of the water (L/L_o = tanh 2 $\pi d/L$).

James [60] has also examined the efficacy of resonators with reduced depth. These were less effective, but could be used with advantage for long wave periods. Because of the reduced resonant length of a rectangular channel with a sloping bottom, 0.192 as against 0.24 for the uniform depth, it would seem advisable to test this shape in any feasibility studies. Besides the shorter length the volume of dredging is lessened by the average half depth.

WATER-BACKED STRUCTURES

Where a marine structure has water bodies on both sides of it there is the possibility of wave transmission on the leeward side. This may result from water oscillating through it or over it. Thus, besides reflection and dissipation of energy, there is a proportion of the wave energy transmitted landward of the structure. This may not necessarily contain the same period as the incident wave train and the length may be further modified by a change in depth on the leeward zone.

Permeable breakwaters

When a breakwater is constructed of rock-fill or fabricated blocks throughout, so permitting oscillation of water throughout its body, waves can be transmitted to the leeward side with reduced height. Kondo [62] has carried out a theoretical analysis from which the transmission coefficient ($K_t = H_t/H_i$) can be derived. This involves an assessment of porosity and tortuosity of the rock body for which laboratory and field tests need to be made. In this respect scale effects arise since the transmissivity of waves appears to depend upon [63] a Reynolds number containing the maximum orbital velocity at SWL, the size of stone and the kinematic viscosity of water. Prototype transmission coefficients could be up to double those obtained in a 1 : 4 model.

Tests carried out by Iwasaki and Numata [64] showed distinct behaviour by breakwaters whose crests were higher than the incident wave and those which were not. For the former the following empirical relationship resulted:

$$H_t/H_i = 1/[1 + K(H_i/L_i)^\beta]^{1/\beta} \tag{7-20}$$

where K and β are factors to be determined from prototype or experimental data, with due regard to scale effects in using the latter.

Submerged breakwaters

For the sake of economy breakwaters may extend from the seabed to SWL or some depth below or slightly above this level. Waves, in overtopping the structure, will be transmitted on the leeward side with reduced amplitude. Many tests have been conducted on this subject and Dick [65] has supplied a critical review of these, together with experimental data. However, Goda's [66, 67] comprehensive tests, with the corrections noted previously, have been reported with different dimensionless parameters. There are many choices in this regard from depth of submergence, breakwater width, wave height and length, and water depth.

Goda [66] has used h/H_i as an abscissa for K_r, K_t and K_e as reproduced in

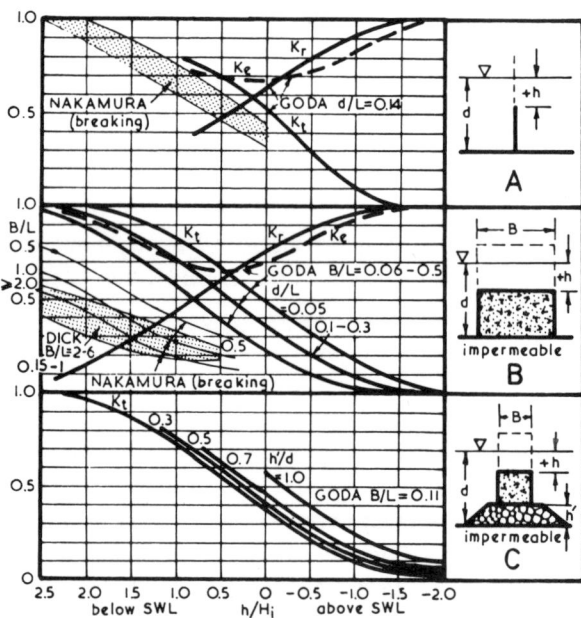

Fig. 7-30. Reflection and transmission ratios plus proportion of energy conserved for: A. thin wall; B. block mound; and C. composite breakwater.

WATER-BACKED STRUCTURES

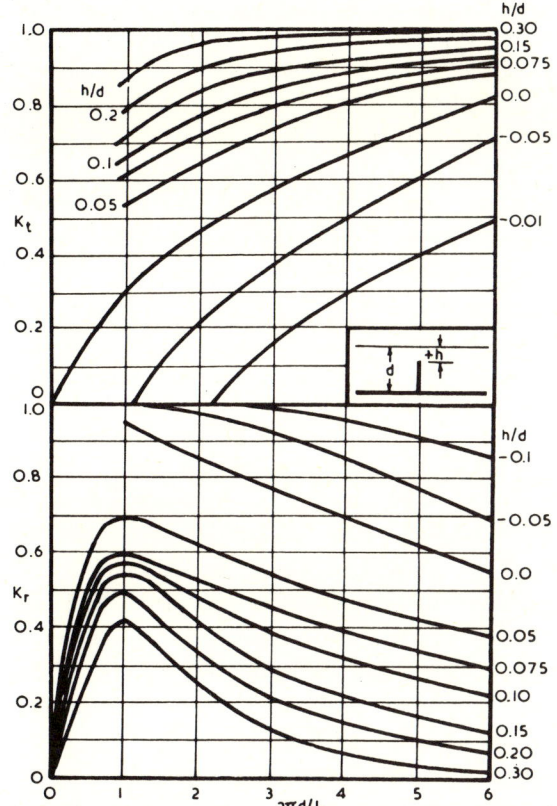

Fig. 7-31. Reflection and transmission ratios for thin-walled submerged breakwater.

Fig. 7-30. The incident wave energy must be accounted for as follows:

$$H_i^2 = H_t^2 + H_r^2 + H_l^2 \tag{7-21}$$

or:

$$1 - K_l^2 = K_t^2 + K_r^2 \tag{7-22}$$

or:

$$K_e^2 = K_t^2 + K_r^2 \tag{7-23}$$

where K_e^2 is the energy conserved in the overtopping process. It should be the aim of the coastal engineer to minimize this proportion because even if zero wave transmission is achieved with high structures, the reflection may be unduly amplified, so aggravating erosion of the sea bed in front of the structure.

In Fig. 7-30A Goda's results [67] for the thin barrier are graphed for $d/L = 0.14$

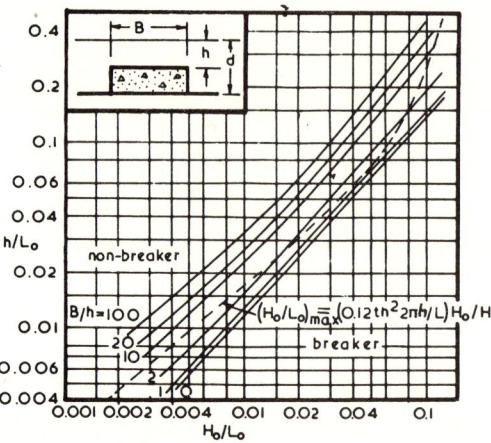

Fig. 7-32. Breaking criterion over submerged breakwater.

with wave steepness not entering as a variable. Also shown are data supplied by Nakamura et al. [68]. The alternate and more comprehensive tests of Dick [65] are presented in Fig. 7-31, where they are combined with the theoretical analyses of Dean [69] and Ogilvie [70]. Also included are the curves for K_t and K_r for $h/d = 0, -0.05$ and -0.10, which have been extrapolated from the remaining curves. For $d/L = 0.14$ and $h/d = 0$ the curve matches Goda's result in Fig. 7-30A for $h/H_i = 0$. Nakamura et al. [68] differentiated between breaking and non-breaking over their submerged dikes. Although they included no data for the thin plate in the latter category, results for other widths indicated that the transmission coefficient was higher than that indicated in Fig. 7-30A.

Where the dike is of finite width B the possibility of wave breaking is greater. From the experimental results of Nakamura et al. Fig. 7-32 has been drawn up, in which breaking is dependent upon H_0/L_0, h/L_0 and B/h. From the curve of $(H_0/L_0)_{max} = 0.12 \tanh^2 2\pi h/L (H_0/H)$ from eq. 4-41 it is seen that premature breaking occurs over the mound for $B/h > 10$, but that for $B/h \leq 1.0$ smaller depths are required than given by the theory for normal bed conditions.

The transmission and reflection coefficients for the finite wide breakwaters are prescribed in Fig. 7-30B. It should be noted that the tests of Nakamura et al. cover different ranges of B/L than those of Goda [67], who derived a general relationship:

$$K_t = \frac{H_t}{H_i} \cdot 0.5 \left[1 - \sin \frac{\pi}{2\alpha}\left(\frac{h}{H_i} + \beta\right)\right] \tag{7-24}$$

for $-2.5 \leq h/H_i \leq 2.5$, where α and β are determined from experiment. The value of $\alpha = 2.2$ and β ranged from 0 to 0.8 in the Goda curves illustrated in Fig. 7-30B.

The results from Nakamura et al. for non-breaking waves have much higher K_t, and for $d/H_i > 2.0$ indicate $K_t > 1.0$. Dick [65] has noted this to be the interaction of transmitted waves with a range of periods. He overcame this problem by measuring the variance of the transmitted wave over one period T and finding the height of the wave with equivalent energy, but having the period of the incident wave, that is:

$$\frac{w}{8} H_{eq}^2 = \frac{w}{T} \int_0^T y^2(t) \, dt \qquad (7\text{-}25)$$

where w is specific weight of water, and y is the water surface deviation from SWL at intervals dt along one wave period of the record.

This procedure was valid since all secondary waves were smaller in length than the basic one which was equal to that of the incident waves. Errors arose which were considered to be less than ± 10%.

The measurements permitted the computation of energy in the secondary waves compared to that of the basic band. This ratio approximated 50% so that the transmission coefficient for the secondary waves was $0.7 K_t$ for the equivalent height band with incident wave period.

Dick's results do not differ greatly from those of Nakamura et al. [68], as illustrated in Fig. 7-30B, where the B/L values from 2 to 6 could not be differentiated. It would seem that, for optimum reduction in transmitted wave height, B should be as large as possible, in fact up to 2 wave lengths. This is unlikely to be an economical proposition, so that some narrower breakwater taken to some greater height will have to be found from Fig. 7-30B.

Composite breakwaters have been tested by Goda [67], the results of which are reproduced in Fig. 7-30C. The same general eq. 7-24 is recommended with factors α and β changing appropriately to suit the new curves. This type of construction is widely used in Japan where a rock mound is placed and shaped to serve as a foundation for monolithic concrete boxes which are floated into position, sunk and filled with sediment or rock.

The effect of permeability on submerged rectangular breakwaters has been studied by Dick who used rows of tubes with changeable orifices to simulate differing flow resistances through a rock-fill structure. Some wave dissipation accrues from the form and viscous drag through the mound, although it was difficult to differentiate between the various percentages of porosity except for the 100% case, which was simulated by a flat plate. The mean curves [65] are reproduced in Fig. 7-33 and 7-34. In the former it is seen that the greater the submersion (greater h/d) the higher is the transmitted wave. Also, for the permeable breakwaters (the flat plate = 100% porosity) a distinct reduction occurs around $L/B = 4$, although this is not evident for the solid mound. This is occasioned by the phase lag of the

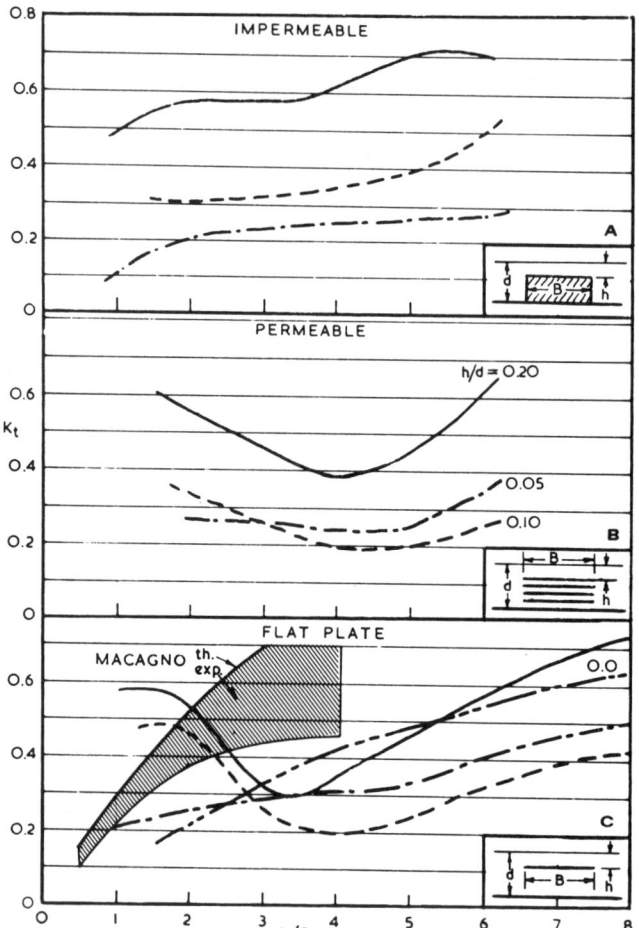

Fig. 7-33. Transmission coefficient for: A. solid mound; B. permeable mound; and C. flat plate.

wave passing across the top of the breakwater and that passing beneath its surface. Dick found this phase lag to be near π radians for $h/d = 0.10$ and 0.20 and about $1.5\,\pi$ for $h/d = 0.05$, which correlates with the imperceptable depression in the 0.05 curve for the flat plate. When the flat plate was placed at the surface ($h/d = 0.0$) no interaction took place so that K_t is a rising curve, giving greater wave transmission as B decreased.

Also presented in Fig. 7-33C are the theoretical and experimental results of Macagno [71] for the flat plate. These have a higher K_t than those just discussed, although the rising trend is similar.

Fig. 7-34 compares the three types of breakwater for similar depths of submergence. For $h/d = 0.05$ there is little difference in the wave attenuation. For the two

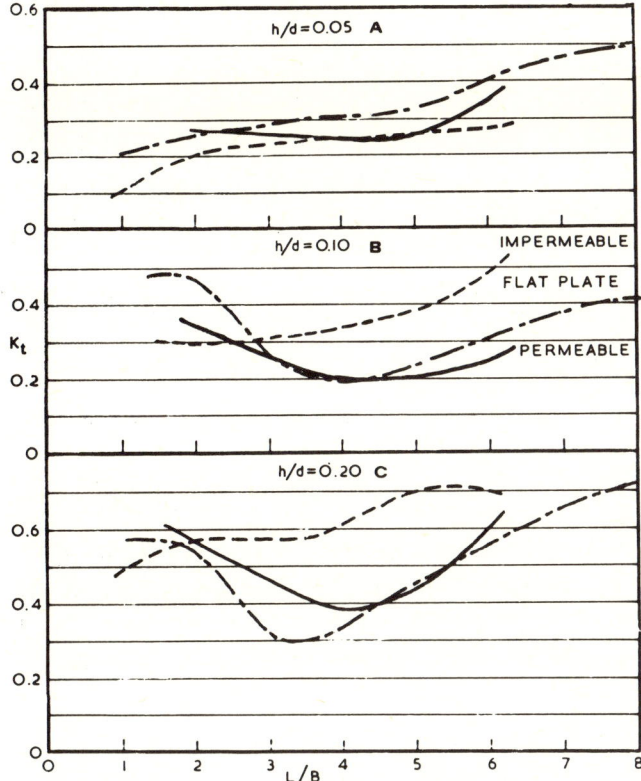

Fig. 7-34. Transmission coefficients as for conditions in Fig. 7-33 grouped for: A. solid mound; B. permeable mound; and C. flat plate. $h/d = 0.10$; and C. $h/d = 0.20$.

other depth ratios the permeable structures are more effective than the solid mound. From overall observation, a flat plate $L/4$ in breadth and submerged to 0.10 of the depth is the best attenuator studied. Although this breakwater is effective through a resonance action, and therefore for optimum performance must be tuned to a wave length, inspection of Fig. 7-34B shows that $K_t < 0.3$ over the range of $B/L = 3$ to 6 for $h/d = 0.10$. It could therefore attenuate waves from $2\,T_{max}/3$ to $4\,T_{max}/3$ which is a large proportion of the spectrum of a FAS nearshore.

A theoretical analysis of the flat plate attenuator has been provided by Ijima et al. [72] who derived curves for K_t and K_r and relationships for pressure forces on the structure. The experimental curves of Dick [65], reproduced in Fig. 7-33 and 7-34, give K_t values much lower than those predicted by the equations. With the apparent efficiency of this type of structure, feasibility studies should be conducted into means of installation and overall costs.

Pile arrays

Closely spaced piles may be used as a protection against waves, or as a support to jetties which will attenuate waves whilst serving this purpose. Hayashi et al. [73] derived a theoretical value for K_t, which was verified closely by laboratory tests, of the form:

$$K_t = 2(2d/H_i) E[-E + \sqrt{E^2 + H_i/2d}] \qquad (7\text{-}26)$$

where:

$$E = C_d \frac{b}{D+b} \bigg/ \sqrt{1 - \left(\frac{b}{D+b}\right)^2}$$

where C_d is the coefficient of discharge through the slot of width b between piles of transverse dimension D.

This relationship is graphed in Fig. 7-35 for variations in b/D and d/H_i. It is applicable to only shallow-water conditions.

Sawaragi and Iwata [74] derived a similar analysis based on the concept of pile drag, rather than orifice flow for the previous case. The results are very similar, the main difference occurring in the term $H_i/2d$ where the $2d$ became a function of d/L_o, so that the following multiplying factor A to the values of H_t/H_i in Fig. 7-35 could be employed:

d/L_o	0.005	0.01	0.02	0.05	0.10
A	1.005	1.010	1.023	1.055	1.123

Values of H_t/H_i in Fig. 7-35 relate to circular piles where $C_d \doteq 0.9$. For square or other sharp-edged piles the coefficient of contraction will make the effective width $\doteq 0.6\,b$. This affects K_t very little as according to eq. 7-26 the following multiplying factor B results:

d/H_i	2	3	4	5
B	0.93	0.97	0.98	0.98

The reflection coefficient $K_r = H_r/H_i$ can be obtained from K_t in a similar manner to eq. 7-23, where $K_e^2 = 0.4$ from the average of experimental results by Hayashi et al. [73].

Eq. 7-26 applies for circular piles in which $1/20 < b/D < 1/4$. Sawaragi and Iwata report tests on wider spacing ($0.65 < b/D < 3.0$), and the influence of 2 and 3 rows of piles. Fig. 7-36 has been prepared from averaging values for the two limiting wave steepnesses tested. It is seen that the provision of two and three rows at a specific spacing reduces the transmitted wave height slightly. The theoretical transmission curve of Sawaragi and Iwata is above the experimental traces, but not so high as the curve for $d/H_i = 3.75$ taken from Fig. 7-35 [73]. For other values of

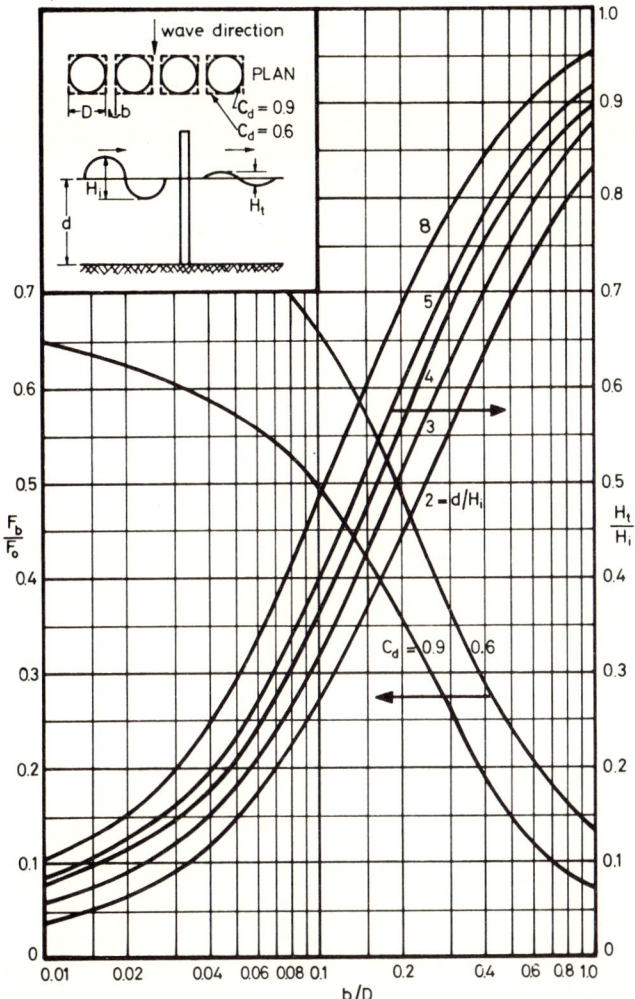

Fig. 7-35. Transmission coefficient of closely spaced piles and ratio of force with spacing b to force with zero spacing.

d/H_i there will be some variation of K_t in Fig. 7-36, which may be interpolated from the theoretical and experimental traces in the two figures.

Costello [75] has tested attenuation of waves by rows of piles either in line or diagonally spaced. There are so many combinations which could be considered that only general trends are worth noting here. Transmission decreases with increasing wave steepness and for swell waves ($H_i/L < 0.03$) $H_t \geqslant 0.8$ for 24 rows of piles with $b/D = 5/3$ in two directions. Doubling the number of rows reduces K_t by about 25% for very steep waves ($H_i/L \doteq 0.10$).

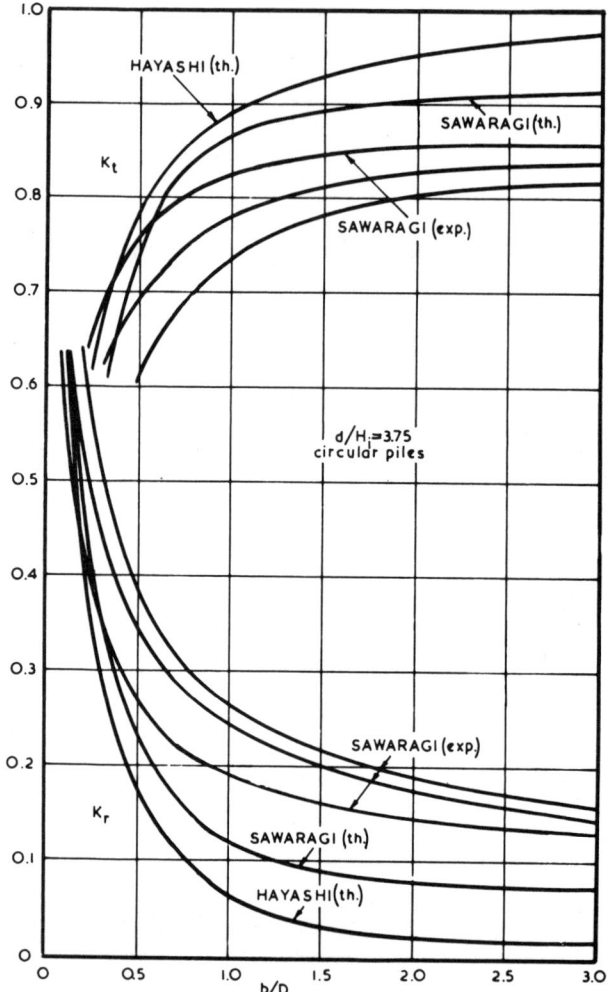

Fig. 7-36. Transmission and reflection coefficients for closely spaced circular piles.

MOBILE BREAKWATERS

There are many advantages in using mobile breakwaters over those fixed in position. Mobile in this sense should take into account designs which allow for ready removal of structures that serve whilst resting on the seabed. However, this static installation has already been treated, so that only floating structures are now considered. Also amongst mobile units workers include the pneumatic and hydraulic breakwater, but these are discussed separately in the next section.

Carr [76] has mentioned the advantages of the floating breakwater system, namely: (*a*) economy of material; (*b*) freedom from foundation problems; (*c*) freedom from erosion problems; (*d*) indifference to tidal changes in site water depth; (*e*) probable relative ease of transportation.

Freedom from erosion depends upon the action in reducing wave transmission. If this is through substantial reflection then scouring beneath the breakwater is to be expected, as noted by Vinjé [77]. But the ease in transportation leaves an extra degree of freedom in planning for the many unknowns in wave incidence on a coast. If the design data is not quite correct a change in position is soon effected.

Like the fixed structure, the floating breakwater reduces wave transmission by dissipation through breaking, reflection, friction (generation of vortices), or by the interaction of waves, so that orbital motions of water particles are reduced. The various types of breakwater attempt to maximize all these effects. They may be divided into pontoon structures, either vertical or horizontal, flexible rafts or membranes, and miscellaneous floats of specific shape.

Pontoons

A floating obstacle which penetrates almost to the seabed relies on reflection for most of its transmission interference in which K_t is given by [76]:

$$K_t = 1/[1 + (\pi W/wLd)^2]^{1/2} \tag{7-27}$$

where W is the weight of the pontoon per unit length of wave crest intercepted and w is the specific weight of seawater.

The ratio W/wLd is the ratio of the barrier weight to the weight of seawater per unit wave length. Eq. 7-27 applies to a freely floating structure. Where some restraint is placed on the vertical motion, the combined system has a natural period of oscillation:

$$T_s = 2\pi/(Kg/W)^{1/2} \tag{7-28}$$

where K is the spring constant of an elastic mooring system. This provides a new transmission coefficient:

$$K'_t = 1/[1 + (\pi W/wLd)^2 (1/(T_s/T)^2 - 1)^2]^{1/2} \tag{7-29}$$

From equations 7-28 and 7-29 the value of K'_t/K_t varies with T_s/T, and when this is unity K'_t/K_t is optimum. For $T_s/T \gtrsim 1.0, K'_t/K_t < 1.0$.

The action of the floating breakwater is not dissimilar to that of a rigid thin barrier penetrating the water surface to some depth. Ursell [78] and Wiegel [79] have derived theoretical relationships of K_t for deep-water and transitional depths respectively which agree reasonably well with experiment as seen in Fig. 7-37. Mobile structures of this character were designed for temporary shelter during the invasion of Normandy during World War II [80, 81].

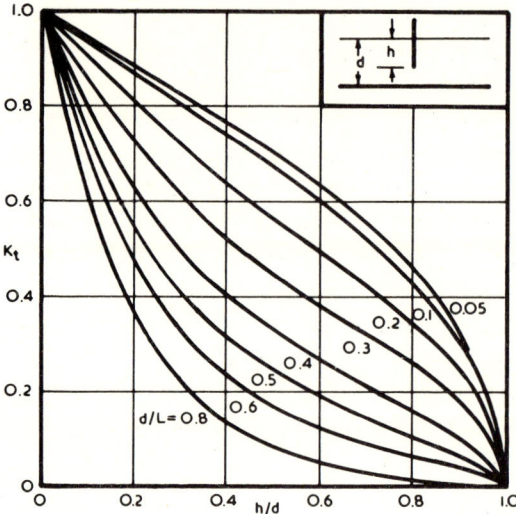

Fig. 7-37. Transmission coefficient for a thin plate penetrating the water surface.

The maximum force per unit length of breakwater experienced by the mooring system in the floating case is:

$$F_{max} = \left| \frac{wdH_r}{1-(T_s/T)^2} \right| = \left| \frac{wd\sqrt{H_i^2 - H_t^2}}{1-(T_s/T)^2} \right| \qquad (7\text{-}30)$$

assuming no loss of energy (i.e., $H_i^2 = H_r^2 + H_t^2$). From eq. 7-30 it is seen that F_{max} can become theoretically infinite when $T_s = \cdot T$. It is advisable to make $T_s > 2T$.

A novel breakwater has been tested at Queens University [82] consisting of two cylindrical floats between which is supported a vertical plate. The floats are parallel to the wave crests and to be effective must be about 2 wave lengths apart. Reduction in transmission is effected mainly by reflection, so substantial mooring forces must be suffered. The wide flat pontoon relies on its reflection along its bottom to reduce the wave transmission. For this purpose, as for the fixed structure discussed previously, the ratio L/B is important, as well as the mass of the floating body to withstand the forces tending to push it upwards. Pontoons tested [83] with differing drafts provided results which are graphed in Fig. 7-38. Also reproduced is the curve for the fixed plate with zero submersion as obtained by Dick [65]. It is seen that the floating case differs little from the fixed platform, especially at the minima where $B/L = 0.54$. This depression in K_t occurs through some out-of-phase action of the pontoon, which possibly reflects or breaks the wave at a critical stage of its approach. For the three d_2/d_1 ratios it occurs at different ratios of T_s/T. The

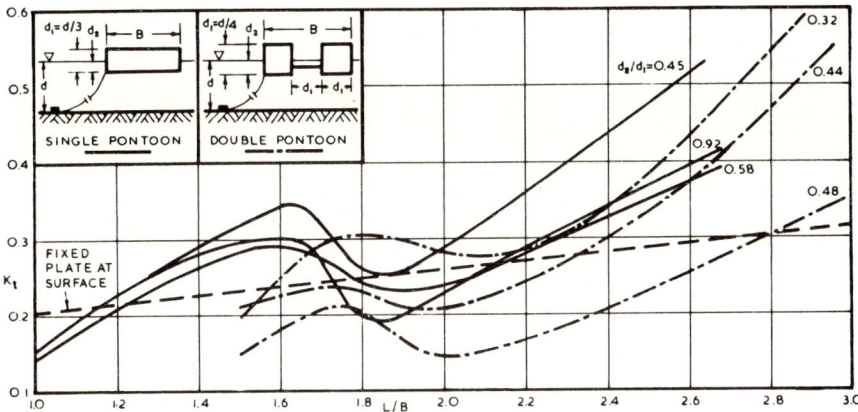

Fig. 7-38. Transmission coefficient of pontoons.

double pontoon tested by Ofuya [83] acts in a similar manner to the single float as seen in Fig. 7-38, the larger draft in this case providing for greater reflection and less transmission.

From the benefit previously seen to accrue from submerging a fixed horizontal flat plate it would appear that a worthwhile attenuation should be achieved by floating such a plate below the surface. Since the action in this case depends upon the phase lag of the wave above the plate to that below it, some amplification of this action by roughness elements on the top surface may prove useful. Also, the transient differential pressures above and below the plate could effect flow through any perforations in the plate and so help wave dissipation. The concept is depicted in Fig. 7-39A. The overall breadth of such a system could be made up of several units since the dissipation is not being achieved by rigidity of the plate but by the differential speeds of the waves above and below it.

Hom-ma et al. [84] tested a two-layer pontoon, the lower one being rigidly fixed to the upper at various spacings. Results of their tests are depicted in Fig. 7-40, which clearly show the advantage of this system. The presence of the surface raft assists in the retardation of the wave motion above the lower plate, so reducing the breadth of the breakwater to accomplish the annulment at the leeward end. Since the rafts in these tests consisted of parallel pipes there would have been exchange of water through the surface which would help in wave dissipation. For the surface raft alone the breadth was made up of 1, 2 or 4 articulated units with little difference being exhibited by this freedom of motion.

The mooring forces on such breakwaters depend greatly on the length of anchor line in respect to water depth. The measurements made by Hom-ma et al. showed the maximum force on the two-layer raft to be of the same magnitude as a pontoon, the difference being in the frequency of such forces. The two-layer raft had

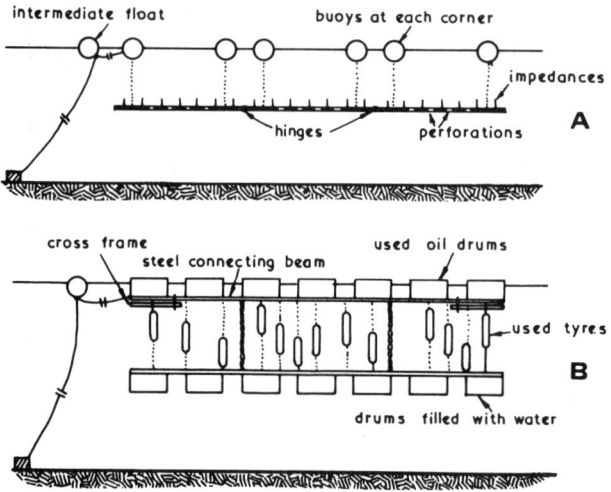

Fig. 7-39. Suggested floating breakwater from: A. submerged articulated platform; B. tow-layer raft.

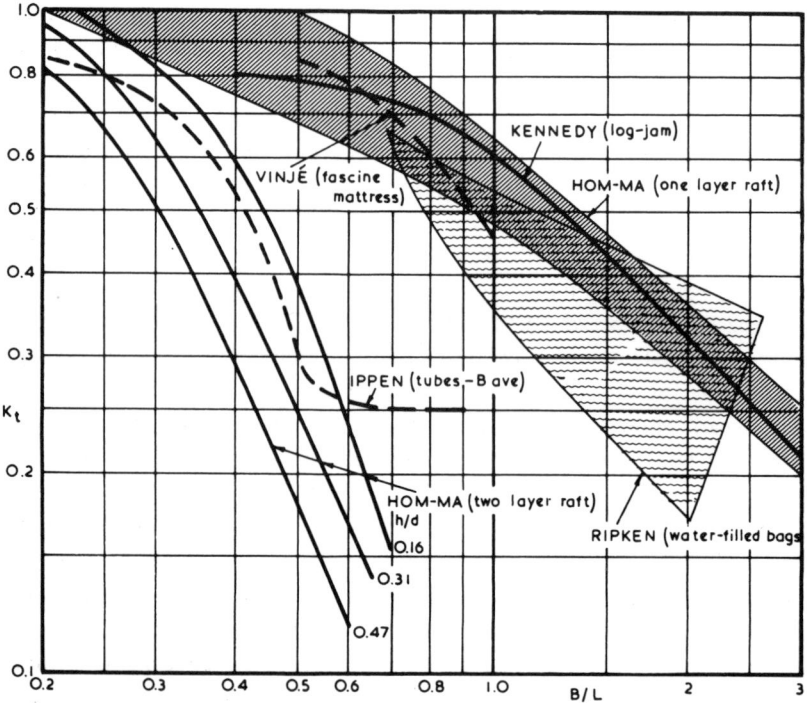

Fig. 7-40. Transmission coefficient for various floating breakwaters and fixed tubes near upper surface.

MOBILE BREAKWATERS

major forces exerted once every 2 to 4 waves whilst the pontoon experienced them almost every wave. In the frequency graph for $L'/d = 4$, which is a reasonable ratio of mooring line length to depth, the double-layer raft showed over 60% of the time with zero force. To provide greater elasticity in the line it is suggested that an intermediate float be incorporated into the anchoring system, so that it can be dragged down on the occasion of large forces which can thus be spread over a longer period of time. Such an addition is depicted in Fig. 7-39 A, B.

An economical breakwater could be envisaged consisting of used oil drums welded end to end and joined together to form rafts (see Fig. 7-39B). The surface raft would provide the buoyancy for the lower one, which would be flooded to provide sufficient inertia in the system. To further impede the wave travelling between the layers, used automobile tyres could be slung from the surface raft. However, this may increase the mooring forces out of proportion to the attenuation gained. The matter deserves further research because of the economies available.

Flexible rafts

As noted in the foregoing, the rafts by Hom-ma et al. were partly flexible, but other tests have been conducted on completely flexible systems which follow the undulations of the longer waves, if not the shorter ones present. Wiegel et al. [85] tested water-filled bags which floated on the surface. They found that $K_t = L/B$, so that large breadths would be involved in attenuating longer waves. They also noted the need to account for wave diffraction behind a floating breakwater, both in the laboratory and in the field.

Kennedy and Marsalek [86] conducted tests on a log-jam, results of which are included in Fig. 7-40. It is seen that they match very well the raft system of Hom-ma et al. They also examined the attenuation of wooden booms in the shape of a U with the sides perforated. These were found to be effective when spaced crestwise at half wave-length intervals. Delft tests on fascine mattresses, as reported by Vinjé [77], are also graphed as a mean curve in Fig. 7-40. It is seen that all these tests with permeable and flexible floats provide similar attenuation, so it is a matter only of choosing the most economical material available in any area.

Miscellaneous

Kato et al. [87] conducted interesting tests on box-type floats which varied in shape from rectangular to triangular, having trapezoidal sections upright and inverted. These had lips at the top edges which assisted in the reflection of waves when in the appropriate position and orientation to receive an incoming crest. For the breadths tested (about one wave length), best attenuation was obtained when $T_s/T = 1.0$. Such pontoon-type structures will have high mooring forces.

Johnston [88] has tested a floating breakwater consisting of two buoyant walls with a water space between to promote resonance. Another form was a pontoon containing water which oscillated back and forth, providing both inertia and resonance effects. Marks [89] has tested similar structures in which the seaward face was perforated. For the floating case he examined the forces involved and in the static case [90] the erosion of the bed beneath the breakwater.

Tanaka [91] has tested two vertical walls as a breakwater, which might be adapted to a floating structure. He concludes that the seaward wall should be of height H_i and its top submerged $H_i/4$. The leeward wall should be $2 H_i$ in height, half of which is above SWL, and a distance $L/4$ behind the seaward wall.

Ippen and Bourodimos [93, 93] have examined the influence of a group of horizontal tubes immersed in the upper zone of water on the attenuation of waves. Whilst they were held steady the application is there for a floating structure. The tubes occupied only 30–40% of the breakwater volume. The action is similar to a submerged plate in that flow through the tubes, caused by differential pressures at front and back, preceeds the pressure flux accompanying the wave. This occurs at both ends of the tubes and as seen in Fig. 7-40 is very effective in attenuating waves. Its economy would have to be examined, especially in the light of barnacle and other marine growth which is likely to block the tubes.

PNEUMATIC AND HYDRAULIC BREAKWATERS

Both these systems rely on the breaking of waves by opposing currents. The pneumatic installation generates this current by air bubbles being released near the bed which drag water to the surface as they rise. This flow divides at the surface, with part being directed against incoming waves. It and the hydraulic system are more effective against short-period waves and waves in deep-water, where the bulk of the energy is contained in the surface region. It should be realized that where short waves are predominant a local wind exists and that under these conditions a strong mass transport of water and shear current will tend to force the bubbles and rising current shorewards, so reducing the effect of a pneumatic system.

Waves arriving which are essentially shallow-water waves will have horizontal water-particle oscillations which are relatively uniform throughout the depth. Thus the vertical current produced by the bubbles will have a very small influence on this motion and hence will attenuate the waves very little.

The volumes of water or of air to run such breakwaters are prodigious and hence no economic solution appears to be at hand. It may be possible to direct the discharge of a river in such a way as to help dissipate waves, but navigational hazards must also be appreciated. Much research has been carried out on the subject but it is not intended to review it here. The selected references provided on pneumatic [94–99] and hydraulic [100–104] breakwaters should assist any reader in commencing a literature survey on the topic.

One final point to note is that pipelines supplying air for a pneumatic breakwater can soon become silted up if placed too close to the bed. The current generated by the bubbles will draw water at the bed towards the line, which, together with the possible reduction of the mass-transport velocity at the bed on the breakwater line, could cause excessive accretion.

Having discussed a great variety of mobile breakwaters it is worth considering for what specific problems they might be used. These are:

(*a*) Military applications where swift protection must be given to small craft discharging men and munitions on an enemy coast.

(*b*) Protection of dredges operating in a fixed locality for extensive periods, but which have a small free board.

(*c*) Protection of marinas on inland lakes or reservoirs where water depths are great close to shore, or levels can vary greatly throughout the year.

(*d*) Protection of vessels being employed in salvage operations in exposed sea areas.

(*e*) Temporary protection of shore facilities whilst certain marine operations are carried out, such as pulling sewage outfalls through the surf zone to the offshore depths, or repairing seawalls or dikes.

(*f*) To serve as temporary headlands for the accumulation of sediment from littoral drift. As accretion proceeds so can the headland be removed seawards until a stable bay has been formed.

Bulson [105] has made a comprehensive survey of transportable breakwaters and made a cost assessment of floating mattresses, caissons and bubble breakwaters. His concluding remarks are worth quoting: "The final conclusion is that all three forms of breakwater are feasible, but expensive. The most attractive form is the caisson. In all three forms there are a number of factors connected with full-scale operation that are difficult to assess, but it does seem that the idea of installing a transportable breakwater anywhere off the coast is not really practical. Its obvious use is to improve the facilities of an already existing natural harbour. Research has produced no obvious breakthrough, and high cost seems bound to be a basic feature of any apparatus designed to combat the energy of the open sea."

One may take hope because Bulson's bibliography contained only one reference to Japanese research on this topic. The submerged platform, either single or multi-layered, would appear to serve as the break-through required to bring such structures in the realms of engineering practicability.

EXAMPLES

1

A long straight breakwater runs into an area of sea where the depth is essentially constant at 35 ft. Swell of 12-sec period and 4 ft. height is approaching the breakwater normally. A berth for ships is planned for a point on a radius of 45° from the breakwater and 0.35 NM from its tip. Find the wave height at the berth with and without reflection taking place.

See Fig. 7-41.

$H = 4$ ft.; $d/L_0 = 35/5.12 \times 12^2 = 0.0475$; $d/L = 0.0915$; $L = 384$ ft.;

$R = 0.35 \times 6080 = 2130$ ft.; $R/L = \dfrac{2130}{384} = 5.5$; $\theta = 90°$; $\alpha = 45°$; $2\theta - \alpha = 135°$

$K = 0.088 + 0.037 = 0.125$; $H = 4 \times 0.125 = 0.5$ ft. with 100% reflection, $H = 0.088 \times 4 = 0.35$ ft. without reflection.

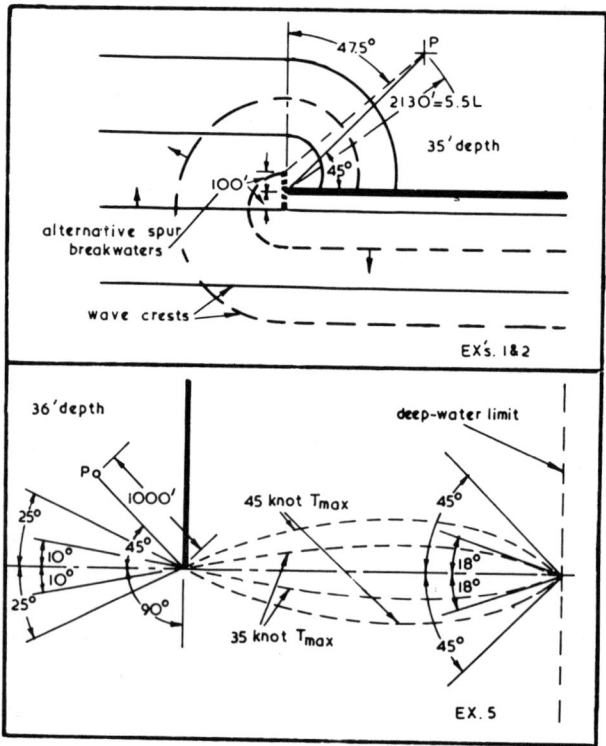

Fig. 7-41. Sketch plans for Examples 1, 2 and 5.

2

In Example 1 it is proposed to add about 100 ft. to the breakwater in the form of an L. Determine whether this should run seawards or into the harbour in order to reduce the wave height at the berth.

Running the extension into the harbour (see Fig. 7-41) maintains the height of the incident wave to the tip apparently making $\theta = 180°$. But this small extension does little to alter the previous situation so that:

EXAMPLES

$\theta = 90°; \alpha = 47.5°; 2\theta - \alpha = 132.5°; R/L = 5;$
$K = 0.588 + 0.039 = 0.127$ similar to Example 1.

Running the extension seawards delays the reflected wave from diffracting into the harbour. The length is about 1/4 wave length, so that when a crest has reflected to the tip X a trough of an incident wave is present. This will create an annulment which will continue into the shadow zone, so that $K = 0.088$ or even less. It should be realized, however, that this extension will be subject to large vortices and high waves in the case of waves arriving at angles θ greater than 90°. The structural units of this extension must be larger than normal and the bed around it protected by stones for a reasonable radius.

3

A breakwater gap 585 ft. wide in 30 ft. of water faces due west. The most persistent swell arrives from the WSW direction and has a period of 10 sec and height of 3 ft. List the wave height experienced at various R/L along the centre-line of the gap, both in the direction of maximum intensity and normal to the breakwaters.

See Fig. 7-10.
$H = 3.0$ ft.; $d/L_0 = 30/512 = 0.0585; d/L = 0.1028; L = 292$ ft.;
$B/L = 585/292 = 2.0; B'/L = 2.0 \cos 22.5° = 1.73; \theta = 67.5°$

Penny and Price solution (Fig. 7-10 with $B'/L = 1.73$):
$K\sqrt{R/L} = 1.75$ and 0.65 for $\alpha = 0$ and $-22.5°$, respectively.
Morse and Rubenstein solution (Fig. 7-11 with $B/L = 2.0$):
$K\sqrt{R/L} = 1.83$ and 0.65, say 1.8 and 0.65.

K should not be computed for $R/L < B/L = 2.0$, see Table 7-III

TABLE 7-III

Diffracted wave heights as per Example 3

R/L	2	5	10	15	20
$H_{\alpha=0}$	2.7	1.08	0.54	0.36	0.27
$H_{\alpha=22.5}$	0.98	0.39	0.195	0.13	0.098

It is seen in this table that, at only 22.5° deviation from the alignment of maximum intensity, the wave height is only 36% of this maximum at equal R/L values.

4

In Example 3, compute the wave height in the same positions when the breakwater gap is 50% greater than before.

$B/L = 3.0; B'/L = 2.59; \theta = 67.5°$.
Penny and Price solution (Fig. 7-10 with $B'/L = 2.59$):
$\qquad\qquad\qquad \alpha = 0 \qquad -22.5°$
for $R/B > 1, K\sqrt{R/L} = 2.55 \qquad 0.65$
$\qquad R/B = 1, K\sqrt{R/L} = 2.1 \qquad 0.95 \qquad$ for values of K see Table 7-IV

TABLE 7-IV

Diffracted wave heights as per Example 4

R/L	2.59	5	10	15	20
$H_{\alpha=0}$	2.43	1.53	0.765	0.51	0.383
$H_{\alpha=22.5°}$	1.1	0.39	0.195	0.13	0.098

356   EFFECT OF STRUCTURES

There is no solution by Morse-Rubenstein or Lacombe for this B/L ratio. It is seen that along the approach alignment the waves are increased, but apart from the zone near the gap the waves in the direction normal to it remain the same as before. As seen in Fig. 7-10 it is only within 15° of the entrance orthogonal that waves increase in height due to widening of the gap. This points to the need for accurate assessment of all possible approach directions for waves.

5

A breakwater protecting a harbour 36 ft. deep runs parallel to the coast, which can be considered uniform in slope to the sea. It is desired to find the significant wave height and period at a point 1000 ft. radius from the tip and 45° to it when two storm conditions exist. One is a FAS sea produced by $U_{19.5}$ = 35 knots and the other is a 45-knot wind blowing for 14 h; in each case the wind direction is normal to the coast and the breakwater.

See Fig. 7-41.

$U_{19.5}$ = 35 knots FAS.
From Table 3-VII, $H_{1/3}$ = 22.6 ft., T_{max} = 11.6 sec
In 36 ft. water d/L_0 = 0.0523
From Fig. 5-20, $H_{1/3}/T_{max}^2$ = 0.18, which is greater than that existing for FAS.
From Fig. 2-3, 11.6-sec waves in 35-knot wind have maximum energy 18° to wind in deep water.

$U_{19.5}$ = 45 knots, t = 14 h
From Table 3-VII, $H_{1/3}$ = 37.5 ft., T_{max} = 14.9 sec, t_{FAS} = 53.4 h.
t/t_{FAS} = 14/53.4 = 26.2%
From Table 3-VIII, $(H/H_{FAS})_{1/3}$ = 59%, $H_{1/3}$ = 22.1 ft., $(T/T_{FAS})_{max}$ = 70%, T_{max} = 10.4 sec.
In 36 ft. water d/L_0 = 0.065, $H_{1/3}/T_{max}^2$ is the same order as 15% FAS in Fig. 5-22.
From Fig. 2-3, 10.4-sec waves in 45-knot wind have maximum energy 45° to wind in deep water.

From Table 5-IV angle of waves to wind after refraction to d/L_0 = 0.0523 is 9.3° (say 10°).
Consider half of the wave energy in each angled component of this period band, so that each contains a height $22.6/\sqrt{2}$ = 16 ft.
Diffraction computation: d/L = 0.0965; L = 373 ft.; R/L = 1000/373 = 2.7.
Assume 50% reflection, or reflected wave = 0.7 incident height;
θ_1 = 80°; α_1 = 35°; $2\theta_1 - \alpha_1$ = 125°.
θ_2 = 100°; α_2 = 55°; $2\theta_2 - \alpha_2$ = 145°.

From Table 7-I:
K_1 = 0.155 + 0.7 × 0.055 = 0.193;
K_2 = 0.105 + 0.7 × 0.053 = 0.142;
H = (0.193 + 0.142)16 = 5.35 ft.

From Table 5-IV angle of waves to wind after refraction to d/L_0 = 0.065 is 25°.
The height of each angled component is $22.1/\sqrt{2}$ = 15.65 ft.
Diffraction computation: d/L = 0.1092; L = 330 ft.; R/L = 1000/330 = 3.0.
Assume 50% reflection, or reflected wave = 0.7 incident height:
θ_1 = 65°; α_1 = 20°; $2\theta_1 - \alpha_1$ = 110°
θ_2 = 115°; α_2 = 70°; $2\theta_2 - \alpha_2$ = 160°.

From Table 7-I:
K_1 = 0.228 + 0.7 × 0.056 = 0.267;
K_2 = 0.080 + 0.7 × 0.047 = 0.113;
H = (0.267 + 0.113)15.65 = 5.95 ft.

Thus the short duration condition for 45-knot wind produces slightly higher wave height than the FAS from 35-knot wind. The relative T_{max} values are 10.4 and 11.6 sec, respectively.

6

Waves were measured seaward of a block-type breakwater in 15 ft. of water when 10-sec swell was striking it normally. By means of a boat the heights of the antinodal and nodal oscillations of the standing wave were measured. These were respectively 8.5 and 4.2 ft. Find the height of the incident wave, the coefficient of reflection and the order of scour at the foot of the structure.

EXAMPLES

From eq. 7-5, $K'_r = \dfrac{H'_r}{H'_i} = \dfrac{H_{max} - H_{min}}{H_{max} + H_{min}} = \dfrac{4.3}{12.7} = 0.34$; $T = 10$ sec; $d/L_0 = 15/512 = 0.0293$; $d/L = 0.0705$; $L = 213$ ft.; $H'_i/L = 12.7/213 = 0.06$.

From Fig. 7-17 (for $d/L = 0.08$), $K_r = 0.8$, $H_i/L = 0.43$ or $H_i/H'_i = (1+K'_r)/(1+K_r)$ or $H_i = 12.7 \left(\dfrac{1+0.34}{1+0.8} \right) = 9.4$ ft. Check $H_i/L = 9.4/213 = 0.044$.

From Fig. 7-20 for zero porosity and $\alpha = 90°$ and $K_r \doteq 0.8$, $S/H_0 \doteq 0.7$. At $d/L_0 = 0.0293$, $H/H_0 = 1.13$ so that $S = 9.4/1.13 = 8.3$ ft. Check $H_0/L_0 = 9.4/1.13 \times 512 = 0.0163$ OK.

Should the waves be angled to the wall, or near the extremity of a section, greater scour must be allowed for.

7

A FAS from a 35-knot wind is directed at a coast and strikes a stone revetment consisting of 3 ft. blocks sloped at 1 : 3 which runs down to the seabed, which is 15 ft. deep at the toe. Considering the $H_{1/3}$ and T_{max} values to constitute a monochromatic wave of equal energy, determine the maximum reach of the waves at the antinodes of the partial standing wave just seaward of the revetment on the basis of: (a) a smooth impermeable wall; (b) a rough impermeable wall; and (c) a permeable rough wall.

From Table 3-VII, $U_{19.5} = 35$ knots, $T_{max} = 11.6$ sec.
From Fig. 5-20, $d/L_0 = 15/5.12 \times 11.6^2 = 0.0217$; $H_{1/3}/T_{max}^2 = 0.08$; $H_{1/3} = 10.8$ ft. in 15 ft. depth. For $d/L_0 = 0.0217$, $d/L = 0.06$, $H/H_0 = 1.2$, $L = 250$, $H_0 = 9.0$ ft.; $H_0/L_0 = 9.0 / 5.12 \times 11.6^2 = 0.013$.

From Fig. 7-18 (using Moraes' results), $K'_r = 0.75$; $H_r = 10.8 \times 0.75 = 8.1$ ft. (nearly complete clapotis). $H_{max} = H_i + H_r = 18.9$ ft. Check for stability $H/L = 18.9/250 = 0.0757$.

From eq. 4-89, $(H/L)_{max} = 0.22 \tanh 2\pi d/L = 0.22 \times 0.35 = 0.077$ wave near breaking.

(a) smooth impermeable wall: from Fig. 4-18 for $d/L = 0.06$ and $H/L = 0.075$, $\Delta H/H = 1.0$; $a_c/H = 0.5 + 1.0 = 1.5$; $a_c = 1.5 \times 18.9 = 28.3$ ft.

(b) rough impermeable wall: from Fig. 7-19, $B/L_0 = 44 \times 10^{-4}$; $H_0/L_0 = 0.013$; $K_r = 0.43$; $H_{max} = 1.43 \times 10.8 = 15.4$ ft.; $H/L = 15.4/250 = 0.0615$; from Fig. 4-18, $\Delta H/H = 0.7$; $a_c = 15.4(0.5 + 0.7) = 18.5$ ft.

(c) rough permeable wall: from Fig. 7-20, assuming void ratio = 45%, $\alpha = 20°$; $H_0/L_0 = 0.013$; $K_r = 0.155$; $H_{max} = 1.155 \times 10.8 = 12$ ft.; $H/L = 12/250 = 0.48$; from Fig. 4-18, $\Delta H/H = 0.58$; $a_c = 12(0.5 + 0.58) = 12.85$ ft.

Thus, great savings can be effected in the height of a structure to be safe from such waves if roughness and permeability can be used to dissipate wave energy. By using both these mechanisms the height is reduced to half of that for a smooth sloping wall.

The scour for case c with void ratio = 45%, from Fig. 7-20, is $S/H_0 = 0.13$, $S = 0.13 \times 10.8/1.2 = 1.17$ ft.

The scour for case c with void ratio = 0, from Fig. 7-20, is $S/H_0 = 0.45$, $S = 0.45 \times 10.8/1.2 = 4.05$ ft.

Although Fig. 7-20 is not directly applicable to the smooth-wall case it can be used by amplifying the H_0 by the ratio of the standing waves in the rough and smooth cases. Thus with $S/H_0 = 0.45$, $S = 0.45 \times 10.8 \times 18.9/(15.4 \times 1.2) = 5.0$ ft. Thus, there is an added saving in reducing the reflected wave by decreasing the scour to about a quarter of that without the effect of porosity.

8

A 12-sec swell whose deep-water height is 8.0 ft. strikes normally on a stone revetment of slope 1 : 2.5, the toe of which is in 10 ft. of water at high tide and zero at low tide. Assuming the surface to be smooth, determine the run-up for HWL and LWL. If the waves are angled at 30° to the wall at high tide what is the run-up? For the high-tide normal approach compute the run-up if the revetment is both permeable and rough.

Smooth-wall HWL: Fig. 7-21, $H_o/L_o = 8/5.12 \times 12^2 = 0.011$; $d/L_o = 10/5.12 \times 12^2 = 0.0136$; $R/H_o = 2.9$; $R = 8 \times 2.9 = 23.2$ ft.

Smooth-wall LWL: Fig. 7-22, $d/L_o = 0$; $R/H_o = 0.65$; $R = 5.2$ ft.

Smooth-wall HWL (30° approach): Table 7-II, $K_\theta = 0.90$; $R = 0.90 \times 23.2 = 20.7$ ft.

Rough-wall HWL (normal approach): since the run-up should be in proportion to the H_{max} existing in front of the wall ($= H_i + H_r$) it should be in proportion to $(1+K_r)$ for the smooth and rough conditions. Fig. 7-18, slope 40%, $K_r = 1.0$; Fig. 7-20, slope 20°, $K_r = 0.16$; $R_R/R_S = (1+0.16)/(1+1) = 0.58$; $R_R = 23.2 \times 0.58 = 13.5$ ft.

9

A FAS from a 30-knot wind ($U_{19.5}$) strikes a coast where a smooth seawall of 1 : 2 slope exists. The sea bed slopes at 1 : 20 in front and is at a depth of 12 ft. at the toe of the wall. Determine the run-up of the significant waves.

Fig. 5-22: $d/L_o = 12/5.12 \times 9.9^2 = 0.024$; $H_{1/3}/T^2_{max} = 0.085$; $H_{1/3} = 8.34$ ft.

From eq. 7-8, $R = C \tan \alpha H_{1/3} (gT^2_{max}/2\pi H_{1/3})^{1/2}$; $R = 0.7 \times 0.5 \times 8.34 (5.12/0.085)^{1/2} = 22.7$ ft.

10

A vertical seawall is in 12 ft. of water and rises 3 ft. above SWL. Waves of 10 ft. height and 9 sec period reach the wall. Calculate the overtopping mean discharge per ft. length of wall. What is the velocity of flow at the top of the wall? What would be the required height of wall for no flooding to take place?

$H = 10$ ft., $h = 3$ ft., $d = 12$ ft., $d/L_o = 12/5.12 \times 9^2 = 0.029$, $d/L = 0.07$, $L = 171$ ft.

From Fig. 7-24 for $R_o/H = 0.6$ (independent of H/L) and $h/R_o = 3/0.6 \times 10 = 0.5$, $q_{ave} 10^2/g^{1/2} H^{3/2} 1.2 = 0.65$; $q_{ave} = 0.65(32.2)^{1/2}(10)^{3/2} 1.2/10^2 = 1.4$ ft.^{3}/sec per ft. length.

From eq. 7-10, the volume discharged per wave period $V_T = q_{ave} T = 12.6$ ft.^{3} per ft. length. The velocity at the wall crest $v = 2q_{ave}/(R_o - h) = 2 \times 1.4/(6-3) = 0.93$ ft./sec, but this value could be doubled if proven in the laboratory. For $d/L_o = 0.029$, $H/H_o = 1.133$; $H_o = 8.8$ ft.; $H_o/L_o = 0.0213$. From Fig. 7-25, for $(H_o/d)_c = 8.8/12 = 0.735$, $h/d = 1.8$; $h = 21.6$ ft.

Thus it is uneconomical to preclude overtopping in these circumstances. This value of h/d in essence $= R/d$, so that $R/H_o = 1.8/0.735 = 2.45$ which should be compared with values from Fig. 7-21 and 7-23.

11

A harbour 2400 ft. square is 40 ft. deep and is connected to the ocean by a breakwater gap. Determine the wave periods that could set up resonance in such a basin with various widths of gap.

Considering the first harbour with very small gap or as fully enclosed. From eq. 7-14:

$$T = \frac{2}{\sqrt{gd}} \left[\left(\frac{m}{a}\right)^2 + \left(\frac{n}{a}\right)^2 \right]^{-1/2} = \frac{2a}{\sqrt{gd}} \left(\frac{1}{m^2 + n^2}\right)^{1/2}$$

A variety of modes in each direction is possible, as listed in Table 7-V, from which the related period can be determined: $\dfrac{2a}{\sqrt{gd}} = \dfrac{2 \times 2400}{\sqrt{32.2 \times 40}} = 134$.

Consider now a finite gap. From Fig. 7-26 for fundamental mode and $W/B = 1.0$, for a very small gap $G/W = 0.01$, $W/L = 0.08$, thus $2400/T\sqrt{g40} = 0.08$; $T = 836$ sec. This fundamental mode has no nodes and is a pumping action over the whole harbour.

For the second mode in Fig. 7-26 with $G/W = 0.01$, $W/L = 0.51$, so that $T = 131.5$ sec which is slightly less than the same mode of the closed basin.

With the fully open harbour $G/W = 1.0$, for the fundamental mode $W/L = 0.15$ and the second mode $W/L = 0.56$, give $T = 4460$ and 119.5 sec, respectively.

EXAMPLES

TABLE 7-V

Parameters for Example 11

m	0	0	1	1	1	2	2	3
n	1	2	1	2	3	2	3	3
$1/\sqrt{m^2+n^2}$	1	1/2	$1/\sqrt{2}$	$1/\sqrt{5}$	$1/\sqrt{10}$	$1/\sqrt{8}$	$1/\sqrt{13}$	$1/\sqrt{18}$
	1	0.5	0.707	0.447	0.317	0.353	0.278	0.236
T (sec)	134	67	90	60	42	47	37	32

Other modes for the finite open case will be in proportions somewhat similar to those for the closed basin listed above. For modes higher than those included the resonant wave periods will decrease. By this stage they may no longer be shallow-water waves. Although the amplification process decreases as higher modes are generated so also does the attenuation due to friction of the bed. Thus, in any rectangular or square harbour dimensions which are multiples of swell wave length for the specific harbour depth (or its tidal variations) should be avoided. The critical periods for ship ranging are 30–120 sec, but shorter-period horizontal oscillations can make tug and barge operation difficult.

If this were a small craft harbour it could be one quarter the size and one quarter the depth. This would cause the resonant periods to be about half those quoted above, which brings them well within the range of long-period swell.

12

A harbour has a dredged rectangular approach channel 500 ft. wide, 45 ft. deep and 4000 ft. in length. It is desired to attenuate 10-sec waves to 1/4 their height at the seaward entrance (= 5 ft.) by means of vertical baffles on the sides. What size and spacing should these be to effect such attenuation?

From eq. 7–19, $H/H_o = 1(1 + \beta H_o x/2)$, so that $1/4 = 1/(1 + \beta 5 \times 4000/2)$, $\beta = 3/10,000$, where $\beta = \frac{8}{3W_c L_o} \frac{C}{C_g} \left(\frac{1}{3} + \frac{1}{\sinh^2 2\pi d/L}\right) C_f$, $T = 12$ sec., $d/L_o = 45/5.12 \times 10^2 = 0.088$, $d/L = 0.1304$, $L = 345$ ft., $C/C_g = 0.831$, $\sinh 2\pi d/L = 0.914$, thus:

$$\frac{3}{10,000} = \frac{8}{3 \times 500 \times 5.12 \times 10^2} 0.89 \left(\frac{1}{3} + \frac{1}{(0.914)^2}\right) C_f$$

$35.9 = (0.33 + 1.2)C_f$: $C_f = 23.5$

From Fig. 4-4, $X = 1.1$ at bed and 1.45 at surface so that:

$$\frac{2x_{max}}{H} = \frac{1.1 + 1.45}{2} \quad \text{or} \quad x_{max} = 5 \times 2.55/4 = 3.18 \text{ ft.}$$

In Fig. 7-28, although C_f exceeds 20, the limit of $d/L_o > 0.5$ and that $d/L_o < 0.1$, these differences are small enough for the graph to be used for initial design purposes. The choices are: $s/k = 2$, $x_{max}/k = 0.08–0.14$ and $s/k = 4x_{max}/k = 0.08–0.25$.

Since the narrowest batten and greatest spacing is desirable, choose $x_{max}/k = 0.25$ or $k = 3.18/0.25 = 12.7$ ft.; $s/k = 4$; $s = 51$ ft.

13

It is required to design a rubble-mound breakwater which will reduce a 10-sec wave with deep-water height of 5.0 ft. to one third its height in front of the structure. The mound can be considered as rectangular and is located in 20 ft. of water.

H_o = 5 ft., d/L_o = 20/5.12 × 10² = 0.039, d/L = 0.0821, L = 244 ft., H/H_o = 1.069, H_i = 5.35 ft.
From Fig. 7-30B for K_t = 0.3, see results in Table 7-VI.

TABLE 7-VI

Experimental ratios for Example 13

Worker	h/H_i	B/L	B	$d-h$	$(d-h)B$
A. Goda (d/L = 0.1–0.3)	−0.25	0.06–0.5	< 122	21.3	< 2600
B. Nakamura	0.5	0.5	122	17.3	2110
C. Nakamura	1.2	1.0	244	13.6	3310
D. Nakamura	1.5	≥ 2.0	488	12.0	5850

From Fig. 7-33A, impermeable (Dick [65]),
(E): h/d = 0.05; L/B = 6.5; B = 37.6; $(d-h)$ = 19; $(d-h)B$ = 714.
(F): h/d = 0.1; L/B = 2.0; B = 122, $(d-h)$ = 18; $(d-h)B$ = 2200.
Check that waves will break, for solutions B, E and F in Fig. 7-32 this is so.
The solution E appears the most economical from the point of view of volume. If the structure could be considered permeable, Fig. 7-33B would indicate that h/d = 0.10, and L/B = 6.5 would provide K_t = 0.3.

14

A submerged horizontal platform 50 ft. wide is held rigidly 4 ft. below the surface in 20 ft. of water. A wave 7 ft. high and 9 sec period traverses it. Determine the height of the wave transmitted shorewards from it, excluding the effects of diffraction at the ends. Calculate the phase lag of the wave passing over the plate and that propagating beneath it. Is this the most ideal length and depth to attenuate waves?

d/L_o = 20/5.12 × 9² = 0.048, d/L = 0.092, L = 217 ft.
Fig. 7-33C: L/B = 4.35; h/d = 4/20 = 0.20; K_t = 0.4. Speed of the wave above the plate = $\sqrt{gd}$ = $\sqrt{g}$ 4 = 11.35 ft./sec; speed of wave beneath plate = C_o tanh $2\pi d/L$ = 5.12 × 9.0 × 0.52 = 24 ft./sec. Lower wave takes 50/25 = 2.08 sec to reach leeward end in which time upper wave has advanced 2.08 × 11.35 = 23.6 ft. This is (23.6/11.35 × 9)2π = 0.461π radians. According to Fig. 7-33C, this is not ideal length for this submergence as lowest K_t occurs at L/B = 3.25 or B = 67 ft. This results in a phase lag of (35.3/11.35 × 9)2π = 0.69π.
A better proposition according to Fig. 7-33C is a plate B/L = 1/4 (B = 54 ft.) at h/d = 0.10 or h = 2 ft. The celerity above the plate $\sqrt{gd}$ = 8 ft./sec, so that whilst lower wave traverses the plate in 2.25 sec the upper wave travels 18 ft., so that the phase lag is (36/8 × 9)2π = π. In this case the trough of the upper wave is synchronous with the crest of the lower wave.
Note that further research is required on this phenomenon.

15

The hulk of a ship is to be anchored in front of a beach to provide a sheltered region behind temporarily. It has a 3 ft. clearance from the bed in 18 ft. of water and has a natural period of oscillation when anchored of 20 sec. The hulk can be considered of rectangular section and beam width of 60 ft. Determine the height of wave reaching the beach when 6-sec waves 4 ft. high are arriving normal to the coast. Compare the cases of rigidly fixed, free floating and anchored.

d/L_o = 18/5.12 × 6² = 0.0978, d/L = 0.139, L = 130 ft. Using Fig. 7-37 for rigid barrier, h/d = 15/18 = 0.83, K_t = 0.35.
For free floating, eq. 7-27 gives:

$$K_t = 1/\sqrt{1 + (\pi W/wLd)^2} = 1/\sqrt{1 + (\pi\, 15 \times 60 \times 64/64 \times 130 \times 18)^2} = 0.64.$$

For anchored condition, eq. 7-29 gives:

$$K_t = 1/\sqrt{1+(\pi W/wLd)^2[(T/T_s)^2 - 1]^2} = 1/\sqrt{1+(\quad)^2[(6/20)^2 - 1]^2} = 0.67.$$

Thus, the waves reaching the beach have heights of 1.4, 2.56 and 2.86 ft., respectively.

The maximum force on the anchor line per unit length of hulk is given by eq. 7-30:

$$F_{max} = |wd\sqrt{H_i^2 - H_t^2}/[1-(T_s/T)^2]|$$

F_{max} per unit length = $64 \times 18\sqrt{4^2 - 2.68^2}/[1 - (20/6)^2] = 9900/10.1 = 981$ lb. The force on the rigid barrier is 9900 lb./ft. It can be seen from the above that use of old ships to provide shelter is not very efficient unless they are sunk in position.

16

In place of the ship hulk in Example 15, a double-layer pontoon as tested by Hom-ma et al. [84] is to be used. Its width will be the same at 60 ft. and the submerged raft is slung at 8.5 ft. depth. Determine the wave height reaching the beach for similar conditions as existing before.

The tests conducted by Hom-ma et al. do not specify d/L or H/L conditions for each series of tests but a wide range of conditions appears to be included. From Fig. 7-40, $h/d = 8.5/18 = 0.47$, and $B/L = 60/130 = 0.46$, $K_t = 0.22$.

If these two layers of pipes were laid end to end at the surface (articulated for convenience) $B/L = 0.92$, $K_t = 0.5$ to 0.65. The advantage of the submerged layer should be apparent. To determine mooring forces, tests would need to be carried out. Because of the attenuation through dissipation rather than reflection these forces should be minimal.

PROBLEMS

1
A harbour has a breakwater to provide sheltered conditions for ships. Determine the wave height reduction for a point 1000 ft. radius from the breakwater tip and 30° to it. The 10-sec waves are arriving at 60° to the breakwater in a depth which can be considered uniform at 35 ft. What is the reduction for 7- and 13-sec waves at the same point when arriving at the same angle?

2
A port is formed by two breakwaters which leave a gap of 936 ft. at the 35 ft. deep entrance. Waves of 10 sec period and 5 ft. height arrive at 90° to the gap. If the whole harbour basin can be considered of uniform 35 ft. depth, determine the wave height on the radius 1200 ft. from the gap centre.

3
The alternatives are available in Fig. 7-42 of constructing a breakwater on AB or BC in order to protect point P from waves arriving as shown. With AB, AC and CP all equal to 4 times the wave length, determine the best location on the basis of the smallest wave height at P.

4
Cyclonic waves are arriving at the site illustrated in Fig. 7-42. They consist of $H_{1/3} = 22$ ft., and $T_{max} = 10$ sec. The area has a uniform depth of 92 ft. except for the small craft harbour, which is uniformly 45 ft. deep. Compute the wave heights at P and Q.

5
A headland runs parallel to the general coastline where the continental shelf is reasonably

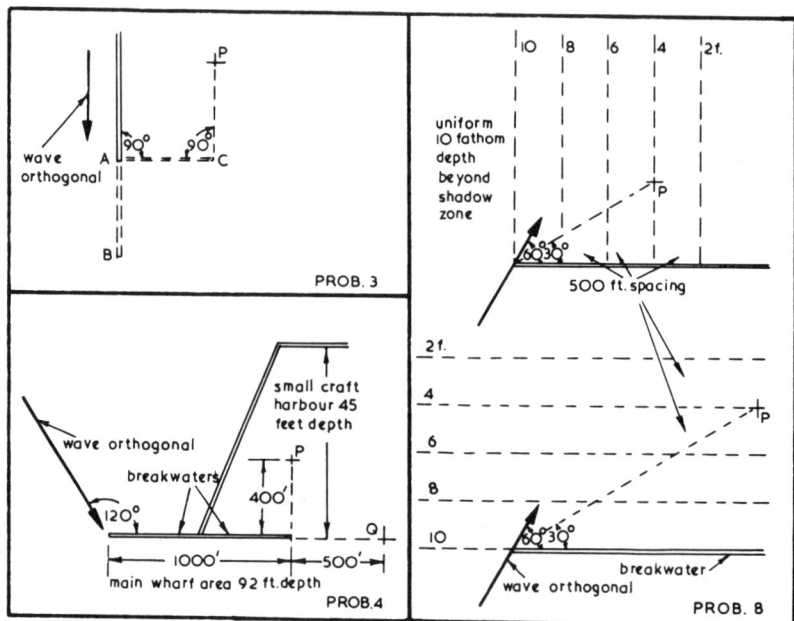

Fig. 7-42. Sketch plans for Problems 3, 4 and 8.

uniform in contour. The zone leeward of the headland can be considered of a uniform depth of 30 ft. and is being considered for a port area. A study of cyclonicity in the adjacent ocean area indicates that the worst wave conditions could arise from 40-knot winds, with a fetch of 400 NM following the waves for two days, at an angle of 45° to the coast, which gives maximum protection to the port. Another wind condition must also be considered, that of a FAS from a 25-knot wind blowing directly onshore. Find the wave height at a point 2000 ft. inside the shadow zone on a radius from the headland extremity at 60° to its inside face. Because of the existence of a rocky beach in front of the headland assume no reflection occurs.

6
An offshore breakwater is to be constructed parallel to the shore to afford protection for small craft. It is 600 ft. long in 15 ft. of water and the storm waves, which arrive at 45° to the breakwater have a period of 9 sec and significant height of 10 ft. Considering these waves to be concentrated in this one direction, draw a plan of the shadow zone, noting the maximum wave heights at the corners of 100 ft. square grids to a distance of 400 ft. from the breakwater. Assume 100% reflection from the seaward side of the structure. Assuming the mass transport of sediment to be proportional to (wave height)^2 in each component wave, draw the vectors and their resultants at each grid point. What conclusions can be drawn about possible silting behind the breakwater and the effect this will have on wave heights in the shadow zone?

7
In designing breakwater lengths and alignments for normal commercial ports and for refuge harbours for small craft, what differing design criteria would you use?

PROBLEMS 363

8
Two conditions of combined refraction and diffraction are depicted in Fig. 7-42, for which comparative heights at point P are required. The waves are 10-sec period and the depths are as illustrated. In each case compare the height at P with that for a uniform depth over the whole shadow zone of 60 ft. Account for diffraction, refraction, shoaling and change of wave length in determining R/L or S/L values.

9
Tests are being conducted in a flume on a marine structure from which waves are being partially reflected. The maxima and minima have been measured by a mobile wave gauge as 12 and 3 cm, respectively when the water depth was 50 cm and the wave period T sec. Determine the apparent and correct reflection coefficient plus the correct height of the incident wave.

10
Compare the reflection coefficients for a 30% smooth slope with that of an impermeable surface with roughness elements of 2 ft. dimension when deep-water waves of 4 ft. high and 10 sec period are striking them. If this slope could be considered to have a void ratio of 20% what would be the K_r value? In this last case what scour could be expected at the toe of the wall?

11
Compute the run-up for a smooth wall of 1 : 2 slope, whose toe is in 6 ft. of water, when waves 5 ft. high and 6 sec period are striking it normally. For this particular wave band what is the largest run-up to be expected? At this optimum condition determine the ratio H/d and comment on the result.

12
If the waves in Problem 11 were angled 45° to the coast in deep water instead of arriving normally, compute the resulting run-up.

13
A 35-knot wind raises a FAS towards the shore where a wall is installed to protect power-house facilities from overtopping. What height should the wall be built if it is to slope at 1 : 2, consist of rock fill of large armour size, and have a rock bed in front at a depth of 10 ft.

14
Protection for a shoreline consists of a stone revetment at 30° slope. It extends from a depth of 15 ft. to a height 4 ft. above SWL. Waves 8 ft. high and 12.5 sec period run up the slope and wash over its crest. Determine the average discharge and the volume to be drained each wave cycle per unit length of revetment. If a vertical wall were constructed at the crest which was 2 ft. high, by what percentage would the overtopping be reduced?

15
Waves of height 10 ft. and period 11 sec must be contended with in an 8 ft. depth where seawalls are to be provided. Compare the average overtopping discharge for the five varieties of wall depicted in Fig. 7-24. Determine the height of each wall to exclude such flooding completely.

16
A harbour basin of constant depth 6 fathoms, is 1500 ft. wide and 7500 ft. long. It is open completely at one end to the sea. What wave periods can create resonance in such a harbour? If the harbour were triangular in plan, with zero width at the head, what changes in these resonant conditions would accrue?

17
A rectangular harbour is connected to the sea through a breakwater gap 600 ft. wide in the centre of one side which is 6000 ft. long. The width of the basin is 4000 ft. If the harbour is dredged for a uniform depth of 40 ft., check which wave periods could cause standing oscillations in it. If such waves were known to exist in the area, what changes should be considered first to overcome the problem?

18
In a harbour layout as in Fig. 7-27 the following dimensions apply: $W_1 = 450$ ft., $W_2 = 200$ ft., $B = 1000$ ft., $d_1 = 60$ ft., $d_2 = 30$ ft. Long waves of 20 sec are known to occur frequently in the area. Test to see if the lay-out chosen will resonate with this input oscillation.

19
A rectangular canal joins a harbour basin to the sea. It is 2 km in length, 20 m deep and 200 m wide. Design vertical battens for the sides to attenuate waves of 8 sec period from their 2 m height at the entrance to 0.5 m at the harbour.

20
A rectangular canal runs at right angles to the coast inland to a harbour basin. A 35-knot FAS is directed up this canal which is 40 ft. deep and it is desired to attenuate the bulk of the spectrum by three resonant structures at the entrance to the canal. Compute the dimensions of these resonators which are of the same depth as the channel. How should these be spaced along its length?

21
A submerged breakwater in 20 ft. of water is meant to break all waves arriving from a FAS generated by a 30-knot wind. Design the structure for the minimum volume of rock-fill assuming a rectangular cross-section.

22
Compare the attenuation of waves by a thin barrier rising from the seabed 0.8 of the depth with that of a barrier from the surface penetrating a similar proportion. List these for the ratios $d/L = 0.5, 0.2, 0.1, 0.05$, and comment on the result.

23
A rock-fill structure is to be used to reduce wave energy arriving at a beach by one quarter. If it is placed in 12 ft. of water and the most persistent swell is 12 sec with deep-water height of 3 ft., design the structure to achieve this goal. Consider it to be rectangular in shape. Find a solution for both impermeable and permeable conditions.

24
A trapezoidal impermeable rock-filled breakwater is built up to SWL where its width is 18 m. Determine the height of the wave being transmitted to the shore if the incident wave has 10 sec period and height of 5 ft.

25
A platform is held rigidly horizontal 5 ft. beneath the surface in 25 ft. of water. Determine its length in order for 10-sec waves to have minimum transmission to leeward of it. What effect does it have on waves of 8 and 12 sec?

26
Piles of 40 cm diameter and 15 cm clearance are driven into the sea bed where the water depth is 10 m. Waves of 3 m height and 12 sec period arrive normal to the pile alignment. Calculate

PROBLEMS

the height of the transmitted wave. What is the effect of adding a second row of piles behind the first? What is the maximum reach of the waves on the seaward side of the pile array in both cases?

27
Flexible or articulated floats all appear to have the same attenuation effects on waves. List the materials and conditions where such solutions may be economical.

28
It has been suggested that a submerged flat plate is an economical breakwater, particularly if it is perforated to permit flow through it. At which end of the floating structure would such perforations be most effective?

29
Compare the attenuation by a double raft as in Fig. 7-40 with rigid flat plate as in Fig. 7-33C. Test for the condition of $h/d = 0.2$ which is common to both, but observe the trends outside this range. What do the results suggest for further research in this topic?

30
Discuss the phenomenon of harbour seiching or resonance and give reasons for avoiding it.

31
If you were called upon to design a structure, either fixed or floating, to protect some offshore operation, what shape in plan would you contemplate? List the possible installations without consideration of cost. List also the various facets of the problem from fabrication to final withdrawal from the site.

32
Discuss the possible uses of offshore breakwaters near the coast. Compare the benefits of fixed or floating structures.

33
It has been noted in this chapter that it is better to dissipate wave energy rather than reflect it seawards. List reasons for this.

34
Discuss the various possibilities of waves running-up and possibly overtopping a sea wall. Whilst such a structure may be designed for one tidal condition, what different water levels need to be taken into account?

35
For what reasons should one need to assess the overtopping discharge of a sea wall? What effects could wind have on such flooding?

36
List the advantages of a trapezoidal channel approach to a harbour over a rectangular shape. What safeguards should be provided at the waterline of such channels?

37
Describe the action of dissipation for a submerged raft consisting of rows of pipes.

REFERENCES

[1] A. Wada, 1965. On a method of solution of diffraction problems. *Coastal Eng. Japan*, 8: 1–19.
[2] W.G. Penney and A.T. Price, 1952. The diffraction theory of sea waves by breakwaters. *Phil. Trans. R. Soc.*, A 244: 236–253.
[3] A. Sommerfeld, 1896. Mathematische Theorie der Diffraction. *Math. Ann.*, 47: 317–374.
[4] T.K. Lim, 1968. *Wave Diffraction*. Thesis, Asian Institute of Technology, Bangkok.
[5] F. Biésel, 1964. Les phénomènes du second ordre rayonnants dans les ondes de gravité. *Houille Blanche*, 17: 403–420.
[6] S.H. Fan, J.E. Cumming and R.L. Wiegel, 1967. Computed solution of wave diffraction by semi-infinite breakwater. *Univ. Calif., Berkeley, Tech. Rep.*, HEL-1-8.
[7] J. Larras, 1966. Diffraction de la houle par les obstacles rectilignes semi-indéfinis sous incidence oblique. *Cah. Océanogr.*, 18: 661–667.
[8] R.L. Wiegel, 1962. Diffraction of waves by semi-infinite breakwater. *Proc. ASCE*, 88 (HY1): 27–44.
[9] R. Silvester and T.K. Lim, 1968. Application of wave diffraction data. *Proc. 11th Conf. Coastal Eng.*, 1: 248–270.
[10] J.A. Putman and R.S. Arthur, 1948. Diffraction of water waves by breakwaters. *Trans. Am. Geophys. Union*, 29: 481–490.
[11] F.L. Blue and J.W. Johnson, 1949. Diffraction of water waves passing through a breakwater gap. *Trans. Am. Geophys. Union*, 30: 705–718.
[12] P.M. Morse and P.J. Rubenstein, 1938. The diffraction of waves by ribbons and slits. *Phys. Rev.*, 54: 895–898.
[13] J.H. Carr and M.E. Stelzriede, 1952. Diffraction of water waves by breakwaters. In: *Gravity Waves – U.S. Natl. Bur. Std., Circ.*, 521: 109–125.
[14] H. Lacombe, 1952. The diffraction of a swell. A practical approximate solution and its justification. In: *Gravity Waves – U.S. Natl. Bur. Std., Circ.*, 521: 129–140.
[15] F.L. Blue, 1948. *Diffraction of Water Waves Passing through a Breakwater Gap*. Thesis, Univ. of California, Berkeley.
[16] I.E. Mobarek and R.L. Wiegel, 1966. Diffraction of wind-generated water waves. *Proc. 10th Conf. Coastal Eng.*, 1966: 185–206.
[17] I.E. Mobarek, 1962. Effects of bottom slope on wave diffraction. *Univ. Calif., Berkeley, Tech. Rep.* HEL-1-1.
[18] S.S. Fan and L.E. Borgman, 1970. Computer modelling of diffraction of wind waves. *Proc. 12th Conf. Coastal Eng.*, 1: 473–488.
[19] A.M. Zhukovets, 1964. The action of wind on diffracted waves or wakes. *Deep-Sea Res.*, 11: 637–645.
[20] J.J. Healy, 1953. Wave damping effect of beaches. *Proc. Minnesota Int. Hydr. Conv.*, 1953: 213–220.
[21] C. Carry, 1953. Clapotis partiel. *Houille Blanche*, 8: 482–494.
[22] Y. Goda and Y. Abe, 1968. Apparent coefficient of partial reflection of finite amplitude waves. *Rep. Port Harbour Res. Inst.*, 7 (3).
[23] R. Miche, 1951. Le pouvoir réflêchissant des ouvrages maritimes exposés à l'action de la houle. *Ann. Ponts Chaussées*, 121: 285–319.
[24] C.C. Moraes, 1970. Experiments of wave reflection of impermeable slopes. *Proc. 12th Conf. Coastal Eng.*, 1: 509–522.
[25] L.G. Straub, C.E. Bowers and J.B. Herbich, 1957. Laboratory tests of permeable wave absorbers. *Proc. 6th Conf. Coastal Eng.*, 729–742.
[26] T. Sawaragi, 1966. Scouring due to wave action at the toe of permeable coastal structures. *Proc. 10th Conf. Coastal Eng.*, 2: 1036–1047.

[27] Y. Tominaga, H. Hashimoto and N. Sakuma, 1966. Wave run-up and overtopping on coastal dikes. *Proc. 10th Conf. Coastal Eng.*, 1: 364–381.
[28] M.S. Longuet-Higgins and R.W. Stewart, 1962. Radiation stress and mass transport in gravity waves with application to "surfbeats". *J. Fluid Mech.*, 13: 481–504.
[29] S. Sato and T. Kishi, 1958. Experimental study of wave run-up on sea wall and shore slope. *Proc. Coastal Eng. Japan*, 1: 39–43.
[30] M. Hosoi and N. Shuto, 1964. Run-up height on a single slope dike due to waves coming obliquely. *Proc. Coastal Eng. Japan*, 7: 95–99.
[31] J.L. Machemehl and J.B. Herbich, 1970. Effects of slope roughness on wave run-up on composite slopes. *Texas A.M. Univ., COE Rep.*, 129.
[32] R.P. Savage, 1959. Laboratory data on wave run-up on roughened and permeable slopes. *Beach Erosion Board, Tech. Mem.*, 109.
[33] J.H. Van Oorschot and K. D'Angremond, 1968. The effects of wave energy spectra on wave run-up. *Proc. 11th Conf. Coastal Eng.*, 2: 888–900.
[34] I.A. Hunt Jr., 1959. Design of seawalls and breakwaters. *Proc. ASCE*, 85 (WW3): 123–152.
[35] T. Ishihara, Y. Iwagaki and H. Mitsui, 1960. Wave overtopping on seawalls. *Proc. Coastal Eng. Japan*, 3: 53–62.
[36] Y. Iwagaki, A. Shima and M. Inoue, 1965. Effects of wave height and sea water level on wave overtopping and wave run-up. *Proc. Coastal Eng. Japan*, 8: 141–151.
[37] T. Saville Jr., 1955. Laboratory data on wave run-up and overtopping on shore structures. *Beach Erosion Board, Tech. Mem.*, 64.
[38] O.J. Sibul and E.G. Tickner, 1956. Model study of overtopping of wind-generated waves on levees with slopes of 1 : 3 and 1 : 6. *Beach Erosion Board, Tech. Mem.*, 80.
[39] H. Kikkawa, H. Shi-Igai and T. Kono, 1968. Fundamental study of wave overtopping on levees. *Proc. Coastal Eng. Japan*, 11: 107–115.
[40] H. Shi-igai and T. Kono, 1970. Analytical approach on wave overtopping on levees. *Proc. 12th Conf. Coastal Eng.*, 1: 563–574.
[41] S. Tsuruta and Y. Goda, 1968. Expected discharge of irregular wave overtopping. *Proc. 11th Conf. Coastal Eng.*, 2: 833–852.
[42] N. Shiraishi, A. Numata and T. Endo, 1968. On the effects of armour block facing on the quantity of wave overtopping. *Proc. 11th Conf. Coastal Eng.*, 2: 853–869.
[43] Y. Iwagaki, Y. Tsuchiya and M. Inoue, 1966. On the effect of wind on wave overtopping on vertical seawalls. *Bull. Disaster Prev., Res. Inst., Kyoto Univ.*, 16 (I): 11–30.
[44] J.S. McNown, 1952. Waves and seiche in idealised ports. In: *Gravity Waves – Natl. Bur. Std., Circ.*, 521: 153–164.
[45] A.T. Ippen and F. Raichlen, 1962. Wave-induced oscillations in harbours: the problem of coupling of highly reflective basins. *Hydr. Lab. MIT, Rep.*, 49.
[46] R. Dorrestein, 1961. Amplification of long waves in bays. *Florida Exp. Sta., Tech. Pap.*, 213.
[47] A.T. Ippen and Y. Goda, 1963. Wave-induced oscillations in harbours: The solution of a rectangular harbour connected to the open sea. *Hydr. Lab. MIT, Rep.*, 59.
[48] J. Miles and W. Munk, 1961. Harbour paradox. *Proc. ASCE*, 87 (WW3): 113–130 (see also discussion in 88 (WW4): 173–195).
[49] F. Biésel and B. Le Méhauté, 1956. Mouvements de résonance à deux dimensions dans une enceinte sous l'action d'ondes incidentes. *Houille Blanche*, 11: 348–374.
[50] B. Le Méhauté, 1961. Theory of wave agitation in a harbour. *Proc. ASCE*, 87 (HY): 31–50.
[51] J.H. Carr, 1953. Long-period waves or surges in harbors. *Trans. ASCE*, 118: 588–603.
[52] K. Horikawa, N. Shuto and H. Nishimura, 1969. Characteristic oscillation of water in an L-shaped basin. *Proc. Coastal Eng. Japan*, 12: 47–56.
[53] S. Ishiguro, 1959. A method of analysis for long-wave phenomena in the ocean, using electronic network models, 1. The earth's rotation ignored. *Phil. Trans. R. Soc.*, A 251: 303–340.

[54] F. Raichlen, 1965. Long-period oscillations in basins of arbitrary shapes. *Proc. Santa Barbara Conf. Coastal Eng., 1965*: 115–145.
[55] C.E. Lee, 1968. On wave damping in harbors. *Proc. ASCE* 94 (WW4): 489–501.
[56] J.A. Battjes, 1965. Wave attenuation on a channel with roughened sides. *Proc. Santa Barbara Conf. Coastal Eng., 1965*: 425–460.
[57] J. Valembois, 1953. Etude de l'action d'ouvrages résonants sur la propagation de la houle. *Proc. Minnesota Int. Hydr. Conv., 1953*: 193–200.
[58] J. Valembois and C. Birard, 1955. Les ouvrages résonants et application à la protection des ports. *Proc. 5th Conf. Coastal Eng., 1955*: 637–641.
[59] W. James, 1968. Rectangular resonators for harbour entrances. *Proc. 11th Conf. Coastal Eng.*, 2: 1512–1530.
[60] W. James, 1971. Two innovations for improving harbour resonators. *Proc. ASCE*, 97 (WW1): 115–122.
[61] W. James, 1970. Spectral response of harbour resonator configurations. *Proc. 12th Conf. Coastal Eng.*, 3: 2181–2194.
[62] H. Kondo, 1970. An analytical approach to wave transmission through permeable structures. *Proc. Coastal Eng. Japan*, 13: 31–42.
[63] J.W. Johnson, H. Kondo and R. Wallihan, 1966. Scale effects in wave action through porous structures. *Proc. 10th Conf. Coastal Eng.*, 2: 1022–1024.
[64] T. Iwasaki and A. Numata, 1970. Experimental studies on wave transmission of a permeable breakwater constructed by artificial blocks. *Proc. Coastal Eng. Japan*, 13: 25–29.
[65] T.M. Dick, 1968. On solid and permeable submerged breakwaters. *Queen's Univ., Civil Eng. Rep.*, 59.
[66] Y. Goda, H. Takeda and Y. Moriya, 1967. Laboratory investigations on wave transmission over breakwaters. *Rep. Port Harbours, Res. Inst.*, 6 (13).
[67] Y. Goda, 1969. Re-analysis of laboratory data on wave transmission over breakwaters. *Rep. Port Harbour, Res. Inst.*, 8 (3).
[68] M. Nakamura, H. Shiraishi and Y. Sasaki, 1966. Wave damping effect of submerged dike. *Proc. 10th Conf. Coastal Eng.*, 1: 254–267.
[69] W.R. Dean, 1945. On the reflection of surface waves by a submerged plane barrier. *Proc. Camb. Phil. Soc.*, 41: 231–236.
[70] T.F. Ogilvie, 1960. Propagation of waves over an obstacle in water of finite depth. *Univ. Calif., IER Ser.* 82, *Rep.*, 14.
[71] E.O. Macagno, 1954. Houle dans un canal présentant en passage en charge. *Houille Blanche*, 9: 10–37.
[72] T. Ijima, S. Ozaki, Y. Eguchi and A. Kobayashi, 1970. Breakwater and quay wall by horizontal plates. *Proc. 12th Conf. Coastal Eng.*, 3: 1537–1556.
[73] T. Hayashi, T. Kano and M. Shirai, 1966. Hydraulic research on the closely spaced pile breakwater. *Proc. 10th Conf. Coastal Eng.*, 2: 873–884.
[74] T. Sawaragi and K. Iwata, 1970. Effect of structural shape on wave run-up and wave damping. *Proc. Coastal Eng. Japan*, 13: 55–74.
[75] R.D. Costello, 1952. Damping of water waves by vertical cylinders. *Trans. Am. Geophys. Union*, 33: 513–519.
[76] J.H. Carr, 1951. Mobile breakwaters. *Proc. 2nd Conf. Coastal Eng., 1951*: 281–295.
[77] J.J. Vinjé, 1966. Increase of effective working-time during operations at sea by means of movable structures. *Delft Hydrol. Lab., Publ.*, 42.
[78] F. Ursell, 1947. The effect of a fixed barrier on surface waves in deep water. *Proc. Camb. Phil. Soc.*, 43: 374–382.
[79] R.L. Wiegel, 1960. Transmission of waves past a rigid vertical thin barrier. *Proc. ASCE*, 86 (WW1): Pap. 2413.
[80] R. Lochner, O. Faber and W. Penney, 1948. The bombardon floating breakwaters. In: *The Civil Engineer in War*, 2. Inst. of Civil Engrs., London,

REFERENCES

[81] C.J.R. Wood, 1948. Phoenix. In: *The Civil Engineer in War, 2.* Inst. of Civil Engrs. London,

[82] A. Brebner and A.O. Ofuya, 1968. Floating breakwaters. *Proc. 11th Conf. Coastal Eng.*, 2: 1055–1094.

[83] A.O. Ofuya, 1968. On floating breakwaters. *Queen's Univ., CE Rep.*, 60.

[84] M. Hom-ma, K. Horikawa and H. Mochizuki, 1964. An experimental study on floating breakwaters. *Proc. Coastal Eng. Japan*, 7: 85–94.

[85] R.L. Wiegel, H.W. Shen and J.D. Cumming, 1962. Hovering breakwater. *Proc. ASCE*, 88 (WW2): 23–50.

[86] R.J. Kennedy and J. Marsalek, 1968. Flexible porous floating breakwaters. *Proc. 11th Conf. Coastal Eng.*, 2: 1095–1103.

[87] J. Kato, S. Hagino and Y. Uekita, 1966. Damping effect of floating breakwater to which anti-rolling system is applied. *Proc. 10th Conf. Coastal Eng.*, 2: 1068–1078.

[88] A.K. Johnston, 1958. Observations on floating breakwaters for reflection of shallow-water waves. *Houille Blanche*, 13: 540–550; 619–638.

[89] W. Marks, 1966. A perforated mobile breakwater for fixed and floating application. *Proc. 10th Conf. Coastal Eng.*, 2: 1079–1129.

[90] W. Marks and G.E. Jarlan, 1968. Experimental studies on fixed perforated breakwater. *Proc. 11th Conf. Coastal Eng.*, 2: 1121–1140.

[91] S. Tanaka, 1966. Researches on double curtain wall breakwater. *Proc. 10th Conf. Coastal Eng.*, 2: 913–931.

[92] A.T. Ippen and E.L. Bourodimos, 1964. Breakwater characteristics of open tube systems. *Hydr. Lab. MIT, Rep.*, 73.

[93] E.L. Bourodimos and A.T. Ippen, 1968. Characteristics of an open tube wave attenuation system. *Proc. ASCE*, 94 (WW4): 465–488.

[94] T.M. Dick and A. Brebner, 1960. A laboratory study of pneumatic breakwaters. *Queen's Univ., Civil Eng., Rep.*, 12.

[95] J.T. Evans, 1955. Pneumatic and similar breakwaters. *Proc. R. Soc. Lond., Ser. A*, 231: 457–466.

[96] J.T. Evans, 1955. Pneumatic and similar breakwaters; model experiments using surface currents. *Dock Harbour Authority*, 36: 251–256.

[97] J.L. Green, 1961. Pneumatic breakwaters to protect dredges. *Dock Harbour Authority*, 42: 41–42.

[98] W. Hensen, 1955. Modellversuche mit pneumatischen Wellenbrechern. *Mitt. Hannoverschen Versuchsanst.*, 7: 179–212.

[99] L.G. Straub, C.E. Bowers and E.S. Tarapore, 1959. Experimental studies of pneumatic and hydraulic breakwaters. *Univ. Minnesota, St. Anthony Falls Lab., Tech. Pap.*, B 25.

[100] L.G. Straub, J.B. Herbich and C.E. Bowers, 1957. An experimental study of hydraulic breakwaters. *Proc. 6th Conf. Coastal Eng., 1957*: 715–728.

[101] G.I. Taylor, 1955. The action of a surface current used as a breakwater. *Proc. R. Soc. Lond., Ser. A*, 231: 466–478.

[102] J.A. Williams and R.L. Wiegel, 1962. Attenuation of wind waves by a hydraulic breakwater. *Proc. 8th Conf. Coastal Eng., 1962*: 500–520.

[103] C.M. Snyder, 1957. Model study of a hydraulic breakwater over a submerged barrier. *Univ. Calif., Berkeley, IER Ser.*, 104 (3).

[104] K. Horikawa, 1958. Three-dimensional model studies of hydraulic breakwaters. *Univ. Calif., Berkeley, IER Ser.*, 104 (8).

[105] P.S. Bulson, 1967. Transportable breakwaters. *Dock Harbour Authority*, 48: 41–46.

Chapter 8

EFFECT OF WAVES ON STRUCTURES

Having discussed the influence of structures on waves it is now necessary to consider the opposite effect of waves on structures. These are interrelated in that reflection will change the size and nature of the wave which, in turn, will alter the action of the wave on the structure. This action is mainly the pressure exerted on the body. Although there are other minor considerations to be taken into account when designing a marine structure, the main one is that of force due to static and dynamic loads from waves, currents and changes in water level.

One important class of structure is the wall or revetment. It can be vertical or sloping, but its common characteristic is the creation of complete or partial clapotis. It is for this condition that pressures must be calculated. When the structure is monolithic, such as concrete or steel fabrication, the known distribution of pressure can be integrated to an overall force acting through some centroid, whose location at some proportion of the depth can be determined. But even in this case, the high localized pressures possible from breaking waves near the water-line should be used in the design of small unit areas of the skin of the structure.

Another important class of structure is that employing relatively slender cylindrical or box-shaped members. Supports for jetties and legs of oil drilling rigs, plus horizontal and sloping braces, fall into this category. Submarine pipelines are also designed on the same basis of progressive wave theory. Where such isolated bodies are large in horizontal dimension, compared to the amplitude of water-particle oscillation, the calculation of forces must differ from that for the slender case. These submerged structures can be of any shape or proportion and unless tests have already been conducted on the one being planned, hydraulic models may have to be used to ascertain the relevant dimensionless ratios.

Where a structure is not monolithic, but consists of units loosely packed together, the forces exerted on individual armour units must be assessed. A certain degree of interlocking and friction between blocks may be included in design, which is based upon some degree of stability against waves of a given energy content when the armour blocks are deposited at a given outer slope.

Composite structures are in use which comprise a block unit seated on sand or a rubble mound resting on the sea-bed. The purpose of the rock-fill foundation is not only to spread the load of the main structure, but also to prevent scour of the sand bed adjacent to the vertical walls. In the design of such units shear forces at the base must be resisted by friction and dead weight of the breakwater. Rocking may help reduce the shear force induced by shock pressures of breaking waves.

It is convenient in most design problems involving forces of waves to bulk the wave energy into a monochromatic wave of height $H_{1/3}$ and period T_{max}. From the discussion on wave shoaling it is evident that energy in even a FAS is concentrated in this band of the spectrum and this energy is a large proportion of the total available when the water is reasonably shallow. Where the water is much deeper, as is the case for many oil rigs operating on the continental shelf, the forces encountered should be treated from a statistical basis.

Many of the relationships between wave characteristics and resulting forces on structures must necessarily revolve on model tests, since the action is not readily amenable to theory. There is a tendency to convert wave heights used in flume tests to deep-water values (H_o) and these are normally associated with L_o or gT^2. This is unfortunate since it introduces an extra parameter of d/L_o in order to bring characteristics to those adjacent to the structure. This transposition normally applies only to a wave shoaling without refraction, so that if this phenomenon is taking place an extra computation is necessary. If the local heights were used with local length then the two major ratios H/L and d/L might be combined to give H/d, HL/d^2 or HL/d^3 as a single dimensionless parameter to relate forces to wave characteristics.

SEAWALLS

Seawalls may be vertical or sloping. The major difference introduced by slope is the action on the wave such as reflection, run-up, and the ability to break the wave. Otherwise the pressure distribution throughout the depth is similar.

Pressures due to standing waves

The procedure to follow is compute the reflection coefficient for the wall in order to determine the height of the clapotis or partial clapotis in front of the wall. Graphs and tables from Chapter 7 should permit this to be done. Then from Fig. 4-18 for near complete clapotis the reach of the wave at the wall above SWL (a_c) can be determined. For conditions where $K_r \leqslant 0.5$ the a_c will be some value between that for a standing wave (Fig. 4-18) and that for a progressive wave (Fig. 4-5).

The pressure head at SWL is then assumed to be a_c, that is a static variation from the limit of wave reach down to this level. The pressure variation from there to the bed is then given to various orders of accuracy by different workers, a comprehensive summary of which has been presented by Rundgren [1]. Experiments have shown that those to first and second order are not extremely accurate. Tadjbaksh and Keller [2] derived a third-order solution for finite height waves in shallow water, whilst Goda [3] has extended this to a fourth-order and presented a comprehensive graph from the computer. This includes the wave set-up, so incorporating a_c, so that the pressure is given in terms of the incident wave height on the assumption of $K_r = 1.0$ for the vertical wall case so treated.

The underwater pressure due to standing waves fluctuates with twice the variation of the water surface oscillation [1]. A peak value occurs just before and just after the antinode at the wall reaches its peak. These are caused by the harmonic components, the second of which predominates near the bed, where the humps of pressure first appear as wave steepness increases or depth ratio decreases. Near the water surface third and higher harmonics assume importance, so that greater extremes of H/L and d/L are required before double humps of pressure appear at the MWL. Goda [3] has graphed the theoretical relationship of H/L and d/L (L being derived from first-order theory) for the first appearance of double humps in pressure at the bed, at the MWL and for the total force per unit length of wall. He also has shown experimental results to confirm this fourth-order theory, but these have been omitted from Fig. 8-1, which has been drawn from Goda's presentation. Also illustrated is the experimental appearance of unsymmetrical humps and the breaking criterion for standing waves. In this condition the first peak exceeds the second peak greatly and tends towards the shock pressure experienced in breaking waves, with a second lower but longer peak following.

Goda has also verified his fourth-order theory experimentally, both for pressure

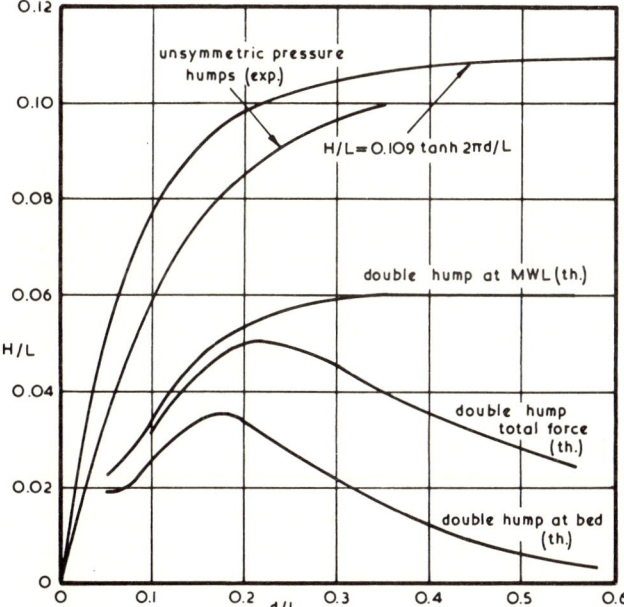

Fig. 8-1. Occurrence of double humps from pressure of standing waves on a vertical wall.

TABLE 8-I

Dimensionless force ratio for standing waves on a vertical wall, F_{max}/wH_id; upper figure: landwards, lower figure: seawards

d/L	H_i/L									
	0.01	0.02	0.03	0.04	0.05	0.06	0.07	0.08	0.09	0.10
0.05	1.40	1.31	1.18	1.11	–	–	–	–	–	–
	0.47	0.38	–	–	–	–	–	–	–	–
0.10	0.98	1.4	1.02	0.98	0.97	0.97	0.99	–	–	–
	0.76	0.65	0.56	0.49	0.43	0.38	0.35	–	–	–
0.12	0.90	0.93	0.93	0.89	0.87	0.88	0.90	0.93	–	–
	0.76	0.68	0.60	0.55	0.50	0.45	0.41	–	–	–
0.14	0.83	0.85	0.84	0.82	0.79	0.79	0.80	0.83	–	–
	0.76	0.70	0.65	0.60	0.56	0.52	0.48	0.44	–	–
0.16	0.78	0.79	0.78	0.75	0.72	0.70	0.71	0.73	0.76	–
	0.73	0.69	0.64	0.60	0.57	0.54	0.50	0.47	–	–
0.18	0.72	0.72	0.71	0.68	0.65	0.64	0.63	0.64	0.66	–
	0.70	0.68	0.64	0.61	0.58	0.55	0.52	0.50	–	–
0.20	0.68	0.67	0.65	0.62	0.60	0.57	0.57	0.57	0.58	–
	0.67	0.65	0.63	0.61	0.59	0.56	0.54	0.52	0.50	–
0.25	0.56	0.57	0.58	0.58	0.59	0.60	0.61	0.64	0.67	0.69
	0.66	0.66	0.65	0.64	0.64	0.63	0.62	0.62	0.61	0.60
0.3	0.51	0.52	0.52	0.52	0.53	0.54	0.55	0.58	0.60	0.62
	0.67	0.66	0.65	0.65	0.64	0.64	0.63	0.63	0.62	0.61
0.5	0.28	0.28	0.28	0.28	0.30	0.31	0.33	0.35	0.37	0.38
	0.67	0.66	0.66	0.66	0.65	0.65	0.65	0.64	0.64	0.64
0.7	0.23	0.23	0.23	0.25	0.26	0.28	0.29	0.30	0.33	0.35
	0.67	0.67	0.66	0.66	0.66	0.66	0.65	0.65	0.65	0.65
1.0	0.19	0.19	0.20	0.22	0.23	0.25	0.27	0.28	0.30	0.31
	0.67	0.67	0.67	0.66	0.66	0.66	0.66	0.66	0.65	0.65

at the bed and MWL and for total force as integrated over the full depth. From his graphical presentation Table 8-I has been drawn up which gives the peak pressure, that prior to the crest reaches the wall. Goda also concluded from his experiments with irregular waves that the theory can be applied to a spectrum of waves, and for the purposes of design suggests H_{max} rather than $H_{1/3}$ or $H_{1/10}$.

The distribution of pressure on a vertical seawall is depicted in Fig. 8-2A. Goda has shown that reflection can be accepted as complete for all conditions of H_i/L and d/L, so that the resultant clapotis has a height $H = 2 H_i$. Therefore the level of

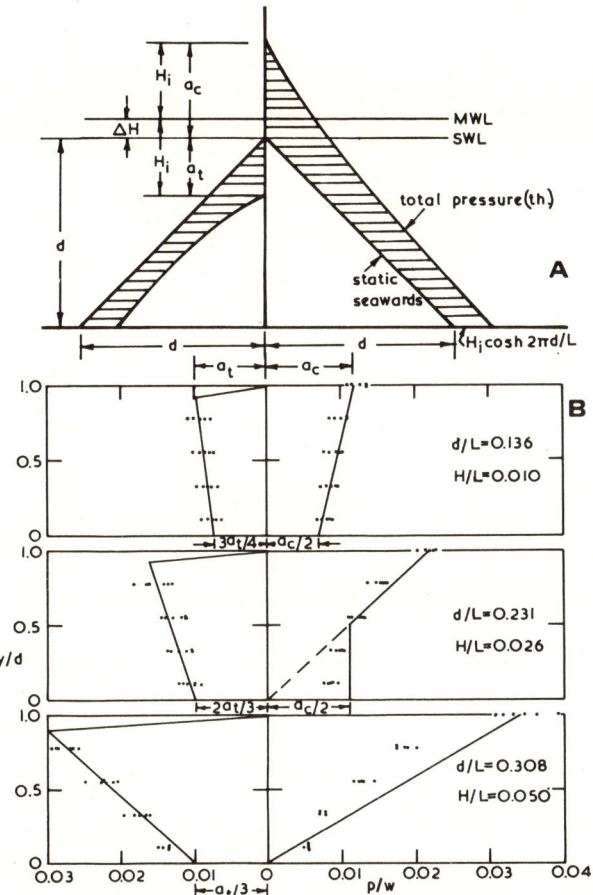

Fig. 8-2. Distribution of pressure on a vertical wall. A. Definition sketch. B. Experimental results from Rundgren [1] and assumed distribution.

the water against the wall alternates between MWL + H_i and MWL − H_i. These heights with respect to SWL are $a_c = H_i + \Delta H$ and $a_t = H_i - \Delta H$. The static distribution of pressure on either side of a structure located in the sea will cancel out, or in the case of a land-backed structure soil pressure plus live load may have to be taken into account. But the fluctuating pressure due to wave action is as shown in Fig. 8-2A, with a positive and negative head of around $H_i/\cosh 2\pi d/L$ at the bed. The distribution from this point to SWL varies with wave steepness and shallowness, and is the issue where most theories differ from experiment.

In Fig. 8-2B are graphed mean curves from Rundgren's [1] tests for a range of d/L, plus an empirical distribution suggested for the determination of the point through which the resultant force previously calculated will act. Three categories

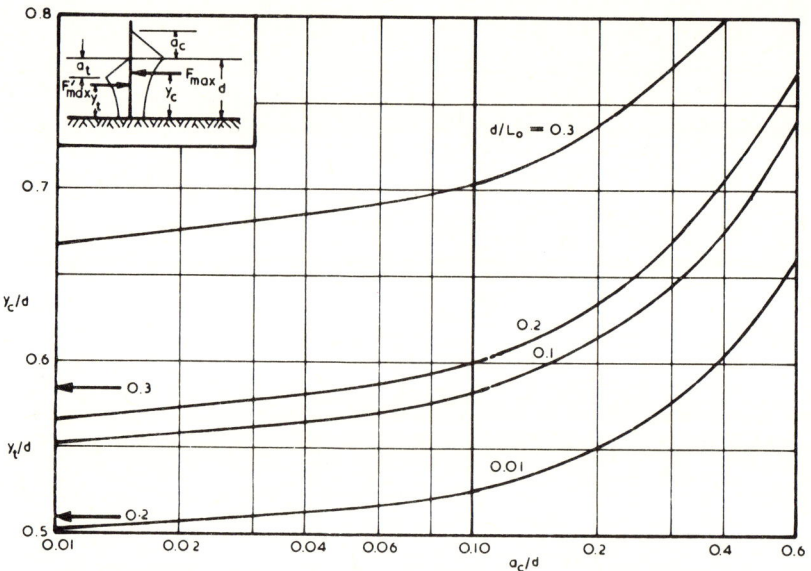

Fig. 8-3. Height of resultant force of standing wave against vertical wall.

have been assumed, from which interpolations can be made for y_c/d from given ratios of a_c/d as in Fig. 8-3. For the seaward net force F'_{max} the values of y_t/d are essentially constant over the full possible range of a_t/d so only the two values of 0.585 and 0.51 are shown for $d/L \geqslant 0.3$ and $\leqslant 0.2$ respectively. Once the force and its line of action are known the bending moment on the wall can be determined.

Tsuchiya and Yamaguchi [4] have examined the limitations of the various orders of theory respecting standing waves, including that applicable to pressures on walls. The perturbation method used for the third and fourth [3] order solutions has no mathematical proof for their convergence, so a numerical test for validity has been carried out similar to that by Dean [5] for progressive waves. The results for kinematic and dynamic conditions did not differ much, permitting a graph as in Fig. 8-4 to indicate the ranges of H_i/d and d/L_o within which the errors from the various theories are reasonably small. It is seen that the fourth-order theory has the widest zone of usage.

Within the bounds of $d/L \leqslant 0.2$ there is a limiting wave steepness for a given d/L beyond which the fourth-order theory should be used with caution. These limits have been marked in Table 8-I by vertical lines.

Since the theory is not recommended for $d/L > 0.2$, values of F_{max}/wHd have been derived by the linear variations of pressure as illustrated in Fig. 8-2B. For deep-water waves this assumes a triangular variation from a_c at the SWL to zero at the bed and at such depth ratios $a_c \rightarrow H_i$. In fact, for $d/L \geqslant 0.5$ experimental evidence [6] indicates that the triangle could be taken to half the depth. Rund-

SEAWALLS

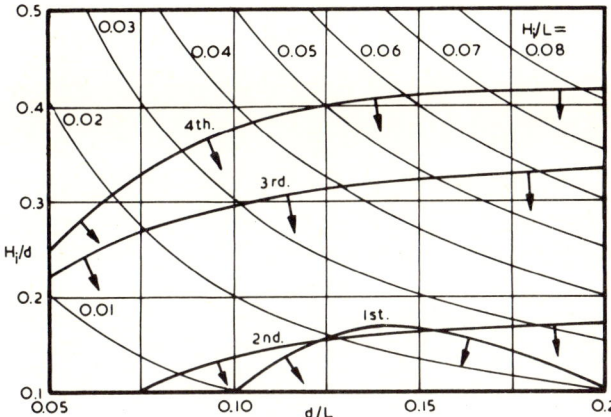

Fig. 8-4. Zones of applicability of various orders of theory for standing waves.

gren's tests showed negative values in the lower regions of the wall for such wave conditions. From the ratios of F_{max}/wHd so obtained, an average has been taken between there and the computed results of Goda [3], cognisance being taken of the comprehensive laboratory tests by Nagai [6]. The greatest difference arises in the seaward force, which involved a doubling and sometimes tripling of Goda's values, since no averaging was carried out in this case. Confirmation of the procedure was based on pressure distributions from Rundgren's tests [1]. They may be somewhat conservative, but should be used for design until further verification is forthcoming.

It is interesting to note that Kuznetsov [7] supplies design curves which give forces in excess of those in Table 8-I. He also reports Russian tests on forces due to angled waves. These can exert forces greatly exceeding those for normal approach. The degree of increase varies with θ, d/L and H/L but the maximum indicated (for large d/L, small H/L and $60° > \theta > 30°$) is an increase of 70%. For $H/L > 0.065$ the force of angled waves is less than that for $\theta = 0°$.

Tsuchiya and Yamaguchi [4] also measured pressures on walls with normal wave approach when overtopping was taking place. They found that for vertical walls the overtopping height was given by the crest height of the wall above SWL (h) and the incident wave height H_i, independently of wave steepness and depth ratio. This is summarized in Table 8-II. If these heights are used in Table 8-I, they will give the F_{max}/wR_od values necessary for design.

A somewhat similar situation to the vertical straight wall is a barrier consisting of upright cylinders on which model tests have been carried out [8]. The vertical pressure distribution could be considered the same as for a straight wall, the intensities at any given level varying around the periphery of the cylinder as per Fig. 8-5, as produced from the relationships supplied. It is seen that on the zone facing the wave (either side of $\beta = 0$) the pressure is less than that for the straight wall,

TABLE 8-II

Overtopping wave height for vertical wall

H_i/h	1.00	1.25	1.50	1.75	2.00	2.25	2.50
R_o/h	0.97	1.18	1.41	1.62	1.83	2.06	2.27

whereas for $\beta > 70°$ the pressure is greater. These latter pressures cancel each other out as far as overturning is concerned, so that the total force acting on the cylinder could be less than on a cubic caisson of the same transverse dimension. The high pressures at the contact point of the cylinders are associated with higher wave heights there, approximately 10% greater according to Fig. 8-5. This should be allowed for in overtopping if this is a crucial issue.

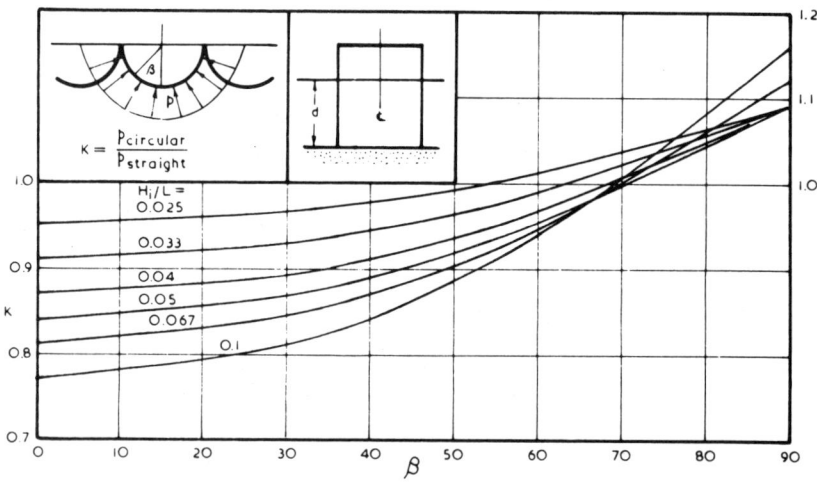

Fig. 8-5. Distribution of pressure around cylindrical wall units.

The closely spaced pile array has been discussed in Chapter 7 in respect to reflection and transmission of waves. From the analysis and experimental verification of Hayashi et al. [9] the force exerted on each pile can be determined from Fig. 7.35 from a ratio of that for a straight wall per pile diameter length of structure. Two curves are supplied, one for circular piles $C_d = 0.9$ and another for square piles $C_d = 0.6$. If triangular piles were being employed the above curves would be used if the apexes faced the incident waves and the bases were to so oppose the attack respectively.

Pressures due to breaking waves

In considering the extreme forces exerted by breaking or broken waves on seawalls it should be appreciated that the depths are small in respect to the wave height. The depth at breaking for progressive waves over a mildly sloped bed is around 1.3 H, but this will decrease with bed slope (see Fig. 5-19 which represents mainly solitary waves). It will be when the wave is at its breaking point, irrespective of the presence of the wall, that the greatest shock force will be experienced.

Where a sedimentary coast is concerned, variability in depth in front of the wall must be considered. Erosion or silting may occur which can bring about the critical conditions at which forces of breaking waves must be allowed for in design. In this context, changes in water level through tides or storm surges (particularly the latter) are extremely important. Under such storm conditions the slope of the bed in front of a structure is well nigh unpredictable, it possibly being undulatory with an offshore bank triggering the large waves to break before or at the wall.

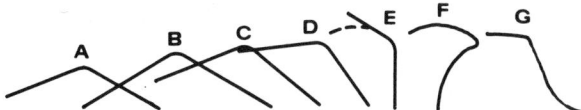

Fig. 8-6. Stages of breaking for waves in shallow water.

The most frequent breaking process is depicted in Fig. 8-6 where steepening is taking place at A, until the contained angle at the crest reaches 120° (B). At this stage the wave will commence to break by the leading face twisting vertically whilst the 120° angle is maintained (C and D), until at this limit (E) the surface particles sweep forward (F) and turbulence is generated whilst the wave advances on an almost vertical front (G). The wave can impact on the seawall at any of these stages but the resulting force is vastly different on each occasion, even for the single wave train.

The pressure experienced is similar to the water-hammer action in closed conduits when the flow is reduced quickly. The abruptness of the stop determines the magnitude of the pressure since:

$$p = \left| \frac{d(mU)}{dt} \right| \tag{8-1}$$

where m is the mass of fluid whose velocity of advance U is brought to zero in time dt.

The time dt within which the bulk of a wave is brought to a stop will depend upon the slope of the leading face of crest when it strikes the wall. It can be appreciated that waves at stages C and D in Fig. 8-6 will have their retardation spread over a longer time than E, F or G.

The shock forces being discussed rise to their peak pressure (p_{max}) in some small interval of time t_{max}. Weggel and Maxwell [10] have shown the product $p_{max} t_{max}$ to be constant for any specific wave train. From eq. 8-1 it can be shown that:

$$p_{max} t_{max} = 2\rho VC/A \tag{8-2}$$

where ρ is seawater density, V is some volume of water in the wave brought to rest, C is the horizontal velocity of the breaking wave face and A is the area over which the mean pressure (p_{max}) is considered to act.

Taking a unit length of crest V/A can be considered some proportion of the wave length which is involved in the production of p_{max}. Weggel and Maxwell [10] have presented a possible relationship, which should serve the present discussion, which can be modified to the form:

$$\frac{V/A}{d_b + (a_c)_b} = 0.0084 (H_o/L_o)^{-1/2} \tag{8-3}$$

where d_b is the depth at breaking, and $(a_c)_b$ is the crest height above SWL at breaking.

Assuming shallow-water conditions to exist on breaking, then: $C = \sqrt{gd_b}$, so that from eq. 8-2 and 8-3:

$$p_{max} t_{max} = 2\rho [d_b + (a_c)_b] 0.0084 (H_o/L_o)^{-1/2} \sqrt{gd_b} \tag{8-4}$$

Rundgren [1] conducted tests in which breaking occurred during a slow rise in water level. At some stage he was able to record the peak pressures on piezometers inserted in the vertical wall, in front of which was a fixed bed sloped at 1 : 9.4. As the water level rose the impact from broken waves was first recorded, after which the peak values occurred when breaking took place right at the wall. Soon after this, complete reflection reduced the possibility of breaking as a progressive wave, and therefore a clapotis was established. The sequence of maximum pressures was contained within an envelope as in Fig. 8-7, where the probable stages of breaking in Fig. 8-6 are related to the intensities measured. Other aspects of this diagram will be discussed later, but the depth d_m at which the peak values occurred (noted as those when breaking took place at the face or d_b) was related to H_o/L_o by:

$$d_b/H_o = 0.25 (H_o/L_o)^{-1/4} \tag{8-5}$$

so that eq. 8-4 can be expressed as:

$$\frac{p_{max} t_{max} g^{1/2}}{w H_o^{3/2}} =$$

$$2 \left[0.25 (H_o/L_o)^{-1/4} + \frac{(a_c)_b}{H_b} \frac{H_b}{H_o} \right] 0.0084 (H_o/L_o)^{-1/2} \sqrt{0.25 (H_o/L_o)^{-1/4}} \tag{8-6}$$

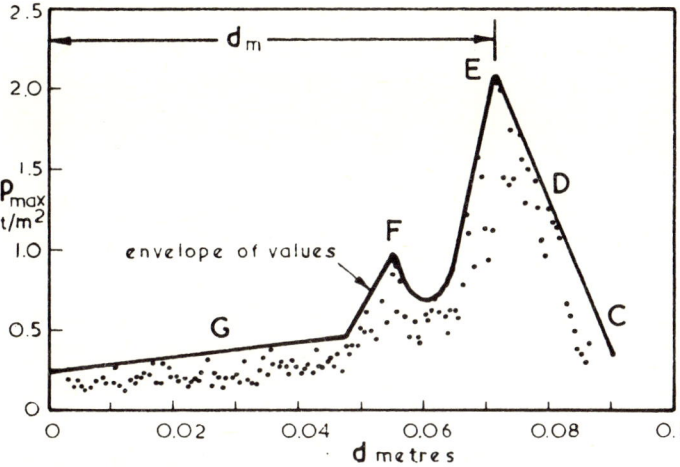

Fig. 8-7. Shock pressures recorded at increasing depths of water [1] with possible stages of breaking from Fig. 8-6.

From Fig. 4-5 it is seen that $(a_c/H)_b$ is a function of bed slope, H_o/L_b and d_b/L_b, but the first variable can be ignored for the present. Also from eq. 4-41 it can be shown that:

$$H_b/H_o = (0.12 \tanh^2 2\pi d_b/L_b) L_o/H_o \tag{8-7}$$

Adeyemo [11] conducted tests with a single wave train ($H_o/L_o = 0.037$) in which he measured horizontal wave asymmetry = crest to preceding trough/crest to following trough. When his curve is extrapolated to zero (i.e., vertical crest face) $d/L = d_b/L_b = 0.055$. Substituting these values plus $(a_c)_b/H_b = 0.75$ from Fig. 4-5 into eq. 8-6 and 8-7 gives:

$$\frac{p_{max} t_{max} g^{1/2}}{w H_o^{3/2}} = 0.092 \tag{8-8}$$

By substituting values of p_{max}/wH_o and H_o from tests conducted by Rundgren, for tests where $H_o/L_o \doteq 0.037$, an assessment can be made of t_{max}. In spite of test conditions being almost identical four values of 1/1150, 1/437, 1/695, 1/847 and 1/1092 sec were obtained for p_{max} = 101.0, 38.9, 107.5, 80.0, 100.0 g/cm², respectively. This detail emphasizes the important role of steepness of the front face of the wave when it strikes the wall. From field measurements of shock pressures at Dieppe [12] identical waves, as far as H_o and H_o/L_o were noted, produced force ratios differing by 3 : 1. If the relative t_{max} values are taken to represent the time difference between the trough and the crest impacting on the wall this implies a difference in angle between the wave face and the wall of 3° to 9°.

According to eq. 8-8 if the wave face is vertical when it strikes the wall the pressure is infinite. However, in this case air will be trapped by the wave, even without the spilling mechanism, which will help cushion the blow. Bagnold [13] analyzed the problem on the basis of a pocket of air being compressed against the surface. This has been extended by Mitsuyasu [14], whose more exact solution differs very little from Bagnold's, giving:

$$p_{max} = p_o + 2.7 \frac{(V/A)wC^2}{dg} \tag{8-9}$$

where p_o is atmospheric pressure, V/A is as defined for eq. 8-2, C is speed of advancing water mass ($\doteq \sqrt{gd}$). The limiting value of p_{max} can also be expressed in terms of the water-hammer expression:

$$p_{lim} = \rho C c \tag{8-10}$$

where c is the speed of sound in water.

For the condition of entrapment the water will contain a certain percentage of air which greatly affects the value of c. For example 1% air reduces this speed from some 4000 ft./sec to around 600 ft./sec. Substitution of representative wave characteristics into eq. 8-10 results in p_{lim} values many orders of magnitude greater than p_{max} in eq. 8-9. This is not to preclude their existence but the determination of air content does permit such calculations to be made.

In the Dieppe tests [12], already mentioned, the average value of p_{max}/wH_o was 20 whilst the maximum was 38. Bagnold's [13] tests gave optimum values of 28, but occasionally visual maxima of 244 were observed. Denny [15], from extensive statistical data, found the most frequent value of the above ratio to be 28, extreme values reaching 110.

The temporal and spatial distribution of shock pressure must be known if the resulting force on the wall is to be assessed. Mitsuyasu [14], from his comprehensive tests with multiple piezometers, has made the following observations:

(a) "Although the measurements were carried out under the almost same conditions by using the uniform part of a train of waves, the measured trace of wave pressure varied greatly wave by wave."

(b) Whilst $p_{max}/wH_o \leq 10$ were simultaneously recorded at two points at the one level $2H_o$ apart, values ≥ 20 were also recorded by one recorder only, indicating a very localized influence of the high intensity pressure. (It can be imagined that whilst one part of a near vertical crest face may strike the wall with no air cushion, other parts nearby may suffer substantial attenuation of pressure from entrapped air.)

(c) When p_{max} was large there was very little negative pressure associated with it, whereas when more moderate peaks were experienced these were followed by negative pressures of almost equal intensity. This latter appears to be due to the

SEAWALLS

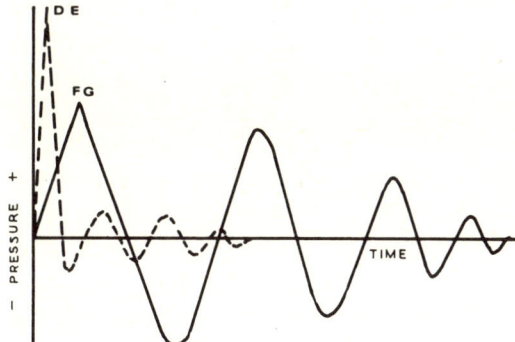

Fig. 8-8. Typical pressure/time curves experienced with breaking, with stages indicated from Fig. 8-6.

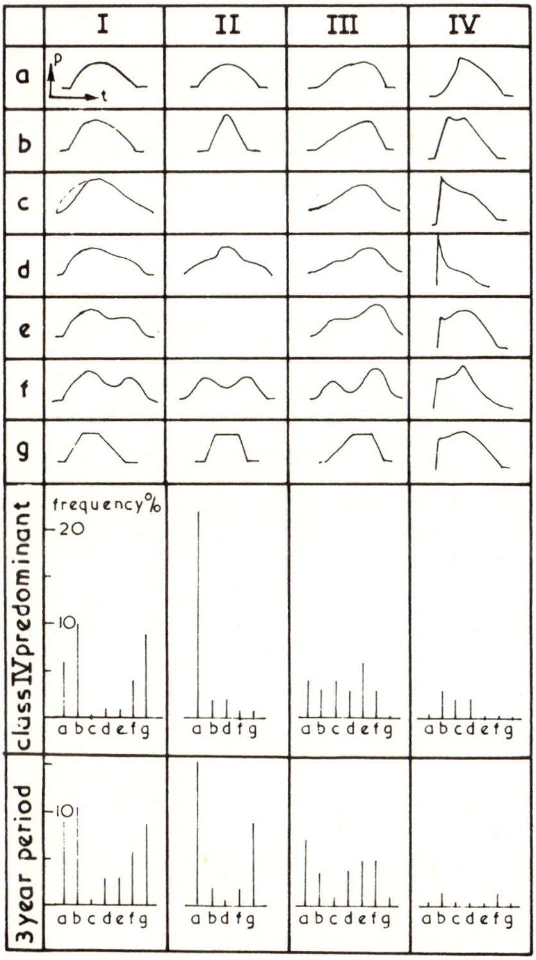

Fig. 8-9. Pressure/time curves at SWL measured in Hoboro harbour [16].

contraction and expansion of the air cushion [14] on reflection of the elastic wave from the sea surface [10]. The possible time pressure curves are depicted in Fig. 8-8, where the likely stages of breaking as illustrated in Fig. 8-6 are indicated on the curves.

Muraki [16] has made field measurements of pressure on a vertical breakwater at Hoboro harbour in Hokkaido. He presents many pressure time curves at SWL divided into four classes as in Fig. 8-9. These consist of a rising dome either asymmetrical forwards or backwards, or symmetrical about the optimum point. The fourth category includes the shock pressure discussed above, which is normally accompanied by a more extended maximum. Larras [17] has termed the shock pressure section the "gifle" and the follow-up the "bourrage". The negative pressure mentioned previously, that can follow a medium shock pressure, may be eliminated by positive pressure from the remainder of the crest being brought to rest. Hence the variety of shapes shown in classification IV, where the "bourrage" may even exceed the "gifle". Groups I to III are the clapotis-type pressure waves [3] for which the previous analysis is applicable, whilst group IV is the type under present discussion.

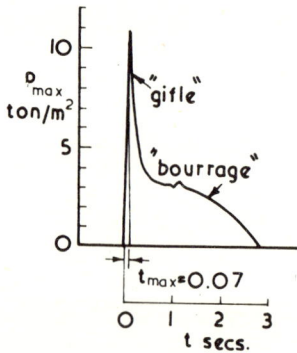

Fig. 8-10. Most typical shock-type wave measured at Hoboro harbour [16].

The most typical shock pressure curve measured [16] is indicated in Fig. 8-10, consisting of a peak reached in 0.07 sec followed by a pressure $p_{max}/3$ approximately, which falls to zero in about 3 sec. However, the frequency of occurrence of all four categories is important in design considerations. For the Hoboro tests these frequencies were determined for the full recording period of 3 years and also for a period in which shock pressures were predominant. These are listed in Table 8-III, from which it can be seen that category IV occurs relatively infrequently. Even within this category, the distributions in which the peak pressure is greatly in excess of the follow-up "were found in only two or three cases in the entire data observed, which involved two thousand and several hundred waves" [16].

SEAWALLS

TABLE 8-III

Frequency of occurrence of wave pressures at Hoboro harbour

	Classification			
	I	II	III	IV
% 1957–1960	45	27	24	4
% when shock pressures predominated	39	28	25	8

Hayashi and Hattori [18] had earlier conducted laboratory tests in which time histories at different depths were taken. Solitary waves were used, the record from one of which is reproduced in Fig. 8-11. From this it is clear that the "gifle" experienced near SWL is not concurrent with the "bourrage" occurring at points lower down. Only when a wave actually arrives at the wall with a near vertical face could simultaneous peak pressures be experienced throughout the full depth.

In the tests by Rundgren [1] with rising water levels, the maxima were likely to be experienced at some stage. A graph of $p_{max}/w H_o$ against H_o/L_o results in much

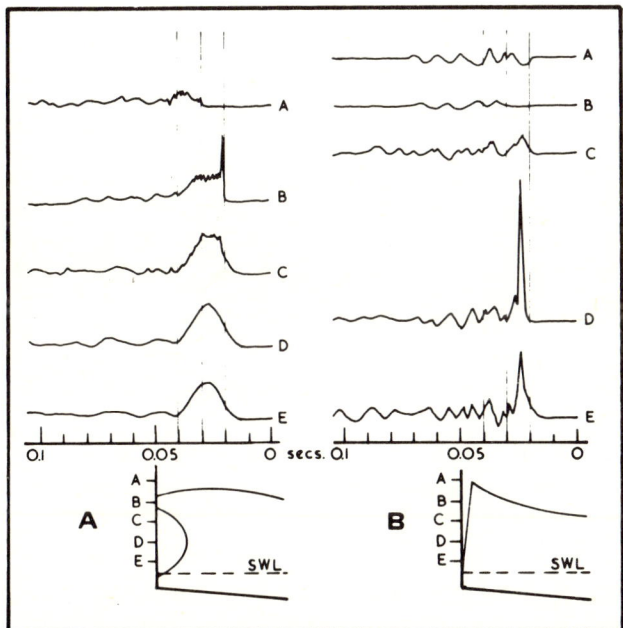

Fig. 8-11. Concurrent recordings of pressure at various heights of a vertical wall with breaking waves [18].

scatter, the top limit of the envelope being expressed by:

$$\frac{p_{max}}{wH_o} = 3.5 \left(\frac{H_o}{L_o}\right)^{1/3} \quad \text{or} \quad \frac{p_{max}}{w} = 3.5(L_o H_o^2)^{1/3} \qquad (8\text{-}11)$$

which appears a reasonable relationship since $L_o H_o^2$ represents the energy per wave in deep water approaching the wall. Eq. 8-11 can be put in the form:

$$p_{max}/wH_o = 1.9(g/H_o)^{1/3} T^{2/3} \qquad (8\text{-}12)$$

which has been graphed in Fig. 8-12 with the constant on RHS raised to 3.8 since Rundgren admitted that his piezometers had high inertia and therefore gave lower values than Bagnold [13] and Denny [15]. It is seen in the figure that for a value of $p_{max}/wH_o = 28$, a reasonable optimum [12], the range of periods 12 to 8 sec is associated with wave heights (H_o) of 12 and 5 ft. respectively. Substitution in eq. 8-8 gives t_{max} values of 0.002 and 0.0013 sec, which are commensurate with times measured for larger shock pressures. However, values of p_m/wH_o four to five times 28 can occur at exceptionally infrequent intervals.

Broken waves arriving at a wall will have a height as determined by procedures in Chapter 5. The force in such a wave is dictated by its momentum so that:

$$p = KwC^2/2g \qquad (8\text{-}13)$$

where C is wave celerity, and K is some factor to be determined by experiment. To the above dynamic pressure must be added the static triangular distribution from the crest level to the bed if this is applicable.

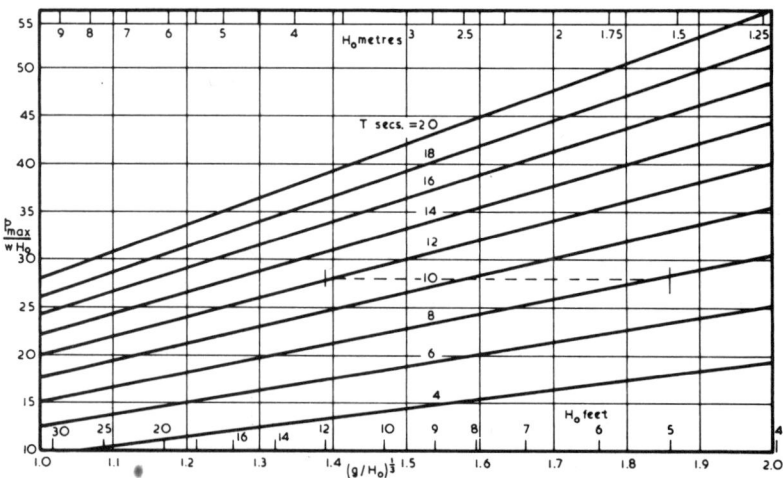

Fig. 8-12. Maximum shock pressures likely at SWL on a vertical wall.

The wave celerity in shallow water can be determined from graphs given by Iwagaki [19] for hyperbolic waves, the breaking condition of which results in the relationship:

$$\frac{C}{C_o} = 0.05 + 2.5 \left(\frac{d}{L_o}\right)^{0.525} \tag{8-14}$$

This provides a value of 0.05 for $d/L_o = 0$, which will reduce further if the wave has to traverse any length of dry bed.

Hayashi and Hattori [18] have listed the variety of K values published, which in the main have ranged from 1.6 to 4. From their tests with solitary waves they conclude that 4 is the more realistic figure, even though their results indicated the possibility of twice this value. Accepting $K = 4$, eq. 8-13 and 8-14 give:

$$\frac{p}{wT^2} = \frac{1}{2\pi^2} [0.05 + 2.5(d/L_o)^{0.525}]^2 \tag{8-15}$$

which has been graphed in Fig. 8-13. This value of p is the maximum, occurring at SWL, whence a triangular reduction occurs to the bed and the reach of the wave $(a_c)_b$. In Fig. 8-13 is traced the curve for $(a_c/H)_b$ as obtained by Iwagaki [19]. When $d/L_o = 0$ the distribution is triangular with the apex at the bed, as indicated in the inset of the figure. From this knowledge the total force (per unit length of crest) and the point of action of the resultant can be ascertained. The height of the broken wave at any depth within the surf zone is given by Fig. 5-21. When waves

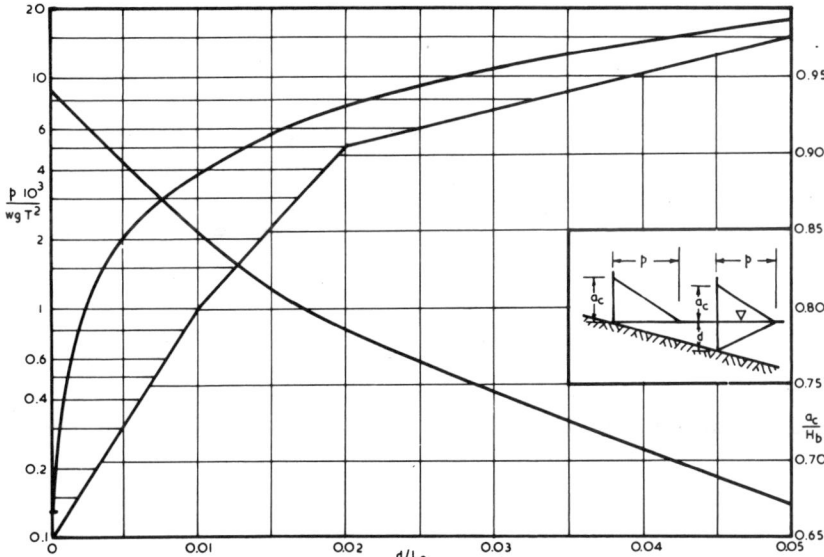

Fig. 8-13. Pressure distribution from broken waves striking vertical walls.

break right at the wall then the p_{max} values presented previously should be utilized. Hom-ma and Horikawa [20] have suggested an armour block mound in front of walls to reduce the possibility of shock pressures.

Pressures on sloping walls

For sloping walls the clapotis-type waves will have similar pressures as derived for vertical walls. The reach and reflection will be modified, but otherwise the computation is the same for the relative depths. However, when the waves are breaking on a sloping revetment, forces are localized on zones that receive the spill of the wave. Skladnev and Popov [21] have conducted tests for such structures and found relationships as reproduced in Fig. 8-14. The values of p_{max}/wH in this case are quite modest as they result only from the fall of water from the wave crest. The authors discuss the minimum model scale to achieve a Froude relationship of pressure. This was found to be $H_m/H_p = 0.5$ or $H_m \geqslant 50$ cm.

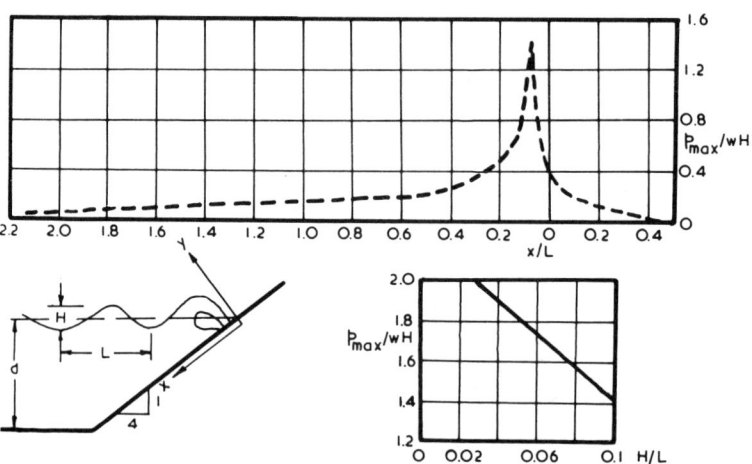

Fig. 8-14. Pressure due to breaking waves on walls with slope of 1 : 4.

Molero [22] has presented data on flexible sloping pavements which are more economical than the massive block protection normally provided. Krylov [23] has provided a theoretical analysis on the "gifle" and the "bourrage".

Pressures on composite structures

These comprise a monolithic structure which is sited on the sea-bed or on a rubble mound which serves to distribute the load and impede erosion of a sedimen-

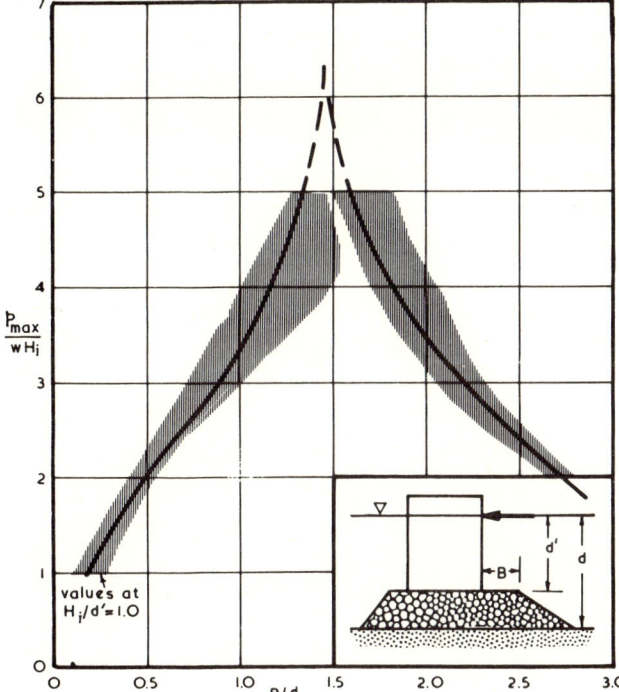

Fig. 8-15. Experimental peak pressures from waves breaking on composite structures.

tary surface bed (see inset of Fig. 8-15). The stability of this rock-fill base is treated later, but it is accepted that it will rise a given proportion (d') of the depth (d) and extend seawards (B) and leewards flat for some distance before sloping at some stable slope to the bed. The proportion of depth (d'/d) utilized for rubble-mound and concrete or steel superstructure is a matter of economics, but Nagai [24] has also indicated that it is a matter of hydrodynamics. Mounds can precipitate breaking at the vertical wall, which has already been seen to cause large shock pressures. Breaking was probable for $0.4 < d'/d < 0.75$ whilst standing waves occurred for $d'/d > 0.75$ and $H/d \leqslant 0.55$. The incidence of peak pressures at SWL is depicted in Fig. 8-15, where it is seen that p_{max}/wH_i is optimum for the breadth ratio (B/d) of 1.3 to 1.6. Seaward widths (B) greater or less than this ratio reduce the pressure, due to greater friction and mal-formation of the breaker respectively. It is possible, as seen from the discussion on normal seawalls, for p_{max}/wH to be much greater than the value of 5 for which Nagai provided the largest isohyets.

Composite breakwaters do not have land-backing to help them resist the horizontal forces. They must therefore be designed to resist them by shear stress at the base. Since it is difficult underwater to provide protrusions on the base of the breakwater, as is the case for foundations of dams, the shear resistance must be developed by friction on the sand or rubble-mound bed. Stability is then normally expressed as:

$$\mu W > P_{max} \tag{8-16}$$

where μ is the coefficient of static friction ($\doteq 0.6$ for sand or rubble material against concrete), W is the submerged weight of the breakwater, P_{max} is the maximum simultaneous horizontal force due to waves (W and P are generally expressed as force per unit length of structure).

Since the maximum force P_{max} is often due to shock pressure from waves breaking right at the breakwater face, the resultant force may be of extremely short duration. Also, since these structures can sway small amounts on the soil or stone foundation, Hayashi and Hattori [25] submit that "the inertia resistance of the structures against the sway may 'absorb' a portion of the shock pressures, and consequently the shearing forces which the structures are called upon the resist may be reduced by that amount".

The forces on and motions of a monolithic structure are illustrated in the inset of Fig. 8-16. The resultant wave force P_{max} acts at some height y_w above the bed. This causes the breakwater of weight W to rotate about some centre O which Hayashi and Hattori [25] have taken to be in the centre of the contact plane. This is more likely to be at some depth in the foundation since a portion of it will move

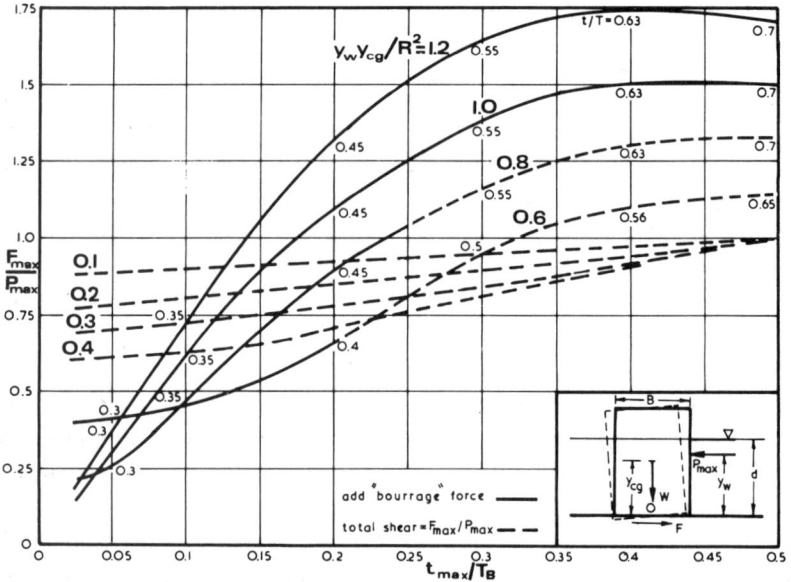

Fig. 8-16. Maximum shear stress due to shock load on composite breakwater.

integrally with the breakwater. As discussed by Lundgren [26], the assessment of this centre of rotation must be made from geotechnical considerations.

Thus the suggested criterion [25] against sliding should be:

$$\mu W > F = P_{max} \pm \text{inertial resistance} \qquad (8\text{-}17)$$

where F is the shearing force, the maximum value of which varies with parameters:

$$\lambda = y_w y_{cg}/R^2 \qquad (8\text{-}18)$$

and:

$$\tau = t_{max}/T_B \qquad (8\text{-}19)$$

where y_w is the distance from the base of the breakwater to the line of action of the resultant force P_{max} (for composite structures with rubble mound base this is assumed to be d'), y_{cg} is the distance from bed to centre of gravity of the breakwater, R is the radius of rotation of the wall around point $O = [\frac{4}{3}y_{cg}^2 + \frac{1}{12}(B^2)]^{1/2}$, B is the base-width of the breakwater, t_{max} is the time for maximum shock force (P_{max}) to be developed, T_B is the natural period of rocking of the wall (this may change as the bed stabilizes). The magnitude of the rocking period T_B is given by:

$$T_B = 2\pi/\sqrt{J/K} \qquad (8\text{-}20)$$

where J is the moment of inertia of the wall around point O, and K is the coefficient of the resisting moment of the foundation ($= kB^3/12$), where k is the coefficient of bearing resistance of the foundation.

For a wall of rectangular cross-section eq. 8-20 can be written:

$$T_B = 2\pi\sqrt{gB_1^2(1+B_1^2)/18\gamma}\sqrt{k/y_{cg}} \qquad (8\text{-}21)$$

where γ is the specific weight of the breakwater, and B_1 is the dimensionless width of the wall ($B/2y_{cg}$).

Using values suggested by Hayashi and Hattori [25] of $B_1 = 0.7$ to 1.3; $2y_{cg} = 5-15$ m; $k = 2-20$ (kg/cm^2)/cm the following ranges of T_B result:

wall height (m)	5	10	15
T_B (sec)	0.07–0.5	0.1–0.8	0.1–0.9

Values of t_{max} as measured in laboratories and in the field [25] range from 0.025 to 0.045 sec, hence:

$$t_{max}/T_B \text{ ranges from } \frac{0.025}{0.9} \text{ to } \frac{0.045}{0.1} = \frac{1}{36} \text{ to } \frac{1}{2.2}.$$

The analysis by Hayashi and Hattori provides graphs of F/P_{max} versus t/T_B (t being a measure of time from first build up of pressure) for a range of t_{max}/T_B and of λ. After a maximum is initially reached, values of F/P_{max} vary sinusoidally, theoretically giving negative values almost equal to the positive force landwards. This has been shown to attenuate quickly [27] due to the damping effect of kinetic energy emission through the foundation which has non-elastic properties. The added attenuation due to wave making by the motion of the breakwater is considered to be slight.

The initial maxima of F/P_{max} have been graphed in Fig. 8-16 against t_{max}/T_B for a range of λ. Hayashi and Hattori have cited breakwaters around the world to indicate that in most cases $1.1 > \lambda > 0.7$. Also indicated in Fig. 8-16 in each λ curve are the time ratios t/T_B to reach the F_{max}/P_{max} values. It should be remembered that the P_{max} and hence the F_{max} are due solely to the "gifle". This builds up in time t_{max} to a maximum (P_{max}) and is zero again by another interval of time t_{max}. Due to the sway of the breakwater the effect of this impulse is longer lasting by which time the momentum of the "bourrage" is exerted against the wall. Where the time ratio (t/T) is greater than $2\,t_{max}/T$ the total shear force is the addition of the "gifle" and "bourrage" forces i.e., $F_{max} + P_b$. Where $\lambda > 0.8$ (which is normally the case) the extension of the "gifle" effect is such that the "bourrage" force must be added on to all values of F_{max} determined from Fig. 8-16. For $0.6 \leq \lambda \leq 0.8$ the lower parts of the curves must have P_b added, but not the upper part where $t/T < 2t_{max}/T$. For $\lambda \leq 0.4$ the maximum shear force is produced by

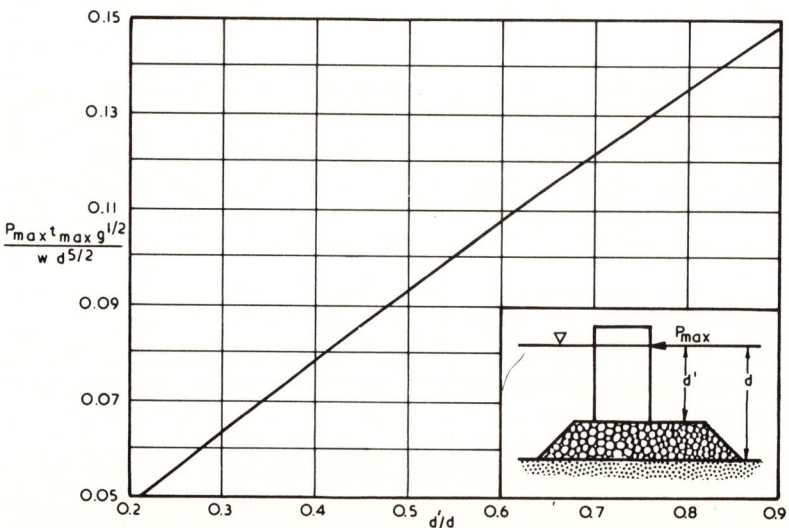

Fig. 8-17. Wave inpulse from waves breaking on composite breakwater.

CYLINDRICAL STRUCTURES

the "gifle" and no addition of P_b is required. These various requirements are indicated in Fig. 8-16 by dotted or full lines respectively.

The product $P_{max}t_{max}$ is the impulse of the shock and although P_{max} and t_{max} may vary individually Denny [15] found that the impulse tends to be constant and have a maximum value of $0.07 M_b$, where M_b is the momentum of the breaking wave per unit length of crest. Using an approximation for the solitary wave [25] and introducing a depth d' to a rubble mound support with a value of $(H/d')_b \doteq 1/1.6$ from tests by Nagai [28] the following relationship can be derived:

$$\frac{P_{max}t_{max}g^{1/2}}{wd^{5/2}} = 0.128 \frac{d'}{d} \sqrt{\frac{1}{2}\left(1+\frac{d}{d'}\right)} + 0.625 \qquad (8\text{-}22)$$

Eq. 8-22 has been graphed in Fig. 8-17. In order to evaluate P_{max} the order of t_{max} must be assessed, which has been indicated as ranging from 0.02 to 0.05.

CYLINDRICAL STRUCTURES

Prior to World War II the main use of cylindrical sections in maritime structures were supports for jetties, generally consisting of timber trunks which could withstand marine borers. However, with the extension of oil exploration to the continental shelf and the need for radar stations and light-houses in deep-water, the demand for steel and concrete structures has assumed great importance. Due to the efficiency of the circular section in withstanding loads for the minimum resistance to wind and wave forces it is the most widely used feature for such structures.

Oil rigs and other large supporting space frames consist of vertical, horizontal and sloping cylinders, each to be designed for its dead and live loads. It is with the latter, in respect to water motions, that the coastal engineer is concerned. In spite of the simplicity of an element, such as a circular pile extending from the bed through the water surface, the analysis is both complex and difficult to verify experimentally. Unlike the drag force experienced in steady flow, the marine pile suffers accelerations in two major directions. In a complex sea, as for FAS conditions, both velocities and accelerations can be considered as three dimensional variables.

The force exerted by maximum horizontal orbital velocity and that due to maximum horizontal orbital acceleration of water particles combine to give a resultant maximum. Depending upon the wave characteristics, one or other may predominate, as also with the ratio of diameter to the amplitude of horizontal oscillation of the water particles. For large diameter cylinders the force due to acceleration, or the virtual mass force, is optimum whilst the reverse is true for the slender pile, although both drag force and inertial force must be determined for the latter.

Slender piles

Where piles are widely spaced they do not interfere with the propagation of the wave so that progressive wave theory is applicable, as distinct from standing wave theory for walls where reflection is predominant. Also the hydrostatic pressure is balanced on either side of the pile, but its existence should not be forgotten in the design of cylinders which exclude water from their insides.

The drag force F_D is produced by the lower pressure in the wake of flow around the pile plus the friction on each side. This can be expressed as:

$$dF_D = \underbrace{C_D \, w \, \frac{u|u|}{2g}}_{\text{pressure}} \underbrace{D \, ds}_{\substack{\text{projected} \\ \text{area}}} \tag{8-23}$$

where dF_D is the instantaneous drag force on an elemental strip of pile of diameter D and length ds when the instantaneous velocity u occurs, and C_D is the drag coefficient which varies with the shape of the pile and the Reynolds number of flow ($R = uD/\nu$) in which ν is the kinematic viscosity of seawater.

The inertia force (F_m) is that exerted by the pile to retain its position and not move with the water particles. It is the force required to accelerate the same body of water as is occupied by the pile. It can be expressed as:

$$dF_m = \underbrace{C_m \, \frac{w}{g} \, \frac{D^2}{4} \, ds}_{\text{mass}} \underbrace{\frac{du}{dt}}_{\text{acceleration}} \tag{8-24}$$

where du/dt is the instantaneous acceleration occurring on an elemental length of pile as defined in eq. 8-23, and C_m is the coefficient of inertia or virtual mass coefficient, which could be equated to conditions of pure acceleration.

As will be understood from wave theory, water particles within waves have unsteady oscillating flow, wherein the velocities and accelerations vary temporally and with depth for a two-dimensional wave. Morison et al. [29, 30] have dealt with this complicated kinematic problem by summing the two forces over the full length of the pile, so that:

$$\Sigma dF_T = \Sigma dF_D + \Sigma dF_M \tag{8-25}$$

The maximum value of ΣdF_T will depend upon variation of ΣdF_D and ΣdF_M throughout the wave cycle, as depicted in Fig. 8-18, where it is seen that the total force does not necessarily occur when the crest of the wave is at the pile. The predominance of F_D or F_M varies with H/D and d/L as in Fig. 8-19 [31].

The computation of forces as in eq. 8-23 and 8-24 depends upon an assessment of C_D and C_M, which in turn relies upon an accurate measurement of velocity and

CYLINDRICAL STRUCTURES

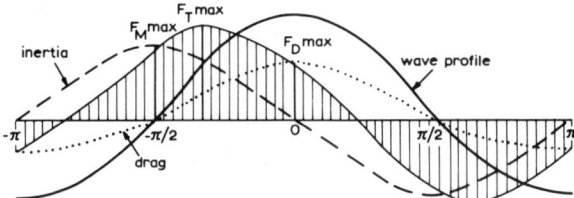

Fig. 8-18. Variation of F_D and F_M throughout wave cycle.

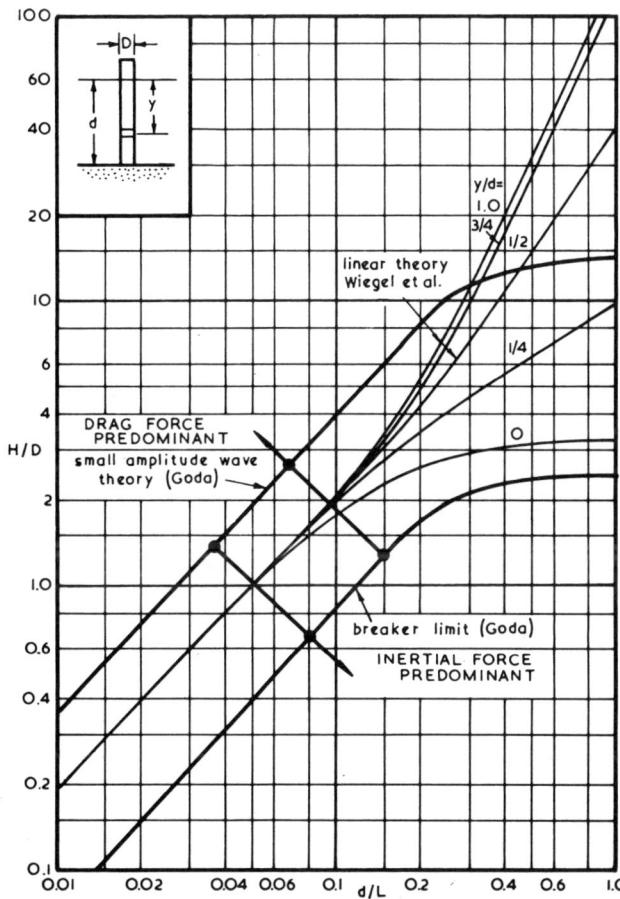

Fig. 8-19. Predominance of F_D and F_M on cylindrical structures.

acceleration over time at all depths whilst simultaneously measuring the total wave force. This procedure is very complicated experimentally, so that velocities and accelerations have generally been calculated from theory, from knowledge of the force on the pile in respect to the position of the wave crest. As noted in Chapter 4 the various wave theories do not predict velocities throughout the water depth very accurately. It could also be accepted that they do not provide good evaluation of accelerations. It is little wonder, therefore, that values of C_D and C_M computed in this way vary so drastically. For this reason it has been proposed to treat these coefficients statistically [32], but this approach has been criticized by Bretschneider [33].

The most realistic correlation of C_D and C_M with wave characteristics requires the continuous recording of velocities, accelerations and wave forces. Such a comprehensive study has been carried out by Goda [31], part of which study has been referred to in the theory of waves. From his data he was able to relate the separate forces F_D and F_M to general wave characteristics and provide a means of adding these together (see Fig. 8-20). From his graphs values of:

$$(F_D)_{max}/wDH^2 = C_D K_D \qquad (8\text{-}26)$$

$$(F_M)_{max}/wD^2H = C_M K_M \qquad (8\text{-}27)$$

have been listed in Table 8-IV for $C_D = 1.00$ and $C_M = 1.66$ as suggested by Goda from his tests. Note the difference in the denominators of these two dimensionless ratios. Goda measured moments on the piles in his tests from which forces could be derived. His graph of ratios $(M_T)_{max}/(M_D)_{max}$ versus $(M_M)_{max}/(M_D)_{max}$ is reproduced in Fig. 8-20 from which the empirical values as in Table 8-V have been drawn up.

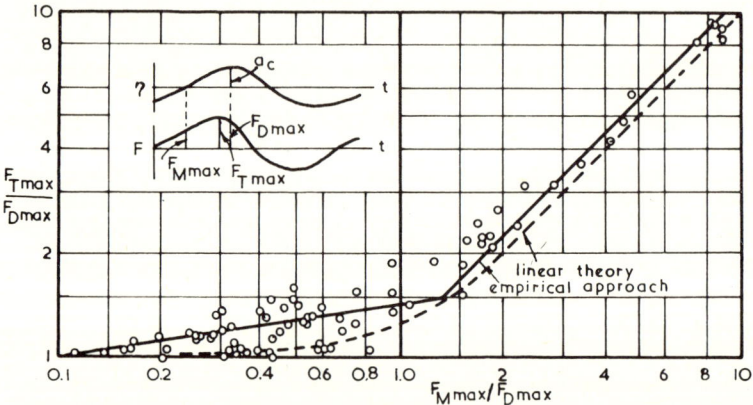

Fig. 8-20. Data from Goda [31] for determining resultant moment or force on slender pile.

CYLINDRICAL STRUCTURES

TABLE 8-IV

Drag and inertial forces on slender piles; upper figure = $(F_D)_{max} 10^2/wDH^2 = K_D$, lower figure = $(F_M)_{max} 10^2/wD^2H = K_M$

d/L	H/d							
	0.1	0.2	0.3	0.4	0.5	0.6	0.7	0.8
0.02*	14.8	17.0	20.0	23.5	28.0	36.0	45.0	62.0
	8.1	8.1	8.1	8.1	8.1	8.1	8.1	8.1
0.03	14.8	17.0	20.0	23.5	28.0	36.0	45.0	62.0
	12.2	12.2	12.2	12.2	12.2	12.2	12.2	12.2
0.05	14.5	16.6	20.0	23.0	28.0	36.0	45.0	62.0
	20.0	20.0	20.0	20.0	20.0	20.0	20.0	20.0
0.10	13.5	15.6	18.8	22.5	28.0	36.0	47.5	62.0
	36.4	36.4	36.4	36.4	36.4	36.4	36.4	36.4
0.20	10.8	13.3	16.2	22.0	28.0	40.0	48.0	62.0
	55.5	55.5	55.5	55.5	55.5	55.5	55.5	55.5
0.30	9.6	12.6	17.0	24.5	35.0	40.0	48.0	62.0
	62.4	62.4	62.4	62.4	62.4	62.4	62.4	62.4
0.50	9.5	14.5	24.5	32.0	35.0	40.0	48.0	62.0
	65.0	65.0	65.0	65.0	65.0	65.0	65.0	65.0
0.70	10.8	22.0	31.0	32.0	35.0	40.0	48.0	62.0
	65.3	65.3	65.3	65.3	65.3	65.3	65.3	65.3
1.00	14.1	31.0	31.0	32.0	35.0	40.0	48.0	62.0
	65.3	65.3	65.3	65.3	65.3	65.3	65.3	65.3

* For $d/L < 0.02$, K_D same as for $d/L = 0.02$, $K_m = 2.5\, d/L$

TABLE 8-V

Resultant force $(F_T)_{max}$ from Fig. 8-20

$\dfrac{(F_M)_{max}}{(F_D)_{max}}$	0.1	0.2	0.3	0.5	0.7	1.0	1.5	2.0	3.0	5.0	7.0	10.0
$\dfrac{(F_T)_{max}}{(F_D)_{max}}$	1.00	1.15	1.20	1.30	1.35	1.45	1.70	2.30	3.30	5.60	7.70	11.0

The point of action of $(F_D)_{max}$ and $(F_M)_{max}$ above the bed is given in Table 8-VI. That for $(F_T)_{max}$ can be determined by the relative values of the force and y_D or y_M.

In choosing a design wave, that for separate members of a structure should be

TABLE 8-VI

Line of action of drag and inertial forces on slender piles; upper figure = y_D/d, lower figure = y_M/d

d/L	H/d							
	0.1	0.2	0.3	0.4	0.5	0.6	0.7	0.8
0.02	0.56	0.63	0.70	0.77	0.85	0.94	1.05	1.19
	0.50	0.50	0.50	0.50	0.50	0.50	0.50	0.50
0.03	0.56	0.63	0.70	0.77	0.85	0.94	1.05	1.20
	0.50	0.50	0.50	0.50	0.50	0.50	0.50	0.50
0.05	0.56	0.63	0.70	0.78	0.86	0.95	1.06	$\overline{1.21}$
	0.51	0.51	0.51	0.51	0.51	0.51	0.51	0.51
0.10	0.60	0.66	0.72	0.81	0.89	1.00	1.11	$\overline{1.21}$
	0.52	0.52	0.52	0.52	0.52	0.52	0.52	0.52
0.20	0.68	0.75	0.82	0.91	1.00	$\overline{1.11}$	$\overline{1.13}$	$\overline{1.21}$
	0.56	0.56	0.56	0.56	0.56	0.56	0.56	0.56
0.30	0.77	0.84	0.92	1.01	$\overline{1.09}$	$\overline{1.11}$	$\overline{1.13}$	$\overline{1.21}$
	0.61	0.61	0.61	0.61	0.61	0.61	0.61	0.61
0.50	0.89	0.96	1.04	$\overline{1.08}$	$\overline{1.09}$	$\overline{1.11}$	$\overline{1.13}$	$\overline{1.21}$
	0.71	0.71	0.71	0.71	0.71	0.71	0.71	0.71
0.70	0.94	1.02	$\overline{1.06}$	$\overline{1.08}$	$\overline{1.09}$	$\overline{1.11}$	$\overline{1.13}$	$\overline{1.21}$
	0.78	0.78	0.78	0.78	0.78	0.78	0.78	0.78
1.00	0.98	$\overline{1.04}$	$\overline{1.06}$	$\overline{1.08}$	$\overline{1.09}$	$\overline{1.11}$	$\overline{1.13}$	$\overline{1.21}$
	0.84	0.84	0.84	0.84	0.84	0.84	0.84	0.84

Note: bar above a figure indicates breaking.

greater than for the structure as a whole. Whilst some peak wave may break a single member, its frequency is unlikely to create a hazard of toppling a whole space frame. Thus, whilst a significant height may serve for the latter the limiting height in the depth of water should be used for the former. Such limiting heights, which are breakers, are given in Table 8-VII [31]. These values represent breaking conditions in flat or mildly sloped beds, for steeper grades the breaker height ratio can be larger. If the meteorological conditions do not possibly allow for such generation the limiting wave may be computed by the formulae in Chapter 3.

Paape and Breusers [34] have also used the concept of F_{max} with some wave

TABLE 8-VII

Limiting wave heights in given depths of water

d/L	0.02	0.05	0.07	0.10	0.20	0.30	0.50	0.70	1.0
H_b/d	0.83	0.79	0.76	0.72	0.59	0.47	0.32	0.24	0.17

characteristic to obviate the use of C_D and C_M. They plotted results from tests on square piles of F_M/wD^2H versus H/D with specific d/L_o curves identified. It can be seen from Table 8-IV that this pilot study covers but a part of the overall picture where drag and inertial forces can predominate with various d/L and H/L conditions.

The effect of roughness of piles has been examined by Burton and Sorensen [35]. Whilst there was some indication of increased resultant force $(F_T)_{max}$, the authors concluded that the accuracy of their empirical approach was insufficient to quantify this influence.

Where piles are in close proximity the scattering of the waves and of the eddies from the oscillatory water motions can influence the drag and lift (transverse) forces exerted on neighbouring piles. Laird has summarized the research on this topic over a number of years [36]. Goda [31] noted that if the centreline spacing of piles is greater than 2 times their diameter little interaction is experienced. When it is less than this there can be an increase in force due to a "blocking effect" and a reduction due to a "sheltering effect". The reader is referred to articles published on this topic since the variety of spacings is infinite. Wiegel [37] has provided a comprehensive survey of this problem.

Where a current exists in the presence of waves it could be expected that the addition of the forward horizontal orbital velocity (u_{max}) and the mean flow (V), will effect a greater force on a pile. Hino [38] has introduced this variable into his spectral analysis of wave forces, from which the ratio of force with current (F_c) to that without (F_o) at any specific depth can be reduced to:

$$\frac{F_c}{F_o} = \frac{AB+2}{A+2} \qquad (8\text{-}28)$$

where $A = u_{rms}^2 T^2/\pi^5 D^2$ (assuming $C_D = 1.0$ and $C_m = 2.0$), where u_{rms} = r.m.s. of the velocity of orbital motion of waves at the specific depth, T = wave period, D = pile diameter, and B = factor dependent upon V/u_{rms} as per Table 8-VIII.

TABLE 8-VIII

Amplification factor B in eq. 8-28 for wave force on pile in presence of current

V/u_{rms}	0.5	1.0	1.5	2.0	2.5	3.0	3.5
B	1.2	2.1	3.8	6.1	9.6	14.2	19.2

Considering a FAS in 100 ft. of water u_{rms} at the SWL from Fig. 4-4 would be around 2 ft./sec. Using this with $T = 12$ sec and $D = 6$ ft., makes $A \doteq 0.05$. If $V/u_{rms} = 3.0$, $B = 14.2$, so that $F_c/F_o = 271/2.05 = 1.32$, or the force at the surface is 32% greater than with zero current. Although a current V is not expected to be uniform throughout the depth it will be optimum near the surface, so enhancing the force at a greater distance from the bed and thus increasing the bending moment on a pile.

The force resulting from a spectrum of waves has been studied analytically by Borgman [39, 40] and Hino [38], who have derived transform functions $f(d/L, T, C_D, C_m)$ from $S_H{}^2(f)$ to $S_F{}^2(f)$ at specific depths. These can be further transformed to total force on a pile by integration and even to a group as a whole by other transfer functions. For this purpose the drag and inertial forces are treated separately and the former is considered to vary linearly with velocity for ease in computation. When C_D and C_m values have been derived from prototype measurements by such processes the scatter is very wide, indicating that further refinements are needed in these procedures. For the case of inertial forces predominating, Jen [41] found the transfer function of Borgman for the total force on a pile to be quite accurate. Once the spectral density function of wave force is thus determined the area under the curve gives the mean square amplitude of force.

In the design of an offshore structure many factors must be taken into account. The intensity of storm centres, their probable and possible proximity to the site, their direction of approach and their ability to create storm surges, all affect the design. However, it is difficult to assess the optimum conditions which imply a certain degree of risk. Borgman [42] has attempted to quantify these variables, based upon certain simplifications, so that probabilities of certain combinations of events can be evaluated.

Pile structures can suffer the force of breaking waves, either in shallow water where breaking is caused by depth, or in deeper water where it is produced by instability. In either case the force is of higher intensity than for a similar sized progressive wave not at this stage. This impact is concentrated over a shorter time and Goda [31] questions whether this is transferred fully to a pile structure which is more flexible than the seawalls previously considered. As an indication Hall [43] has found C_B = 1.2 to 3.0 in the relationship:

$$F_B = C_B w D H_b^2 \tag{8-29}$$

where F_B is the total force on the pile of diameter D.

In the case of shallow-water conditions $d_b = 0.8 H_b$ and $a_c = H_b$ so that the load per foot height of affected area can be computed. Comparison with Table 8-IV for $d/L = 0.02$ and $H_b/d_b > 0.8$, the optimum value of $(F_D)_{max}/wDH^2 = 0.7$, so that the breaking wave exerts a force 2 to 4 times that of a progressive wave of equal height. Goda's experiments have indicated an amplification of 2 but further work is required on this topic.

Although it may not be possible for the larger components of the spectrum to have reached instability in deeper water, it is part of the generating process for smaller-period waves to be breaking near the crests of these longer waves. At the point of breaking the water particles within the shorter wave are being carried forward at the orbital velocity of the longer wave. With the pressure of the wind on the breaking wave and the down hill slope on which it travels the forward velocity of these particles will exceed the orbital speed of the carrying wave.

It is worth substituting some figures by way of examples. Consider a FAS produced by a 35-knot wind. From Table 3-VII, $H_{1/3} = 22.6$ ft., $T_{max} = 11.6$ sec, $T_U = 18.8$ sec, $T_L = 4.1$ sec, with a triangular distribution of spectral energy as exemplified in Fig. 3-13. Consider that the components from T_{max} to T_U combine to carry waves of smaller period, they will have an average period of $T = (11.6 + 18.8)/2 = 15.2$ sec and height $H = 22.6\sqrt{0.62/1.27} = 15.8$ ft. (proportional area under EDC). In a depth of 100 ft. $d/L_o = 100/(5.12 \times 15.2^2) = 0.0845$, $d/L = 0.128$, $L = 787$ ft., $HL/d^2 = 15.8 \times 787/100^2 = 1.25$. From Fig. 4-16, $u_{max}/\sqrt{gd}$ at surface = 0.18, $u_{max} = 10.2$, say 10 ft./sec. Consider the shorter-period components to have an average period $T = (4.1 + 7.7)/2 = 5.9$ sec and a height $H = 22.6\sqrt{0.65/(1.27 \times 4)} = 8.1$ ft. These combined waves will have their own maximum orbital velocity which will be added to that of the carrier wave group. For this smaller batch $d/L_o = 100/(5.12 \times 5.9^2) = 0.56$, $d/L = 0.56$, $L = 179$, $HL/d^2 = (8.1 \times 179)/100^2 = 0.145$, $u_{max} = 0.105\sqrt{g100} = 6.0$ ft./sec. This combined wave crest will be forced forward by the 36 ft./sec wind, but no allowance will be made for this, nor the fact that the broken wave will travel down the leading face of the crest of the carrier wave system. But this near-vertical body of water, some 6–10 ft. high, will advance to the pile at a speed around 16 ft./ sec. This is commensurate with the speed of a similar sized breaking wave travelling in shallow water ($= \sqrt{g(4/3)\, 8.1} = 18.5$) so that the force of the breaking wave will be of the same order as that given by eq. 8-29.

Such breaking waves are the concern of every ship's master, since their impact broadside on to a vessel is sufficient to dislodge cargo and give the ship a dangerous set, besides possibly overstressing plates on the hull. Within a fetch breaking waves will normally follow the line of the wind, but a ship may be turned side-on to it at times as the shorter steeper waves approach from angles up to 30° either side of this alignment.

In the design of large space structures, such as oil rigs, there are many factors to be considered, including that of the interaction of the structure and the waves. These have been discussed by Dean and Harleman [44]. Harleman et al. [45] have provided an example on a four-legged platform cantilevered from the bed. Such a structure implies a degree of fixity as provided by piles driven into the sandy base (see Fig. 8-21A). Another situation is where the legs rest on the sea bed and the structure can rock to and fro [46]. The motion is periodic but not simple harmonic, the natural period (T_s) of which is given by:

$$T_s = 8\sqrt{\frac{(h^2 + B^2/4)\alpha}{Bg}} \qquad (8\text{-}30)$$

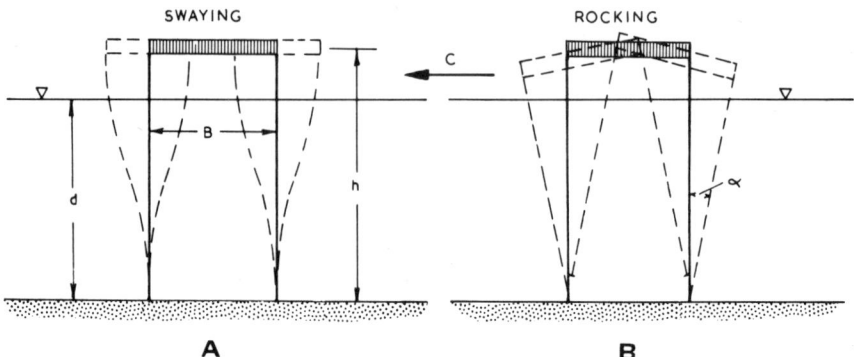

Fig. 8-21. Possible modes of oscillation for oil rigs. A. With fixed ends. B. With free ends.

where α is the extreme angle of tilt (see Fig. 8-21B) in radians, h is the height of the deck above the sea bed where the mass of the structure is concentrated, and B is the spacing of the legs in line with wave advance.

Howe [47] has given typical dimensions for deep-water jack-up rigs as $h = 100$ m, $B = 50$ m, giving $T = 37.3\sqrt{\alpha}$. If $\alpha = 0.5°$ (i.e., leg lifts about 1.5 ft. from bed), $T = 3.5$ sec. This is the same order of resonant period as for structures fixed in the bed and swaying horizontally [45].

It may be thought that with such periodicities, which are much smaller than those encountered in the larger sea waves, that resonant oscillations could not be initiated. However, impacts at intervals which are multiples of these can feed such resonant rocking. Also, waves advancing through a pile structure strike one set of legs and then the rear set some small period of time later. This is illustrated in Fig. 8-22 where the celerity C of waves with given period (T) in depths (d) of water result in impact intervals as shown for leg spacings from 50 to 200 ft. The T_{max}

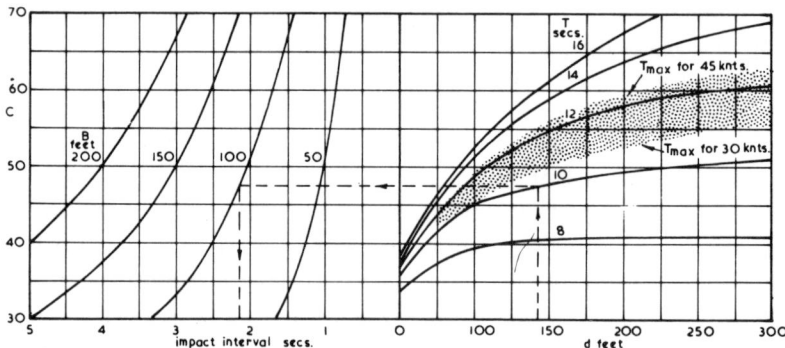

Fig. 8-22. Intervals of consecutive impact of progressive waves on legs of marine structures (see definition sketch Fig. 8-21).

CYLINDRICAL STRUCTURES

values for FAS from 30 to 45-knot winds fall within the stippled area. It is seen that as rigs go into deeper water and the legs become more widely spread so the impact interval grows, as does the resonant period of oscillation of the structure.

It was for prototype periods of 2.5 – 10 sec that model tests on scour beneath legs oscillating up and down on a sandy bed were conducted at the Asian Institute of Technology in Bangkok [46]. These showed a pronounced pumping effect, which resulted in scour to a depth equal approximately to the diameter of the leg. Another type of overall scour around a group of legs cross-connected by horizontal and angled members has been reported by Posey [48]. He reports a cloud of sediment being generated as each long period wave passed through the complex frame. This was probably caused by the macro-turbulence generated as vortices formed either side of legs and braces. Material so thrown into suspension could be removed from the site by currents or mass-transport due to wave action. The scour depth indicated [48] was about 1/3 of the leg spacing, but extended a considerable distance from the structure. Posey has suggested a protective cover of larger-sized material to prevent such action. There is certainly need for much more research into foundation conditions, as well as design loads, for such marine structures.

Horizontal and sloping members

The force on horizontal members can be considered uniform since water-particle velocities and accelerations are similar at the one depth for uni-directional waves. This is not so for short-crested systems, which fact may have to be taken into account with structures near cliffs or vertical walls of nearby facilities. The distribu-

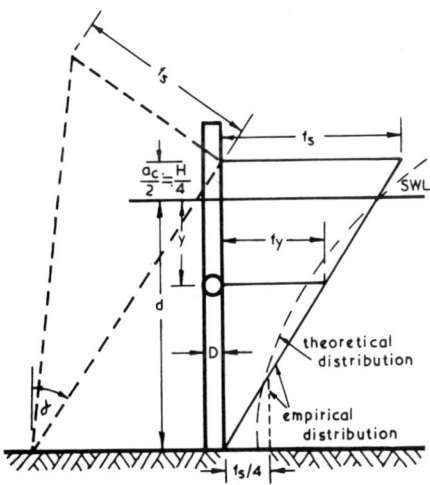

Fig. 8-23. Pressure distribution on vertical and sloping members of marine space frames.

tion of force up a vertical pile in a progressive wave could be considered triangular, as depicted in Fig. 8-23. This provides a conservative pressure at mid-depths, but an under-estimate near the surface and at the bed. To cope with the latter a value of 1/4 the surface pressure is suggested.

The force on a hypothetical vertical pile, the same diameter as the horizontal member, is computed as outlined previously. In the triangular distribution of this an equivalent surface load per unit length of pile (f_s) is given by:

$$F_{max} = f_s(d + a_c/2)/2 \qquad (8\text{-}31)$$

where a_c is the crest height above SWL.

It is assumed that the maximum force (F_{max}) occurs when the water level is half-way between SWL and this maximum crest level. For sinusoidal waves this extrapolated length of affected pile would be $H/4$.

As seen in Fig. 8-23 the per unit length load (f_y) at some depth (y) below SWL is given by:

$$f_y = \frac{f_s(d-y)}{(d+a_c/2)} = \frac{F_{max}(d-y)2}{(d+a_c/2)^2} \qquad (8\text{-}32)$$

It is assumed that the horizontal forces per unit length are the same for the horizontal orientation as for the vertical. However, horizontal members must withstand the forces of vertical velocities and accelerations. For both motions:

$$\frac{\text{vertical value}}{\text{horizontal value}} = \tanh 2\pi(d-y)/L \qquad (8\text{-}33)$$

If the force is predominantly inertial the ratio vertical to horizontal is as given in eq. 8-33. If it is predominantly drag (varying as u^2 or v^2) the force ratio is given by:

$$F_v/F_H = \tanh^2 2\pi(d-y)/L \qquad (8\text{-}34)$$

In the event of the total force comprising both inertia and drag components, a ratio between those from eq. 8-33 and 8-34 should be selected. The predominance of these forces can be checked from the phase angle at which the maximum force is experienced in model tests or by theory. Wiegel et al. [49] have shown that for horizontal forces to be purely inertial:

$$\left(\frac{C_M \, 2V}{C_D \, HA}\right) \frac{\sinh 2\pi d/L}{\cosh 2\pi(d-y)/L} > 1 \qquad (8\text{-}35)$$

and for vertical forces to be predominantly inertial:

$$\left(\frac{C_M \, 2V}{C_D \, HA}\right) \frac{\sinh 2\pi d/L}{\sinh 2\pi(d-y)/L} > 1 \qquad (8\text{-}36)$$

CYLINDRICAL STRUCTURES 405

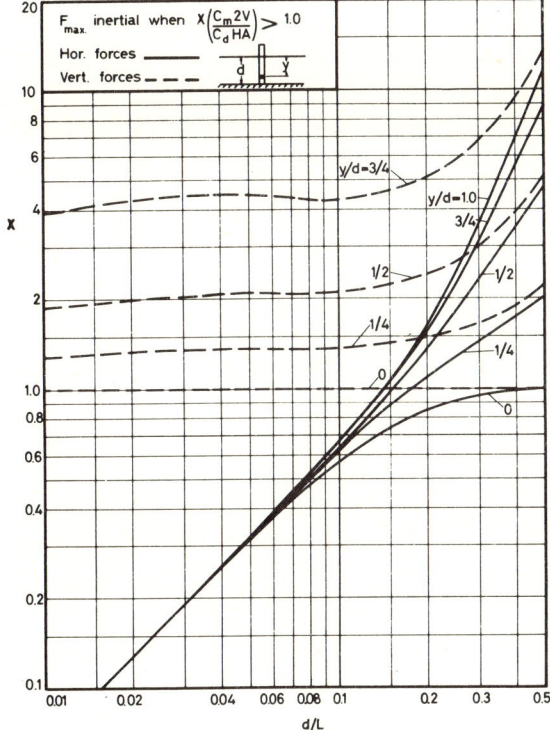

Fig. 8-24. Criterion for predominance of inertial or drag force for horizontal and vertical motions of water particles.

where V is the volume of pile per unit length, A is the area of pile per unit length transverse to the flow, and y is the depth from SWL.

Eq. 8-35 and 8-36 have been graphed in Fig. 8-24 where X replaces the hyperbolic functions. For circular piles the bracketed term may be replaced as follows:

$$\frac{C_M\, 2V}{C_D\, HA} = \frac{2 \times 2 \times \pi D^2}{1 \times H \times D \times 4} = \frac{\pi D}{H} \tag{8-37}$$

When applied in Fig. 8-24 for horizontal forces the result as drawn in Fig. 8-19 results, where it is seen that the H/L ratio influences the criterion for predominance of inertial or drag forces. Fig. 8-24 also indicates that vertical forces are more likely to be inertial than horizontal forces, particularly for small depth ratios. It is possible that the predominant force can change from drag to inertial from SWL to the bed. However, this does not influence the total force as computed by Table 8-V, but will dictate the proportion of vertical force to horizontal force at any given level.

Where a brace is sloping the distribution of force per unit length vertically is as before, but there is now a longer length over which this acts. Assuming the triangular distribution as before (see Fig. 8-23), the unit force at the surface is a given by eq. 8-31. This applies to the surface section of the sloping member per unit length so that the total force on the member sloping at angle α to the vertical is:

$$(F_{max})_\alpha = \frac{f_s(d+a_c/2)}{2 \cos \alpha} = \frac{2(F_{max})_{90}(d+a_c/2)}{(d+a_c/2) 2 \cos \alpha} = \frac{(F_{max})_{90}}{\cos \alpha} \quad (8\text{-}38)$$

Submerged pipelines

Pipelines are a special case of horizontal members, normally of circular cross-section, the major difference being in their proximity to the bed. This distorts the flow around the top surface of the pipe and prevents flow beneath it. This increased velocity is associated with a decreased pressure so that a lift force is exerted on the pipe which must be resisted by the buoyant weight of the pipe and its contents. This lift force is optimum concurrently with the maximum horizontal velocity, or when the drag force is at its peak. The overall force condition is depicted in Fig. 8-25A where F_D = drag force, F_L = lift force, F_W = weight of pipe and contents, F_B = buoyant force, and F_F = friction force at bed.

Employing the concepts of uni-directional flow, these forces per unit length of pipe are given by:

$$F_D = C_D D_o w u_{max}^2/2g \quad (8\text{-}39)$$

$$F_L = C_L D_o w u_{max}^2/2g \quad (8\text{-}40)$$

$$F_W - F_B = (D_o^2 - D_i^2)\gamma_p \pi/4 + D_i^2 \gamma_c \pi/4 - D_o^2 w \pi/4$$

$$= \frac{\pi}{4}[D_o^2(\gamma_p - w) + D_i^2(\gamma_c - \gamma_p)] \quad \text{41)}$$

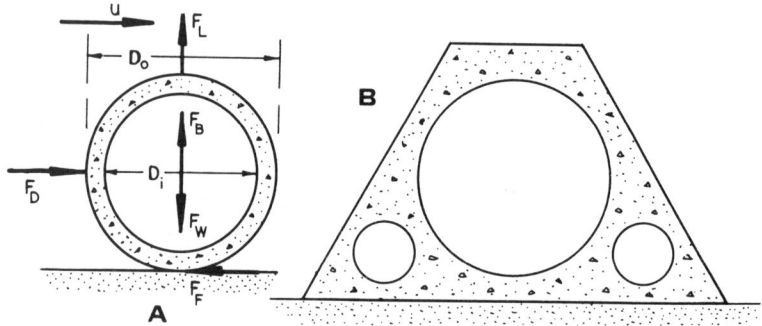

Fig. 8-25. A. Resultant wave forces on a submerged circular pipeline. B. Possible layout for recommended stable trapezoidal cross-section.

$$F_F = f(F_W - F_B - F_L) \tag{8-42}$$

where C_D is the drag coefficient, C_L is the lift coefficient, D_o, D_i are outside and inside diameters of pipeline, w is the specific weight of fresh or seawater, u is the free velocity of flow at the bed (maximum orbital value normally used), γ_p is specific weight of pipe material, γ_c is specific weight of contents of pipe, and f is friction factor between pipe and bed.

Appropriate equations can be derived for conduits of other shapes and combinations of materials. As noted by Beckmann and Thibodeaux [50] a trapezoidal outer shape can minimize the drag and lift forces whilst supplying sufficient weight to hold down a circular tube. As seen in Fig. 8-25A, a torsional force is exerted on the pipe which is resisted by the broad base of the suggested trapezoid (see Fig. 8-25B).

In eq. 8-39 and 8-40 the unknowns to be determined by theory or experiment are C_D and C_L. There would appear to be much more research needed on this topic for oscillatory flow, so that only the range of values can be noted here. Beckmann and Thibodeaux conclude that for oscillatory flow over an obstacle on a bed the drag coefficient is reduced by two factors: (a) the streamlining effect of the flow towards the obstacle by the plane of the bed; and (b) by the development of turbulence due to the oscillation of vortices developed each half wave period. This is equivalent to very high Reynolds number conditions. They conclude that $C_D = C_L = 0.5$ is sufficient guarantee of safety.

Wilson and Reid [50], in their discussion of this paper, dispute the above coefficients and cite the results of many workers to show that more realistic values are $C_D = C_L = 1.0$. Alterman [50] also notes the variety of wave conditions compared to results obtained with a uni-directional flow, where depth is of no concern.

Beckmann and Thibodeaux also discussed the inertial force which is out of phase with the concurrent drag and lift forces. This must therefore be computed separately and compared to the resultant of the former. The inertial force (F_M) per unit length of pipeline will have a maximum value given by:

$$(F_M)_{max} = C_M \frac{w}{g} \frac{\pi}{4} D_o^2 (du/dt)_{max} \tag{8-43}$$

where $(du/dt)_{max}$ is the maximum orbital acceleration in the horizontal direction, and other terms as previously defined.

The value of C_M recommended by Wilson and Reid is 2.5. Assuming sinusoidal motion near the bed:

$$(du/dt)_{max} = 2\pi u_{max}/T \tag{8-44}$$

so that Fig. 4-16 can be used from experimental tests to determine the particle acceleration. The force $(F_M)_{max}$ will occur when the water level is passing through the SWL value, which for a sinusoidal motion is at $T/4$ and $3T/4$, with reversals at these respective times. It will become predominant when the amplitude of horizontal particle motion (*x*) given by:

$$x = uT/2\pi \qquad (8\text{-}45)$$

is less than 3 times the vertical dimension of the object, or the diameter in the case of a pipe [50]. It should be noted that in eq. 8-42 the $-F_L$ term will not be present if the friction force F_F is due to inertial force (F_M) only.

Brown [51] conducted full-scale tests by inserting piezometers in a pipe suffering a unidirectional current. Suction forces were experienced over the top two quadrants plus the leeward lower quadrant. The coefficient of lift was found to be greater than that of drag. Longitudinal web members projecting 1/8 diameter near the top of the pipe, termed "spoilers", influenced the lift and drag forces greatly.

Ralston and Herbich [52] have summarized many articles on the effects of waves and currents on submerged pipelines. These deal mainly with oil or gas lines which may rest on the sea bed or be buried. The crucial zone of such pipes is that adjacent to the riser at the oil platform.

A report on the research needs in pipeline engineering has been prepared [53], which expresses the necessity for better analytical and experimental results for predicting forces. Another unknown is the scour which is prevalent along pipes resting on the sea bed, even at depths near the edge of the continental shelf. It would appear that the most costly damage during some hurricanes on the Gulf coast of the United States was sustained by movement or failure of underwater pipe lines or risers [54].

Reid [52, 55] has listed some of the hazards in marine pipeline construction as follows:

(*a*) destruction or damage of pipe risers or platform structures by wave and wind action; (*b*) shifting of pipelines by wave action or currents; (*c*) damage to pipeline by sag into soft sediments; (*d*) damage by anchors of ships or dredgers when buried lines are exposed by scour; (*e*) damage during pipe-laying operations.

The many aspects of design and construction of submarine pipelines has been outlined by Miller [56], who discusses the application of marine geology, oceanography and survey requirements for oil installations and sewage outfalls. Coastal engineers concerned with such projects should read this reference and use its comprehensive bibliography.

SUBMERGED OBJECTS OF LARGE DIMENSIONS

It has been noted already that when the horizontal dimensions of submerged tanks or piles are large in respect to wave height or the amplitude of orbital motion of the water particles, the force experienced is predominantly inertial. Besides this, the wave motion itself is influenced by the object, so its accelerations and pressure fluctuations are in part dictated by its shape and size as well as the characteristics of the incident wave. The problem of assessing forces on such shapes as upright cylinders, spheres, hemispheres and boxes has become increasingly important with the need to store petroleum products from offshore wells and construct undersea habitats. It is also anticipated that in the future food may be stored in such submarine containers to obviate rodent attack.

The upright cylinder has been analyzed by Lebreton and Cormault [57] who verified experimentally the relationship:

$$\frac{(F_M)_{max}}{wD^2H} = \frac{C_M \pi \sinh 2\pi h/L}{8 \cosh 2\pi d/L} \qquad (8\text{-}46)$$

where D is the diameter of an upright cylinder of height h from the bed, and C_M is the inertial coefficient (= 1.0 for $h/d \leqslant 0.8$ and = 1.66 for $h/d > 0.8$)

When $h/d > 1.0$ the RHS of equation reduces to $C_M(\pi/8) \tanh 2\pi d/L$, which value was derived by Goda [31] and used for $(F_M)_{max}$ in Table 8-IV. Eq. 8-46 has been graphed in Fig. 8-26, where it is seen that for cylinder heights less than the water depth a maximum in the dimensionless force occurs at d/L_o from 0.17 to 0.25. Since the accelerations are directly related to u_{max} from eq. 8-44 they will have a similar vertical distribution throughout the depth as the velocities which have previously been accepted as triangular. Also, as the force is proportional to acceleration, the pressure exerted on the cylinder can be approximated to a triangular distribution, giving the height of the resultant force as $2h/3$ from the bed. As cylinders of smaller height are employed so should this height be reduced to $h/2$, but the former figure is conservative for computing overturning moments.

For the particular case of large cylinders penetrating through the water surface, MaCamy and Fuchs [58] provided a diffraction theory based upon linear theory or waves of small height. Their relationships can be put in the form graphed in Fig. 8-27 where F_{max} represents the total force experienced up to the SWL. The turning moment, which includes forces to the water surface, can be obtained from Fig. 8-28 by using $M_{SWL} = F_{max} \times y_c$ from Fig. 8-27. The equations of MaCamy and Fuchs do not include any inertial coefficient and a comparison of forces in Fig. 8-26 and 8-27 indicates the latter to be low by a factor of 1.4, indicating the need for a C_M of this order. The curve for $h/d > 1.0$ in Fig. 8-26 contains a value of $C_M = 1.66$.

Tanks of box shape have been studied by Brater et al. [59] for mid-depth submersion and by Herbich and Shank [60, 61] for location on the bed. The experimental results of the former have been combined with the theoretical analysis of Riabouchinski [62] to give values of C_M in:

$$\frac{(F_M)_{max}}{wAD} = \frac{C_M}{g} (du/dt)_{max} \qquad (8\text{-}47)$$

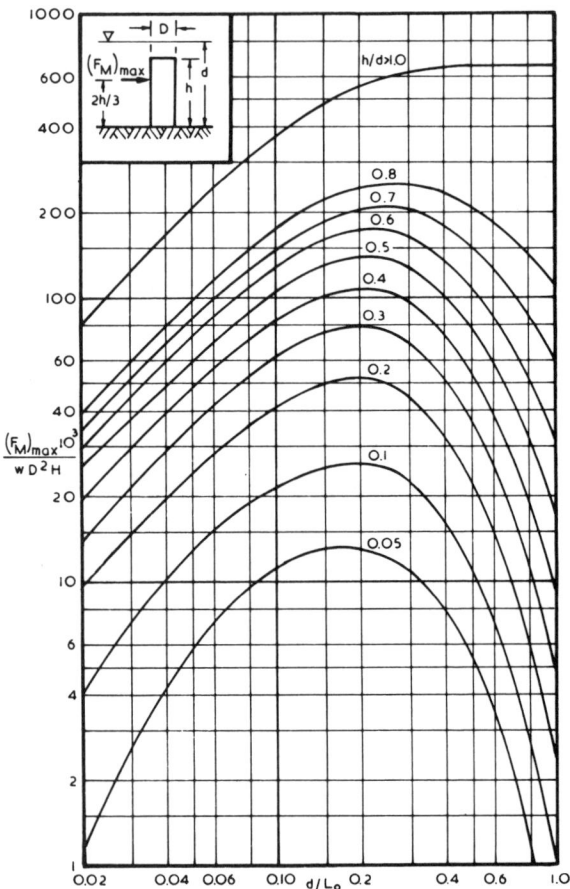

Fig. 8-26. Inertial forces on submerged upright cylinders of large diameter.

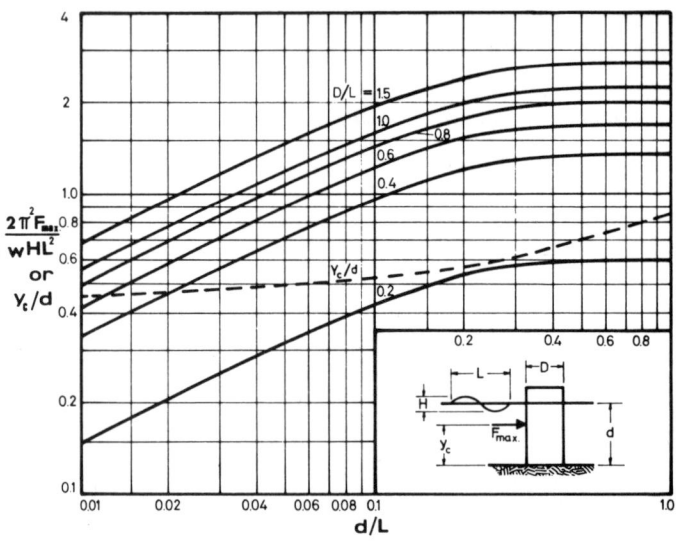

Fig. 8-27. Maximum horizontal force on upright cylinder exerted up to SWL.

SUBMERGED OBJECTS OF LARGE DIMENSIONS

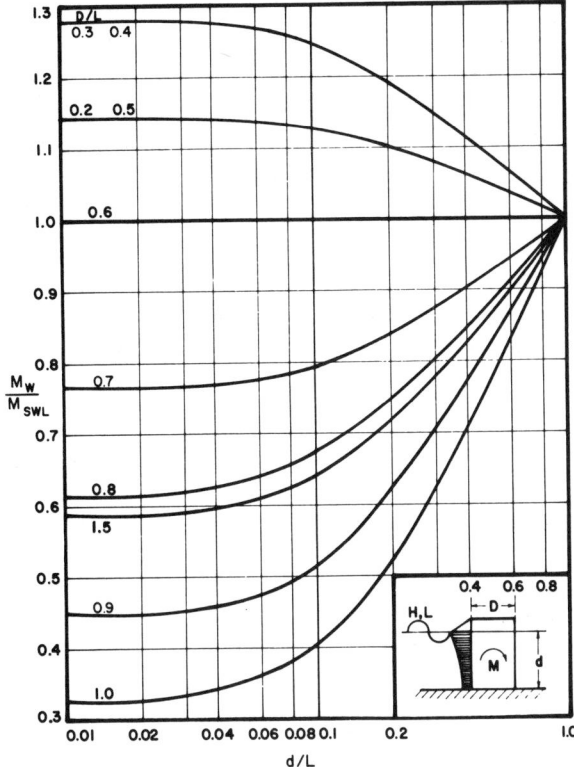

Fig. 8-28. Maximum bending moment on upright cylinder produced by forces up to water surface.

where $(F_M)_{max}$ is the force per unit width of box measured crest-wise, A is the length of box measured in direction of wave advance, and D is the height of box measured vertically.

These values of C_M are listed in Table 8-IX and are seen to vary with wave

TABLE 8-IX

Values of C_M for box-type structures submerged at mid-depths

H/L	A/D										
	0.1	0.2	0.4	0.6	0.8	1.0	2	4	6	8	10
0.02	12.2	7.0	4.3	3.5	3.1	2.7	2.1	1.75	1.6	1.55	1.45
0.03	11.9	6.7	4.1	3.4	3.0	2.6	2.0	1.8	1.55	1.5	1.4
0.04	11.3	6.4	3.9	3.3	2.8	2.5	1.9	1.7	1.45	1.4	1.35
0.05	10.0	5.7	3.5	2.9	2.5	2.2	1.7	1.45	1.3	1.25	1.2

steepness and ratio A/D. No differentiation is made in this table for the depth at which the box is held rigidly, but there would be a tendency for greater values of C_M the nearer the box was to the surface.

The experimental results obtained for forces on such structures resting on the bed [60, 61] were graphed against a wave parameter HL/d^2 and distinct relationships obtained for each A/D ratio. From these the added volume V' required to be added to $V = AD$ was obtained, in order that ratios $(F_M)_{max}/w(V + V')$ versus HL/d^2 fell essentially along one curve. This volume V' was found to be constant for all A/D ratios, indicating a similar wake structure for all proportions of right parallelepiped. The results of half-cylinders oriented side-on and end-on to the waves were similarly analyzed and finally combined in one graph where the following equation defines the lower envelope of the force ratio:

$$\frac{(F_H)_{max}}{w(V+V')} = 0.055 (HL/d^2)^{1.133} \tag{8-48}$$

or:

$$\frac{(F_H)_{max}}{wV} = C_M \, 0.055 (HL/d^2)^{1.133} \tag{8-49}$$

An alternative relationship which averages the results is:

$$\frac{(F_H)_{max}}{w(V+V')} = 0.07 \frac{HL}{d^2} \pm 0.02 \tag{8-50}$$

or:

$$\frac{(F_H)_{max}}{wV} = C_M \, 0.07 \frac{HL}{d^2} \pm 0.02 \tag{8-51}$$

The ratio $(V + V')/V$ equals the coefficient C_M in eq. 8-49 and 8-51. Values obtained for the box-type structures at the bed are listed in Table 8-X for a range of H/L from 0.01 to 0.156. These are graphed with values for mid-depth submersion

TABLE 8-X

Values of C_M for box-type structures and half cylinders resting on the bed

A/D	0.445	0.89	1.78
C_M	4.0	2.5	1.75
half-cylinder		end-on	side-on
A/D (equiv. rect.)		2.18	2.55
C_M		1.85	1.69

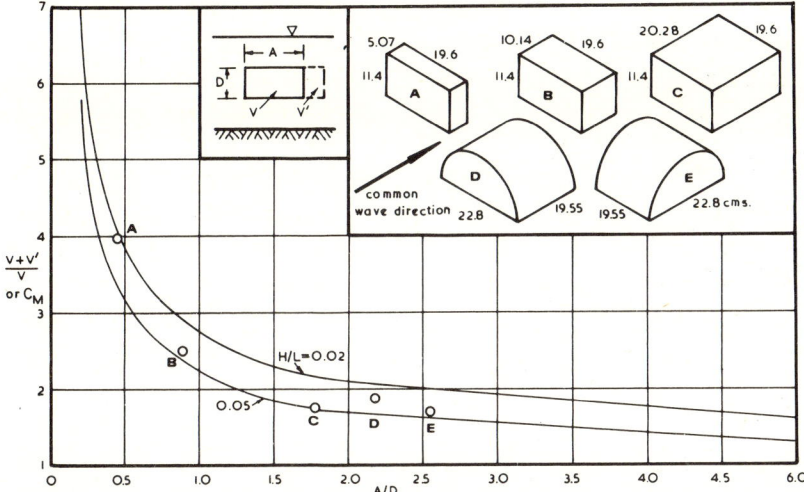

Fig. 8-29. Coefficient of inertia for submerged block structures in mid-depths and at the bed.

(Table 8-IX) in Fig. 8-29, where it is seen that values of C_M for bed location are commensurate with those for mid-depth suspension. It would appear that a similar approach to the above might be suitable in analyzing tests on submarine pipelines.

Submerged tanks of hemi-spherical shape resting on the bottom have been studied by Garrison [63, 64]. A theory is derived, based upon the assumptions of wave length large compared to object size and viscous effects negligible, which gives horizontal and lifting forces. These two dimensionless force ratios are:

$$\frac{8(F_H)_{max}}{wD^2 H} = \frac{\pi^2 D/L}{\cosh(d/D)(2\pi D/L)} \tag{8-52}$$

and:

$$\frac{8(F_V)_{max}}{wD^2 H} = 2\pi \left[\frac{(\pi D/L) \sinh \pi D/L - \cosh \pi D/L + 1}{(\pi D/L)^2 \cosh(d/D)(2\pi D/L)} \right] \tag{8-53}$$

where D is the diameter of the hemi-sphere.

Eq. 8-52 and 8-53 have been reproduced in Fig. 8-30, together with curves from experimental results provided [63]. Results for F_H experimental for $d/D = 1.0, 1.5$ and 2.5 were essentially the same as for the theory. The F_V experimental curves were not quite so close to the theoretical traces, but Fig. 8-30 may serve for design purposes until further data are available. The equation for F_V has an implied $C_M = 1.5$ which could be increased to 1.7 in order to coincide with the diffraction theory [63]. For small $\pi D/L$ the horizontal force is in phase with the horizontal particle acceleration whilst the vertical force is maximum as the trough passes

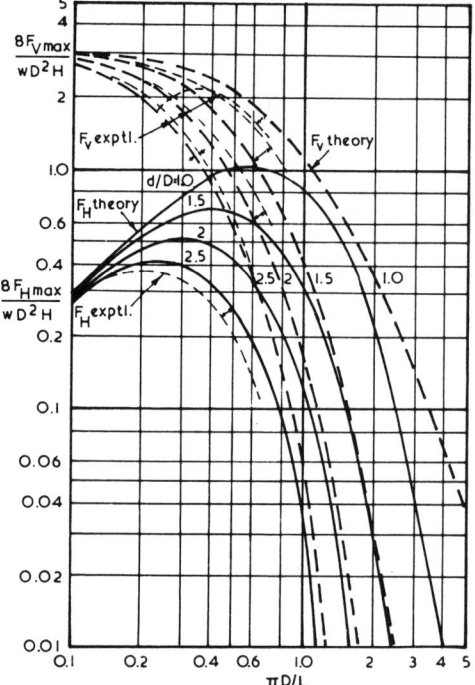

Fig. 8-30. Horizontal and vertical wave forces on a hemisphere resting on the sea bed.

across the hemisphere. At no values of $\pi D/L$ do these forces synchronize so they should not be added vectorially. There is no overturning moment on the hemisphere since all forces are normal to the surface, hence the resultant passes through the centre.

The force and motion on a sphere supported at a given depth have been analyzed by Shi-igai and Kono [65]. The maximum horizontal force is given by:

$$\frac{(F_H)_{max}}{wV} = \frac{\pi H}{L} \frac{\cosh 2\pi(d-h)/L}{\cosh 2\pi d/L} \qquad (8\text{-}54)$$

where V is the volume of the sphere ($= \pi D^3/6$), and h is the depth of the centre of the sphere below SWL.

Fig. 8-31 contains the definition sketch of this system plus the graph of eq. 8-54. The reader is referred to the original paper [65] for the motions in the two horizontal directions with varying degrees of fixity. The motion of such an object can influence the water particle motion and this can have a greater effect than viscous resistance or added mass. Grace and Casciano [66] have attempted to correlate forces on a sphere in the ocean with characteristics of waves passing overhead. In this case the sphere was small enough for the forces of drag to be

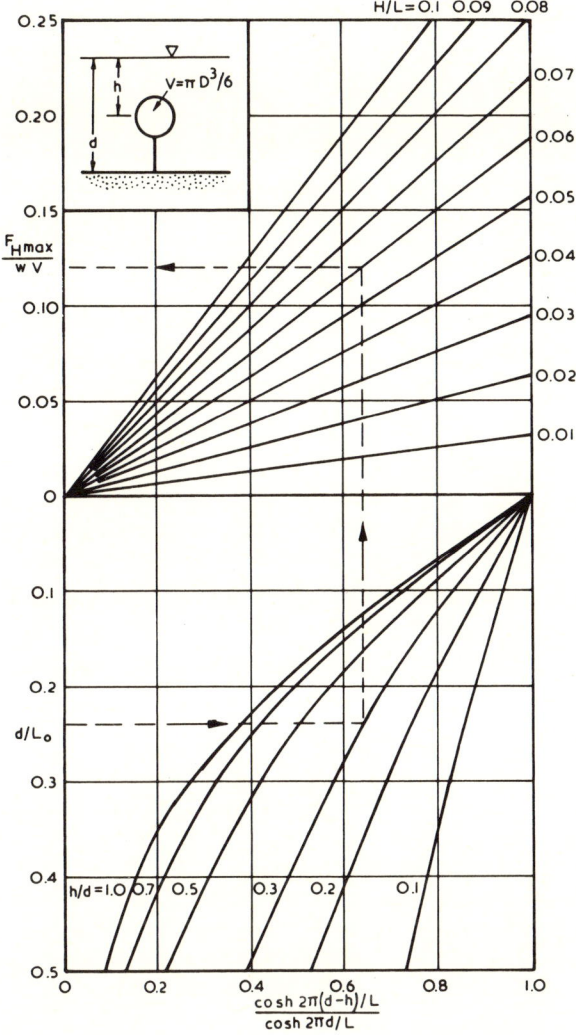

Fig. 8-31. Wave force on a sphere submerged at mid-depths.

predominant whereas the analysis and tests of Shi-igai and Kono [65] apply strictly to inertial forces.

The forces and resulting motions of a sphere, a triaxial ellipsoid and other shapes has been analyzed for complex wave systems by Sommet and Vignat [67]. Application is possible to fixed, moored or free bodies; in the latter case the heaving force, rolling, pitching and yawing moment coefficients are obtainable in the various directions.

RUBBLE-MOUND BREAKWATERS

Where rock is available conveniently for quarrying, the rubble-mound breakwater is generally the most economical structure for protection. It can dissipate steep storm waves by turbulence, as water penetrates the voids between the rocks. Most of this attenuation is effected near the water surface, although oscillatory water motions at depth are reduced somewhat. The longer swell waves of moderate height have a high reflectivity from even rock-fill structures.

The stability of armour units on the seaward face depends upon many factors, including the size and density of units and their capacity to lock with each other. For given wave conditions there will be a slope of the breakwater face which will be in equilibrium for the specific armour units used. Tests have shown that the stone characteristics and stable slope vary with H^3, but further tests may indicate that wave length should enter such a relationship, perhaps as energy arriving per wave length (proportional to $H^2 L$).

Svee [68] has analyzed the forces acting on armour units of a breakwater as depicted in Fig. 8-32. These consist of weight W, uplift force P_u or P_d with components P_x and P_y, friction force F_1 at bed and F_2 between a unit and its neighbours. Assuming that movement directly downslope is prevented by stability of armour units at depth, stability of any single block is determined by the balance of forces normal to the face.

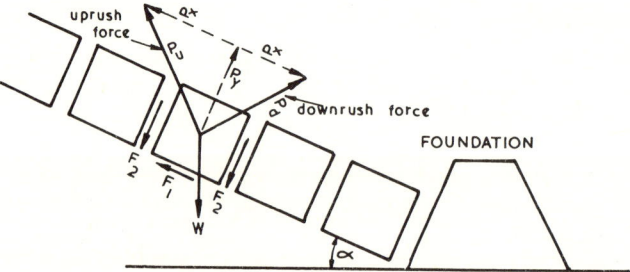

Fig. 8-32. Forces acting on armour units of a breakwater.

Svee distinguishes two conditions where blocks might be plucked from the surface, one when the uprush occurs from the incoming wave (either breaking or non-breaking), and the other when downrush occurs prior to the arrival of the next wave. In the former case the drag force of the tangential flow P_u will be concentrated at the top surface of the armour unit, probably effecting an interlocking of a unit with its upslope neighbour, so enhancing the friction component of the force F_2. The balance of forces normal to the face is then:

$$W \cos\alpha - P_y + \mu'(P_x - W \sin\alpha) \geq 0 \qquad (8\text{-}55)$$

where α is the angle of the face to the horizontal, and μ' is a friction plus interlocking coefficient.

In the case of the downrush, water is pouring through the voids of the armour units and issuing to the sea at, or just below, SWL as the trough exists at the face. This flow between blocks, Svee contends, can nullify the friction between blocks, so that the balance of forces for stability is given by:

$$W \cos\alpha - P_y \geqslant 0 \tag{8-56}$$

By putting the various forces in eq. 8-55 and 8-56 into terms of specific weights, dimensions of rock and kinematic functions, Svee reduces the uprush stability criterion to:

$$V = \frac{KH^3}{(SG_s - 1)^3 (\cos\alpha - \mu' \sin\alpha)^3} \tag{8-57}$$

where V is the volume of armour unit required for stability, H is the wave height at the toe of the breakwater, SG_s is the specific gravity of the stone ($= \gamma_s/w$), μ' is a coefficient of friction and interlocking ($\doteq 1.05$), K is a factor which is dependent upon: (a) the nature and placement of the armour units (it is least for rough and porous layers since these qualities reduce velocities; where blocks are flattish in shape they should be placed with their long dimensions normal to the breakwater face); (b) the steepness of the waves, which affects the breaker height for any given incident amplitude (a long wave of a given height is more destructive than a short one; aspects of the wave spectrum will be discussed later).

As noted already, the factor μ' in the downrush situation can well be zero, so that eq. 8-57 reduces to:

$$V = \frac{KH^3}{(SG_s - 1)^3 \cos^3\alpha} \tag{8-58}$$

In this case the "quick-sand" effect is likely to be greatest for steeper slopes, where a larger hydrostatic head from the prior uprush is available. For milder slopes the uprush velocities are of greater importance. Thus there must be a limiting value of α when uprush or downrush is decisive. Hedor [68, 69] has shown by extensive model tests that $\alpha_{crit} = 21°$ or a slope of 1 : 2.5.

Eq. 8-57 and 8-58 have been graphed in Fig. 8-33, with a change from the former to the latter at α_{crit}. This has brought curves for $\alpha - 20°$ and $30°$ close together. Values of K or its reciprocal are available for various types of armour unit [70, 71] and for various positions in a breakwater [72].

The volume V versus wave height H graphs have been drawn in Fig. 8-33 for angular rubble stone of specific weight 165 lb/ft.^{3} (SG = 2.65) in a shallow trunk section of a structure. Thus the inset curves of "percentage V" are 100% for these conditions.

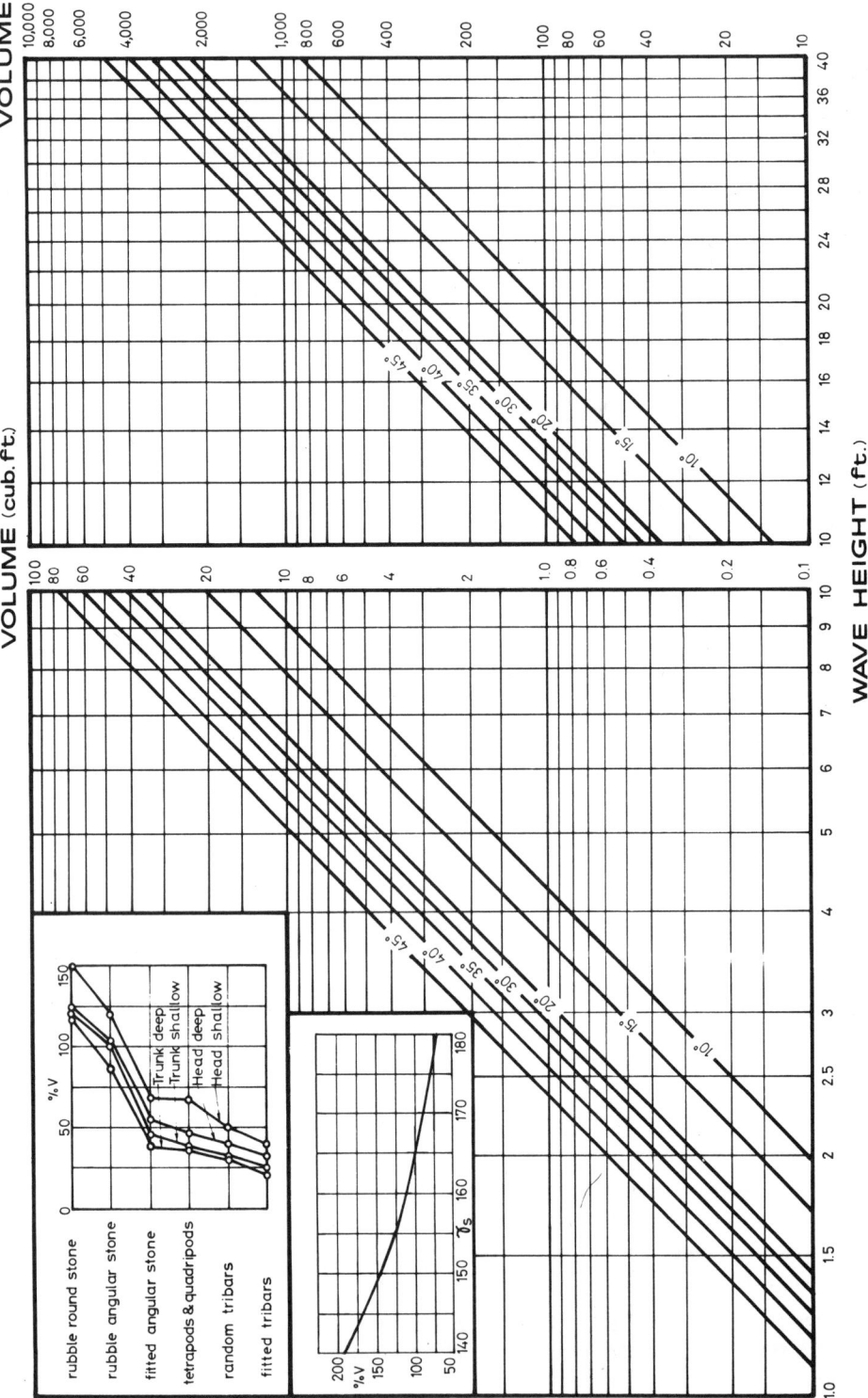

Fig. 8-33. Design graph to determine size of armour units for breakwater stability.

The top inset of Fig. 8-33 contains the following curves: (*a*) trunk deep: implies non-breaking waves; (*b*) trunk shallow: implies breaking waves; (*c*) head deep: conical shape with non-breaking waves; (*d*) head shallow: conical slope with breaking waves.

The lower inset gives the different volume required for units with specific weights different from 165 lb./ft.^{3} or SG = 2.65, e.g., concrete of 150 lb./ft.^{3} must have volume 46% greater than that derived from the main graph. The advantage of using volume rather than weight of armour units is that it obviates the mis-use of tons of various magnitudes (1 ton = 2240 lb. U.K., 2000 lb. U.S., 1000 kg = 2205 lb. c.g.s.).

Hudson [70] has introduced a stability number N_s given by:

$$N_s = \frac{H}{V^{1/3}(SG_s - 1)} = (K \cot \alpha)^{1/3} \tag{8-59}$$

in which the difference of the angular functions of α has been illustrated by Svee [68]. Values of H are designated in eq. 8-59 for zero amount of damage. At one wave height armour units will just commence to move and at a greater height more units will be dislodged, which can be identified as a percentage of complete failure. Font [73] has shown that damage is a function of wave duration. If a certain degree of damage could be entertained during a storm, with the possibility of repair before the next storm arrives, the size of stone could be reduced somewhat.

Since most design criteria will be based upon storm conditions and most model studies based upon monochromatic waves, a correlation between effects of $H_{1/3}$ and T_{max} of the spectrum and H and T of the model tests needs to be proven. Discussion of this issue [74] does not seem to be resolved, as sufficient tests, with a variety of height, period, depth, slope and size of armour units, have not been conducted. At present it would appear that a wide spectrum with $H_{1/3}$ is slightly less severe on a breakwater than a monochromatic wave with the same height and $T = T_{max}$. However, a narrower spectrum concentrated around T_{max} can be more damaging than a single train with $H = H_{1/3}$. The implication here is that the infrequent higher waves are sufficient to dislodge one armour unit and so open the way for the average wave to further the attack. Where aprons or submarine mounds are provided to precipitate breaking in front of a breakwater face the location of such action is of prime importance in respect to run-up and destruction of the main structure. Model testing should be conducted either with a spectrum of waves or a large range of wave periods separately.

Certain fabricated armour units are contained in the design graph of Fig. 8-33. Other shapes have been developed [75–77] the economics of which should be investigated for any site. In establishing a size for a specific slope of breakwater face, not only does the handling equipment required during construction need to be considered, but also the capacity of crane units for repair work at a later date.

Since friction or interlocking between armour units appears to be of considerable value in providing stability it is understandable that attempts to increase these have succeeded. The various shapes developed are aimed at interlocking whereas increase of friction may be provided by asphaltic materials grouted into the breakwater face after placement. A comprehensive report on successful application of such methods used in The Netherlands has been presented by d'Argremond et al. [78]. Model tests have indicated that stones can be virtually glued together, both above and below SWL, so that they act very much as larger blocks. This permits smaller rock units to be employed for a given wave climate, or a larger factor of safety for an existing structure. The upgrading factor so derived is in the order of 5, so that great economies would appear possible if the bitumen products were readily available. The authors provide relevant references in the literature respecting mixtures and techniques developed in the use of bitumen in hydraulic and marine engineering.

Kagawa [79] has recommended an alsphalt mat between a concrete block breakwater and its rubble-mound mat. Such composite breakwaters, as noted already, must be designed to resist sliding, overturning and rocking, all of which should be aided by the cohesive effect of bitumen products.

The rock-fill material used at the base of composite breakwaters or as partial protection for vertical walls will have wave forces tending to remove or shift it, even though it is submerged well below the surface. Rubble-mound foundations for concrete structures are necessary where a sand bed is unable to withstand the pressures of the breakwater itself plus the shock loads from breaking waves, or where the sand will be eroded on the seaward side, particularly at extremities of the structure. The alternative of a complete rock-fill mound can be too expensive when depths become great.

A model study of stability has been made by Brebner and Donnelly [80] on rock fill serving as foundation for a superstructure, or as protection at the toe of a block structure sitting on the sea bed. In their tests stones of uniform size were used, so that for the former case water could flow beneath the structure. The seaward flow when the trough of the wave was at the wall was probably instrumental in removing units of the outer layer, because larger sizes were required for a given wave height than was the case of the structure blocking such flow. The tests indicated that stability was not greatly affected by wave steepness, nor width of mat in front of the superstructure. Results for the top width used of $0.4\,d$ were expressed as

$$V(SG_s - 1)/H^3 = f(h/d, d/L) \tag{8-60}$$

where V is the volume of stone of specific gravity SG_s which remained stable for an incident wave height of H, and h is the height of the stone mound from the bed.

Eq. 8-60 is presented in Fig. 8-34, where it is seen that the permeable mound

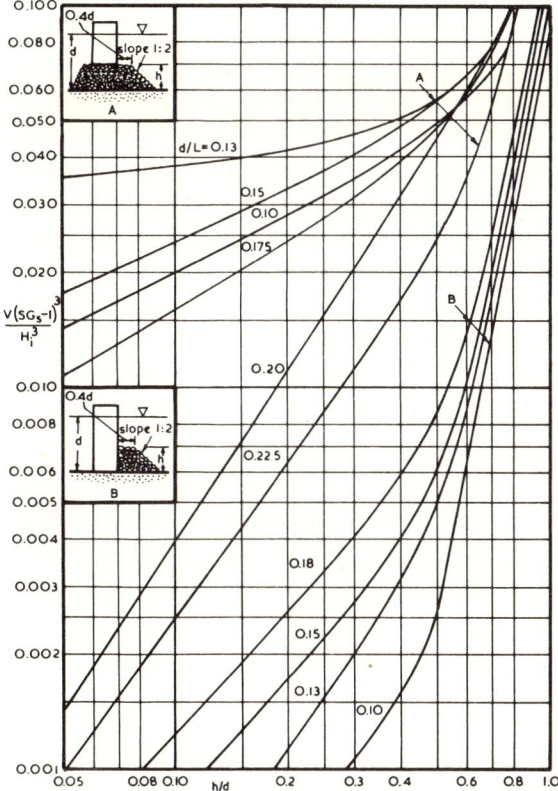

Fig. 8-34. Design graph for armour units of a composite breakwater.

requires larger blocks. For this case $d/L = 0.13$ provides the severest conditions, with smaller stones being possible for greater and smaller depth ratios. Where smaller stones are used inside such a mound they impede flow so that condition A in Fig. 8-34 will tend to that of B. As the mounds rise to SWL the stability requirements of each become similar.

Nagai [81] found that pressures exerted on steel pipes used as a breakwater increased significantly when rubble-mound "protection" was used to reduce scour. This was due to the propensity for breaking provided by the mat. The total load on the pipes was increased up to 36%. This aspect of toe protection for any given site should be examined by a model study.

EXAMPLES

1

Waves of 10 sec period and 6 ft. height arrive to strike a vertical wall normally. Find the bending moment about the bottom of the wall due to the wave forces in both the landward and seaward directions, if the water depth is 25 ft. Make the same calculations for the maximum standing wave possible at this depth.

$T = 10$ sec; $d = 25$ ft.; $d/L_o = 25/512 = 0.049$; $d/L = 0.093$; $L = 272$ ft., $H = 6$ ft., $H/L = 6/512 = 0.0117$.

From Table 8-I, $F_{max}/wH_i d = 0.98$ and 0.76; $F_{max} = 0.96 \times 64 \times 6 \times 25 = 9250$ lb./ft. landwards and $F'_{max} = 0.76 \times 64 \times 6 \times 25 = 7320$ lb./ft. seawards.

To use Fig. 8-3 determine a_c from Fig. 4-18 for $H/L = 2 \times 6/272 = 0.0234$, $d/L = 0.093$. Thus $\Delta H/H = 0.09$, $a_c/H = 0.59$, $a_c = 0.59 \times 12 = 7.06$ ft.

Fig. 8-3 for $a_c/d = 7.06/25 = 0.283$, $d/L_o = 0.049$, gives: $y_c/d = 0.6$ and $y_t/d = 0.5$.

Landward moment $= F_{max} \times y_c = 9250 \times 0.6 \times 25 = 139,000$ ft./lb., seaward moment $= F'_{max} \times y_t = 7320 \times 0.5 \times 25 = 91,500$ ft./lb.

From eq. 4-89 maximum possible standing wave given by $H/L = 0.22 \tanh 2\pi d/L$ where H is total wave height: $H/272 = 0.22 \times 0.526$, $H = 31.5$ ft., $H_i = 15.8$ ft., $H_i/L = 15.8/272 = 0.058$. Table 8-I gives $F_{max}/wH_i d = 0.97$ and 0.38.

Test accuracy of 4th-order solution with Fig. 8-4, $H_i/d = 15.8/25 = 0.63$; this is beyond the limit of accurate application of the 4th-order theory, so that values in Table 8-I should be used with caution, otherwise the procedure for finding moment is as before.

2

If the wall in Example 1 extended 10 ft. above SWL, find the horizontal force exerted landwards on the wall for the two wave conditions.

For $H_i = 6$ ft., $a_c = 7.06$ ft., so that wave crest does not reach to top of wall, so force is the same as before.

For $H_i = 15.8$ ft., $a_c > 10$ ft., so overtopping will occur. In Table 8-II for $H_i/h = 15.8/10 = 1.58$, $R_o/h = 1.48$, $R_o = 14.8$ ft.

In Table 8-I for $H_i/L = 15.8/272 = 0.058$, $d/L = 0.093$, $F_{max}/wR_o d = 0.97$, $F_{max} = 0.97 \times 64 \times 14.8 \times 25 = 22,950$ lb./ft.

3

Compare the horizontal force exerted on a 10-ft. square caisson with a circular unit of 10 ft. diameter when they stand in 12 ft. of water and suffer waves 5 ft. high and 9 sec period.

$T = 9$ sec; $d/L_o = 12/(5.12 \times 9^2) = 0.029$; $d/L = 0.07$; $L = 172$ ft.; $H_i/L = 5/172 = 0.029$; from Table 8-I, $F_{max}/wH_i d = 1.1$. Total force on 10-ft. square caisson $= 1.1 \times 64 \times 5 \times 12 \times 10 = 42,200$ lb. Consider $15°$-elements of circular caisson each of width $\pi d(15/360) = 1.31$ ft. Component of horizontal force on each element $= F_{max} 1.31 K \cos \beta$ where $K = f(\beta)$ as in Fig. 8-5 and listed in Table 8-XI.

TABLE 8-XI

Functions of circular elements in Example 3

Element no.	1	2	3	4	5	6	Total
K	0.94	0.95	0.96	0.98	1.02	1.06	
$\cos \beta$	0.991	0.924	0.793	0.609	0.383	0.130	
$K \cos \beta$	0.932	0.877	0.760	0.596	0.391	0.138	3.694

EXAMPLES   423

Total force = $1.1 \times 64 \times 5 \times 12 \times 3.694 \times 1.31 \times 2$ = 40,900 lb., which is very similar to that for the square caisson.

Even if $H_i/L = 0.1$ the force on the circular units would still be 86% that for the square caissons.

4

Piles 14 inches diameter are driven with 3" spacing in 15 ft. of water. Find the force exerted on a pile when 12-sec waves of 4 ft. height are impacting on them normally. Determine also the height of the wave transmitted on the leeward side of the barrier. What care should be taken with the sedimentary bed into which the piles are driven?

The force exerted on a plain wall must be computed first. $T = 12$ sec; $d/L_o = 15/(5.12 \times 12^2)$ = 0.0203; $d/L = 0.058$; $L = 259$ ft.; $H_i/L = 4/259 = 0.0155$.

From Table 8-I, $F_{max}/wH_id = 1.35$, $F_{max} = 1.35 \times 64 \times 4 \times 15 = 5190$ lb./ft. Force over 14 inches width of pile = $5190 \times 14/12 = 6060$ lb. Since $d/L = 0.058$, close enough to shallow-water conditions for the application of Fig. 7-35, $b/D = 3/14 = 0.214$, $d/H_i = 15/4 = 3.75$, $C_D = 0.9$, $F_b/F_o = 0.34$. Force on pile = $6060 \times 0.34 = 2060$ lb. Since the velocities are practically uniform throughout the depth, the resultant acts about mid-depth.

From Fig. 7-35 also, $H_t/H_i = 0.57$, $H_t = 2.3$ ft. Scouring is likely to occur on the bed in front of and between the piles. Stones of reasonable size could be used to serve as an apron over the bed. If this is built up to any great extent the waves may be forced to break, so producing large shock forces on the piles.

5

Deep-water wave heights and related swell wave periods likely to emerge from a 35-knot FAS are 6 sec (1.5), 8 (2.1), 10 (2.7), 12 (2.8), 14 (2.4), 16 (1.7) (combining $S_H^2(T)$ over 2 sec in each period band). If these were all likely to impact on a wall on breaking, compare the maximum pressure developed at SWL from each.

Eq. 8-12 gives: $p_{max}/wH_o = 1.9(g/H_o)^{1/3} T^{2/3}$, factors of which are listed in Table 8-XII. The influence of both period and wave height is observed and the range of $p_o/wH_o = 20-38$, found to be most frequent in the field [12], fits the wave conditions assumed.

TABLE 8-XII

Variables as in Example 5

TH_o	6(1.5)	8(2.1)	10(2.7)	12(2.8)	14(2.4)	16(1.7)
$(g/H_o)^{1/3}$	2.78	2.48	2.30	2.26	2.37	2.67
$T^{2/3}$	3.30	4.00	4.65	5.25	5.80	6.36
p_{max}/wH_o	17.50	18.90	20.30	22.60	26.10	32.30
p_{max} (lb./inches^2)	11.70	17.70	24.40	28.20	27.90	24.40

6

A 12-sec wave, which is 4 ft. high in deep-water, arrives normally to a shoreline whose bed in front of a seawall averages a slope of 1:30. If the water depth at the wall is 3 ft., find the force exerted by the wave per ft. length of crest.

With such a small depth it is probable that the wave will break before reaching the wall. $T = 12$ sec; $H_o = 4$ ft., $H_o/L_o = 4/(5.12 \times 12^2) = 0.0054$. From Table 5-V, with $\alpha_o = 0$, H_b/H_o = 1.82, $H_b = 1.82 \times 4 = 7.28$ ft.; from Fig. 5-19 for $s = 1/30$, $H_b/d_b = 1.5$, $d_b = 7.28/1.5 = 5$ ft.; from Fig. 5-21 for $s = 1/30$ and $d'_b/d_b = 3/5 = 0.6$, $H'_b/H_b = 0.5$, $H'_b = 7.28 \times 0.5 = 3.64$ ft.

424　　　　　　　　　　　　　　　　　　　　　　　EFFECT OF WAVES ON STRUCTURES

From Fig. 8-13 for d/L_0 at wall = $3/(5.12 \times 12^2)$ = 0.004, $p \, 10^3/wgT^2$ = 0.73, p = $(0.73 \times 64 \times 32.2 \times 12^2)/10^3$ = 216.5 lb./ft.^{2}, a_c/H_b = 0.9, a_c = 0.9×3.64 = 3.3 ft. Total force from triangular distribution = $216.5 \times (3 + 3.3)/2$ = 682 lb./ft.

7

A wave train of 8.25 sec period and 4 m height breaks as it rises over a rubble mound on which a caisson-type breakwater is founded. The water depth beyond the mound is 6 m and the depth to the mound is 4 m. Compute the maximum shear stress for which the composite breakwater must be designed if t_{max} = 0.025 sec and the natural period of rocking of the structure is 0.2 sec. Assume the breakwater characteristic λ = 0.8. What cross-section does this imply for safety against shift?

T = 8.25 sec; d/L_0 = $6/1.56 \, T^2$ = 0.056; d/L = 0.1; L = 60 m.

From Fig. 8-16, for λ = 0.8, t_{max}/T = 0.025/0.2 = 0.125, F_{max}/P_m = 0.6; from Fig. 8-17, for d'/d = 4/6 = 0.67, $P_{max}t_{max}g^{1/2}/wd^{5/2}$ = 0.117, P_{max} = $(0.117 \times 1.025 \times 6^{5/2})/0.025(9.81)^{1/2}$ = 135 tons/m: F_{max} = 135×0.6 = 81 tons/m. Total shear force is F_{max} plus "bourrage" force; from Table 8-I, for d/L = 6/60 = 0.1, H_i/L = 4/60 = 0.067; F_{max}/wH_id' = 0.98 (to be used with caution): F_{max} = $0.98 \times 1.025 \times 4 \times 4$ = 16 tons/m. Thus total shear force = 135 + 16 = 151 tons/m.

If average S.G. of concrete caisson plus filling is 2.3 and a coefficient of friction between breakwater and mound = 0.6, the area of cross-section of the breakwater = $151/(2.3-1)1$ = 116 m^2.

If the overall height of the breakwater above the mound is 10 m, the width should be at least 12 m. The safety factor used should take cognisance of the wave height (e.g., $H_{1/3}$ once in 50 years) and the infrequent nature of the peak loads. From Fig. 8-15 it can be seen that mound extensions B in front of the breakwater should not be in the vicinity of 1 to 2 times the depth of 6 m.

8

The legs of an oil rig in 100 ft. of water are 6 ft. diameter. Find the maximum force exerted on them by waves from a 35-knot FAS and its point of action from the bed. Compare the influence of the significant wave of such a storm and the maximum wave possible at this depth. What allowance should be made for barnacles 1.5 inches high? If a current of 3 knots should develop at the site, due to storm surge water passing towards the coast, compute the extra force exerted by the waves in this condition.

A 35-knot FAS gives $H_{1/3}$ = 22.6 ft., H_{max} = 42.0 ft., T_{max} = 11.6 sec from Table 3-VII. Since the strength of the legs should be designed for the largest possible wave in the storm, H_{max} should be used in preference to $H_{1/3}$.

d/L_0 = $100/5.12 \times (11.6)^2$ = 0.145; d/L = 0.179; L = 560 ft. Maximum wave height in this depth = $0.12 \, L \tanh 2\pi d/L$ = $0.12 \times 560 \times 0.81$ = 54.5 ft. Observation of Table 3-VII indicates that this is similar to H_{max} for 40-knot FAS and hence is quite possible. The design wave could therefore be 50 ft.

From Table 8-IV, for H/d = 50/100 = 0.5, and d/L = 0.18, $(F_D)_{max}10^2/wDH^2$ = 28 and $(F_M)_{max}10^2/wD^2H$ = 43.6: $(F_D)_{max}$ = $28 \times 64 \times 6 \times 50^2/10^2$ = 269,000 lb.; $(F_M)_{max}$ = $43.6 \times 64 \times 6^2 \times 50/10^2$ = 50,200 lb.

From Fig. 8-20 or Table 8-V for $(F_M)_{max}/(F_D)_{max}$ = 50.2/269 = 0.187, $(F_T)_{max}/(F_D)_{max}$ = 1.15: $(F_T)_{max}$ = $1.15 \times 269,000$ = 309,000 lb.

From Table 8-VI, y_D/d = 0.98, y_M/d = 0.55. Thus $309 \times y_T/d$ = $269 \times 0.98 + 50.2 \times 0.55$, y_T/d = 0.94.

For barnacles with ϵ = 1.5 inches, ϵ/D = 1.5/72 = 0.02. Table 8-VIII gives increase of $(F_T)_{max}$ $\doteq$ 10%.

For a current of V = 3 knots = 5.06 ft./sec, eq. 8-28 gives: F_c/F_0 = $(AB+2)/(A+2)$ where A = $u_{rms}^2 T^2/\pi^5 D^2$.

From Fig. 4-16, for HL/d^2 = $50 \times 560/100^2$ = 2.8, $u_{rms}/\sqrt{gd}$ = $0.18 : u_{rms}$ = $0.18\sqrt{32.3 \times 100}$ = 10.25 ft./sec.

EXAMPLES

From Table 8-IX, for $V/u_{rms} = 0.5$, $B = 1.2$, $A = (10.25)^2(11.6)^2/\pi^2 6^2 = 40$, $F_c/F_o = (40 \times 1.2 + 2)/42 = 1.19$.

Hence to allow for roughness and current the maximum force could be increased to 309,000 × 1.1 × 1.19 = 405,000 lb.

9

If the rig in Example 8 had 16 legs in two rows spaced 50 ft. apart what force should be used in the determination of stability. If a maximum inclination of 0.5° is contemplated, what wave band in the spectrum is likely to promote resonant rocking if the legs are just resting on the bed, and the platform is 40 ft. above SWL?

In the case of overall stability the more frequent $H_{1/3} = 22.6$ ft. could be used for design. However, from the "optimum" section of Table 3-VII it is seen that $H_{1/3} = 30$ ft. would cope with a 45-knot FAS and may be a safer design value.

From Table 8-IV, for $H/d = 30/100 = 0.3$, $d/L = 0.18$, $(F_D)_{max} = 17.0 \times 64 \times 6 \times 22.6^2/10^2 = 33,300$ lb., $(F_M)_{max} = 51.7 \times 64 \times 6^2 \times 22.6/10^2 = 27,000$ lb.

From Fig. 8-20, for $(F_M)_{max}/(F_D)_{max} = 27/33.3 \doteq 0.8$, $(F_T)_{max} = 1.4 \times 33,300 = 46,600$ lb. Total force = 16 × 46,600 = 746,000 lb.

This total force on two rows of legs is not much in excess of twice that for a single leg with $H_{max} = 50$ ft. However, such a peak wave as H_{max} occurs at one point in the ocean and is not likely to exist for long or extend far transverse to the wind. The inertia of the flexible structure should be able to withstand a single impact of H_{max} magnitude, even on two legs simultaneously, without overturning. In practice there will be brace members also involved in this calculation.

The resonant period for the structure from eq. 8-30 is:

$$T = 8\sqrt{\frac{(h^2 + B^2/4)\alpha}{Bg}} = 8\sqrt{\frac{(140^2 + 50^2/4)0.5\pi}{50 \times 32.2 \times 180}} = 2.65 \text{ sec.}$$

The wave band with an impact interval of 2.65/2 = 1.32 sec could create resonant rocking.

From Fig. 8-22 this could be caused by waves just over 8 sec period. From Table 3-VII a 25-knot FAS has $T_{max} = 8.25$ sec. Also 11.6 × 70/100 = 8.1 sec, which from Table 3-VIII is provided by 35-knot wind with $F/F_{FAS} = 15\%$ or $t/t_{FAS} = 26\%$. Even within the EDC of a 35-knot FAS the band at $(T_{max} + T_U)/2 = 7.8$ sec. Thus there is ample opportunity for a resonant wave input.

10

A horizontal member of the rig in Example 8 is 4 ft. diameter and is located at mid-depth. Compute the horizontal and vertical forces due to the water oscillations.

For the single member the design wave should be $H_{max} = 50$ ft.; as in Example 8, $H/d = 0.5$, $d/L = 0.18$. The approach suggested is to find the force on a vertical pile of diameter 4 ft. and then convert this to a load per unit height, in order to determine this load at the level of the horizontal member, in this case at $y = 50$ ft. from the surface.

From Table 8-IV, $(F_D)_{max} = 28 \times 64 \times 4 \times 50^2/10^2 = 179,000$ lb., $(F_M)_{max} = 43.6 \times 64 \times 4^2 \times 50/10^2 = 22,300$ lb., $(F_M/F_D)_{max} = 0.125$, $(F_T)_{max}/(F_D)_{max} = 1.05$, $(F_T)_{max} = 1.05 \times 179,000 = 188,000$ lb.

From Fig. 8-23 and eq. 8-32, $f_y = \dfrac{F_{max}(d-y)2}{(d+a_c/2)^2}$.

From Fig. 4-5, for $H/L = 50/560 = 0.009$, $a_c/H = 0.605$, $a_c = 50 \times 0.605 = 30.25$ ft., $f_y = \dfrac{188,000(100-50)2}{(100+30.25/2)^2} = 1420$ lb./ft.

It is seen from above that this horizontal force is predominantly drag, which will be accompanied by a lift force of equal magnitude. The resultant force is thus $\sqrt{2} f_y = 2000$ lb./ft. For a case in which the horizontal force comprises both drag and inertial forces the lift force should be taken equal to the former only. Check to see if force due to vertical water-particle motions are predominantly inertial.

From Fig. 8-24, for $d/L = 0.18$, $X = 2.3$. Using eq. 8-37, $2.3\pi D/H = 2.3\pi 4/50 = 0.58 < 1.0$, therefore force is both drag and inertial. Ratio of (vert./hor.) force = $\tanh 2\pi(d-y)/L$ (= 0.5) for inertia and $\tanh^2 2\pi(d-y)/L$ (= 0.25) for drag, use a value between these, say 0.375. Thus vertical $(F_T)_{max} = 0.375 \times 1420 = 532$ lb./ft.

11

Find the distribution of the resultant force on a brace member of the rig in Example 8 which is 4 ft. in diameter and is angled at 45° between legs 50 ft. apart, with its top end 25 ft. from SWL. What is the total dynamic load on the brace?

Using $H_{max} = 50$ ft., the force on the equivalent vertical pile is first assessed, i.e., 188,000 lb., from which f_y at 25 ft. and 75 ft. depths can be determined from eq. 8-32: $f_{25} = \dfrac{188,000(100-25)2}{(100+15.12)^2} = 2120$ lb./ft., $f_{75} = 710$ lb./ft. Since these forces are predominantly drag the resultant forces are $f_{25} = 2120\sqrt{2} = 3000$ lb./ft., $f_{75} = 1000$ lb./ft. Total load F_T = length × $(f_{25}+f_{75})/2 = 50\sqrt{2}(4000)/2 = 142,000$ lb.

12

A sewage outfall consists of a pipe 3 ft. internal diameter, whose material has an average specific gravity of 2.5. It is to resist horizontal and lifting forces when resting on the sea bed at a depth of 10 fathoms in waves 20 ft. high and 10 sec period. Determine the thickness of the pipe material.

$T = 10$ sec; $d/L_0 = 60/512 = 0.118$; $d/L = 0.156$; $L = 384$ ft.

From Fig. 4-16, for $HL/d^2 = (20 \times 384)/60^2 = 2.135$, $u_{max}/\sqrt{gd} = 0.11$, $u_{max} = 4.82$ ft./sec. From eq. 8-44, $(du/dt)_{max} = 2\pi u_{max}/T = 2\pi\, 4.82/10 = 3.04$ ft./sec^2.

Assume an outside diameter of 3.5 ft., so that from eq. 8-39, 8-40 and 8-43 we have: $F_D = C_D D_0 w u_{max}^2/2g = 1.0 \times 3.5 \times 64 \times 4.82^2/2 \times 32.2 = 81.5$ lb./ft., $F_L = C_L D_0 w u_{max}^2/2g = 81.5$ lb./ft., $F_T = 81.5\sqrt{2} = 115$ lb./ft., $F_M = C_M w \pi D_0^2 (du/dt)_{max}/4g = 2.5 \times 64 \times \pi \times 3.5^2 \times 3.04/4 \times 32.2 = 146$ lb./ft.

From eq. 8-41, $F_W - F_B = [D_0^2(\gamma_p-w) + D_i^2(\gamma_c-\gamma_p)]\pi/4$; $F_W - F_B = 62.4[D_0^2(2.5-1.025) + 3.0^2(1-2.5)]\pi/4 = 62.4[1.475D_0^2 - 13.5]\pi/4$.

From eq. 8-42, $F_D = f(F_W - F_B - F_L)$ or $F_M = f(F_W - F_B)$. Assuming $f = 0.6$, we have either $81.5 = 0.6\{[1.475D_0^2 - 13.5]62.4\pi/4 - 81.5\}$, $D_0 = 3.49$ ft., or $146 = 0.6[1.475D_0^2 - 13.5] \times 62.4\pi/4$, $D_0 = 3.55$ ft.

Using $D_0 = 3.55$ ft. does not provide any safety factor against sliding or additional forces due to a current. From eq. 8-45 $x = u_{max}T/2\pi = 4.82 \times 10/2\pi = 7.66$ ft., which is less than 3×3.05, the criterion stated in the text for the predominance of inertial force.

13

Compare the wave forces exerted on a submerged tank 40 ft. wide, 40 ft. long and 18 ft. high, resting on the bed, with those on an upright cylinder, a half cylinder and a hemisphere, all equal volume. The half cylinder may be side-on or end-on to the waves. The depth of water is 100 ft., the wave height is 40 ft. and the period 12 sec.

$T = 12$ sec; $d/L_0 = 100/(5.12 \times 12^2) = 0.136$; $d/L = 0.171$; $L = 585$ ft.

For a box-type tank, Fig. 8-29, for $A/D = 40/18 = 2.22$ and $H/L = 0.07$, gives $C_M = 1.6$. Eq. 8-51 gives: $(F_H)_{max}/wV = C_M\, 0.07(HL/d^2) \pm 0.02$. $HL/d^2 = 40 \times 585/100^2 = 2.34$, so that $(F_H)_{max}/64 \times 40 \times 40 \times 18 = 1.6 \times 0.07 \times 2.34 \pm 0.02 = 0.262 + 0.02 = 0.282$, $(F_H)_{max} = 520,000$ lb.

For a half cylinder of equal volume $B\pi A^2/8 = 40 \times 40 \times 18$, if $B = 40$ ft., $A = 42.8$ ft., A/D(eq.) $= 42.8/(21.4\pi/4) = 2.55$, $C_M = 1.7$ from Fig. 8-29 for side-on.

$A/D = 40/(21.4\pi/4) = 2.38$, $C_M = 1.8$ from Fig. 8-29 for end-on. For side-on $(F_H)_{max}/wV = 1.7 \times 0.07 \times 2.34 + 0.02 = 0.299$, for end-on $(F_H)_{max}/wV = 1.8 \times 0.07 \times 2.34 + 0.02 = 0.315$, which are increases of 6% and 12% over the box-type tank.

EXAMPLES   427

For upright cylinder with height 18 ft., $D^2 = 40 \times 40 \times 18 \times 4/(\pi \times 18)$, $D = 45.2$ ft. From Fig. 8-26, for $h/d = 0.18$ and $d/L_o = 0.136$, $(F_M)_{max}/wD^2H = 0.043$, $(F_M)_{max} = 64 \times 45^2 \times 40 \times 0.043 = 224{,}000$ lb.

For hemisphere equal volume requires $\pi D^3/6 = 40 \times 40 \times 18$, $D = 47.9$ ft. From Fig. 8-30 for $\pi D/L = \pi \times 47.9/585 = 0.25$ and $d/D = 100/47.9 = 2.1$, 8 $(F_H)_{max}/wD^2H = 0.48$, $(F_H)_{max} = 0.48 \times 64 \times 47.9^2 \times 40/8 = 352{,}000$ lb.

Thus, the upright cylinder seems to provide the least shear force on the bed. The lift forces acting on all these structures could be determined from the drag force assuming $C_L = C_D$, or in the case of the hemisphere from Fig. 8-30 where it is seen in the present case that the lift force is about 4 times the horizontal thrust.

14

A 10-ft. diameter sphere is held 20 ft. above the bed in 60 ft. of water. Find the horizontal force exerted on it by 20 ft. high waves with 11 sec period.

$T = 11$ sec; $d/L_o = 60/(5.12 \times 11^2) = 0.0967$; $d/L = 0.138$; $L = 435$ ft., $H/L = 20/435 = 0.046$, $h/d = 40/60 = 0.67$. From Fig. 8-31, $(F_H)_{max}/wV = 0.11$, $(F_H)_{max} = 0.11 \times 64\pi 10^3/6 = 3{,}690$ lb.

15

Find the weight of various armour units that will remain stable on a breakwater slope of 1:2.5. For concrete fabricated units, use a specific weight of 150 lb./ft.^{3}. Determine weights for 10 and 30 ft. depths for trunk and head sections when $T_{max} = 11$ sec.

The design wave in each section should be the maximum possible at the respective depths, if these can possibly be generated in the area.

In 10 ft. depth, $d/L_o = 10/5.12 \times 11^2 = 0.0161$, $d/L = 0.051$, $L = 196$ ft., $(H/L)_{max} = 0.12 \tanh 2\pi d/L = 0.12 \times 0.312$, $H_{10} = 7.35$ ft.

In 30 ft. depth, $d/L_o = 30/5.12 \times 11^2 = 0.048$, $d/L = 0.092$, $L = 326$ ft., $(H/L)_{max} = 0.12 \times 0.52$, $H_{30} = 20.4$ ft. This is $H_{1/10}$ for a 30-knot FAS, or $H_{1/3}$ for a 35-knot FAS, both considered probable. Slope 1:2.5 gives $\alpha = 22°$. From Fig. 8-33 volume of angular rubble stone units (SG = 2.65) = 15 and 300 ft.^{3} for 10 and 30 ft. depths, respectively. If the breakwater head were located at either of these depths the volumes would need to be increased by 20%. (In both cases the curve for trunk shallow applies.) For fabricated units with specific weight of 150 lb./ft.^{3} the volume must be increased by 46%.

Fabricated units require on the average 40% of the volume of rubble angular stone according to the top inset of Fig. 8-33. Other types of armour block on which information has been published could be inserted into this inset.

16

A composite breakwater at the deepest part of its trunk in 6 fathoms of water has a rubble-mound base (SG = 2.65), 12 ft. high on which the concrete caisson is founded. If the mound extends 15 ft. in front of the caisson face and then slopes at 1:2, find the volume of blocks required for stability when waves 20 ft. high and 11 sec period are striking the structure. Consider cases for uniform size of mound material and an impervious condition for this foundation.

$T = 11$ sec; $d/L_o = 36/5.12 \times 11^2 = 0.0581$; $d/L = 0.102$; $h/d = 12/36 = 0.33$; $B/d = 15/36 = 0.415$. From Fig. 8-34, $V(SG_s-1)^3/H_i^3 = 0.037$, $V = 0.037 \times 20^3/1.65^3 = 66$ ft.^{3}.

In Fig. 8-34 the critical $d/L = 0.13$, in which case the volume requires to be $0.047/0.037 = 27\%$ greater.

If the mound can be considered impervious the condition B in Fig. 8-34 exists, so that for $d/L = 0.1$ and $h/d = 0.33$, $V(SG_s-1)^3/H_i^3 = 0.0013$. For $d/L = 0.13$ this parameter is 0.0025, which is probably the more realistic design figure, in which case $V = 4.5$ ft.^{3}.

There is thus a large range from which to choose an armour-unit size. If bitumastic grouting equipment were available it should not be difficult to seal the mound against flow of water from the harbour side, but even a mixture of sizes should effect blockage due to sand fairly readily.

PROBLEMS

1

Waves of 10 sec period are striking a wall in 20 ft. of water. As the waves increase, the pressure exerted on the wall by the resulting standing waves will contain humps either side of instant when the crest is there. Determine the incident wave heights when such humps are likely to occur at the bed, at MWL, and for the total force. At what heights will unsymmetric humps and final breaking occur?

2

For 12-sec waves in 30 ft. of water, what are the limiting wave heights for which the various order theories can give an accurate picture of the pressures exerted on a vertical wall from the standing waves developed?

3

A jetty contains a vertical wall which must withstand the force of waves generated across a 60 NM wide bay where winds of 55 knots can be experienced for 3 h. The wall is in 30 ft. of water and extends 20 ft. above SWL. Find the magnitude of the wave forces and the depth of their resultants below SWL. If the jetty extended only 8 ft. above SWL, determine the resultant forces.

4

A breakwater consists of 15 ft. circular caissons sitting on the sea bed in 12 ft. of water. Determine the maximum height of a non-breaking wave train with 10 sec period against which these can resist horizontal shift and over-turning. Assume the mean SG of the breakwater is 2.5 and the coefficient of friction with the bed is 0.6. Discuss other criteria of design for such a structure.

5

Closely spaced piles are used to attenuate waves approaching a harbour by 50%. If the piles are 18 inches in diameter, find their spacing and force to be resisted if the incident waves are 10 ft. high and 11 sec period, and the water depth 25 ft. What safeguards respecting the sedimentary bed would you advise?

6

A wave train 4 ft. in height and 12 sec period approaches normally to a coast where a wall receives the impact of these breaking waves. Find the maximum pressure exerted for waves whose leading faces of their crests are sloping at $15°$, $10°$, $5°$, and $1°$ to the vertical. Compare your results with values derived from eq. 8-8 and explain any differences. The bed slope in front of the wall averages 1 : 20.

7

Why should the peak pressures experienced from waves breaking against seawalls not be used for overall stability of the structure?

8

Discuss the similarities and differences in the pressure records A and B in Fig. 8-11, considering each piezometer level in turn.

9

A sea revetment consists of a sand-dune slope of 1 : 4 sealed with a bitumen compound on a compacted base. Find the maximum pressure for which this cover must be designed when deep-water waves of height 4 ft. and period 12 sec are arriving at $45°$ to the coast. The water

EXAMPLES

level varies with the tide and the slope runs down to low water spring level (MLWS). To what height above SWL of the highest tide should the sealed area be taken in order to avoid a build up of ground water pressure from wave swash?

10
A seawall is so placed that at MHWS tide plus a 4 ft.-storm surge the water level is 6 ft. above its toe, the sea bed sloping away at 1 : 30. Waves 10 ft. high and 10 sec period are directed against the wall. Find the maximum force for design purposes.

11
A vertical cliff faces the open sea on which the fiercest extra-tropical storms can exist. The submarine slope is very steep and runs to 20 fathoms very close to shore. Calculate the peak pressures most likely to be experienced at SWL considering all components of the spectrum and the spectrum as a whole.

12
A composite breakwater consists of square caissons sitting on a rubble-mound foundation 6 m high in 12 m of water. Find the maximum shear force for which the concrete units should be designed if they must withstand any wave conditions. Assume t_{max} = 0.025 sec, the rocking period 0.1 sec and the breakwater parameter λ = 1.0.

13
An oil rig is designed to operate in 20 fathoms of water. Its legs are circular and are 3 m in diameter. Compute the horizontal force on each leg and the point of action of the resultant when the worst extra-tropical cyclone likely to be experienced generates its FAS. Find also the overturning moment on the structure which has eight legs, omitting any allowance for bracing members.

14
If the rig in Problem 13 has its legs spaced 20 m apart and its deck 8 m above SWL, determine the wave bands which could cause a resonant rocking problem. Are these conditions likely? A maximum inclination of 1° is permitted.

15
If in Problem 14 allowances were made for barnacle growth (normal size 2 inches high) and a mean current of 1 m/sec, what increase in force on individual legs and the rig as a whole should be made?

16
Piles of a jetty are in a tidal zone where the water depth varies from 6 to 15 ft. Find the forces on each 2 ft. diameter pile for the case of a 35-knot FAS.

17
A horizontal member of a marine structure is 5 ft. diameter. It is at a depth of 30 ft. in water 100 ft. deep and is designed to withstand a 30-knot FAS. Find the horizontal and vertical forces per unit length, allowing for barnacle growth (2 inches dimension) and a current of 3 ft./sec.

18
An angled brace consists of a 3.5 ft. diameter tube rising from a depth of 30 ft. to 60 ft. in an 80 ft. depth of water, between legs that are 40 ft. apart. Determine the distribution of horizontal and vertical forces on this member for a 20 ft. high wave with 12 sec period.

19
A submarine pipeline made of steel (SG = 7.8) and coated with plastic carries oil of specific gravity 0.85. Its outer diameter is 15 inches. Determine the thickness of the steel required for stability of the pipe, when experiencing waves 33 ft. high and period 12 sec, and resting on the bed in 80 ft. of water.

20
As a coastal engineering consultant you are required to report on the necessary surveys and other investigations for a submarine pipeline. List the headings and make any necessary succinct comments to elucidate your choice of topics.

21
If the pipeline in Problem 19 were to be buried in a trench jetted for the purpose, at first it could be suspended in a fluid mud, whose specific gravity was around 1.7. Is the pipe likely to "float" on the contents of the trench, if filled with seawater during this operation?

22
The piers of a bridge crossing an estuary are 12 ft. in diameter and they stand in 20 ft. of water. The estuary is 40 miles wide, across which waves can be generated by tropical cyclones whose maximum sustained winds are 55 knots. Compute the bending moment of the piers on their foundations due to the waves alone. If a current of 6 knots can occur simultaneously, what increase in moment is exerted?

23
During the construction of the piers in Problem 22 the strength of the concrete and the bond between it and the steel reinforcement takes time to develop to its full strength. It is therefore necessary to know the bending moment exerted under the same wave conditions when the piers are 1/4, 1/2 and 3/4 depth.

24
A barge-type structure is to rest on the sea-bed for storage of oil with SG = 0.79. Determine its weight in order to prevent upward lift, with a safety factor of 2, when waves 33 ft. high and 11.5 sec occur in the 75 ft. deep water. The barge is 50 ft. wide, 50 ft. long and 20 ft. high. What would be the required weight for a hemisphere containing the same weight of oil?

25
A 20-ft. diameter sphere is supported 15 ft. clear of the bottom in a 50 ft. depth. Find the horizontal forces exerted on it by 20-ft. waves with 9 sec period. What other forces should be considered on such a structure?

26
A breakwater runs from shore out to a depth of 35 ft. and is broadside on to the largest storm waves produced by a 30-knot wind over a 200-mile fetch. Find the weight of tetra-pod units to be used on a 1 : 2 slope at depths of 10, 20 and 30 ft. in the trunk and at 35 ft. at the head of the breakwater. Assume specific weight of concrete as 150 lb./ft.^{3}

27
A composite breakwater consists of square concrete caissons filled with sand founded on a rubble-mound foundation rising 20 ft. from the bed in 40 ft. of water. Determine the size of armour units required if 20-ft. waves of 11-sec period are experienced. It can be assumed that the foundation is impervious, and that the specific gravity of the stone is 2.65. What width of mound seawards of the wall should be avoided?

28
Many breakwaters of a composite type have collapsed seawards, particularly near their heads. Give reasons for this in terms of forces exerted, or of bed scour.

29
Abstract articles in the technical literature dealing with scour under and around oil rigs or similar marine space frames. What effect should this have on the stability of the structure?

30
Compare the approaches used to compute the force on a vertical seawall and that on a vertical circular pile. The closely spaced pile is partly one and partly the other; discuss the approach to this problem.

31
A concrete block breakwater stands in 30 ft. of water and has waves 12 ft. high and 13 sec period approaching it normally. Compute the force exerted on the wall and the overturning moment about the base for each lineal foot of the breakwater.

32
Determine the weights of rock units to be used on a rubblemound breakwater when a 1 : 1.5 slope is to withstand waves of 15 ft. height and rock of SG = 2.65 is to be used. If the rock is considered cubic in form and two layers are to be used on the face of the breakwater, compare the volume of rock required for the above slope with that necessary for a 1 : 3 slope. What is the unit weight of tribars (SG = 2.4) for these two side slopes? Differentiate in each case between the head and trunk of the breakwater, referring all values to shallow-water conditions.

33
A composite breakwater comprises a rectangular caisson filled with sand and gravel which stands on the sea bed. It has been stated that the factor λ of eq. 8-18 is normally greater than 0.8. If the breakwater is of square cross-section, what does this imply regarding the relative height of impact force to the centre of gravity of the structure.

34
If a composite breakwater resting on the sea bed contains a major rectangular section of height h and width B and a superstructure of height h' and width B', derive an expression for the radius of rotation of the wall about the centre of the base.

REFERENCES

[1] L. Rundgren, 1958. Water wave forces. *R. Inst. Technol., Stockholm, Bull.*, 54.
[2] I. Tadjbaksh and J.B. Keller, 1960. Standing surface waves of finite amplitude. *J. Fluid Mech.*, 8: 442–451.
[3] Y. Goda, 1967. The fourth-order approximation to the pressure of standing waves. *Coastal Eng. Japan*, 10: 1–11.
[4] Y. Tsuchiya and M. Yamaguchi, 1970. Limiting conditions for standing wave theories by perturbation method. *Proc. 12th Conf. Coastal Eng.*, 1: 523–542.
[5] R.G. Dean, 1970. Relative validities of water wave theories. *Proc. ASCE*, 96 (WW1): 105–119.
[6] S. Nagai, 1969. Pressures of standing waves on vertical walls. *Proc. ASCE*, 95 (WW1): 53–76.

[7] A. Kuznetsov, 1970. Wave loads against the vertical wall at a random angle of approaching waves. *Abstr. 12th Conf. Coastal Eng.*, 227–232.

[8] G.D. Khaskhachikh and D.M. Vanchagov, 1971. Effect of waves on walls made of circular cells. *Proc. ASCE*, 97 (WW4): 735–754.

[9] T. Hayashi, T. Kano and M. Shirai, 1966. Hydraulic research on the closely spaced pile breakwater. *Proc. 10th Conf. Coastal Eng.*, 2: 873–884.

[10] R. Weggel and W.H.C. Maxwell, 1970. Numerical model for wave pressure distributions. *Proc. ASCE*, 96 (WW3): 623–642.

[11] M.D. Adeyemo, 1968. Effect of beach slope and shoaling on wave asymmetry. *Proc. 11th Conf. Coastal Eng.*, 1: 145–172.

[12] A. De Rouville, P. Besson and P. Petry, 1938. Etat actuel des études internationales sur les efforts dus aux lames. *Ann. Ponts Chaussées*, 108: 5–113.

[13] R.A. Bagnold, 1939. Interim report on wave pressure research. *J. Inst. Civil Eng.*, 12: 201–226.

[14] H. Mitsuyasu, 1966. Shock pressure of breaking wave. *Proc. 10th Conf. Coastal Eng.*, 1: 268–293.

[15] D.F. Denny, 1951. Further experiments on wave pressures. *J. Inst. Civil Eng.*, 35: 330–345.

[16] Y. Muraki, 1966. Field observations of wave pressure, wave run-up and oscillation of breakwater. *Proc. 10th Conf. Coastal Eng.*, 1: 302–321.

[17] J. Larras, 1957. Le déferlement des lames sur les jetées verticales. *Ann. Ponts Chaussées*, 127: 643–680.

[18] T. Hayashi and M. Hattori, 1958. Pressure of the breaker against a vertical wall. *Proc. Coastal Eng. Japan*, 1: 25–37.

[19] Y. Iwagaki, 1968. Hyperbolic waves and their shoaling. *Proc. 11th Conf. Coastal Eng.*, 1: 124–144.

[20] M. Hom-ma and K. Horikawa, 1965. Experimental study on total wave force against sea wall. *Proc. Coastal Eng. Japan*, 8: 119–129.

[21] M.F. Skladnev and I.Ya. Popov, 1969. Studies of wave loads on concrete slope protections of earth dams. *Proc. Symp. Res. Wave Action, Delft*, 2: Paper 7.

[22] F. Molero, 1966. Protection against wave action based on hydro-elastic effect. *Proc. 10th Conf. Coastal Eng.*, 2: 932–952.

[23] V.V. Krylov, 1966. On design of wave pressure acting on structures of sloping type. *Proc. 10th Conf. Coastal Eng.*, 2: 953–957.

[24] S. Nagai, 1968. Pressures of breaking waves on composite-type breakwaters. *Proc. 10th Conf. Coastal Eng.*, 2: 920–933.

[25] T. Hayashi and M. Hattori, 1964. Thrusts exerted upon composite-type breakwaters by the action of breaking waves. *Coastal Eng. Japan*, 7: 65–84.

[26] H. Lundgren, 1969. Wave shock forces: an analysis of deformations and forces in the wave and in the foundation. *Proc. Symp. Res. Wave Action, Delft*, 2: Paper 4.

[27] T. Hayashi, 1965. Virtual mass and the damping factor of the breakwater during rocking, and the modification by their effect of the expression of the thrusts exerted upon breakwaters by the action of breaking waves. *Coastal Eng. Japan*, 8: 105–117.

[28] S. Nagai, 1961. Shock pressures exerted by breaking waves on breakwaters. *Trans. ASCE*, 126: 772–809.

[29] J.R. Morison, M.P. O'Brien, J.W. Johnson and S.A. Schaaf, 1950. The force exerted by surface waves on piles. *Petrol. Trans. Am. Inst. Mining Metal. Eng.*, 189: 149–154.

[30] J.R. Morison, 1951. Design of piling. *Proc. 1st Conf. Coastal Eng.*, 254–258.

[31] Y. Goda, 1964. Waves forces on a vertical circular cylinder: experiments and a proposed method of wave force computation. *Port Harbour Tech. Res. Inst., Rep. 8*.

[32] H.A. Agerschou and J.J. Edens, 1965. Fifth- and first-order wave-force coefficients for cylindrical piles. *Proc. Santa Barbara Conf. Coastal Eng., 1965*: 219–248.

[33] C.L. Bretschneider, 1965. On the probability distribution of wave force and an introduc-

tion to the correlation drag coefficient and the correlation inertial coefficient. *Proc. Santa Barbara Conf. Coastal Eng., 1965*: 183–217.
[34] A. Paape and H.N.C. Breusers, 1966. The influence of pile dimension on forces exerted by waves. *Proc. 10th Conf. Coastal Eng.*, 2: 840–849.
[35] W.J. Burton and R.M. Sorensen, 1970. The effects of roughness on the wave forces on a circular cylindrical pile. *Texas A M Univ., Rep.,* COE 121.
[36] A.D.K. Laird, 1965. Forces on a flexible pile. *Proc. Santa Barbara Conf. Coastal Eng., 1965*: 249–268.
[37] R.L. Wiegel, 1969. Waves and their effects on pile-supported structures. *Proc. Symp. Res. Wave Action, Delft*, 1.
[38] M. Hino, 1969. Spectrum of wave drag on a pile. *Coastal Eng. Japan*, 12: 17–27.
[39] L.E. Borgman, 1965. The spectral density for ocean wave forces. *Proc. Santa Barbara Conf. Coastal Eng., 1965*: 147–182.
[40] L.E. Borgman, 1969. Ocean wave simulation for engineering design. *Proc. ASCE*, 95 (WW4): 557–583.
[41] Y. Jen, 1968. Laboratory study of inertia forces on a pile. *Proc. ASCE*, 94 (WW1): 59–76.
[42] L.E. Borgman, 1970. Maximum wave force probabilities for a random number of random intensity storms. *Proc. 12th Conf. Coastal Eng.*, 1: 53–64.
[43] M.A. Hall, 1958. Laboratory study of breaking wave forces on piles. *Beach Erosion Board, Tech. Mem.*, 116.
[44] R.G. Dean and D.R.F. Harleman, 1966. Interaction of structures and waves. In: A.T. Ippen (Editor), *Estuarine and Coastline Hydrodynamics*. McGraw-Hill, New York, N.Y., pp. 341–403.
[45] D.R.F. Harleman, W.C. Nolan and V.C. Honsinger, 1963. Dynamic analysis of offshore structures. *Proc. 8th Conf. Coastal Eng.*, 482–499.
[46] J.M. Barradell-Smith, 1970. *Scour under Offshore Mobile Jack-up Rig Legs*. Thesis, Asian Inst. Technol., Bangkok.
[47] R.J. Howe, 1968. Offshore mobile drilling units. *Ocean Ind.*, 3: 38–61.
[48] C.J. Posey, 1971. Protection of offshore structures against underscour. *Proc. ASCE*, 97 (HY7): 1011–1016.
[49] R.L. Wiegel, K.E. Beebe and J. Moon, 1957. Ocean wave forces on circular cylindrical piles. *Proc. ASCE*, 83 (HY2): *Pap.* 1199.
[50] H. Beckmann and M.H. Thibodeaux, 1962. Wave force coefficients for offshore pipelines. *Proc. ASCE*, 88 (WW2): 125–138. (See also discussions: 88 (WW4): 149–150; 89 (WW1): 61–65; 89 (WW3): 53–55.)
[51] R.J. Brown, 1967. Hydrodynamic forces on a submarine pipeline. *Proc. ASCE*, 93 (PL1): *Pap.* 5143.
[52] D.O. Ralston and J.B. Herbich, 1968. The effects of waves and currents on submerged pipelines. *Texas A M Univ., Rep.*, COE 101.
[53] Anonymous, 1967. Research needs in pipeline engineering for the decade 1966–1975. *Proc. ASCE*, 93 (PL2): 19–50.
[54] R. Blumberg, 1964. Hurricane winds, waves and currents test marine pipe line design. *J. Pipe Line Ind.*, 1 (20): 42–45; 2 (21): 70–74; 3 (21): 34–39; 4 (21): 67–72; 5 (21): 35–41; 6 (21): 85–88.
[55] R.O. Reid, 1955. Oceanographic considerations in marine pipe line construction. In: *Oceanography and Meteorology, 1950–1954*. Texas A M Univ., pp. 9–14.
[56] D.R. Miller, 1965. Marine studies for the design and construction of offshore pipelines. *Proc. Santa Barbara Conf. Coastal Eng., 1965*: 991–1003.
[57] J.C. Lebreton and P. Cormault, 1969. Wave action on slightly immersed structures: some theoretical and experimental considerations. *Proc. Symp. Res. Wave Action, Delft*, 4: *Pap.* 12A.

[58] R.C. MaCamy and R.A. Fuchs, 1954. Wave forces on piles: a diffraction theory. *Beach Erosion Board, Tech. Mem.*, 69.
[59] E.F. Brater, J.S. McNown and L.D. Stair, 1958. Wave forces on submerged structures. *Proc. ASCE*, 84 (HY6): 1–26.
[60] J.B. Herbich and G.E. Shank, 1971. Forces due to waves on submerged structures. *Proc. ASCE*, 97 (WW1): 57–71.
[61] G.E. Shank and J.B. Herbich, 1970. Forces due to waves on submerged structures. *Texas A M Univ., Rep.*, COE 123.
[62] P. Riabouchinski, 1920. Sur la résistance des fluides. *Int. Congr. Math., Strasbourg*, 568–585.
[63] C.J. Garrison and R.H. Snider, 1970. Wave forces on large submerged tanks. *Texas A M Univ., Rep.*, COE 117.
[64] C.J. Garrison and V.S. Rao, 1970. Interaction of waves with submerged objects. *Proc. ASCE*, 97 (WW2): 259–277.
[65] H. Shi-igai and T. Kono, 1969. Study on vibration of submerged spheres caused by surface waves. *Coastal Eng. Japan*, 12: 29–40.
[66] R.A. Grace and F.M. Casciano, 1969. Ocean wave forces on a subsurface sphere. *Proc. ASCE*, 95 (WW3): 291–317.
[67] J. Sommet and Ph. Vignat, 1969. Complex wave action on submerged bodies. *Proc. Symp. Res. Wave Action, Delft*, 4: Pap. 12.
[68] R. Svee, 1962. Formulas for design of rubble-mound breakwaters. *Proc. ASCE*, 88 (WW2): 11–21.
[69] A. Hedar, 1960. *Stability of Rock-fill Breakwaters*. Thesis, Chalmers Tek. Hog., Goteborg.
[70] R.Y. Hudson, 1959. Laboratory investigation of rubble-mound breakwaters. *Proc. ASCE*, 85 (WW3): 93–121.
[71] Anonymous, 1966. Shore protection, planning and design. *Coastal Eng. Res. Centre, Wash. D.C., Tech. Rep.*, 4.
[72] R.Q. Palmer, 1960. Breakwaters in the Hawaiian Islands. *Proc. ASCE*, 86 (WW3): 39–67. (Discussion, 88 (WW3): 161–169.)
[73] J.B. Font, 1970. Damage functions for a rubble-mound breakwater under the effect of swells. *Proc. 12th Conf. Coastal Eng.*, 3: 1567–1585.
[74] H. Berge and A. Traetteberg, 1969. Stability tests of the Europoort breakwater. *Proc. Symp. Res. Wave Action, Delft*, 3: Pap. 10.
[75] S. Nagai, 1961. Experimental studies of specially shaped concrete blocks for absorbing wave energy. *Proc. 7th Conf. Coastal Eng., 1961*: 659–673.
[76] K.Y. Singh, 1968. Stabit – a new armour block. *Proc. 11th Conf. Coastal Eng.*, 2: 797–814.
[77] E.M. Merrifield and J.A. Zwamborn, 1966. The economic value of a new breakwater armour unit "Dolos". *Proc. 10th Conf. Coastal Eng.*, 2: 885–912.
[78] K. D'Angremond, M.J.Th. Span, J. Van der Weide and A.J. Woestenenk, 1970. Use of asphalt in breakwater construction. *Proc. 12th Conf. Coastal Eng.*, 3: 1601–1627.
[79] M. Kagawa, 1965. Increase of sliding resistance of gravity-type breakwaters by means of an asphalt mat. *Coastal Eng. Japan*, 8: 131–140.
[80] A. Brebner and P. Donnelly, 1962. Laboratory study of rubble foundations for vertical breakwaters. *Queen's Univ, Civil. Eng., Rep.*, 23.
[81] S. Nagai, 1966. Researches on steel-pipe breakwater. *Proc. 10th Conf. Coastal Eng.*, 2: 850–872.

APPENDIX

TABLE OF FUNCTIONS OF d/L_o

d/L_o	d/L	$2\pi d/L$	tanh $2\pi d/L$	sinh $2\pi d/L$	cosh $2\pi d/L$	$4\pi d/L$	sinh $4\pi d/L$	cosh $4\pi d/L$	H/H'_o
0	0	0	0	0	1	0	0	1	
.0001000	.003990	.02507	.02506	.02507	1.0003	.05014	.05016	1.001	4.467
.0002000	.005643	.03546	.03544	.03547	1.0006	.07091	.07097	1.003	3.757
.0003000	.006912	.04343	.04340	.04344	1.0009	.08686	.08697	1.004	3.395
.0004000	.007982	.05015	.05011	.05018	1.0013	.1003	.1005	1.005	3.160
.0005000	.008925	.05608	.05602	.05611	1.0016	.1122	.1124	1.006	2.989
.0006000	.009778	.06144	.06136	.06148	1.0019	.1229	.1232	1.008	2.856
.0007000	.01056	.06637	.06627	.06642	1.0022	.1327	.1331	1.009	2.749
.0008000	.01129	.07096	.07084	.07102	1.0025	.1419	.1424	1.010	2.659
.0009000	.01198	.07527	.07513	.07534	1.0028	.1505	.1511	1.011	2.582
.001000	.01263	.07935	.07918	.07943	1.0032	.1587	.1594	1.013	2.515
.001100	.01325	.08323	.08304	.08333	1.0035	.1665	.1672	1.014	2.456
.001200	.01384	.08694	.08672	.08705	1.0038	.1739	.1748	1.015	2.404
.001300	.01440	.09050	.09026	.09063	1.0041	.1810	.1820	1.016	2.357
.001400	.01495	.09393	.09365	.09407	1.0044	.1879	.1890	1.018	2.314
.001500	.01548	.09723	.09693	.09739	1.0047	.1945	.1957	1.019	2.275
.001600	.01598	.1004	.1001	.1006	1.0051	.2009	.2022	1.020	2.239
.001700	.01648	.1035	.1032	.1037	1.0054	.2071	.2086	1.022	2.205
.001800	.01696	.1066	.1062	.1068	1.0057	.2131	.2147	1.023	2.174
.001900	.01743	.1095	.1091	.1097	1.0060	.2190	.2207	1.024	2.145
.002000	.01788	.1123	.1119	.1125	1.0063	.2247	.2266	1.025	2.119
.002100	.01832	.1151	.1146	.1154	1.0066	.2303	.2323	1.027	2.094
.002200	.01876	.1178	.1173	.1181	1.0069	.2357	.2379	1.028	2.070
.002300	.01918	.1205	.1199	.1208	1.0073	.2410	.2433	1.029	2.047
.002400	.01959	.1231	.1225	.1234	1.0076	.2462	.2487	1.031	2.025
.002500	.02000	.1257	.1250	.1260	1.0079	.2513	.2540	1.032	2.005
.002600	.02040	.1282	.1275	.1285	1.0082	.2563	.2592	1.033	1.986
.002700	.02079	.1306	.1299	.1310	1.0085	.2612	.2642	1.034	1.967
.002800	.02117	.1330	.1323	.1334	1.0089	.2661	.2692	1.036	1.950
.002900	.02155	.1354	.1346	.1358	1.0092	.2708	.2741	1.037	1.933
.003000	.02192	.1377	.1369	.1382	1.0095	.2755	.2790	1.038	1.917
.003100	.02228	.1400	.1391	.1405	1.0098	.2800	.2837	1.040	1.902
.003200	.02264	.1423	.1413	.1427	1.0101	.2845	.2884	1.041	1.887
.003300	.02300	.1445	.1435	.1449	1.0104	.2890	.2930	1.042	1.873
.003400	.02335	.1467	.1456	.1472	1.0108	.2934	.2976	1.043	1.860
.003500	.02369	.1488	.1477	.1494	1.0111	.2977	.3021	1.045	1.847
.003600	.02403	.1510	.1498	.1515	1.0114	.3020	.3065	1.046	1.834
.003700	.02436	.1531	.1519	.1537	1.0117	.3061	.3109	1.047	1.822
.003800	.02469	.1551	.1539	.1558	1.0121	.3103	.3153	1.049	1.810
.003900	.02502	.1572	.1559	.1579	1.0124	.3144	.3196	1.050	1.799
.004000	.02534	.1592	.1579	.1599	1.0127	.3184	.3238	1.051	1.788
.004100	.02566	.1612	.1598	.1619	1.0130	.3224	.3280	1.052	1.777
.004200	.02597	.1632	.1617	.1639	1.0133	.3263	.3322	1.054	1.767
.004300	.02628	.1651	.1636	.1659	1.0137	.3302	.3362	1.055	1.756
.004400	.02659	.1671	.1655	.1678	1.0140	.3341	.3403	1.056	1.746

APPENDIX (Continued)

d/L_o	d/L	$2\pi d/L$	tanh $2\pi d/L$	sinh $2\pi d/L$	cosh $2\pi d/L$	$4\pi d/L$	sinh $4\pi d/L$	cosh $4\pi d/L$	H/H'_o
.004500	.02689	.1690	.1674	.1698	1.0143	.3380	.3444	1.058	1.737
.004600	.02719	.1708	.1692	.1717	1.0146	.3417	.3483	1.059	1.727
.004700	.02749	.1727	.1710	.1736	1.0149	.3454	.3523	1.060	1.718
.004800	.02778	.1745	.1728	.1754	1.0153	.3491	.3562	1.062	1.709
.004900	.02807	.1764	.1746	.1773	1.0156	.3527	.3601	1.063	1.701
.005000	.02836	.1782	.1764	.1791	1.0159	.3564	.3640	1.064	1.692
.005100	.02864	.1800	.1781	.1809	1.0162	.3599	.3678	1.066	1.684
.005200	.02893	.1818	.1798	.1827	1.0166	.3635	.3715	1.067	1.676
.005300	.02921	.1835	.1815	.1845	1.0169	.3670	.3753	1.068	1.669
.005400	.02948	.1852	.1832	.1863	1.0172	.3705	.3790	1.069	1.662
.005500	.02976	.1870	.1848	.1880	1.0175	.3739	.3827	1.071	1.654
.005600	.03003	.1887	.1865	.1898	1.0178	.3774	.3864	1.072	1.647
.005700	.03030	.1904	.1881	.1915	1.0182	.3808	.3900	1.073	1.640
.005800	.03057	.1921	.1897	.1932	1.0185	.3841	.3937	1.075	1.633
.005900	.03083	.1937	.1913	.1949	1.0188	.3875	.3972	1.076	1.626
.006000	.03110	.1954	.1929	.1967	1.0192	.3908	.4008	1.077	1.620
.006100	.03136	.1970	.1945	.1983	1.0195	.3941	.4044	1.079	1.614
.006200	.03162	.1987	.1961	.2000	1.0198	.3973	.4079	1.080	1.607
.006300	.03188	.2003	.1976	.2016	1.0201	.4006	.4114	1.081	1.601
.006400	.03213	.2019	.1992	.2033	1.0205	.4038	.4148	1.083	1.595
.006500	.03238	.2035	.2007	.2049	1.0208	.4070	.4183	1.084	1.589
.006600	.03264	.2051	.2022	.2065	1.0211	.4101	.4217	1.085	1.583
.006700	.03289	.2066	.2037	.2081	1.0214	.4133	.4251	1.087	1.578
.006800	.03313	.2082	.2052	.2097	1.0217	.4164	.4285	1.088	1.572
.006900	.03338	.2097	.2067	.2113	1.0221	.4195	.4319	1.089	1.567
.007000	.03362	.2113	.2082	.2128	1.0224	.4225	.4352	1.091	1.561
.007100	.03387	.2128	.2096	.2144	1.0227	.4256	.4386	1.092	1.556
.007200	.03411	.2143	.2111	.2160	1.0231	.4286	.4419	1.093	1.551
.007300	.03435	.2158	.2125	.2175	1.0234	.4316	.4452	1.095	1.546
.007400	.03459	.2173	.2139	.2190	1.0237	.4346	.4484	1.096	1.541
.007500	.03482	.2188	.2154	.2205	1.0240	.4376	.4517	1.097	1.536
.007600	.03506	.2203	.2168	.2221	1.0244	.4406	.4549	1.099	1.531
.007700	.03529	.2218	.2182	.2236	1.0247	.4435	.4582	1.100	1.526
.007800	.03552	.2232	.2196	.2251	1.0250	.4464	.4614	1.101	1.521
.007900	.03576	.2247	.2209	.2265	1.0253	.4493	.4646	1.103	1.517
.008000	.03598	.2261	.2223	.2280	1.0257	.4522	.4678	1.104	1.512
.008100	.03621	.2275	.2237	.2295	1.0260	.4551	.4709	1.105	1.508
.008200	.03644	.2290	.2250	.2310	1.0263	.4579	.4741	1.107	1.503
.008300	.03666	.2304	.2264	.2324	1.0266	.4607	.4772	1.108	1.499
.008400	.03689	.2318	.2277	.2338	1.0270	.4636	.4803	1.109	1.495
.008500	.03711	.2332	.2290	.2353	1.0273	.4664	.4834	1.111	1.491
.008600	.03733	.2346	.2303	.2367	1.0276	.4691	.4865	1.112	1.487
.008700	.03755	.2360	.2317	.2381	1.0280	.4719	.4896	1.113	1.482
.008800	.03777	.2373	.2330	.2396	1.0283	.4747	.4927	1.115	1.478
.008900	.03799	.2387	.2343	.2410	1.0286	.4774	.4957	1.116	1.474
.009000	.03821	.2401	.2356	.2424	1.0290	.4801	.4988	1.118	1.471
.009100	.03842	.2414	.2368	.2438	1.0293	.4828	.5018	1.119	1.467
.009200	.03864	.2428	.2381	.2452	1.0296	.4855	.5049	1.120	1.463
.009300	.03885	.2441	.2394	.2465	1.0299	.4882	.5079	1.122	1.459
.009400	.03906	.2455	.2407	.2479	1.0303	.4909	.5109	1.123	1.456
.009500	.03928	.2468	.2419	.2493	1.0306	.4936	.5138	1.124	1.452
.009600	.03949	.2481	.2431	.2507	1.0309	.4962	.5168	1.126	1.448
.009700	.03970	.2494	.2443	.2520	1.0313	.4988	.5198	1.127	1.445
.009800	.03990	.2507	.2456	.2534	1.0316	.5014	.5227	1.128	1.442
.009900	.04011	.2520	.2468	.2547	1.0319	.5040	.5257	1.130	1.438

Table of functions of d/L_0

APPENDIX (Continued)

d/L_o	d/L	$2\pi d/L$	tanh $2\pi d/L$	sinh $2\pi d/L$	cosh $2\pi d/L$	$4\pi d/L$	sinh $4\pi d/L$	cosh $4\pi d/L$	H/H_o'
.01000	.04032	.2533	.2480	.2560	1.0322	.5066	.5286	1.131	1.435
.01100	.04233	.2660	.2598	.2691	1.0356	.5319	.5574	1.145	1.403
.01200	.04426	.2781	.2711	.2817	1.0389	.5562	.5853	1.159	1.375
.01300	.04612	.2898	.2820	.2938	1.0423	.5795	.6125	1.173	1.350
.01400	.04791	.3010	.2924	.3056	1.0456	.6020	.6391	1.187	1.327
.01500	.04964	.3119	.3022	.3170	1.0490	.6238	.6651	1.201	1.307
.01600	.05132	.3225	.3117	.3281	1.0524	.6450	.6906	1.215	1.288
.01700	.05296	.3328	.3209	.3389	1.0559	.6655	.7158	1.230	1.271
.01800	.05455	.3428	.3298	.3495	1.0593	.6856	.7405	1.244	1.255
.01900	.05611	.3525	.3386	.3599	1.0628	.7051	.7650	1.259	1.240
.02000	.05763	.3621	.3470	.3701	1.0663	.7242	.7891	1.274	1.226
.02100	.05912	.3714	.3552	.3800	1.0698	.7429	.8131	1.289	1.213
.02200	.06057	.3806	.3632	.3898	1.0733	.7612	.8368	1.304	1.201
.02300	.06200	.3896	.3710	.3995	1.0768	.7791	.8603	1.319	1.189
.02400	.06340	.3984	.3786	.4090	1.0804	.7967	.8837	1.335	1.178
.02500	.06478	.4070	.3860	.4184	1.0840	.8140	.9069	1.350	1.168
.02600	.06613	.4155	.3932	.4276	1.0876	.8310	.9310	1.366	1.159
.02700	.06747	.4239	.4002	.4367	1.0912	.8478	.9530	1.381	1.150
.02800	.06878	.4322	.4071	.4457	1.0949	.8643	.9760	1.397	1.141
.02900	.07007	.4403	.4138	.4546	1.0985	.8805	.9988	1.413	1.133
.03000	.07135	.4483	.4205	.4634	1.1021	.8966	1.022	1.430	1.125
.03100	.07260	.4562	.4269	.4721	1.1059	.9124	1.044	1.446	1.118
.03200	.07385	.4640	.4333	.4808	1.1096	.9280	1.067	1.462	1.111
.03300	.07507	.4717	.4395	.4894	1.1133	.9434	1.090	1.479	1.104
.03400	.07630	.4794	.4457	.4980	1.1171	.9588	1.113	1.496	1.098
.03500	.07748	.4868	.4517	.5064	1.1209	.9737	1.135	1.513	1.092
.03600	.07867	.4943	.4577	.5147	1.1247	.9886	1.158	1.530	1.086
.03700	.07984	.5017	.4635	.5230	1.1285	1.0033	1.180	1.547	1.080
.03800	.08100	.5090	.4691	.5312	1.1324	1.018	1.203	1.564	1.075
.03900	.08215	.5162	.4747	.5394	1.1362	1.032	1.226	1.582	1.069
.04000	.08329	.5233	.4802	.5475	1.1401	1.047	1.248	1.600	1.064
.04100	.08442	.5304	.4857	.5556	1.1440	1.061	1.271	1.617	1.059
.04200	.08553	.5374	.4911	.5637	1.1479	1.075	1.294	1.636	1.055
.04300	.08664	.5444	.4964	.5717	1.1518	1.089	1.317	1.654	1.050
.04400	.08774	.5513	.5015	.5796	1.1558	1.103	1.340	1.672	1.046
.04500	.08883	.5581	.5066	.5876	1.1599	1.116	1.363	1.691	1.042
.04600	.08991	.5649	.5116	.5954	1.1639	1.130	1.386	1.709	1.038
.04700	.09098	.5717	.5166	.6033	1.1679	1.143	1.409	1.728	1.034
.04800	.09205	.5784	.5215	.6111	1.1720	1.157	1.433	1.747	1.030
.04900	.09311	.5850	.5263	.6189	1.1760	1.170	1.456	1.766	1.026
.05000	.09416	.5916	.5310	.6267	1.1802	1.183	1.479	1.786	1.023
.05100	.09520	.5981	.5357	.6344	1.1843	1.196	1.503	1.805	1.019
.05200	.09623	.6046	.5403	.6421	1.1884	1.209	1.526	1.825	1.016
.05300	.09726	.6111	.5449	.6499	1.1926	1.222	1.550	1.845	1.013
.05400	.09829	.6176	.5494	.6575	1.1968	1.235	1.574	1.865	1.010
.05500	.09930	.6239	.5538	.6652	1.2011	1.248	1.598	1.885	1.007
.05600	.1003	.6303	.5582	.6729	1.2053	1.261	1.622	1.906	1.004
.05700	.1013	.6366	.5626	.6805	1.2096	1.273	1.646	1.926	1.001
.05800	.1023	.6428	.5668	.6880	1.2138	1.286	1.670	1.947	.9985
.05900	.1033	.6491	.5711	.6956	1.2181	1.298	1.695	1.968	.9958
.06000	.1043	.6553	.5753	.7033	1.2225	1.311	1.719	1.989	.9932
.06100	.1053	.6616	.5794	.7110	1.2270	1.3231	1.744	2.011	.9907
.06200	.1063	.6678	.5834	.7187	1.2315	1.336	1.770	2.033	.9883
.06300	.1073	.6739	.5874	.7256	1.2355	1.348	1.795	2.055	.9860
.06400	.1082	.6799	.5914	.7335	1.2402	1.360	1.819	2.076	.9837

APPENDIX (Continued)

d/L_o	d/L	$2\pi d/L$	tanh $2\pi d/L$	sinh $2\pi d/L$	cosh $2\pi d/L$	$4\pi d/L$	sinh $4\pi d/L$	cosh $4\pi d/L$	H/H_o'
.06500	.1092	.6860	.5954	.7411	1.2447	1.372	1.845	2.098	.9815
.06600	.1101	.6920	.5993	.7486	1.2492	1.384	1.870	2.121	.9793
.06700	.1111	.6981	.6031	.7561	1.2537	1.396	1.896	2.144	.9772
.06800	.1120	.7037	.6069	.7633	1.2580	1.408	1.921	2.166	.9752
.06900	.1130	.7099	.6106	.7711	1.2628	1.420	1.948	2.189	.9732
.07000	.1139	.7157	.6144	.7783	1.2672	1.432	1.974	2.213	.9713
.07100	.1149	.7219	.6181	.7863	1.2721	1.444	2.000	2.236	.9694
.07200	.1158	.7277	.6217	.7937	1.2767	1.455	2.026	2.260	.9676
.07300	.1168	.7336	.6252	.8011	1.2813	1.467	2.053	2.284	.9658
.07400	.1177	.7395	.6289	.8088	1.2861	1.479	2.080	2.308	.9641
.07500	.1186	.7453	.6324	.8162	1.2908	1.490	2.107	2.332	.9624
.07600	.1195	.7511	.6359	.8237	1.2956	1.502	2.135	2.357	.9607
.07700	.1205	.7569	.6392	.8312	1.3004	1.514	2.162	2.382	.9591
.07800	.1214	.7625	.6427	.8386	1.3051	1.525	2.189	2.407	.9576
.07900	.1223	.7683	.6460	.8462	1.3100	1.537	2.217	2.432	.9562
.08000	.1232	.7741	.6493	.8538	1.3149	1.548	2.245	2.458	.9548
.08100	.1241	.7799	.6526	.8614	1.3198	1.560	2.274	2.484	.9534
.08200	.1251	.7854	.6558	.8687	1.3246	1.571	2.303	2.511	.9520
.08300	.1259	.7911	.6590	.8762	1.3295	1.583	2.331	2.537	.9506
.08400	.1268	.7967	.6622	.8837	1.3345	1.594	2.360	2.563	.9493
.08500	.1277	.8026	.6655	.8915	1.3397	1.605	2.389	2.590	.9481
.08600	.1286	.8080	.6685	.8989	1.3446	1.616	2.418	2.617	.9469
.08700	.1295	.8137	.6716	.9064	1.3497	1.628	2.448	2.644	.9457
.08800	.1304	.8193	.6747	.9141	1.3548	1.639	2.478	2.672	.9445
.08900	.1313	.8250	.6778	.9218	1.3600	1.650	2.508	2.700	.9433
.09000	.1322	.8306	.6808	.9295	1.3653	1.661	2.538	2.728	.9422
.09100	.1331	.8363	.6838	.9372	1.3706	1.672	2.568	2.756	.9411
.09200	.1340	.8420	.6868	.9450	1.3759	1.684	2.599	2.785	.9401
.09300	.1349	.8474	.6897	.9525	1.3810	1.695	2.630	2.814	.9391
.09400	.1357	.8528	.6925	.9600	1.3862	1.706	2.662	2.843	.9381
.09500	.1366	.8583	.6953	.9677	1.3917	1.717	2.693	2.873	.9371
.09600	.1375	.8639	.6982	.9755	1.3970	1.728	2.726	2.903	.9362
.09700	.1384	.8694	.7011	.9832	1.4023	1.739	2.757	2.933	.9353
.09800	.1392	.8749	.7039	.9908	1.4077	1.750	2.790	2.963	.9344
.09900	.1401	.8803	.7066	.9985	1.4131	1.761	2.822	2.994	.9335
.1000	.1410	.8858	.7093	1.006	1.4187	1.772	2.855	3.025	.9327
.1010	.1419	.8913	.7120	1.014	1.4242	1.783	2.888	3.057	.9319
.1020	.1427	.8967	.7147	1.022	1.4297	1.793	2.922	3.088	.9311
.1030	.1436	.9023	.7173	1.030	1.4354	1.805	2.956	3.121	.9304
.1040	.1445	.9076	.7200	1.037	1.4410	1.815	2.990	3.153	.9297
.1050	.1453	.9130	.7226	1.045	1.4465	1.826	3.024	3.185	.9290
.1060	.1462	.9184	.7252	1.053	1.4523	1.837	3.059	3.218	.9282
.1070	.1470	.9239	.7277	1.061	1.4580	1.848	3.094	3.251	.9276
.1080	.1479	.9293	.7303	1.069	1.4638	1.858	3.128	3.284	.9269
.1090	.1488	.9343	.7327	1.076	1.4692	1.869	3.164	3.319	.9263
.1100	.1496	.9400	.7352	1.085	1.4752	1.880	3.201	3.353	.9257
.1110	.1505	.9456	.7377	1.093	1.4814	1.891	3.237	3.388	.9251
.1120	.1513	.9508	.7402	1.101	1.4871	1.902	3.274	3.423	.9245
.1130	.1522	.9563	.7426	1.109	1.4932	1.913	3.312	3.459	.9239
.1140	.1530	.9616	.7450	1.117	1.4990	1.923	3.348	3.494	.9234
.1150	.1539	.9670	.7474	1.125	1.5051	1.934	3.385	3.530	.9228
.1160	.1547	.9720	.7497	1.133	1.5108	1.944	3.423	3.566	.9223
.1170	.1556	.9775	.7520	1.141	1.5171	1.955	3.462	3.603	.9218
.1180	.1564	.9827	.7543	1.149	1.5230	1.966	3.501	3.641	.9214
.1190	.1573	.9882	.7566	1.157	1.5293	1.977	3.540	.3.678	.9209

Table of functions of d/L_0

APPENDIX (Continued)

d/L_o	d/L	$2\pi d/L$	tanh $2\pi d/L$	sinh $2\pi d/L$	cosh $2\pi d/L$	$4\pi d/L$	sinh $4\pi d/L$	cosh $4\pi d/L$	H/H_o'
.1200	.1581	.9936	.7589	1.165	1.5356	1.987	3.579	3.716	.9204
.1210	.1590	.9989	.7612	1.174	1.5418	1.998	3.620	3.755	.9200
.1220	.1598	1.004	.7634	1.182	1.5479	2.008	3.659	3.793	.9196
.1230	.1607	1.010	.7656	1.190	1.5546	2.019	3.699	3.832	.9192
.1240	.1615	1.015	.7678	1.198	1.5605	2.030	3.740	3.871	.9189
.1250	.1624	1.020	.7700	1.207	1.5674	2.041	3.782	3.912	.9186
.1260	.1632	1.025	.7721	1.215	1.5734	2.051	3.824	3.952	.9182
.1270	.1640	1.030	.7742	1.223	1.5795	2.061	3.865	3.992	.9178
.1280	.1649	1.036	.7763	1.231	1.5862	2.072	3.907	4.033	.9175
.1290	.1657	1.041	.7783	1.240	1.5927	2.082	3.950	4.074	.9172
.1300	.1665	1.046	.7804	1.248	1.5990	2.093	3.992	4.115	.9169
.1310	.1674	1.052	.7824	1.257	1.6060	2.104	4.036	4.158	.9166
.1320	.1682	1.057	.7844	1.265	1.6124	2.114	4.080	4.201	.9164
.1330	.1691	1.062	.7865	1.273	1.6191	2.125	4.125	4.245	.9161
.1340	.1699	1.068	.7885	1.282	1.6260	2.135	4.169	4.288	.9158
.1350	.1708	1.073	.7905	1.291	1.633	2.146	4.217	4.334	.9156
.1360	.1716	1.078	.7925	1.300	1.640	2.156	4.262	4.378	.9154
.1370	.1724	1.084	.7945	1.308	1.647	2.167	4.309	4.423	.9152
.1380	.1733	1.089	.7964	1.317	1.654	2.177	4.355	4.468	.9150
.1390	.1741	1.094	.7983	1.326	1.660	2.188	4.402	4.514	.9148
.1400	.1749	1.099	.8002	1.334	1.667	2.198	4.450	4.561	.9146
.1410	.1758	1.105	.8021	1.343	1.675	2.209	4.498	4.607	.9144
.1420	.1766	1.110	.8039	1.352	1.681	2.219	4.546	4.654	.9142
.1430	.1774	1.115	.8057	1.360	1.688	2.230	4.595	4.663	.9141
.1440	.1783	1.120	.8076	1.369	1.696	2.240	4.644	4.751	.9140
.1450	.1791	1.125	.8094	1.378	1.703	2.251	4.695	4.800	.9139
.1460	.1800	1.131	.8112	1.388	1.710	2.261	4.746	4.850	.9137
.1470	.1808	1.136	.8131	1.397	1.718	2.272	4.798	4.901	.9136
.1480	.1816	1.141	.8149	1.405	1.725	2.282	4.847	4.951	.9135
.1490	.1825	1.146	.8166	1.415	1.732	2.293	4.901	5.001	.9134
.1500	.1833	1.152	.8183	1.424	1.740	2.303	4.954	5.054	.9133
.1510	.1841	1.157	.8200	1.433	1.747	2.314	5.007	5.106	.9133
.1520	.1850	1.162	.8217	1.442	1.755	2.324	5.061	5.159	.9132
.1530	.1858	1.167	.8234	1.451	1.762	2.335	5.115	5.212	.9132
.1540	.1866	1.173	.8250	1.460	1.770	2.345	5.169	5.265	.9132
.1550	.1875	1.178	.8267	1.469	1.777	2.356	5.225	5.320	.9131
.1560	.1883	1.183	.8284	1.479	1.785	2.366	5.283	5.376	.9130
.1570	.1891	1.188	.8301	1.488	1.793	2.377	5.339	5.432	.9129
.1580	.1900	1.194	.8317	1.498	1.801	2.387	5.398	5.490	.9130
.1590	.1908	1.199	.8333	1.507	1.809	2.398	5.454	5.544	.9130
.1600	.1917	1.204	.8349	1.517	1.817	2.408	5.513	5.603	.9130
.1610	.1925	1.209	.8365	1.527	1.825	2.419	5.571	5.660	.9130
.1620	.1933	1.215	.8381	1.536	1.833	2.429	5.630	5.718	.9130
.1630	.1941	1.220	.8396	1.546	1.841	2.440	5.690	5.777	.9130
.1640	.1950	1.225	.8411	1.555	1.849	2.450	5.751	5.837	.9130
.1650	.1958	1.230	.8427	1.565	1.857	2.461	5.813	5.898	.9131
.1660	.1966	1.235	.8442	1.574	1.865	2.471	5.874	5.959	.9132
.1670	.1975	1.240	.8457	1.584	1.873	2.482	5.938	6.021	.9132
.1680	.1983	1.246	.8472	1.594	1.882	2.492	6.003	6.085	.9133
.1690	.1992	1.251	.8486	1.604	1.890	2.503	6.066	6.148	.9133
.1700	.2000	1.257	.8501	1.614	1.899	2.513	6.130	6.212	.9134
.1710	.2008	1.262	.8515	1.624	1.907	2.523	6.197	6.275	.9135
.1720	.2017	1.267	.8529	1.634	1.915	2.534	6.262	6.342	.9136
.1730	.2025	1.272	.8544	1.644	1.924	2.544	6.329	6.407	.9137
.1740	.2033	1.277	.8558	1.654	1.933	2.555	6.395	6.473	.9138

APPENDIX (Continued)

d/L_o	d/L	$2\pi d/L$	tanh $2\pi d/L$	sinh $2\pi d/L$	cosh $2\pi d/L$	$4\pi d/L$	sinh $4\pi d/L$	cosh $4\pi d/L$	H/H_o
.1750	.2042	1.282	.8572	1.664	1.941	2.565	6.465	6.541	.9139
.1760	.2050	1.288	.8586	1.675	1.951	2.576	6.534	6.610	.9140
.1770	.2058	1.293	.8600	1.685	1.959	3.586	6.603	6.679	.9141
.1780	.2066	1.298	.8614	1.695	1.968	2.597	6.672	6.747	.9142
.1790	.2075	1.304	.8627	1.706	1.977	2.607	6.744	6.818	.9144
.1800	.2083	1.309	.8640	1.716	1.986	2.618	6.818	6.891	.9145
.1810	.2092	1.314	.8653	1.727	1.995	2.629	6.890	6.963	.9146
.1820	.2100	1.320	.8666	1.737	2.004	2.639	6.963	7.035	.9148
.1830	.2108	1.325	.8680	1.748	2.013	2.650	7.038	7.109	.9149
.1840	.2117	1.330	.8693	1.758	2.022	2.660	7.113	7.183	.9150
.1850	.2125	1.335	.8706	1.769	2.032	2.671	7.191	7.260	.9152
.1860	.2134	1.341	.8718	1.780	2.041	2.681	7.267	7.336	.9154
.1870	.2142	1.346	.8731	1.791	2.051	2.692	7.345	7.412	.9155
.1880	.2150	1.351	.8743	1.801	2.060	2.702	7.421	7.488	.9157
.1890	.2159	1.356	.8755	1.812	2.070	2.712	7.500	7.566	.9159
.1900	.2167	1.362	.8767	1.823	2.079	2.723	7.581	7.647	.9161
.1910	.2176	1.367	.8779	1.834	2.089	2.734	7.663	7.728	.9163
.1920	.2184	1.372	.8791	1.845	2.099	2.744	7.746	7.810	.9165
.1930	.2192	1.377	.8803	1.856	2.108	2.755	7.827	7.891	.9167
.1940	.2201	1.383	.8815	1.867	2.118	2.765	7.911	7.974	.9169
.1950	.2209	1.388	.8827	1.879	2.128	2.776	7.996	8.059	.9170
.1960	.2218	1.393	.8839	1.890	2.138	2.787	8.083	8.145	.9172
.1970	.2226	1.399	.8850	1.901	2.148	2.797	8.167	8.228	.9174
.1980	.2234	1.404	.8862	1.913	2.158	2.808	8.256	8.316	.9176
.1990	.2243	1.409	.8873	1.924	2.169	2.819	8.346	8.406	.9179
.2000	.2251	1.414	.8884	1.935	2.178	2.829	8.436	8.495	.9181
.2010	.2260	1.420	.8895	1.947	2.189	2.840	8.524	8.583	.9183
.2020	.2268	1.425	.8906	1.959	2.199	2.850	8.616	8.674	.9186
.2030	.2277	1.430	.8917	1.970	2.210	2.861	8.708	8.766	.9188
.2040	.2285	1.436	.8928	1.982	2.220	2.872	8.803	8.860	.9190
.2050	.2293	1.441	.8939	1.994	2.231	2.882	8.897	8.953	.9193
.2060	.2302	1.446	.8950	2.006	2.242	2.893	8.994	9.050	.9195
.2070	.2310	1.451	.8960	2.017	2.252	2.903	9.090	9.144	.9197
.2080	.2319	1.457	.8971	2.030	2.263	2.914	9.187	9.240	.9200
.2090	.2328	1.462	.8981	2.042	2.274	2.925	9.288	9.342	.9202
.2100	.2336	1.468	.8991	2.055	2.285	2.936	9.389	9.442	.9205
.2110	.2344	1.473	.9001	2.066	2.295	2.946	9.490	9.542	.9207
.2120	.2353	1.479	.9011	2.079	2.307	2.957	9.590	9.642	.9210
.2130	.2361	1.484	.9021	2.091	2.318	2.967	9.693	9.744	.9213
.2140	.2370	1.489	.9031	2.103	2.329	2.978	9.796	9.847	.9215
.2150	.2378	1.494	.9041	2.115	2.340	2.989	9.902	9.952	.9218
.2160	.2387	1.500	.9051	2.128	2.351	2.999	10.01	10.06	.9221
.2170	.2395	1.506	.9061	2.142	2.364	3.010	10.12	10.17	.9223
.2180	.2404	1.511	.9070	2.154	2.375	3.021	10.23	10.28	.9226
.2190	.2412	1.516	.9079	2.166	2.386	3.031	10.34	10.38	.9228
.2200	.2421	1.521	.9088	2.178	2.397	3.042	10.45	10.50	.9231
.2210	.2429	1.526	.9097	2.192	2.409	3.052	10.56	10.61	.9234
.2220	.2438	1.532	.9107	2.204	2.421	3.063	10.68	10.72	.9236
.2230	.2446	1.537	.9116	2.218	2.433	3.074	10.79	10.84	.9239
.2240	.2455	1.542	.9125	2.230	2.444	3.085	10.91	10.95	.9242
.2250	.2463	1.548	.9134	1.244	2.457	3.095	11.02	11.07	.9245
.2260	.2472	1.553	.9143	2.257	2.469	3.106	11.15	11.19	.9248
.2270	.2481	1.559	.9152	2.271	2.481	3.117	11.27	11.31	.9251
.2280	.2489	1.564	.9161	2.284	2.493	3.128	11.39	11.44	.9254
.2290	.2498	1.569	.9170	2.297	2.506	3.138	11.51	11.56	.9258

Table of functions of d/L_0

APPENDIX (Continued)

d/L_o	d/L	$2\pi d/L$	tanh $2\pi d/L$	sinh $2\pi d/L$	cosh $2\pi d/L$	$4\pi d/L$	sinh $4\pi d/L$	cosh $4\pi d/L$	H/H'_o
.2300	.2506	1.575	.9178	2.311	2.518	3.149	11.64	11.68	.9261
.2310	.2515	1.580	.9186	2.325	2.531	3.160	11.77	11.81	.9264
.2320	.2523	1.585	.9194	2.338	2.543	3.171	11.90	11.93	.9267
.2330	.2532	1.591	.9203	2.352	2.556	3.182	12.03	12.07	.9270
.2340	.2540	1.596	.9211	2.366	2.569	3.192	12.15	12.19	.9273
.2350	.2549	1.602	.9219	2.380	2.581	3.203	12.29	12.33	.9276
.2360	.2558	1.607	.9227	2.393	2.594	3.214	12.43	12.47	.9279
.2370	.2566	1.612	.9235	2.408	2.607	3.225	12.55	12.59	.9282
.2380	.2575	1.618	.9243	2.422	2.620	3.236	12.69	12.73	.9285
.2390	.2584	1.623	.9251	2.436	2.634	3.247	12.83	12.87	.9288
.2400	.2592	1.629	.9259	2.450	2.647	3.257	12.97	13.01	.9291
.2410	.2601	1.634	.9267	2.464	2.660	3.268	13.11	13.15	.9294
.2420	.2610	1.640	.9275	2.480	2.674	3.279	13.26	13.30	.9298
.2430	.2618	1.645	.9282	2.494	2.687	3.290	13.40	13.44	.9301
.2440	.2627	1.650	.9289	2.508	2.700	3.301	13.55	13.59	.9304
.2450	.2635	1.656	.9296	2.523	2.714	3.312	13.70	13.73	.9307
.2460	.2644	1.661	.9304	2.538	2.728	3.323	13.85	13.88	.9310
.2470	.2653	1.667	.9311	2.553	2.742	3.334	14.00	14.04	.9314
.2480	.2661	1.672	.9318	2.568	2.755	3.344	14.15	14.19	.9317
.2490	.2670	1.678	.9325	2.583	2.770	3.355	14.31	14.35	.9320
.2500	.2679	1.683	.9332	2.599	2.784	3.367	14.47	14.51	.9323
.2510	.2687	1.689	.9339	2.614	2.798	3.377	14.62	14.66	.9327
.2520	.2696	1.694	.9346	2.629	2.813	3.388	14.79	14.82	.9330
.2530	.2705	1.700	.9353	2.645	2.828	3.399	14.95	14.99	.9333
.2540	.2714	1.705	.9360	2.660	2.842	3.410	15.12	15.15	.9336
.2550	.2722	1.711	.9367	2.676	2.856	3.421	15.29	15.32	.9340
.2560	.2731	1.716	.9374	2.691	2.871	3.432	15.45	15.49	.9343
.2570	.2740	1.722	.9381	2.707	2.886	3.443	15.63	15.66	.9346
.2580	.2749	1.727	.9388	2.723	2.901	3.454	15.80	15.83	.9349
.2590	.2757	1.732	.9394	2.739	2.916	3.465	15.97	16.00	.9353
.2600	.2766	1.738	.9400	2.755	2.931	3.476	16.15	16.18	.9356
.2610	.2775	1.744	.9406	2.772	2.946	3.487	16.33	16.36	.9360
.2620	.2784	1.749	.9412	2.788	2.962	3.498	16.51	16.54	.9363
.2630	.2792	1.755	.9418	2.804	2.977	3.509	16.69	16.73	.9367
.2640	.2801	1.760	.9425	2.820	2.992	3.520	16.88	16.91	.9370
.2650	.2810	1.766	.9431	2.837	3.008	3.531	17.07	17.10	.9373
.2660	.2819	1.771	.9437	2.853	3.023	3.542	17.26	17.28	.9377
.2670	.2827	1.776	.9443	2.870	3.039	3.553	17.45	17.45	.9380
.2680	.2836	1.782	.9449	2.886	3.055	3.564	17.64	17.67	.9383
.2690	.2845	1.788	.9455	2.904	3.071	3.575	17.84	17.87	.9386
.2700	.2854	1.793	.9461	2.921	3.088	3.587	18.04	18.07	.9390
.2710	.2863	1.799	.9467	2.938	3.104	3.598	18.24	18.27	.9393
.2720	.2872	1.804	.9473	2.956	3.120	3.610	18.46	18.49	.9396
.2730	.2880	1.810	.9478	2.973	3.136	3.620	18.65	18.67	.9400
.2740	.2889	1.815	.9484	2.990	3.153	3.631	18.86	18.89	.9403
.2750	.2898	1.821	.9490	3.008	3.170	3.642	19.07	19.10	.9406
.2760	.2907	1.826	.9495	3.025	3.186	3.653	19.28	19.30	.9410
.2770	.2916	1.832	.9500	3.043	3.203	3.664	19.49	19.51	.9413
.2780	.2924	1.837	.9505	3.061	3.220	3.675	19.71	19.74	.9416
.2790	.2933	1.843	.9511	3.079	3.237	3.686	19.93	19.96	.9420
.2800	.2942	1.849	.9516	3.097	3.254	3.697	20.16	20.18	.9423
.2810	.2951	1.854	.9521	3.115	3.272	3.709	20.39	20.41	.9426
.2820	.2960	1.860	.9526	3.133	3.289	3.720	20.62	20.64	.9430
.2830	.2969	1.866	.9532	3.152	3.307	3.731	20.85	20.87	.9433
.2840	.2978	1.871	.9537	3.171	3.325	3.742	21.09	21.11	.9436

APPENDIX (Continued)

d/L_0	d/L	$2\pi d/L$	tanh $2\pi d/L$	sinh $2\pi d/L$	cosh $2\pi d/L$	$4\pi d/L$	sinh $4\pi d/L$	cosh $4\pi d/L$	H/H_0'
.2850	.2987	1.877	.9542	3.190	3.343	3.754	21.33	21.35	.9440
.2860	.2996	1.882	.9547	3.209	3.361	3.765	21.57	21.59	.9443
.2870	.3005	1.888	.9552	3.228	3.379	3.776	21.82	21.84	.9446
.2880	.3014	1.893	.9557	3.246	3.396	3.787	22.05	22.07	.9449
.2890	.3022	1.899	.9562	3.264	3.414	3.798	22.30	22.32	.9452
.2900	.3031	1.905	.9567	3.284	3.433	3.809	22.54	22.57	.9456
.2910	.3040	1.910	.9572	3.303	3.451	3.821	22.81	22.83	.9459
.2920	.3049	1.916	.9577	3.323	3.471	3.832	23.07	23.09	.9463
.2930	.3058	1.922	.9581	3.343	3.490	3.843	23.33	23.35	.9466
.2940	.3067	1.927	.9585	3.362	3.508	3.855	23.60	23.62	.9469
.2950	.3076	1.933	.9590	3.382	3.527	3.866	23.86	23.88	.9473
.2960	.3085	1.938	.9594	3.402	3.546	3.877	24.12	24.15	.9476
.2970	.3094	1.944	.9599	3.422	3.565	3.888	24.40	24.42	.9480
.2980	.3103	1.950	.9603	3.442	3.585	3.900	24.68	24.70	.9483
.2990	.3112	1.955	.9607	3.462	3.604	3.911	24.96	24.98	.9486
.3000	.3121	1.961	.9611	3.483	3.624	3.922	25.24	25.26	.9490
.3010	.3130	1.967	.9616	3.503	3.643	3.933	25.53	25.55	.9493
.3020	.3139	1.972	.9620	3.524	3.663	3.945	25.82	25.83	.9496
.3030	.3148	1.978	.9624	3.545	3.683	3.956	26.12	26.14	.9499
.3040	.3157	1.984	.9629	3.566	3.703	3.968	26.42	26.44	.9502
.3050	.3166	1.989	.9633	3.587	3.724	3.979	26.72	26.74	.9505
.3060	.3175	1.995	.9637	3.609	3.745	3.990	27.02	27.04	.9509
.3070	.3184	2.001	.9641	3.630	3.765	4.002	27.33	27.35	.9512
.3080	.3193	2.007	.9645	3.651	3.786	4.013	27.65	27.66	.9515
.3090	.3202	2.012	.9649	3.673	3.806	4.024	27.96	27.98	.9518
.3100	.3211	2.018	.9653	3.694	3.827	4.036	28.28	28.30	.9522
.3110	.3220	2.023	.9656	3.716	3.848	4.047	28.60	28.62	.9525
.3120	.3230	2.029	.9660	3.738	3.870	4.058	28.93	28.95	.9528
.3130	.3239	2.035	.9664	3.760	3.891	4.070	29.27	29.28	.9531
.3140	.3248	2.041	.9668	3.782	3.912	4.081	29.60	29.62	.9535
.3150	.3257	2.046	.9672	3.805	3.934	4.093	29.94	29.96	.9538
.3160	.3266	2.052	.9676	3.828	3.956	4.104	30.29	30.31	.9541
.3170	.3275	2.058	.9679	3.851	3.978	4.116	30.64	30.65	.9544
.3180	.3284	2.063	.9682	3.873	4.000	4.127	30.99	31.00	.9547
.3190	.3294	2.069	.9686	3.896	4.022	4.139	31.35	31.37	.9550
.3200	.3302	2.075	.9690	3.919	4.045	4.150	31.71	31.72	.9553
.3210	.3311	2.081	.9693	3.943	4.068	4.161	32.07	32.08	.9556
.3220	.3321	2.086	.9696	3.966	4.090	4.173	32.44	32.46	.9559
.3230	.3330	2.092	.9700	3.990	4.114	4.185	32.83	32.84	.9562
.3240	.3339	2.098	.9703	4.014	4.136	4.196	33.20	33.22	.9565
.3250	.3349	2.104	.9707	4.038	4.160	4.208	33.60	33.61	.9568
.3260	.3357	2.110	.9710	4.061	4.183	4.219	33.97	33.99	.9571
.3270	.3367	2.115	.9713	4.085	4.206	4.231	34.37	34.38	.9574
.3280	.3376	2.121	.9717	4.110	4.230	4.242	34.77	34.79	.9577
.3290	.3385	2.127	.9720	4.135	4.254	4.254	35.18	35.19	.9580
.3300	.3394	2.133	.9723	4.159	4.277	4.265	35.58	35.59	.9583
.3310	.3403	2.138	.9726	4.184	4.301	4.277	35.99	36.00	.9586
.3320	.3413	2.144	.9729	4.209	4.326	4.288	36.42	36.43	.9589
.3330	.3422	2.150	.9732	4.234	4.350	4.300	36.84	36.85	.9592
.3340	.3431	2.156	.9735	4.259	4.375	4.311	37.25	37.27	.9595
.3350	.3440	2.161	.9738	4.284	4.399	4.323	37.70	37.72	.9598
.3360	.3449	2.167	.9741	4.310	4.424	4.335	38.14	38.15	.9601
.3370	.3459	2.173	.9744	4.336	4.450	4.346	38.59	38.60	.9604
.3380	.3468	2.179	.9747	4.361	4.474	4.358	39.02	39.04	.9607
.3390	.3477	2.185	.9750	4.388	4.500	4.369	39.48	39.49	.9610

APPENDIX (Continued)

d/L_0	d/L	$2\pi d/L$	tanh $2\pi d/L$	sinh $2\pi d/L$	cosh $2\pi d/L$	$4\pi d/L$	sinh $4\pi d/L$	cosh $4\pi d/L$	H/H'_0
.3400	.3468	2.190	.9753	4.413	4.525	4.381	39.95	39.96	.9613
.3410	.3495	2.196	.9756	4.439	4.550	4.392	40.40	40.41	.9615
.3420	.3504	2.202	.9758	4.466	4.576	4.404	40.87	40.89	.9618
.3430	.3514	2.208	.9761	4.492	4.602	4.416	41.36	41.37	.9621
.3440	.3523	2.214	.9764	4.521	4.630	4.427	41.85	41.84	.9623
.3450	.3532	2.220	.9767	4.547	4.656	4.439	42.33	42.34	.9626
.3460	.3542	2.225	.9769	4.575	4.682	4.451	42.83	42.84	.9629
.3470	.3551	2.231	.9772	4.602	4.709	4.462	43.34	43.35	.9632
.3480	.3560	2.237	.9775	4.629	4.736	4.474	43.85	43.86	.9635
.3490	.3570	2.243	.9777	4.657	4.763	4.486	44.37	44.40	.9638
.3500	.3579	2.249	.9780	4.685	4.791	4.498	44.89	44.80	.9640
.3510	.3588	2.255	.9782	4.713	4.818	4.509	45.42	45.43	.9643
.3520	.3598	2.260	.9785	4.741	4.845	4.521	45.95	45.96	.9646
.3530	.3607	2.266	.9787	4.770	4.873	4.533	46.50	46.51	.9648
.3540	.3616	2.272	.9790	4.798	4.901	4.544	47.03	47.04	.9651
.3550	.3625	2.278	.9792	4.827	4.929	4.556	47.59	47.60	.9654
.3560	.3635	2.284	.9795	4.856	4.957	4.568	48.15	48.16	.9657
.3570	.3644	2.290	.9797	4.885	4.987	4.579	48.72	48.73	.9659
.3580	.3653	2.296	.9799	4.914	5.015	4.591	49.29	49.30	.9662
.3590	.3663	2.301	.9801	4.944	5.044	4.603	49.88	49.89	.9665
.3600	.3672	2.307	.9804	4.974	5.072	4.615	50.47	50.48	.9667
.3610	.3682	2.313	.9806	5.004	5.103	4.627	51.08	51.09	.9670
.3620	.3691	2.319	.9808	5.034	5.132	4.638	51.67	51.67	.9673
.3630	.3700	2.325	.9811	5.063	5.161	4.650	52.27	52.28	.9675
.3640	.3709	2.331	.9813	5.094	5.191	4.661	52.89	52.90	.9677
.3650	.3719	2.337	.9815	5.124	5.221	4.673	53.52	53.53	.9680
.3660	.3728	2.342	.9817	5.155	5.251	4.685	54.15	54.16	.9683
.3670	.3737	2.348	.9819	5.186	5.281	4.697	54.78	54.79	.9686
.3680	.3747	2.354	.9821	5.217	5.312	4.708	55.42	55.43	.9688
.3690	.3756	2.360	.9823	5.248	5.343	4.720	56.09	56.10	.9690
.3700	.3766	2.366	.9825	5.280	5.374	4.732	56.76	56.77	.9693
.3710	.3775	2.372	.9827	5.312	5.406	4.744	57.43	57.44	.9696
.3720	.3785	2.378	.9830	5.345	5.438	4.756	58.13	58.14	.9698
.3730	.3794	2.384	.9832	5.377	5.469	4.768	58.82	58.83	.9700
.3740	.3804	2.390	.9834	5.410	5.502	4.780	59.52	59.53	.9702
.3750	.3813	2.396	.9835	5.443	5.534	4.792	60.24	60.25	.9705
.3760	.3822	2.402	.9837	5.475	5.566	4.803	60.95	60.95	.9707
.3770	.3832	2.408	.9839	5.508	5.598	4.815	61.68	61.68	.9709
.3780	.3841	2.413	.9841	5.541	5.631	4.827	62.41	62.42	.9712
.3790	.3850	2.419	.9843	5.572	5.661	4.838	63.13	63.14	.9714
.3800	.3860	2.425	.9845	5.609	5.697	4.851	63.90	63.91	.9717
.3810	.3869	2.431	.9847	5.643	5.731	4.862	64.66	64.67	.9719
.3820	.3879	2.437	.9848	5.677	5.765	4.875	65.45	65.46	.9721
.3830	.3888	2.443	.9850	5.712	5.798	4.885	66.20	66.21	.9724
.3840	.3898	2.449	.9852	5.746	5.833	4.898	67.00	67.01	.9726
.3850	.3907	2.455	.9854	5.780	5.866	4.910	67.80	67.81	.9728
.3860	.3917	2.461	.9855	5.814	5.900	4.922	68.61	68.62	.9730
.3870	.3926	2.467	.9857	5.850	5.935	4.934	69.45	69.46	.9732
.3880	.3936	2.473	.9859	5.886	5.970	4.946	70.28	70.29	.9735
.3890	.3945	2.479	.9860	5.921	6.005	4.958	71.12	71.13	.9737
.3900	.3955	2.485	.9862	5.957	6.040	4.970	71.97	71.98	.9739
.3910	.3964	2.491	.9864	5.993	6.076	4.982	72.85	72.86	.9741
.3920	.3974	2.497	.9865	6.029	6.112	4.993	73.72	73.72	.9743
.3930	.3983	2.503	.9867	6.066	6.148	5.005	74.59	74.59	.9745
.3940	.3993	2.509	.9869	6.103	6.185	5.017	75.48	75.48	.9748

APPENDIX (Continued)

d/L_o	d/L	$2\pi d/L$	tanh $2\pi d/L$	sinh $2\pi d/L$	cosh $2\pi d/L$	$4\pi d/L$	sinh $4\pi d/L$	cosh $4\pi d/L$	H/H_o'
.3950	.4002	2.515	.9870	6.140	6.221	5.029	76.40	76.40	.9750
.3960	.4012	2.521	.9872	6.177	6.258	5.041	77.32	77.32	.9752
.3970	.4021	2.527	.9873	6.215	6.295	5.053	78.24	78.24	.9754
.3980	.4031	2.532	.9874	6.252	6.332	5.065	79.19	79.19	.9756
.3990	.4040	2.538	.9876	6.290	6.369	5.077	80.13	80.13	.9758
.4000	.4050	2.544	.9877	6.329	6.407	5.089	81.12	81.12	.9761
.4010	.4059	2.550	.9879	6.367	6.445	5.101	82.08	82.08	.9763
.4020	.4069	2.556	.9880	6.406	6.483	5.113	83.06	83.06	.9765
.4030	.4078	2.562	.9882	6.444	6.521	5.125	84.07	84.07	.9766
.4040	.4088	2.568	.9883	6.484	6.561	5.137	85.11	85.11	.9768
.4050	.4098	2.575	.9885	6.525	6.601	5.149	86.14	86.14	.9777
.4060	.4107	2.581	.9886	6.564	6.640	5.161	87.17	87.17	.9772
.4070	.4116	2.586	.9887	6.603	6.679	5.173	88.20	88.20	.9774
.4080	.4126	2.592	.9889	6.644	6.718	5.185	89.28	89.28	.9776
.4090	.4136	2.598	.9890	6.684	6.758	5.197	90.39	90.39	.9778
.4100	.4145	2.604	.9891	6.725	6.799	5.209	91.44	91.44	.9780
.4110	.4155	2.610	.9892	6.766	6.839	5.221	92.55	92.55	.9782
.4120	.4164	2.616	.9894	6.806	6.879	5.233	93.67	93.67	.9784
.4130	.4174	2.623	.9895	6.849	6.921	5.245	94.83	94.83	.9786
.4140	.4183	2.629	.9896	6.890	6.963	5.257	95.96	95.96	.9788
.4150	.4193	2.635	.9898	6.932	7.004	5.269	97.13	97.13	.9790
.4160	.4203	2.641	.9899	6.974	7.046	5.281	98.30	98.30	.9792
.4170	.4212	2.647	.9900	7.018	7.088	5.294	99.52	99.52	.9794
.4180	.4222	2.653	.9901	7.060	7.130	5.305	100.7	100.7	.9795
.4190	.4231	2.659	.9902	7.102	7.173	5.317	101.9	101.9	.9797
.4200	.4241	2.665	.9904	7.146	7.215	5.329	103.1	103.1	.9798
.4210	.4251	2.671	.9905	7.190	7.259	5.341	104.4	104.4	.9800
.4220	.4260	2.677	.9906	7.234	7.303	5.353	105.7	105.7	.9802
.4230	.4270	2.683	.9907	7.279	7.349	5.366	107.0	107.0	.9804
.4240	.4280	2.689	.9908	7.325	7.392	5.378	108.3	108.3	.9806
.4250	.4289	2.695	.9909	7.371	7.438	5.390	109.7	109.7	.9808
.4260	.4298	2.701	.9910	7.412	7.479	5.402	110.9	110.9	.9810
.4270	.4308	2.707	.9911	7.457	7.524	5.414	112.2	112.2	.9811
.4280	.4318	2.713	.9912	7.503	7.570	5.426	113.6	113.6	.9812
.4290	.4328	2.719	.9913	7.550	7.616	5.438	115.0	115.0	.9814
.4300	.4337	2.725	.9914	7.595	7.661	5.450	116.4	116.4	.9816
.4310	.4347	2.731	.9915	7.642	7.707	5.462	117.8	117.8	.9818
.4320	.4356	2.737	.9916	7.688	7.753	5.474	119.2	119.2	.9819
.4330	.4366	2.743	.9917	7.735	7.800	5.486	120.7	120.7	.9821
.4340	.4376	2.749	.9918	7.783	7.847	5.499	122.2	122.2	.9823
.4350	.4385	2.755	.9919	7.831	7.895	5.511	123.7	123.7	.9824
.4360	.4395	2.762	.9920	7.880	7.943	5.523	125.2	125.2	.9826
.4370	.4405	2.768	.9921	7.922	7.991	5.535	126.7	126.7	.9828
.4380	.4414	2.774	.9922	7.975	8.035	5.547	128.3	128.3	.9829
.4390	.4424	2.780	.9923	8.026	8.088	5.560	129.9	129.9	.9830
.4400	.4434	2.786	.9924	8.075	8.136	5.572	131.4	131.4	.9832
.4410	.4443	2.792	.9925	8.124	8.185	5.584	133.0	133.0	.9833
.4420	.4453	2.798	.9926	8.175	8.236	5.596	134.7	134.7	.9835
.4430	.4463	2.804	.9927	8.228	8.285	5.608	136.3	136.3	.9836
.4440	.4472	2.810	.9928	8.274	8.334	5.620	137.9	137.9	.9838
.4450	.4482	2.816	.9929	8.326	8.387	5.632	139.6	139.6	.9839
.4460	.4492	2.822	.9930	8.379	8.438	5.644	141.4	141.4	.9841
.4470	.4501	2.828	.9930	8.427	8.486	5.657	143.1	143.1	.9843
.4480	.4511	2.834	.9931	8.481	8.540	5.669	144.8	144.8	.9844
.4490	.4521	2.840	.9932	8.532	8.590	5.681	146.6	146.6	.9846

Table of functions of d/L_0

APPENDIX (Continued)

d/L_o	d/L	$2\pi d/L$	tanh $2\pi d/L$	sinh $2\pi d/L$	cosh $2\pi d/L$	$4\pi d/L$	sinh $4\pi d/L$	cosh $4\pi d/L$	H/H_o'
.4500	.4531	2.847	.9933	8.585	8.643	5.693	148.4	148.4	.9847
.4510	.4540	2.853	.9934	8.638	8.695	5.705	150.2	150.2	.9848
.4520	.4550	2.859	.9935	8.693	8.750	5.717	152.1	152.1	.9849
.4530	.4560	2.865	.9935	8.747	8.804	5.730	154.0	154.0	.9851
.4540	.4569	2.871	.9936	8.797	8.854	5.742	155.9	155.9	.9852
.4550	.4579	2.877	.9937	8.853	8.910	5.754	157.7	157.7	.9853
.4560	.4589	2.883	.9938	8.910	8.965	5.766	159.7	159.7	.9855
.4570	.4599	2.890	.9938	8.965	9.021	5.779	161.7	161.7	.9857
.4580	.4608	2.896	.9939	9.016	9.072	5.791	163.6	163.6	.9858
.4590	.4618	2.902	.9940	9.074	9.129	5.803	165.6	165.6	.9859
.4600	.4628	2.908	.9941	9.132	9.186	5.815	167.7	167.7	.9860
.4610	.4637	2.914	.9941	9.183	9.238	5.827	169.7	169.7	.9862
.4620	.4647	2.920	.9942	9.242	9.296	5.840	171.8	171.8	.9863
.4630	.4657	2.926	.9943	9.301	9.354	5.852	173.9	173.9	.9864
.4640	.4666	2.932	.9944	9.353	9.406	5.864	176.0	176.0	.9865
.4650	.4676	2.938	.9944	9.413	9.466	5.876	178.2	178.2	.9867
.4660	.4686	2.944	.9945	9.472	9.525	5.888	180.4	180.4	.9868
.4670	.4695	2.951	.9946	9.533	9.585	5.900	182.6	182.6	.9869
.4680	.4705	2.957	.9946	9.586	9.638	5.912	184.8	184.8	.9871
.4690	.4715	2.963	.9947	9.647	9.699	5.925	187.2	187.2	.9872
.4700	.4725	2.969	.9947	9.709	9.760	5.937	189.5	189.5	.9873
.4710	.4735	2.975	.9948	9.770	9.821	5.949	191.8	191.8	.9874
.4720	.4744	2.981	.9949	9.826	9.877	5.962	194.2	194.2	.9875
.4730	.4754	2.987	.9949	9.888	9.938	5.974	196.5	196.5	.9876
.4740	.4764	2.993	.9950	9.951	10.00	5.986	199.0	199.0	.9877
.4750	.4774	2.999	.9951	10.01	10.07	5.999	201.4	201.4	.9878
.4760	.4783	3.005	.9951	10.07	10.12	6.011	203.9	203.9	.9880
.4770	.4793	3.012	.9952	10.13	10.18	6.023	206.5	206.5	.9881
.4780	.4803	3.018	.9952	10.20	10.25	6.036	209.0	209.0	.9882
.4790	.4813	3.024	.9953	10.26	10.31	6.048	211.7	211.7	.9883
.4800	.4822	3.030	.9953	10.32	10.37	6.060	214.2	214.2	.9885
.4810	.4832	3.036	.9954	10.39	10.43	6.072	216.8	216.8	.9886
.4820	.4842	3.042	.9955	10.45	10.50	6.085	219.5	219.5	.9887
.4830	.4852	3.049	.9955	10.52	10.57	6.097	222.2	222.2	.9888
.4840	.4862	3.055	.9956	10.59	10.63	6.109	225.0	225.0	.9889
.4850	.4871	3.061	.9956	10.65	10.69	6.121	228.3	228.3	.9890
.4860	.4881	3.067	.9957	10.71	10.76	6.134	230.6	230.6	.9891
.4870	.4891	3.073	.9957	10.78	10.83	6.146	233.5	233.5	.9892
.4880	.4901	3.079	.9958	10.85	10.90	6.159	236.4	236.4	.9893
.4890	.4911	3.086	.9958	10.92	10.96	6.171	239.6	239.6	.9895
.4900	.4920	3.092	.9959	10.99	11.03	6.183	242.3	242.3	.9896
.4910	.4930	3.098	.9959	11.05	11.09	6.195	245.2	245.2	.9897
.4920	.4940	3.104	.9960	11.12	11.16	6.208	248.3	248.3	.9898
.4930	.4950	3.110	.9960	11.19	11.24	6.220	251.3	251.3	.9899
.4940	.4960	3.117	.9961	11.26	11.31	6.232	254.5	254.5	.9899
.4950	.4969	3.122	.9961	11.32	11.37	6.245	257.6	257.6	.9900
.4960	.4979	3.128	.9962	11.40	11.44	6.257	260.8	260.8	.9901
.4970	.4989	3.135	.9962	11.47	11.51	6.269	264.0	264.0	.9902
.4980	.4999	3.141	.9963	11.54	11.59	5.282	267.3	267.3	.9903
.4990	.5009	3.147	.9963	11.61	11.65	6.294	270.6	270.6	.9904
.5000	.5018	3.153	.9964	11.68	11.72	6.306	274.0	274.0	.9905
.5010	.5028	3.159	.9964	11.75	11.80	6.319	277.5	277.5	.9906
.5020	.5038	3.166	.9964	11.83	11.87	6.331	280.8	280.8	.9907
.5030	.5048	3.172	.9965	11.91	11.95	6.343	284.3	284.3	.9908
.5040	.5058	3.178	.9965	11.98	12.02	5.356	287.9	287.9	.9909

APPENDIX (Continued)

d/L_o	d/L	$2\pi d/L$	tanh $2\pi d/L$	sinh $2\pi d/L$	cosh $2\pi d/L$	$4\pi d/L$	sinh $4\pi d/L$	cosh $4\pi d/L$	H/H_o'
.5050	.5067	3.184	.9966	12.05	12.09	6.368	291.4	291.4	.9909
.5060	.5077	3.190	.9966	12.12	12.16	6.380	295.0	295.0	.9910
.5070	.5087	3.196	.9967	12.20	12.24	6.393	298.7	298.7	.9911
.5080	.5097	3.203	.9967	12.28	12.32	6.405	302.4	302.4	.9912
.5090	.5107	3.209	.9968	12.35	12.39	6.417	306.2	306.2	.9913
.5100	.5117	3.215	.9968	12.43	12.47	6.430	310.0	310.0	.9914
.5110	.5126	3.221	.9968	12.50	12.54	6.442	313.8	313.8	.9915
.5120	.5136	3.227	.9969	12.58	12.62	6.454	317.7	317.7	.9915
.5130	.5146	3.233	.9969	12.66	12.70	6.467	321.7	321.7	.9916
.5140	.5156	3.240	.9970	12.74	12.78	6.479	325.7	325.7	.9917
.5150	.5166	3.246	.9970	12.82	12.86	6.491	329.7	329.7	.9918
.5160	.5176	3.252	.9970	12.90	12.94	6.504	333.8	333.8	.9919
.5170	.5185	3.258	.9971	12.98	13.02	6.516	337.9	337.9	.9919
.5180	.5195	3.264	.9971	13.06	13.10	6.529	342.2	342.2	.9920
.5190	.5205	3.270	.9971	13.14	13.18	6.541	346.4	346.4	.9921
.5200	.5215	3.277	.9972	13.22	13.26	6.553	350.7	350.7	.9922
.5210	.5225	3.283	.9972	13.31	13.35	6.566	355.1	355.1	.9923
.5220	.5235	3.289	.9972	13.39	13.43	6.578	359.6	359.6	.9924
.5230	.5244	3.295	.9973	13.47	13.51	6.590	364.0	364.0	.9924
.5240	.5254	3.301	.9973	13.55	13.59	6.603	368.5	368.5	.9925
.5250	.5264	3.308	.9973	13.64	13.68	6.615	373.1	373.1	.9926
.5260	.5274	3.314	.9974	13.73	13.76	6.628	377.8	377.8	.9927
.5270	.5284	3.320	.9974	13.81	13.85	6.640	382.5	382.5	.9927
.5280	.5294	3.326	.9974	13.90	13.94	6.652	387.3	387.3	.9928
.5290	.5304	3.333	.9975	13.99	14.02	6.665	392.2	392.2	.9929
.5300	.5314	3.339	.9975	14.07	14.10	6.677	397.0	397.0	.9930
.5310	.5323	3.345	.9975	14.16	14.19	6.690	402.0	402.0	.9931
.5320	.5333	3.351	.9976	14.25	14.28	6.702	406.9	406.9	.9931
.5330	.5343	3.357	.9976	14.34	14.37	6.714	412.0	412.0	.9932
.5340	.5353	3.363	.9976	14.43	14.46	6.727	417.2	417.2	.9933
.5350	.5363	3.370	.9976	14.52	14.55	7.639	422.4	422.4	.9933
.5360	.5373	3.376	.9977	14.61	14.64	6.752	427.7	427.7	.9934
.5370	.5383	3.382	.9977	14.70	14.73	6.764	433.1	433.1	.9935
.5380	.5393	3.388	.9977	14.79	14.82	6.776	438.5	438.5	.9935
.5390	.5402	3.394	.9977	14.88	14.91	6.789	444.0	444.0	.9936
.5400	.5412	3.401	.9978	14.97	15.01	6.801	449.5	449.5	.9936
.5410	.5422	3.407	.9978	15.07	15.18	6.814	455.1	455.1	.9937
.5420	.5432	3.413	.9978	15.16	15.19	6.826	460.7	460.7	.9938
.5430	.5442	3.419	.9979	15.25	15.29	6.838	466.4	466.4	.9938
.5440	.5452	3.426	.9979	15.35	15.38	6.851	472.2	472.2	.9939
.5450	.5461	3.432	.9979	15.45	15.48	6.863	478.1	478.1	.9940
.5460	.5471	3.438	.9979	15.54	15.58	6.876	484.3	484.3	.9941
.5470	.5481	3.444	.9980	15.64	15.67	6.888	490.3	490.3	.9941
.5480	.5491	3.450	.9980	15.74	15.77	6.901	496.4	496.4	.9942
.5490	.5501	3.456	.9980	15.84	15.87	6.913	502.5	502.5	.9942
.5500	.5511	3.463	.9980	15.94	15.97	6.925	508.7	508.7	.9942
.5510	.5521	3.469	.9981	16.04	16.07	6.937	515.0	515.0	.9942
.5520	.5531	3.475	.9981	16.14	16.17	6.950	521.6	521.6	.9943
.5530	.5541	3.481	.9981	16.24	16.27	6.962	528.1	528.1	.9944
.5540	.5551	3.488	.9981	16.34	16.37	6.975	534.8	534.8	.9944
.5550	.5560	3.494	.9982	16.44	16.47	6.987	541.4	541.4	.9945
.5560	.5570	3.500	.9982	16.54	16.57	7.000	548.1	548.1	.9945
.5570	.5580	3.506	.9982	16.65	16.68	7.012	554.9	554.9	.9946
.5580	.5590	3.512	.9982	16.75	16.78	7.025	562.0	562.0	.9947
.5590	.5600	3.519	.9982	16.85	16.88	7.037	569.1	569.1	.9947

Table of functions of d/L_0

APPENDIX (Continued)

d/L_0	d/L	$2\pi d/L$	tanh $2\pi d/L$	sinh $2\pi d/L$	cosh $2\pi d/L$	$4\pi d/L$	sinh $4\pi d/L$	cosh $4\pi d/L$	H/H_0'
.5600	.5610	3.525	.9983	16.96	16.99	7.050	576.1	576.1	.9947
.5610	.5620	3.531	.9983	17.06	17.09	7.062	583.3	583.3	.9948
.5620	.5630	3.537	.9983	17.17	17.20	7.074	590.7	590.7	.9949
.5630	.5640	3.543	.9983	17.28	17.31	7.087	598.0	598.0	.9949
.5640	.5649	3.550	.9984	17.38	17.41	7.099	605.0	605.0	.9950
.5650	.5659	3.556	.9984	17.49	17.52	7.112	613.2	613.2	.9950
.5660	.5669	3.562	.9984	17.60	17.63	7.124	620.8	620.8	.9951
.5670	.5679	3.568	.9984	17.71	17.74	7.136	628.5	628.5	.9951
.5680	.5689	3.575	.9984	17.82	17.85	7.149	636.4	636.4	.9952
.5690	.5699	3.581	.9985	17.94	17.97	7.161	644.3	644.3	.9952
.5700	.5709	3.587	.9985	18.05	18.08	7.174	652.4	652.4	.9953
.5710	.5719	3.593	.9985	18.16	18.19	7.186	660.5	660.5	.9953
.5720	.5729	3.600	.9985	18.28	18.31	7.199	668.8	668.8	.9954
.5730	.5738	3.606	.9985	18.39	18.42	7.211	677.2	677.2	.9954
.5740	.5748	3.612	.9985	18.50	18.53	7.224	685.6	685.6	.9955
.5750	.5758	3.618	.9986	18.62	18.64	7.236	694.3	694.3	.9955
.5760	.5768	3.624	.9986	18.73	18.76	7.249	703.2	703.2	.9956
.5770	.5778	3.630	.9986	18.85	18.88	7.261	711.9	711.9	.9956
.5780	.5788	3.637	.9986	18.97	19.00	7.274	720.8	720.8	.9957
.5790	.5798	3.643	.9986	19.09	19.12	7.286	729.9	729.9	.9957
.5800	.5808	3.649	.9987	19.21	19.24	7.298	739.0	739.0	.9957
.5810	.5818	3.656	.9987	19.33	19.36	7.311	748.1	748.1	.9955
.5820	.5828	3.662	.9987	19.45	19.48	7.323	757.5	757.5	.9958
.5830	.5838	3.668	.9987	19.58	19.60	7.336	767.0	767.0	.9959
.5840	.5848	3.674	.9987	19.70	19.73	7.348	776.7	776.7	.9959
.5850	.5858	3.680	.9987	19.81	19.84	7.361	786.5	786.5	.9960
.5860	.5867	3.686	.9987	19.94	19.96	7.373	796.4	796.4	.9960
.5870	.5877	3.693	.9988	20.06	20.09	7.386	806.5	806.5	.9960
.5880	.5887	3.699	.9988	20.19	20.21	7.398	816.5	816.5	.9961
.5890	.5897	3.705	.9988	20.32	20.34	7.411	826.7	826.7	.9961
.5900	.5907	3.712	.9988	20.45	20.47	7.423	837.1	837.1	.9962
.5910	.5917	3.718	.9988	20.57	20.60	7.436	847.6	847.6	.9962
.5920	.5927	3.724	.9988	20.70	20.73	7.448	858.2	858.2	.9963
.5930	.5937	3.730	.9989	20.83	20.86	7.460	868.9	868.9	.9963
.5940	.5947	3.737	.9989	20.97	20.99	7.473	879.8	879.8	.9963
.5950	.5957	3.743	.9989	21.10	21.12	7.485	890.8	890.8	.9964
.5960	.5967	3.749	.9989	21.23	21.25	7.498	901.9	901.9	.9964
.5970	.5977	3.755	.9989	21.35	21.37	7.510	913.4	913.4	.9964
.5980	.5987	3.761	.9989	21.49	21.51	7.523	925.0	925.0	.9965
.5990	.5996	3.767	.9989	21.62	21.64	7.535	936.5	936.5	.9965
.6000	.6006	3.774	.9990	21.76	21.78	7.548	948.1	948.1	.9965
.6100	.6106	3.836	.9991	23.17	23.19	7.673	1,074	1,074	.9969
.6200	.6205	3.899	.9992	24.66	24.68	7.798	1,217	1,217	.9972
.6300	.6305	3.961	.9993	26.25	26.27	7.923	1,379	1,379	.9975
.6400	.6404	4.024	.9994	27.95	27.97	8.048	1,527	1,527	.9977
.6500	.6504	4.086	.9994	29.75	29.77	8.173	1,771	1,771	.9980
.6600	.6603	4.149	.9995	31.68	31.69	8.298	2,008	2,008	.9982
.6700	.6703	4.212	.9996	33.73	33.74	8.423	2,275	2,275	.9983
.6800	.6803	4.274	.9996	35.90	35.92	8.548	2,579	2,579	.9985
.6900	.6902	4.337	.9997	38.23	38.24	8.674	2,923	2,923	.9987
.7000	.7002	4.400	.9997	40.71	40.72	8.799	3,314	3,314	.9988
.7100	.7102	4.462	.9997	43.34	43.35	8.925	3,757	3,757	.9989
.7200	.7202	4.525	.9998	46.14	46.15	9.050	4,258	4,258	.9990
.7300	.7302	4.588	.9998	49.13	49.14	9.175	4,828	4,828	.9991
.7400	.7401	4.650	.9998	52.31	52.32	9.301	5,473	5,473	.9992

APPENDIX (Continued)

d/L_o	d/L	$2\pi d/L$	tanh $2\pi d/L$	sinh $2\pi d/L$	cosh $2\pi d/L$	$4\pi d/L$	sinh $4\pi d/L$	cosh $4\pi d/L$	H/H_o'
.7500	.7501	4.713	.9998	55.70	55.71	9.426	6,204	6,204	.9993
.7600	.7601	4.776	.9999	59.30	59.31	9.552	7,034	7,034	.9994
.7700	.7701	4.839	.9999	63.15	63.16	9.677	7,976	7,976	.9995
.7800	.7801	4.902	.9999	67.24	67.25	9.803	9,042	9,042	.9996
.7900	.7901	4.964	.9999	71.60	71.60	9.929	10,250	10,250	.9996
.8000	.8001	5.027	.9999	76.24	76.24	10.05	11,620	11,620	.9996
.8100	.8101	5.090	.9999	81.19	81.19	10.18	13,180	13,180	.9996
.8200	.8201	5.153	.9999	86.44	86.44	10.31	14,940	14,940	.9997
.8300	.8301	5.215	.9999	92.05	92.05	10.43	17,340	17,340	.9997
.8400	.8400	5.278	1.000	98.01	98.01	10.56	19,210	19,210	.9997
.8500	.8500	5.341	1.000	104.4	104.4	10.68	21,780	21,780	.9998
.8600	.8600	5.404	1.000	111.1	111.1	10.81	24,690	24,690	.9998
.8700	.8700	5.467	1.000	118.3	118.3	10.93	28,000	28,000	.9998
.8800	.8800	5.529	1.000	126.0	126.0	11.06	31,750	31,750	.9998
.8900	.8900	5.592	1.000	134.2	134.2	11.18	36,000	36,000	9998
.9000	.9000	5.655	1.000	142.9	142.9	11.31	40,810	40,810	.9999
.9100	.9100	5.718	1.000	152.1	152.1	11.44	46,280	46,280	.9999
.9200	.9200	5.781	1.000	162.0	162.0	11.56	52,470	52,470	.9999
.9300	.9300	5.844	1.000	172.5	172.5	11.69	59,500	59,500	.9999
.9400	.9400	5.906	1.000	183.7	183.7	11.81	67,470	67,470	.9999
.9500	.9500	5.969	1.000	195.6	195.6	11.94	76,490	76,490	.9999
.9600	.9600	6.032	1.000	203.5	203.5	12.06	86,740	86,740	.9999
.9700	.9700	6.095	1.000	222.8	222.8	12.19	98,350	98,350	.9999
.9800	.9800	6.158	1.000	236.1	236.1	12.32	111,500	111,500	.9999
.9900	.9900	6.220	1.000	251.4	251.4	12.44	126,500	126,500	1.000
1.000	1.000	6.283	1.000	267.7	267.7	12.57	143,400	143,400	1.000

INDEX

Above-surface wave recorders, 274
Acceleration, water particle, 153, 161, 394, 407, 409
Age of wave, 72, 89, 90, 101, 102
Air-sea temperature effect, 45, 47
Airy wave, 146
Amplitude, wave, 12, 67–70, 76, 78, 265–269
–, orbital water motions, 152–154, 162, 163, 172
Analysis of wave record, 277, 289
Angular dispersion of waves, 116, 117, 132, 133
Anticyclones, 44, 49, 50, 56
Antinodes, 179, 180, 320, 356, 357
Armour units, stability of, 371, 416–421, 427
– –, composite breakwater, 420, 421, 427
– –, rubble-mound breakwater, 371, 416–421, 427
Asphaltic concrete, 420
Asymmetry, wave, 34, 157–159, 165, 166, 171, 172, 182, 183, 209, 210, 244, 245, 248, 249, 357, 373, 376, 379
–, progressive wave, 34, 157–159, 165, 166, 171, 172, 209, 244, 245, 248, 249, 379
–, – –, horizontal, 34, 248, 249, 379
–, – –, vertical, 157–159, 165, 166, 171, 172, 209, 244, 245
–, standing wave, vertical, 182, 183, 210, 357, 373, 376
Atmospheric pressure, 122, 124
Attenuation, 16, 278–285, 294, 295, 339
– of pressure, 278–285, 294, 295
– of waves, 16, 339
Average spectra, 105–108
Average wave height (*see also* Height, wave), 68–70, 267

Beaches, reflection from, 282, 283
Bed contours, 220–224, 242, 243
Below-surface wave recorders, 275, 294, 295
Bottom, sea, 195, 196, 211, 287, 295
–, non-rigid, 196, 287, 295
–, permeability, 195, 211
Bourrage force, 384, 385, 388, 392, 393

Breakers and surf, 240, 246, 248, 324, 325, 423
–, uprush height, 240, 246, 324, 325, 423
Breaking waves, 31–34, 157–159, 242–249, 255, 324, 325, 340, 400, 401, 424
– during generation, 31–33, 401
Breakwater head, 417–419
– trunk, 417–419
Breakwaters, 323, 337–353, 360, 361, 371, 388–392, 416–421, 424, 427
–, composite, 371, 388–392, 420, 421, 424
–, hydraulic, 352, 353
–, mobile, 346–351, 360, 361
–, permeable, 337
–, pile arrays, 344–346
–, pneumatic, 352, 353
–, rubble-mound, 323, 371, 416–421, 427
–, submerged, 338–343, 360

Canyon, submarine, 232, 233
Celerity, wave, 13, 71, 146–149, 160, 164, 166, 168, 171, 207, 209
–, cnoidal, 164
–, finite height, 160, 209
–, hyperbolic, 166, 168
–, linear theory, 146–149, 207, 209
–, solitary, 164, 171
Characteristics, wave, *see* Wave characteristics
Charts, 16, 43–50, 54–58, 78, 82, 228, 263–265, 276, 290, 310
–, cyclonicity, 54–56
–, hydrographic, 16, 228, 276, 310
–, sea and swell, 56, 58
–, synoptic, 43–50, 78, 82, 263–265, 290
Clapotis, *see* Standing wave
Clapotis gaufré, *see* Short-crested wave
Classification, wave, 21, 248, 249
–, breaking, 248, 249
–, progressive, 21
Climate, wave, 21, 53, 56, 61, 269, 271, 277
Cnoidal theory, 15, 163–165, 209
– celerity, 164
– profile, 165

Coastal engineering, 1–11, 18
– –, costs in, 7, 8
– –, definition of, 2, 3
– –, development of, 5
– –, education in, 5, 8–10
– –, interest in, 4, 5
– –, opportunities in, 1, 6, 11
– –, sources of information, 18
Co-cumulative spectrum, 76, 77
Coefficient, 229–231, 301, 306–316, 337–350, 360, 378, 394, 395, 399, 400, 404–413
–, diffraction, 306–316
–, discharge, 344, 345, 378
–, drag, 394, 395, 399, 400, 404–407
–, inertia, 394, 395, 399, 400, 404–406, 409–413
–, lift, 406–408
–, refraction, 230, 231
–, shoaling, 229
Cold front, 46, 48, 50
Comparisons, 95, 96, 173–177, 377
– of forecasting formulae, 95, 96
– of wave theories, 173–177, 377
Composite breakwaters, see Breakwaters
Concrete, 323, 418, 420
–, armour units, 323, 418
–, asphaltic, 420
Conditions, meteorological, 43, 44, 49, 54–58, 61, 84
Contours, bed, 220–224, 242, 243
Convection solution, 197, 198
Coriolis force, 44
Costs in ocean engineering, 7, 8
Crest, wave, 12, 159, 186, 233, 234, 373
–, length of, 186, 233, 234
–, secondary, 159, 373
Crossings, zero, 108, 265–271, 281, 291–293
Cumulative wave height distribution, 269, 270, 291–293
Currents, 82, 237–239, 255, 285, 286
–, effect on waves, 237–239, 255, 285
–, errors in wave recording, 82, 286
Curved bed contours, 224
Cyclones, 44–62, 121–125, 134
–, distribution of, 54–59
–, extra tropical, 44–49, 56, 57, 60, 62
–, tropical, 50–60, 62, 121–125, 134
Cyclonicity chart, 54, 55, 56
Cylindrical structures, 371, 337, 378, 403–412, 425, 426
– –, large-diameter, 371, 377, 378, 408–412, 426

– –, slender, 403–408, 425, 426
– –, –, horizontal and sloping, 403–405, 425, 426
– –, –, submerged pipe line, 406–408, 426

Damping, see Viscosity, effects of
Darbyshire method, 82, 83, 96, 129
Decay area, see Dispersion of waves
Deep water, 13, 145, 147, 151, 152, 158
Definition of wave record, 265, 266, 293
Depth increments, for refraction, 226
Design data, 109–115, 178, 221, 231, 242, 272–274, 307, 308, 374, 397, 398, 418
– –, forces on slender piles, 397, 398
– –, forces on vertical walls, 374
– – from wave records, 272–274
– –, horizontal orbital velocities, 178
– –, rubble-mound breakwaters, 418
– –, wave diffraction, 307, 308
– –, wave forecasting, 109–115
– –, – – developing sea, 111–113
– –, – – fully arisen sea (FAS), 109–111
– –, – – optimum sea, 113–115
– –, refraction, 221, 231, 242
Developing sea, 73, 76, 77, 83, 98–104, 111–115, 121, 125
Deviation, rms, 67–70, 76, 78, 265–268
Difficulties, wave recording, 276
Diffraction, see Wave diffraction
Dimensionless, 72–74, 79, 88–90, 101–105, 119
– EDC, 88, 90, 101, 105, 119
– SMB relationships, 72–74
– spectrum, 79, 89, 101
Direction of waves, recording, 276
Directional energy distribution, 34, 35, 39, 41, 236, 237, 316, 317
Discharge, coefficient of, 344, 345, 378
Dispersion, wave, 15, 39–42, 115–120, 132, 133
–, angular, 116, 117, 132, 133
–, radial, 117–120, 132, 133
Displacements, water particles, 152–154, 162, 181, 184–191
Dissipation, wave, 16, 246, 248, 338, 339
Distribution, of cyclones, 54–59
Drag, coefficient of, 394
– force, 394–407, 424
Duration, storm, 15, 72, 73, 77, 94, 97, 110, 113–115

INDEX

Edge waves, 252
Education in coastal engineering, 5, 8–10
Effect of current on waves, 237–239, 255, 285
Effects of viscosity, 191–206, 211, 287, 295
Energy, wave, 35, 115, 154–156
–, directional distribution, 35, 115
–, kinetic, 154, 156
–, potential, 154, 155
– dissipation, 16, 246, 248, 338, 339
– distribution-curve (EDC), 36, 37, 40, 85, 87–91, 101–105
– –, dimensionless, 88–90, 101–105
– –, equivalent, 90, 91, 104
Errors arising, wave record (see also Wave recording), 278, 284–289, 295
Estuarine conditions, wave recording errors, 287
Eulerian presentation, 149–151, 160, 161, 179–183, 192
Extra-tropical cyclones, 44–49, 56, 57, 60, 62
Extreme wave height (see also Height wave), 271, 272
Extremes of EDC (T_U, T_L), see Period, wave
– of spectrum (f_U, f_L), see Frequency, wave
Eye, cyclone, 50–53

Fathom, 16
Fetch, 15, 34, 39, 43, 72, 73, 77, 93, 94, 97–105, 110–115
Finite height, 14, 156–162, 209
– celerity, 160, 209
– surface profile, 157, 158, 209
– water particle motion (see also Progressive wave and Standing wave), 160–162
First-order theory, see Linear wave theory
Flexible rafts (see also Water-backed structures), 350, 351
Floating breakwaters, see Breakwaters
Forces (see also Seawalls), 347, 349, 351, 379–405, 423, 424
–, drag, 394–405, 424
–, inertia, 394–405, 424
–, mooring, 347, 349, 351
–, shock, 379–393, 423
Forecasting, wave, 65–97, 109–115, 127–132
–, comparison of formulae, 95, 96
–, Darbyshire method, 82, 83, 96, 129
–, design data, 109–115
–, history of, 70
–, PM method, 71, 78–80, 84–92, 96, 127–132
–, PNJ method, 75–77, 93, 96, 97, 129
–, SMB method, 66, 70–75, 93, 96, 97, 127, 129

–, wave statistics, 65–69
–, – –, Gaussian probability distribution, 66, 69
–, – –, Rayleigh probability distribution, 67, 69
Frequency, radian, 144
–, wave (see also Period, wave), 13, 76–79, 86, 87, 100
–, extremes of spectrum (f_U), (f_L), 76–79, 86
– of maximum energy band (f_{max}), 76–79, 86, 87, 100
Fronts, 46, 48, 50
–, cold, 46, 48, 50
–, occluded, 46, 48
–, warm, 46, 48
Fully arisen sea (FAS), 15, 22, 53, 66, 74, 84–92, 103, 109–111, 127–133, 247, 248, 267, 268, 401

Gaussian probability distribution, 66, 69
Generation, wave, 21–33, 102, 103
– resonance, 22–29
– shear flow, 25–29
– sheltering effect, 29–31
– wave breaking, 31–33, 102, 103
Geometric shadow line, 302, 303
Geostrophic wind, 45, 46
Gerstner, 156
Gifle, force, 384, 385, 388, 392, 393
Global wind pattern, 56, 57
Gradient wind, 45, 47
Great circle, 38
Group velocity, wave, 94, 95, 154–156

Harbour, seiching, 301, 330–337, 358, 359
–, circular, 331
–, complex harbour shapes, 334
–, elliptical, 331
–, exclusion of resonant waves, 335–337, 359
–, rectangular, 331–334, 358
Harbour, surging, see Harbour, seiching
Harmonic analysis, wave record, 277, 289
Head, breakwater, 417–419
Height, uprush, see Breakers and surf
–, wave, 12, 67–74, 88, 96, 105–113, 119, 121, 125–132, 266–272
– –, average, 68–70, 267
– –, highest tenth, 68–70, 110, 111, 266–271
– –, maximum, 271, 272
– –, optimum, 105–109, 113
– –, probable maximum, 268–271
– –, significant, 67, 73, 74, 88, 96, 106, 110–113, 119, 121, 125–132, 266–271

High pressure systems, *see* Anticyclones
Highest tenth waves (*see also* Height, wave), 68–70, 110, 111, 266–271
History of wave forecasting, 70
Hong Kong, typhoons at, 52, 134
Horizontal slender members, *see* Cylindrical structures
Horizontal velocities, *see* Orbital water motions
Hurricanes (*see also* Tropical cyclones), 50–52
Hydraulic breakwaters, 352, 353
Hydrographic charts, 16, 228, 276, 310
Hyperbolic functions, 144, 145, 435–448
Hyperbolic wave, 15, 166–169, 209
– celerity, 166, 168
– profile, 166, 169

Impact forces, *see* Forces
Inertia force, 394–406, 409, 411–415, 424
Information, sources for ocean engineering, 18
Interaction, wave, *see* Breakwater, submerged, and Standing wave
Interest in ocean engineering, 4, 5
Isobars, 44, 48, 50

Jetty, *see* Cylindrical structures, and Seawalls

Kinetic energy, 154, 156
King waves, 155
Knot, 16

Lagrangian presentation, 149, 152, 162, 163, 181, 184–191
Laminar damping, *see* Viscosity, effects of
Land-backed structures, *see* Structures, land-backed
Length, crest, 186, 233, 234
–, fetch, 15, 34, 39, 43, 72, 73, 77, 93, 94, 97–105, 110–115
–, wave, 13, 146, 165, 167, 187, 217, 218, 238, 239
–, wave record, 288
Lift, coefficient of, 406–408
Linear wave theory, 146–154, 207, 209
– celerity, wave, 146–149, 207, 209
– water-particle motion, 149–154
Low-pressure systems (*see also* Extra-tropical cyclones, and Tropical cyclones), 44–62, 122

Major difficulties, wave recording, 276
Map projections (*see also* Weather maps), 38
Mass transport, wave, 163, 171, 191, 192, 196–206, 211

– –, cnoidal, 171
– –, progressive, 198–202, 211
– –, short-crested, 204, 205
– –, solitary, 171
– –, standing, 202, 203
– –, stratified fluid, 205, 206
Maximum energy band, *see* Height, wave, and Period, wave
Maximum wave height (*see also* Height, wave), 110, 111
Meteorological conditions, 43, 44, 49, 54–58, 61
– seasonal pattern, 54–58, 61
Microseisms, 282
Millibars, 52, 122, 124, 134
Miscellaneous, mobile breakwaters (*see also* Water-backed structures), 351
Mobile breakwaters, 346–351, 360, 361
Moment, from wave forces, 376, 390, 397, 398, 411
Monsoons (*see also* Tropical cyclones), 50
Mooring forces, 347, 349, 351
Motion, water particle, *see* Water-particle motion
Mud, 196, 205, 206, 287, 295
Multi-direction, waves, 34, 35, 236, 237, 316, 317

Narrow spectrum, 67–69
Natural frequency, basins, 331–334
– –, structures, 391, 392, 401, 402
Nautical mile, 16
Need for wave recording, 263
Nodes (*see also* Standing wave), 179, 180, 320
Non-parallel bed contours, refraction, 222, 223
Non-rigid bottom, 196, 287, 295
Normal atmospheric pressure, 122, 124
Number, wave, 144, 145

Occluded front, 46, 48
Ocean engineering, 1–11, 18
– –, costs in, 1, 6–8
– –, definition of, 2, 3, 4
– –, education in, 8–10
– –, interest in, 4, 5
– –, opportunities in, 1, 6, 11
– –, sources of information, 18
Oceanography, 2, 11
–, definition of, 2
–, difference from engineering, 11
–, institutes of, 2
Oceans, 1, 4–8
–, costs of, 1, 6–8

INDEX

–, interest in, 4, 5
–, rewards from, 1, 6–8
Offshore structures, *see* Cylindrical structures, and Submerged objects, large diameter
Oil rigs (*see also* Cylindrical structures), 401–405, 424, 425
Opportunities in ocean engineering, 1, 6, 11
Orbital water motions, 149–154, 160–163, 171, 172, 178
– accelerations, 153, 161, 162
– amplitudes, 152–154, 162, 163, 172
– velocities, 150–154, 160–162, 171, 172, 178
Orthogonal (*see also* Refraction, wave), 217, 218, 225, 230, 303, 306
Oscillations, floating bodies, 351
–, structures, 391, 392, 401, 402
–, water basins, 331–334
–, water particles, *see* Water-particle motion
Overtopping (*see also* Structures), 301, 319, 327–329, 358, 389

Parallel bed contours, refraction, 220–222, 242, 243
Parameter, Ursell, 165
Partial clapotis, *see* Standing wave, partial
Particle motion, water, *see* Water-particle motion
Patterns, 43, 44, 49, 54–58, 61
–, global, wind, 56, 57
–, weather, 43, 44, 49, 54–58, 61
Period, recording, 265, 268, 288
–, wave, 67, 71–78, 87–94, 99, 101–121, 125–134, 265–271, 281, 291–293
–, extremes of EDC (T_U, T_L), 87, 90–92, 99, 105, 110, 111, 117, 119, 128, 133
–, maximum energy band (T_{max}), 78, 87–91, 94, 99, 101–104, 107–121, 126–134, 267–271, 281, 291–293
–, reciprocal of f_{max} (T_f), 76, 107, 108
–, significant ($T_{1/3}$), 67, 71–74, 110, 111, 121, 125–127, 134
–, zero crossing (T_Z), 108, 265–271, 281, 291–293
Permeability, bottom, 195, 211
Permeable breakwaters, *see* Breakwaters
Phase velocity, *see* Wave celerity
Pile array, 344–346, 378, 421, 423
– –, attenuation by, 344–346
– –, forces on, 345, 378, 421, 423
Piles, *see* Cylindrical structures
Pipeline, submerged, 406–408, 426
PM method, forecasting, 71, 78–80, 84–92, 96, 127–132
Pneumatic breakwaters, *see* Breakwaters
PNJ method, forecasting, 75–77, 93, 96, 97, 129
Pontoons (*see also* Structures, water-backed), 347–350, 360, 361
Port oscillations, *see* Harbour seiching
Potential energy, wave, 154, 155
Power, wave, 154, 155, 229, 230
Pressure attenuation, wave, 278–285, 294, 295
– –, progressive, 278–280, 294
– –, short-crested, 283–285, 295
– –, standing, 282
–, on structures, *see* Structures, forces on
Probable maximum wave height (*see also* Height, wave), 268–271
Probability distribution, 66–69, 271, 272, 382–384
– –, extreme waves, 271, 272
– –, Gaussian, 66, 69
– –, Rayleigh, 67, 69
– –, shock pressures, 382–384
Profile, wave, 14, 66, 143, 157, 158, 165, 166, 209
–, cnoidal, 165
–, finite height, 157, 158, 209
–, hyperbolic, 166
–, linear, 143
–, trochoidal, 14, 66
–, wind, 79–81
Progressive wave, 14–16, 66, 146–171, 207, 209, 217–255, 278–280, 285, 294, 318, 319, 356, 373, 381
– –, cnoidal, 15, 163–165, 209
– –, finite height, 156–162, 209, 373, 381
– –, celerity, 160, 209
– –, pressure attenuation, 280, 294
– –, surface profile, 157–159, 209, 373, 381
– –, water-particle motion, 160–162
– –, hyperbolic, 15, 166–169, 209
– –, linear theory, 146–154, 207, 209
– –, – –, celerity, 146–149, 207, 209
– –, – –, pressure attenuation, 278, 279
– –, – –, refraction of, 217–255, 285, 318, 319, 356
– –, – –, water-particle motion, 149–154
– –, solitary, 15, 171
– –, stream function theory, 169, 170
– –, trochoidal, 14, 66
Propagation, wave, 16

Quadripods, 418

Radial dispersion, wave, 117–120, 132, 133
Radian frequency, 144
Rafts, flexible (*see also* Structures, water-backed), 350, 351
Rayleigh probability distribution, 67, 69
Recommendations on wave recording, 289
Record, wave, *see* Wave record
Recorders, wave, *see* Wave recorders
Recording interval, 265, 268–272
– period, 265, 268, 288
–, wave, *see* Wave recording
Reflection, 282, 283, 338, 339, 344–346
– from breaches, 282, 283
– from structures, 338, 339, 344–346
Refraction, wave, 82, 217–255, 285, 286, 318, 319, 356
– as $\alpha_o \to 90°$, 225
– by currents, 237–239, 255, 285
–, coefficient of, 230, 231
–, curved bed contours, 224
–, depth increments, 226
–, non-parallel bed contours, 222, 223
–, parallel bed contours, 220–222, 242, 243
–, short-crested waves, 233–235
– spectrum, 236, 237
–, wave recording errors, 82, 286
– with diffraction, 318, 319, 356
Resonance, *see* Natural frequency
Resonators, wave, 335–337, 359
Rewards from oceans, 1, 6, 7
Reynolds number, 202, 337
Ridge, submarine (*see also* Shoaling water) 232
Rock-fill breakwater, *see* Breakwaters
Root mean square deviation, wave, 67–70, 76, 78, 265–268
Rubble-mound breakwater, *see* Breakwaters
Run-up (*see also* Overtopping), 240–246, 254, 301, 324–328, 357, 358

Scour, 203, 204, 323, 325, 339, 353, 403, 408, 421
– at seawalls, 203, 204, 323, 325, 339, 421
– at submerged pipelines, 353, 408
– under offshore structures, 403
Sea and swell charts, 56, 58
– breeze, 49, 50
– state, 58
Seasonal variations, wind, 54–58, 61
Seawalls, pressures, 371–393, 422, 423
– breaking waves, 379–393, 423
–, overtopping with, 377, 378, 422
–, standing waves, 372–375, 422
Seawards tracing of orthogonal, 225

Secondary crest (*see also* Progressive wave), 159, 373
Sediment movement, *see* Scour
–, suspended, 205, 206
Seiches, harbour, 301, 330–337, 358
Set-up, wave, 157, 158, 182, 183, 325
Shadow line, 302, 303
– region, 302, 303
Shallow water, 13, 145–148, 151, 152, 158
Shear stress, composite breakwater, 389–391
– –, sea surface, 25–29, 170
Ship-borne wave recorder, 82, 275, 291–294
Shoaling coefficient, 229
– water, 217–249, 254, 255
– –, refraction coefficient, 230, 231
– –, submarine canyon, 232
– –, submarine ridge, 232
– –, wave spectrum, 247, 248
Shock pressures, 379–393, 423
Short-crested wave (*see also* Standing wave), 17, 32, 186–192, 210, 233–235, 283–285, 295, 303, 304, 323, 325
Side channel resonators, 335–337
Significant wave, 67–74, 88, 96, 106, 109–113, 119, 121, 125–131, 134, 266–271
– design data, 109, 110
– height, 67, 69, 70, 73, 74, 88, 96, 106, 110–113, 119, 121, 125–131, 266–271
– period, 67, 71–74, 110, 111, 121, 125–127, 134
Sinusoidal (*see also* Progressive wave), 14, 66, 143, 144, 146
Slender structures, *see* Cylindrical structures
Sloping members, *see* Cylindrical structures
– walls, 321–327, 357, 358, 388
– –, forces on, 388
Sloping walls, overtopping of, 327
– –, reflection from, 321–323, 357
– –, run-up on, 324, 326, 358
SMB method, forecasting, 66, 70–75, 93, 96, 97, 127, 129
Snell's law (*see also* Refraction, wave), 218
Solitary wave (*see also* Progressive wave), 15, 171
Sommerfeld's solution, diffraction, 305, 311
Sources of information, ocean engineering, 18
Spectral analysis of wave record, 277, 289
– energy, 36, 37, 40, 42, 76, 77, 84–86, 100
Spectrum, wave (*see also* PM method, PNJ method), 67–69, 76–79, 88, 89, 101, 105–108, 236, 237, 247, 248, 263, 266, 267, 280, 285, 286, 316, 317

INDEX   455

–, average, 105–108
–, co-cumulative, 76, 77
–, diffraction of. 316, 317
–, dimensionless, 79, 88, 89, 101
–, narrow, 67–69
–, recording of, 263, 266, 267, 280, 285, 286
–, refraction of, 236, 237
–, shoaling of, 247, 248
–, wide, 68, 69
–, width of, 68, 69, 267
Standing wave, 16, 17, 32, 179–192, 210, 233–235, 282–285, 295, 303, 304, 320–325, 338, 356, 357, 372, 376, 377, 384
– –, linear theory, 179–181, 210
– –, – –, surface profile, 179
– –, – –, water-particle motion, 179–181, 210
– –, partial, 184, 185, 320–323, 338, 356, 357, 372
– –, –, water particle motion, 185
– –, second-order, 182, 183, 321, 372, 376, 377, 384
– –, –, pressure attenuation, 282
– –, –, surface profile, 182, 183
– –, –, water-particle motion, 182, 183
– –, short-crested, 17, 32, 186–192, 210, 283–285, 295, 303, 304, 323, 325
– –, –, celerity, 186, 210
– –, –, pressure attenuation, 283–285, 295
– –, –, refraction of, 233–235
– –, –, water-particle motion, 187–192, 210
State of sea, 58
Statistics of waves (see also Forecasting, wave), 65–69
Stokes I wave, see Linear wave theory
– II or III, see Finite height
Storm surge, 3, 52, 54, 379, 400, 424
– waves (see also Developing sea, and Fully arisen sea), 33–38, 56–59
– zones of oceans, 41, 55–59
Stream function, 169, 170
Structures, effects of, 301–360
–, diffraction, 302–319, 353–356
–, overtopping, 301, 319, 327–329, 358
–, reflection, 338, 339, 344–346
–, resonant basins, 301, 331–334
–, run-up, 301, 319, 324–328, 357, 358
–, seiches, 301, 330, 358
–, wave transmission, 301, 337–339, 340–350, 360
Structures, forces on, 371–393, 403–412, 425–427

– composite breakwaters, 388–393
–, cylindrical, 371, 377, 378, 403–412, 425, 426
– –, large diameter, 371, 377, 378, 408–412, 426
– –, slender, 403–408, 425, 426
–, rubble-mound breakwaters, 371, 416–421, 427
–, seawalls, 372–376, 422
– –, breaking waves, 379–393, 423
– sloping walls, 388
– submerged large objects, 371, 377, 378, 408–412, 426, 427
Structures, land-backed, 203, 204, 301, 320–339, 356–358, 389, 421
–, harbour seiching, 301, 330–337, 358
–, overtopping, 327–329, 358, 389
– – of spectrum, 329
–, run-up, 324–327, 357, 358
–, – of spectrum, 326, 327
–, reflection and dissipation, 320–322, 356
–, scour, 203, 204, 323, 325, 339, 421
Structures, water-backed, 337–351, 360, 361, 378, 421, 423
–, mobile breakwaters, 346–351, 360, 361
–, – –, flexible rafts, 350, 351
–, – –, miscellaneous, 351
–, – –, pontoons, 347–350, 360, 361
–, permeable breakwaters, 337
–, pile arrays, 344–346, 378, 421, 423
–, submerged breakwaters, 338–343, 360
Submarine canyon, (see also Shoaling water) 232, 233
– ridge, (see also Shoaling water) 232
Submerged breakwaters, (see also Structures, water-backed) 338–343, 360
– objects, large dimensions, 371, 377, 378, 408–415, 426, 427
– pipeline, 406–408, 426
Summation of wave records, 269, 270, 291–293
Surf, see Breakers and surf
Surface profile, wave, 157–159, 165, 166, 172, 179, 209, 373, 381
– –, cnoidal, 165
– –, finite height, 157–159, 209, 373, 381
– –, hyperbolic, 166
– –, secondary crest, 159, 373
– –, solitary, 172
– –, standing, 179
– wave recorders, 274, 275, 291–294
– wind, 15, 45, 47, 79–81
Surge, storm, 3, 52, 54, 379, 400, 424

Suspended sediment, 205, 206
Swell, (*see also* Dispersion, wave) 38–42, 56, 58
Synoptic chart, *see* Weather map

Tetrapods, 323, 418
Theory, wave (*see also* Progressive wave, Standing wave), 13, 143–152, 158, 173–177, 191–206, 211, 287, 295, 377
–, comparisons of, 173–177, 377
–, deep water, 13, 145, 147, 151, 152, 158
–, radian frequency, 144
–, shallow water, 13, 145–148, 151, 152, 158
–, transitional water, 13, 145, 150, 152, 158
–, viscosity, effects of, 191–206, 211, 287, 295
–, wave number, 144, 145
Tidal currents, 82, 237–239, 255, 285
Tracing orthogonals seawards, (*see also* Refraction, wave) 225
Train, wave, 13, 94, 95, 154–156
Transformation, wave, (*see also* Refraction, wave) 217–249, 254, 255
Transitional water, 13, 145, 150, 152, 158
Transmission, coefficient of, 301, 337–350, 360
Transport, sediment, *see* Scour
–, wave, *see* Mass transport
Tribars, 418
Trochoidal wave, 14, 66
Tropical cyclones, 50–60, 62, 121–125, 134
– –, waves from, 121–125, 134
– – wind velocities in, 122–124
Trough, wave, 12, 157–159
Trunk, breakwater, 417–419
Types of recorder, 274, 275, 291–295
Typhoons (*see also* Tropical cyclones), 50–52, 131
–, at Hong Kong, 52, 131

Uprush height, *see* Breakers and surf
Ursell parameter, 165

Valley, submarine, (*see also* Refraction, wave) 232, 233
Velocities, water-particle, 150–154, 160–162, 171, 172, 178–183, 192, 207, 208
Velocity, wave, 146–149, 160, 164, 166, 168, 171, 207, 209
–, wind, 15, 45, 47, 79–81, 122–124
Vertical wall, 320, 321, 325–329, 356, 371–393, 422, 423
– –, overtopping, 327–329

– –, pressures, 371–393, 422, 423
– –, reflection, 320, 321, 356
– –, run-up, 325
Viscosity, effects of, 163, 171, 191–206, 211, 287, 295
–, bottom permeability, 195, 211
–, mass transport, 163, 171, 191, 192, 196–206, 211
–, non-rigid bottom, 196, 287, 295
–, suspended sediment, 205, 206
–, viscous damping, 193–195, 211
Viscous conduction, (*see also* Mass transport), 197, 198
– damping, (*see also* Viscosity, effects of) 193, 195, 211

Wall, 320–329, 356–358, 371–393, 422, 423
–, sloping, 321–327, 357, 358, 388
–, vertical, 320, 321, 325–329, 356, 371–393, 422, 423
Warm front, 46, 48
Water depth contours, 220–224, 242, 243
Water particle motion, 149–154, 160–162, 171, 172, 178–192, 207, 208, 394, 406–409
– accelerations, 153, 161, 394, 407, 409
– displacements, 152–154, 162, 181, 184–191
– velocities, 150–154, 160–162, 171, 172, 178–183, 192, 207, 208, 394, 406, 408
Water-backed structures, *see* Structures, water-backed
Wave age, 72, 89, 90, 101, 102
– amplitude (*see* Amplitude wave)
– celerity (*see* Celerity, wave)
– characteristics, 33–43
– –, storm, 33–38
– –, swell, 38–43
– classification (*see* Classification, wave)
– climate, 21, 53, 56, 61, 269, 271, 277
– crest (*see* Crest, wave)
–, definition, 12
– diffraction, 286, 302–319, 353–356
– –, breakwater gap, 302, 303, 311–316, 353
– –, coefficient of, 306–316
– –, semi-infinite breakwater, 302–310, 316–319, 353–356
– – of spectrum, 316, 317
– –, with refraction, 318, 319, 356
– –, recording errors, 286
– direction, recording of, 276
– energy, *see* Energy, wave
– forecasting, *see* Forecasting, wave
– frequency, *see* Frequency, wave

– generation, *see* Generation, wave
– group velocity, 94, 95, 154–156
– height, *see* Height, wave
– hindcasting, *see* Forecasting, wave
– length, 13, 146, 165, 167, 187, 217, 218, 238, 239
– number, 144, 145
– orthogonal, *see* Orthogonal
– period, *see* Period wave
– prediction, *see* Forecasting, wave
– ray, *see* Orthogonal
– record, 108, 265–271, 277, 278, 281, 288–295
Wave record definition, 265, 266, 293
– –, length, 288
– –, pressure attenuation, 278, 294, 295
– –, spectral analysis, 277, 289
– –, summation, 269, 270, 291–293
– –, width of spectrum, 266–270, 291
– –, zero crossings, 108, 265–271, 281, 291–293
– recorders, 82, 274, 275, 291–295
– – above surface, 274
– – at surface, 82, 274, 275, 291–294
– – below surface, 275, 294, 295
– recording, 82, 263, 266–280, 284–295
– –, errors arising, 82, 278, 284–289, 295
– –, – –, currents, 82, 286
– –, – –, diffraction, 286
– –, – –, estuarine conditions, 287
– –, – –, length of record, 288
– –, – –, non-rigid bottom, 287, 295
– –, – –, analysis, 289
– –, – –, refraction, 286
– –, – –, steepness, 284, 285
– –, extreme values, 271, 272
– –, design waves, 272–274
– –, major difficulties, 276

– –, need for, 263
– –, probable maximum wave, 268–271
– –, recommendations, 289
– –, spectrum, 263, 266, 267, 280, 285, 286
– –, types of recorder, 274, 275, 291–295
– –, wave direction, 276
– resonators, 335–337, 359
– set-up, 157, 158, 182, 183, 325
– spectrum, *see* Spectrum, wave
– statistics, (*see also* Forecasting, wave), 65–69
– steepness, (*see also* Breaking waves), 14, 33, 41, 75, 102, 103, 157–159, 182, 183, 238, 241, 284, 285, 321, 340, 373, 377
– train, 13, 94, 95, 154–156
– transformation, (*see also* Refraction, wave), 217–249, 254, 255
– trough, 12, 157, 158, 159
Weather map, 43–50, 78, 82, 263–265, 290
Weather patterns, 43, 44, 49, 54–58, 61
White horses, (*see also* Breaking waves), 48
Wide spectrum, (*see also* Spectrum, wave), 68, 69
Width of spectrum, (*see also* Spectrum, wave), 68, 69, 267
Wind, (*see also* Fetch), 15, 45–47, 54–58, 61, 79–81, 122–124
–, geostrophic, 45, 46
–, global patterns, 56, 57
–, gradient, 45, 47
– profile, 79–81
–, seasonal variations, 54–58, 61
– velocity, 15, 45, 47, 79–81, 122–124
– pattern, global, 56, 57
Wind-wave parameters, *see* Forecasting, wave

Zero crossings, 108, 265–271, 281, 291–293
Zone, surf, *see* Breakers and surf
Zones, storm, 41, 55–59

A3